T0329834

An Introduction to the Global Circulation of the Atmosphere

An Introduction to the Global Circulation of the Atmosphere

DAVID RANDALL

Princeton University Press
Princeton and Oxford

Published by Princeton University Press, 41 William Street, Princeton, New Jersey 08540
In the United Kingdom: Princeton University Press, 6 Oxford Street, Woodstock, Oxfordshire OX20 1TW

press.princeton.edu

Library of Congress Cataloging-in-Publication Data

Randall, David A. (David Allan), 1948–
An introduction to the global circulation of the atmosphere / David Randall.
pages cm
Includes bibliographical references and index.
ISBN 978-0-691-14896-0 (hardcover : alk. paper)—ISBN 0-691-14896-1 (hardcover : alk. paper)
1. Atmospheric circulation—Study and teaching. 2. Meteorology—Study and teaching. I. Title.
QC880.4.A8R35 2015
551.51'7—dc23
2014043658

British Library Cataloging-in-Publication Data is available

This book has been composed in Minion Pro and Cronos Pro

Printed on acid-free paper. ∞

Printed in the United States of America

10 9 8 7 6 5 4 3 2 1

Contents

Preface

This is a graduate-level introductory overview of the global circulation of the atmosphere, a subject closely tied to atmospheric dynamics. A course on dynamics tends to focus on basic physical concepts and methods for their analysis, however, while *a course on the global circulation must focus on what the atmosphere actually does, and why.*

Graduate-level studies in atmospheric dynamics are an essential prerequisite for reading this book. It is assumed that the reader is familiar with basic concepts of atmospheric dynamics such as the equation of motion, the approximate hydrostatic and geostrophic balances, potential temperature, vorticity, pressure coordinates, and planetary waves. More advanced dynamical concepts are introduced as needed. Chapter 4, in particular, gives a brief but fairly detailed overview of the concepts used throughout the book. Some instructors may choose to refer to chapter 4 only as needed to explain concepts used in the later chapters.

It is difficult to draw a line between the global circulation and climate. The two subjects are growing closer together as the roles of heating and dissipation in the global circulation emerge as key issues. Such topics as monsoons, the hydrologic cycle, and the planetary energy budget can be included under either "climate" or "global circulation," although perhaps with different slants. Climate is the bigger subject. This book skirts the edges of physical climatology.

Our understanding of the global circulation has advanced enormously in recent decades, and the subject is rapidly becoming both broader and deeper. There is far too much to cover in one book, so I have had to make choices. Several parts of the book stress the role of cloud systems and other small-scale processes in the global circulation. Isentropic coordinates are used extensively, and energetics are discussed in some detail. There is a chapter on the global circulation as turbulence, including an extended discussion of predictability.

I have chosen to discuss many of the topics in terms of their original sources, rather than the latest papers. This quasi-historical approach gives credit to the pioneers of our field and highlights the human aspects of the research.

Some of the end-of-chapter problems involve working with observations. The Internet has made it unnecessary to include the data with the book. Many of the figures in this book utilized the Interim Reanalysis of the European Centre for Medium Range Weather Forecasts, which is one suitable source of global atmospheric data for use with the problems, although there are many others.

This book is based on classes I have taught at Colorado State University over the past 26 years. Mike Kelly, Cara-Lyn Lappen, Katherine Harris, Stefan Tulich, Anning Cheng, Mike Toy, Kyle Wiens, Cristiana Stan, Jason Furtado, Maike Ahlgrim, Luke Van Roekel, Levi Silvers, and Matt Masarik performed superbly as teaching assistants for the course, and both the students and I learned as a result of their efforts. Mick Christi, Kate Musgrave, and Kevin Mallen pointed out numerous typos and other errors in the text. The many students who have taken the course over the years asked questions and made suggestions that taught me a lot.

Thomas Birner, Charlotte DeMott, Scott Denning, Celal Konor, and Wayne Schubert of Colorado State University commented on drafts of various chapters and helped me decide what to keep in and what to leave out. George Kiladis of NOAA's Earth System Research Laboratory and Richard Somerville of the Scripps Institution for Oceanography also made valuable suggestions. Any and all errors are my responsibility.

Mark Branson and Don Dazlich ably produced many of the figures. I could not have finished without their help. Mark in particular endured a multiweek crunch as the book neared completion.

Michelle Beckman and Valerie Hisam helped with early versions of the manuscript. Connie Hale helped to obtain permission to use various figures taken from the literature.

Princeton University Press (PUP) has been a pleasure to work with. I am especially grateful to Ingrid Gnerlich for her encouragement, patience, and advice. She recruited four readers who contributed valuable reviews of a preliminary draft of the book. One of them, in particular, wrote a lengthy and detailed set of extremely helpful comments. Karen Fortgang and Karen Carter of PUP efficiently managed production of the book. Barbara Liguori improved the manuscript through an impressive job of copyediting.

Finally, I thank my wife, Mary Kay, for patiently cheering me on as I wrestled this project to the ground. She read several of the chapters, asked thoughtful questions, and enjoyed finding some of my mistakes.

Handy Numbers

Radius of the Earth	6.37×10^6 m
Angular velocity of the Earth's rotation	7.29×10^{-5} s^{-1}
Acceleration of Earth's gravity	9.81 m s^{-2}
Globally averaged surface pressure	984 hPa
Density of air near sea level	1.2 kg m^{-3}
Annual mean incident solar radiation	340 W m^{-2}
Global albedo	0.30
Outgoing longwave radiation	240 W m^{-2}
Globally averaged surface air temperature	288 K
Globally averaged precipitable water	25 mm (= 25 kg m^{-2})
Globally averaged precipitation rate	3 mm day^{-1}
Latent heat of condensation at 0 °C	2.52×10^6 J kg^{-1}
Stefan-Boltzmann constant	5.67×10^{-8} W m^{-2} K^{-4}
c_p for dry air	1000 J kg^{-1} K^{-1}
Gas constant for dry air	287 J kg^{-1} K^{-1}
Molecular viscosity of air	1.5×10^{-5} m^2 s^{-1}

An Introduction to the Global Circulation of the Atmosphere

CHAPTER 1

Perpetual Motion

The atmosphere *circulates*. The circulation is global in extent (see fig. 1.1). The circulating mass consists of "dry air" and three phases of water. Energy and momentum are carried with the air but evolve in response to various processes along the way. Many of those same processes add or remove moisture.

The circulation is sustained by thermal forcing, which ultimately comes from the Sun. On the average, the Earth absorbs about 240 W m^{-2} of incoming or "incident" solar energy, of which roughly 2% is converted to maintain the kinetic energy of the global circulation against frictional dissipation. Additional, "primordial" energy leaks out of the Earth's interior but at the relatively tiny rate of 0.1 W m^{-2} (Sclater et al., 1980; Bukowinski, 1999). The thermal forcing of the global circulation is strongly influenced by the circulation itself—for example, as clouds form and disappear. The interactions between the circulation and the heating are fascinating but complicated.

Averaged over time, the global circulation has to satisfy various balance requirements: for example, the infrared radiation emitted at the top of the atmosphere must balance the solar radiation absorbed, precipitation must balance evaporation, and angular momentum exchanges between the atmosphere and the ocean–solid Earth system must sum to zero. We will discuss the global circulation from this classical perspective. We will also supplement this discussion with descriptions and analyses of the many and varied but interrelated phenomena of the circulation, such as the Hadley and Walker circulations, monsoons, stratospheric sudden warmings, the Southern Oscillation, subtropical highs, and extratropical storm tracks. In addition, we will discuss the diabatic and frictional processes that maintain the circulation, and the ways in which these processes are affected by the circulation itself.

The circulations of energy and water are closely linked. It takes about 2.5×10^6 J of energy to evaporate 1 kg of water from the oceans, and the same amount of energy is released when the water vapor condenses to form a cloud. The energy released through condensation drives thunderstorm updrafts that in one hour or less can penetrate a layer of the atmosphere 10 or even 20 km thick. The cloudy outflows from such storms reflect sunlight to space and block infrared radiation from the warm surface below. Shallower clouds cast shadows over vast expanses of the oceans. One of the aims of this book is to give appropriate emphasis to the role of moisture in the global circulation of the atmosphere.

Figure 1.1. A full-disk image of the Earth on July 27, 2009, looking down on the equator, with North and South America in view. Many elements of the global circulation can be seen in this picture, including the "intertropical" rain band in the eastern North Pacific, swirling midlatitude storms, and the low clouds associated with the high-pressure systems over the eastern subtropical oceans. From http://cimss.ssec.wisc.edu/goes/blog/wp-content/uploads/2009/07/FIRST_IMAGE_G14_V_SSEC.gif.

It is conventional and useful, although somewhat arbitrary, to divide the atmosphere into parts. For purposes of this quick sketch, we will divide the atmosphere vertically and meridionally, only briefly mentioning the longitudinal variations. Let's start at the bottom.

Most of the solar radiation that the Earth absorbs is captured by the surface rather than within the relatively transparent atmosphere. Several processes act to transfer the absorbed energy upward from the ocean and land surface into the lower portion of the atmosphere.

The layer of air that is closely coupled with the Earth's surface is, by definition, the *planetary boundary layer*, or PBL. The top of the PBL is often very sharp and well defined (see fig. 1.2). The depth of the PBL varies considerably in space and time, but a ballpark value to remember is 1 km. The air in the PBL is turbulent, and the turbulence is associated with rapid exchanges or "fluxes" of *sensible heat* (essentially temperature), moisture, and momentum between the atmosphere and the surface. These exchanges are produced by the turbulence, and also promote the turbulence, through

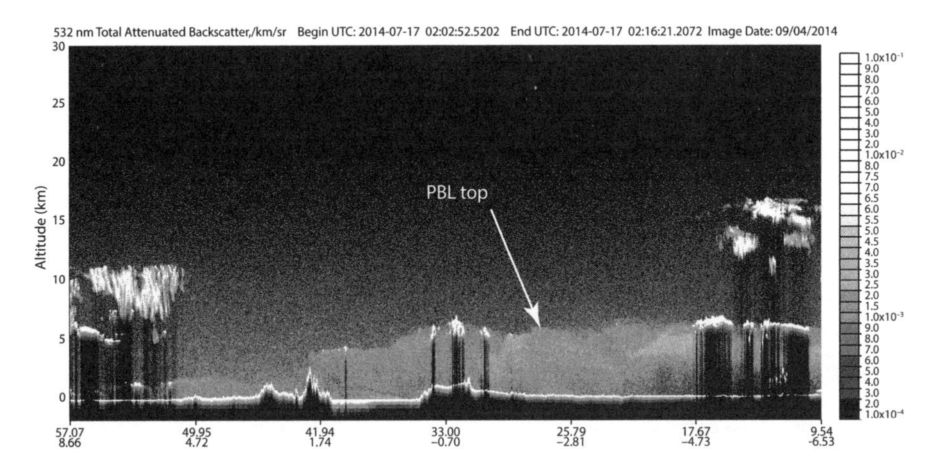

Figure 1.2. This figure shows lidar backscatter from aerosols and clouds. The wavelength of the beam is 532 nm, which is in the green portion of the visible spectrum. The figure was created using data from CALIPSO. The lidar beam cannot penetrate thick clouds, which explains the vertical black stripes in the figure. The PBL is visible because the aerosol concentration decreases sharply with height at the PBL top. The data shown represent observations extending from over the North Atlantic Ocean on the left toward the southeast, over Africa, on the right. Longitudes and latitudes are given along the bottom of the figure. The data also show the Cantabrian Mountains of northern Spain (at about 42° N) and the Atlas Mountains of northern Africa (at about 33° N). The image was kindly provided by Dr. David Winker and the CALIPSO team, of the NASA Langley Research Center.

mechanisms that will briefly be discussed later. The most important exchanges are of moisture, upward into the atmosphere via evaporation from the surface, and of momentum, via friction. The *latent heat* associated with the surface moisture flux is a key source of energy for the global circulation, and surface friction strongly influences the ocean currents.

Above the PBL is the *free troposphere*. Because the troposphere includes the PBL, we add the adjective "free" to distinguish the part of the troposphere that lies above the PBL. The free troposphere is characterized by positive *static stability*, which means that buoyancy forces resist vertical motion. The depth of the troposphere varies strongly with latitude and season.

A turbulent process called *entrainment* gradually incorporates free-tropospheric air into the PBL. Over the oceans, entrainment is, with a few exceptions, relatively slow but steady. Over land, entrainment is promoted by strong daytime heating of the surface, which helps generate turbulence. As a result, the turbulent PBL rapidly deepens during the day. When the Sun goes down, the processes that promote turbulence and entrainment are abruptly weakened, and the PBL reorganizes itself into a much shallower nocturnal configuration, leaving behind a layer of air that was part of the PBL during the afternoon. This diurnal deepening and shallowing of the PBL acts as a kind of "pump" that captures air from the free troposphere and adds it to the PBL starting shortly after sunrise, modifies the properties of that air during the day through strong turbulent exchanges with the surface, and then releases the modified air back into the free troposphere at sunset. This diurnal pumping is one way that the PBL exerts an influence on the free troposphere.

In addition, moisture and energy are carried upward from the PBL into the free troposphere by several mechanisms. Throughout the tropics and the summer-hemisphere middle latitudes the most important of these mechanisms is *cumulus*

Figure 1.3. A space shuttle photograph of tropical thunderstorms. The storms are topped by thick anvil clouds. Much shallower convective clouds can be seen in the foreground. From http://eol.jsc.nasa.gov/sseop/EFS/lores.pl?PHOTO=STS41B-41-2347.

convection. Cumulus clouds typically grow upward from the PBL. The updrafts inside the clouds carry PBL air into the free troposphere, where it is left behind when the clouds decay (see fig. 1.3). One of the effects of this process is to remove air from the PBL and add it to the free troposphere.

Frontal circulations also can carry air from the PBL into the free atmosphere, essentially by "peeling" the PBL from the Earth's surface, like the rind from an orange, and lofting the detached air toward the tropopause. This process is especially active in the middle latitudes in winter.

Figure 1.4 shows somewhat idealized observed midlatitude vertical distributions of temperature, pressure, density, and ozone mixing ratio, from the surface to the 70 km level. In the lowest 12 km, the *troposphere,* the temperature decreases (almost) monotonically with height. The troposphere is cooled radiatively, because it emits infrared radiation much more efficiently than it absorbs solar radiation. The net radiative cooling is balanced mainly by the release of the latent heat of water vapor as clouds form and precipitate.

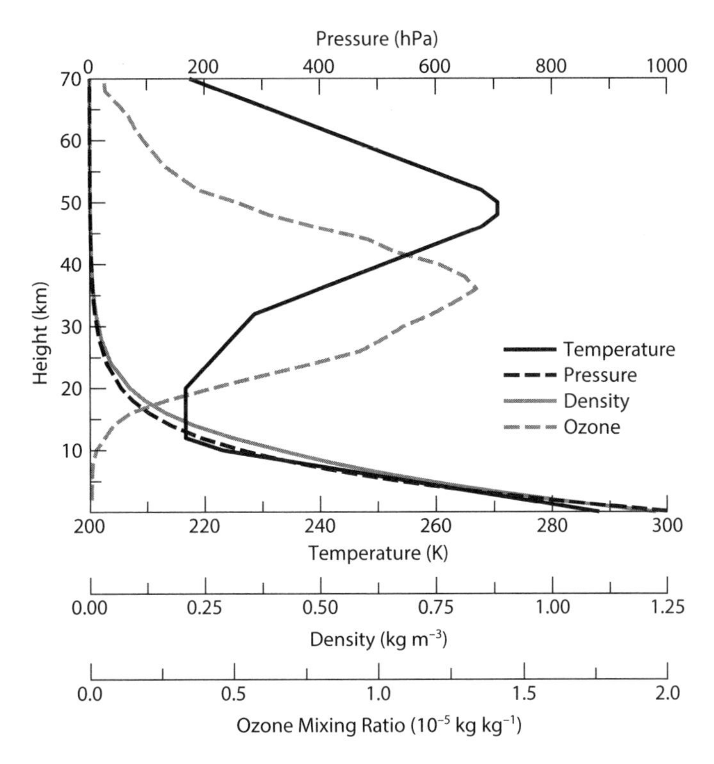

Figure 1.4. Idealized midlatitude temperature, pressure, density, and ozone profiles, for the lowest 70 km of the atmosphere. The temperature, pressure, and density profiles are based on the U.S. Standard Atmosphere (1976). The ozone profile is from Krueger and Minzner (1976).

The upper boundary of the troposphere is called the *tropopause*. The height of the tropopause varies from 17 km or so in some regions of the tropics to about half that near the poles. Above the tropopause, the temperature becomes uniform with height and then begins to increase with altitude in the region known as the *stratosphere*. The temperature increase is due to the absorption of solar radiation by ozone, which is created in the stratosphere by photochemical processes. Without ozone there would be no stratosphere. The summer-hemisphere stratosphere is almost devoid of active weather and has warm air over the pole. The winds of the summer-hemisphere stratosphere are predominantly *easterly*; that is, they blow from east to west. In contrast, the winter-hemisphere stratosphere experiences much more active weather, mainly owing to waves propagating upward from the troposphere below; has very cold air over the pole; and experiences strong *westerly* winds. During winter, the polar stratosphere is occasionally disturbed by *sudden stratospheric warmings*, which are dramatic changes in temperature (and wind) that occur sporadically in the Northern Hemisphere and much less frequently in the Southern Hemisphere.

Even though the stratosphere is very dry, its moisture budget is quite interesting. It receives small amounts of moisture from the troposphere and also gains some moisture through the oxidation of methane. The upper boundary of the stratosphere, called the *stratopause*, occurs near the 1 hPa (~50 km) level. In this book we focus mainly on the tropospheric circulation. We discuss selected aspects of the stratospheric circulation, but we do not discuss the portion of the atmosphere that resides above the stratopause.

For meteorological purposes, the tropics can be defined as the region from about 20° S to 20° N. Although the tropical temperature and surface pressure are remarkably uniform in space and temporally monotonous, the winds and rainfall are quite variable. In many parts of the tropics deep cumulus and cumulonimbus clouds—that is, thunderstorms—produce lots of rain and transport energy, moisture, and momentum vertically, essentially continuing the job begun closer to the surface by the turbulence of the PBL. The convective clouds often produce strong exchanges of air between the PBL and the free troposphere, in both directions: positively buoyant PBL air "breaks off" and drifts upward to form the cumuli, while negatively buoyant downdrafts associated with the evaporation of falling rain can inject free-tropospheric air into the PBL. In the convectively active parts of the tropics, the air is slowly rising in an area-averaged sense.

The mean flow in the tropical PBL is easterly. This is the *trade wind regime*. The tropical temperature and surface pressure distributions are generally very flat and monotonous, for simple dynamical reasons (discussed in chapter 3) that are connected to the smallness of the Coriolis parameter in the tropics. The tropical moisture and wind fields are more variable than the temperature, however. The tropics is home to a variety of distinctive traveling waves and vortices that organize the convective clouds on scales of hundreds to thousands of kilometers. Finally, the tropics is dominated by powerful and very large-scale monsoon systems that extend into the subtropics and even middle latitudes.

The tropical atmosphere acquires the angular momentum of the Earth's rotation from the continents and oceans. The global atmospheric circulation carries the angular momentum to higher latitudes, where it is "put back" into the continents and oceans.

The tropics is home to some circulation phenomena that do not occur in higher latitudes. Most famously, *tropical cyclones* produce tremendous amounts of rainfall and strong winds. They are relatively small in scale and highly seasonal. In contrast, *monsoons* are driven by seasonally varying continental-scale land-sea contrasts. The *Madden-Julian Oscillation*, or MJO, is a powerful tropical weather system that influences rainfall across about half of the tropics. *El Niño*, *La Niña*, and the *Southern Oscillation*, collectively known as ENSO, make up a strong, quasi-regular oscillation of the ocean-atmosphere system, with a period of a few years (Philander, 1990). In an El Niño, the sea-surface temperatures warm in the eastern tropical Pacific, while in a La Niña they cool. The Southern Oscillation is a shift in the pressure and wind fields of the tropical Pacific region that occurs in conjunction with El Niño and La Niña. The tropical stratosphere features an amazing periodic reversal of the zonal (i.e., west-to-east) wind, called the *Quasi-Biennial Oscillation*, or QBO. Its period is slightly longer than two years.

The subtropical portion of each hemisphere is roughly the region between 20° and 30° from the equator. In many parts of the subtropical troposphere, the air is sinking in large anticyclonic circulation systems called, appropriately enough, *subtropical highs*. The subsidence suppresses precipitation, which is why the major deserts of the world are found in the subtropics. Surface evaporation is very strong over the subtropical oceans, which have extensive systems of weakly precipitating shallow clouds. The subtropical upper troposphere is home to powerful *subtropical jets*, which are westerly currents that are particularly strong in the winter hemisphere (see fig. 1.5).

Figure 1.5. This photograph taken from the space shuttle shows cirrus clouds associated with the subtropical jet stream. From http://earth.jsc.nasa.gov/sseop/efs/lores.pl?PHOTO=STS039-601-49.

The tropical rising motion and subtropical sinking motion can be seen as the vertical branches of a "cellular" circulation in the latitude-height plane. This *Hadley circulation* transports energy and momentum poleward, and it transports moisture toward the equator. The Hadley circulation interacts strongly with the monsoons.

The region that we call the *middle latitudes* extends, in each hemisphere, from about 30° to 70° from the equator. There the surface winds are primarily westerly. The midlatitude free troposphere is filled with vigorous weather systems called *baroclinic eddies*, which have scales of a few thousand kilometers and grow through a process in which warm air shifts upward and poleward and is replaced by colder air that descends as it slides toward the equator (see fig. 1.6). These storms transport energy and moisture poleward and upward, primarily in the winter but also to some extent during summer. They transport westerly momentum both meridionally and downward. The downward momentum flux drives currents in the oceans and rustles the leaves on trees. The energy of the storms is derived from horizontal temperature differences that can be sustained only outside the tropics. The storms produce massive cloud systems and heavy precipitation.

On the average, the polar troposphere is characterized by sinking motion and radiative cooling to space. The North Pole is in the Arctic Ocean, which is covered with sea ice (see fig. 1.7) and is often blanketed by extensive cloudiness, while the South Pole is in the middle of a dry, mountainous continent (see fig. 1.8). Near the surface, the polar winds tend to be easterly, but weak.

The polar regions and middle latitudes are home to prominent *annular modes*, which by definition are visible even when the data are averaged over all longitudes. The annular modes fluctuate on a variety of timescales, almost uniformly in longitude. They are seen in both the stratosphere and troposphere, and make major contributions to the variability of the global circulation.

Overall, the atmosphere cools radiatively, and this cooling is balanced primarily by the release of latent heat, which in turn is made possible by surface evaporation. The net flow of energy is upward and toward the poles, carried by thunderstorms and the Hadley circulation in the tropics and by baroclinic eddies in middle latitudes. The energy escapes to space via infrared radiation at all latitudes, but especially in the subtropics.

Figure 1.6. A pair of beautiful winter storms over the North Atlantic Ocean in winter. From http://earthobservatory.nasa.gov/IOTD/view.php?id=7264.

Figure 1.7. An ice station used to collect data in and above the Arctic Ocean during 1997/98. From http://en.wikipedia.org/wiki/CCGS_Des_Groseilliers. Photo supplied by Don Perovich.

This book is organized as follows. Chapter 2 provides an overview of the upper and lower boundary conditions on the global circulation. At the "top of the atmosphere," the observed pattern of radiation implies net poleward energy transports by the atmosphere and ocean together. Lower boundary conditions include the distributions of oceans, continents, and mountains; the pattern of sea-surface temperature and the

Figure 1.8. Cold-looking clouds over the mountains of Antarctica. Photo courtesy of Dr. Byron Adams, Brigham Young University.

directly related pattern of sea-surface saturation vapor pressure; the heat capacity of the surface; the distribution of vegetation on the land surface; and the distributions of sea ice, continental glaciers, and ice sheets. These lower boundary conditions strongly affect the flows of energy and moisture across the Earth's surface. The chapter closes with a brief overview of the vertically integrated energy and moisture budgets of the surface and atmosphere, and the connections between them.

Chapter 3 introduces some basic aspects of the global circulation from an observational perspective, starting with the global distribution of atmospheric mass, then progressing to winds, temperature, and moisture. The amount of interpretation is deliberately kept to a minimum in the chapter.

While chapter 3 focuses on the observations, chapter 4 presents a brief but intensive review of theory to be used later in the book, beginning with a review of the dynamics of fluid motion on a rotating sphere. Angular momentum conservation is derived, and energy transports and transformations are discussed in detail. The key subject of potential vorticity is introduced near the end of the chapter. The quasi-geostrophic approximation and the shallow water equations are also presented.

Chapter 5 discusses how the zonally averaged circulation is influenced by sources and sinks of moisture, energy, and momentum. The chapter also takes a first look at the effects of *eddies* that transport dry air, moisture, energy, and momentum along *isentropic* surfaces, that is, on which the entropy is uniform.

Chapter 6 introduces the effects of convective energy sources and sinks, beginning with a brief review of the nature of buoyant convection in both dry and moist atmospheres. The famous observational study of convective energy transports by Riehl and Malkus is outlined, followed by a discussion of the idealized but important concept of radiative-convective equilibrium and an overview of convective cloud regimes, ranging from deep convection in the tropics to shallow stratocumulus convection in higher latitudes. The concept of convective mass flux is introduced, along with a theory that explains why cumulus updrafts tend to be separated by broad expanses of relatively dry, sinking air. The chapter shows how the convective mass flux can be used to understand the heating and drying associated with cumulus clouds and what determines the intensity of the convection as a function of the large-scale weather regime. Finally, the concept of conditional symmetric instability is introduced, which is particularly relevant in middle latitudes.

Chapter 7 presents the energetics of the global circulation, beginning with an in-depth discussion of *available potential energy* and the related concept of the *gross static stability*. These ideas are developed following Lorenz, through the use of isentropic coordinates, although it is also shown how they can be expressed using pressure coordinates. Discussion of the mechanisms by which vertical and meridional gradients of the zonally averaged circulation can be converted into zonally varying features associated with *eddy variances* is followed by an examination of how available potential energy is generated and converted into *eddy kinetic energy*. The chapter closes with a discussion of the observed *energy cycle* of the global atmosphere.

Chapter 8 introduces various types of eddies, beginning with a brief discussion of the *Laplace tidal equations*, omitting the details of the mathematics. It presents observations of the distribution of energy with scale for midlatitude eddies. The theory of *Rossby waves* forced by flow over topography and Matsuno's theory of *equatorial waves* are examined. Discussion of the monsoons, the east-west *Walker circulation* of the tropical Pacific, and the energy balance of the tropics and subtropics follows. The chapter closes with a discussion of the Madden-Julian Oscillation.

Chapter 9 deals with interactions of eddies with the mean flow beginning with the original *noninteraction theorem* of Eliassen and Palm as applied to gravity waves, followed by a brief discussion of the importance of *gravity-wave drag* for the global circulation. Examination of the quasi-geostrophic wave equation and its use by Charney, Dickinson, Matsuno, and others to interpret the interactions of Rossby waves with the zonally averaged circulation leads to a discussion of sudden stratospheric warmings.

Next, an analysis of the poleward and upward fluxes of sensible heat and momentum associated with winter storms is followed by a discussion of the Eliassen-Palm theorem as it relates to balanced flows. The theorem is first developed in pressure coordinates and then in isentropic coordinates to show the additional insights and greater simplicity gained, including the relationship between the divergence of the Eliassen-Palm flux and the flux of potential vorticity. The chapter closes with discussions of the quasi-biennial oscillation, blocking, and the Brewer-Dobson circulation as additional examples of the interactions between eddies and the zonally averaged circulation.

Chapter 10 discusses the global circulation as a kind of large-scale *turbulence*, starting with an examination of the nature of turbulence, framed in terms of vorticity dynamics. The very different flows of energy between scales in two- and three-dimensional turbulence are analyzed and connected with a description of the distribution of atmospheric kinetic energy with spatial scale. Examination of mechanisms that allow dissipation of enstrophy (squared vorticity) without dissipation of kinetic energy is followed by a discussion of mixing along isentropic surfaces. Finally, the predictability of the weather in a chaotic circulation regime is discussed, and the difference between climate prediction and weather prediction is explained.

The closing chapter briefly summarizes current trends and projected future changes in the global circulation due to increasing concentrations of atmospheric greenhouse gases.

The preceding outline shows that the study of the thermally driven global circulation of the atmosphere brings together concepts from all areas of atmospheric science. We touch on large-scale dynamics, convection, turbulence, cloud processes, and radiative transfer, with an emphasis on their interactions. Often, these various topics are presented as if they were somehow neatly separated from one another. This book will help you see how they fit together.

Because the global circulation spans all seasons, the concepts of atmospheric science encompass a wide range of conditions and contexts. For example, surface friction occurs everywhere: over the convectively disturbed tropical oceans, in the tumultuous storm track north of Antarctica, above the Himalayas, and over the tropical jungles. Discussions of boundary-layer meteorology are typically confined to relatively simple, horizontally uniform, cloud-free conditions, such as might be encountered on a summer morning in Kansas. Welcome to *An Introduction to the Global Circulation of the Atmosphere*; you are not in Kansas anymore. When we try to understand the global circulation, we quickly run up against the limits of knowledge in all the subdisciplines of atmospheric science, and so we are led to push those limits outward. This makes the study of the global circulation a particularly challenging and exciting field—as you are about to see for yourselves.

CHAPTER 2

What Makes It Go?

The Earth's Radiation Budget: An "Upper Boundary Condition" on the Global Circulation

Radiation is (almost) the only mechanism by which the Earth can exchange energy with the rest of the Universe. The most important upper boundary condition on the global circulation of the atmosphere is the incident solar radiation, also called the *insolation*. The insolation varies with geographic location and with time. It is determined by the energy output of the Sun, the spherical geometry of the Earth, and the geometry of the Earth's orbit (see fig. 2.1), which can be described in terms of the obliquity (the angle that the Earth's equatorial plane makes with the plane of the Earth's orbit around the sun), the eccentricity (a measure of the degree to which the shape of the Earth's orbit differs from a perfect circle), and the dates of the equinoxes. These all vary over geologic time (e.g., Crowley and North, 1991).

The solar energy flux at the mean radius of the Earth's orbit is about 1365 W m^{-2}. One way to get an intuitive grasp of this number is to imagine fourteen 100 W lightbulbs per square meter. Another is to consider that 1365 W m^{-2} is equivalent to 1.365 GW km^{-2}, which is the energy output of a large power plant.

The globally averaged top-of-the-atmosphere radiation budget is summarized in table 2.1. The numbers given in the table are now known to three significant digits. The Earth's albedo is near 0.30, independent of season; this number has been known to better than 10% accuracy only since the 1970s. The energy absorbed by the Earth is

$$\begin{aligned} S_{abs} &= S\left(\frac{\pi a^2}{4\pi a^2}\right)(1-\alpha) \\ &= \frac{1}{4}S(1-\alpha). \end{aligned} \tag{1}$$

Here S_{abs} is the globally averaged absorbed solar energy per unit area (given in table 2.1), S is the globally averaged insolation, a is the radius of the Earth, and α is the planetary albedo. The global average used to compute S_{abs} includes the zero values on the night side of the Earth. That is why S_{abs} is multiplied by πa^2, the area of the absorbing disk, and divided by $4\pi a^2$, the area of the sphere.

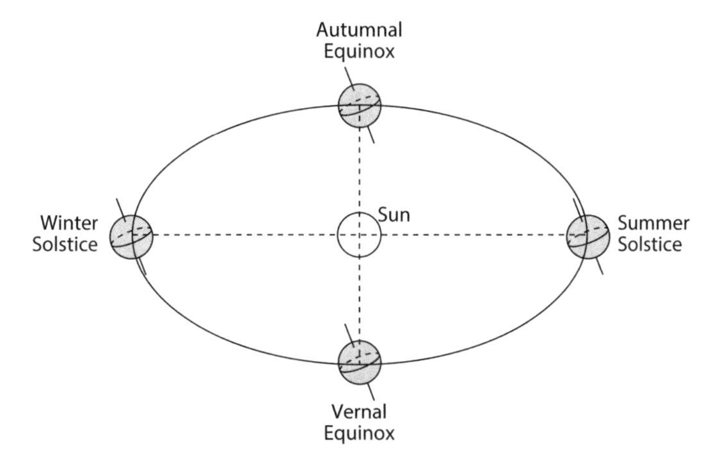

Figure 2.1. March of the seasons. As the tilted Earth revolves around the Sun, changes in the distribution of sunlight cause the succession of seasons. Used with permission of the University of Michigan Department of Astronomy.

Table 2.1. Summary of the Annually Averaged Top-of-the-Atmosphere Radiation Budget

Incident solar radiation	341 W m^{-2}
Absorbed solar radiation	239 W m^{-2}
Planetary albedo	0.30
Outgoing longwave radiation	239 W m^{-2}

Source: After Trenberth et al. (2009).

As a matter of common experience, the insolation varies both diurnally and seasonally. At a given moment, it also varies strongly with longitude. Because a year is much longer than a day, the daily-mean insolation is (almost) independent of longitude, but it varies strongly with latitude in a way that depends on the season, as summarized in figure 2.2. As we move from the solar equator (i.e., the latitude immediately "under" the Sun) to the summer pole, the insolation initially decreases, because at a given local time (e.g., local noon) the Sun appears to be lower in the sky. However, the increase in length of day at high latitudes in summer leads to an increase in the daily-mean insolation. Near the poles, the length-of-day effect dominates, so that at high latitudes in summer the insolation actually increases toward the pole. Thus there is a minimum of the insolation about 23° away from the pole in the summer hemisphere, as shown in the figure.

Seasonal and, to a lesser extent, diurnal cycles are clearly evident in the circulation patterns. Around the time of the solstices, there is no insolation at all near the winter pole (the polar night), but at the same time, near the summer pole the daily mean insolation is very strong despite low sun angles, simply because the sun never sets (the polar day). As is well known, these effects arise from the Sun-Earth geometry shown in figure 2.1. In addition, the distance from the Sun to the Earth varies slightly with time of year, resulting in about 7% percent more globally averaged insolation in January than in July, in the current epoch. The month of maximum insolation varies over geologic time. According to the widely accepted astronomical theory of the ice

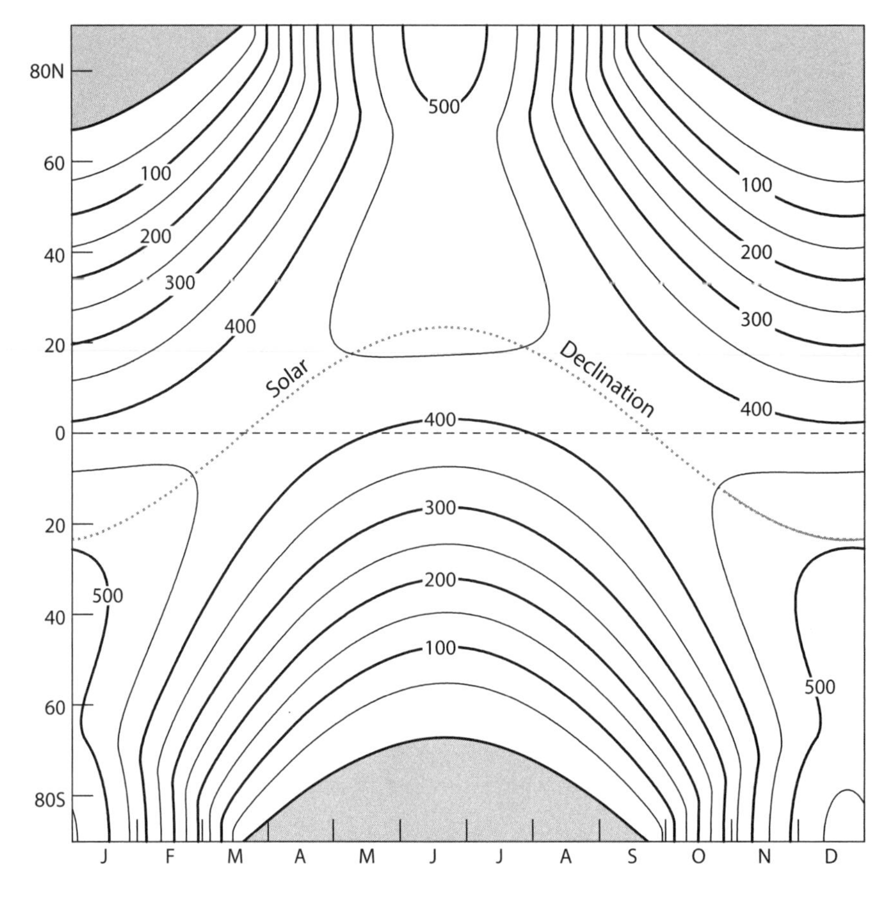

Figure 2.2. The seasonal variation of the zonally (or diurnally) averaged insolation at the top of the atmosphere.

ages, extensive glaciation is favored when the minimum insolation occurs during the Northern Hemisphere summer, because the Northern Hemisphere contains about twice as much land as the Southern Hemisphere (e.g., Crowley and North, 1991).

The Earth emits about as much infrared radiation as required to balance the absorbed solar radiation; both energy flow rates are about 240 W m^{-2}. This near balance has been directly confirmed by analysis of satellite data and has been observed to hold within a few watts per square meter, which is comparable to the uncertainty of the measurements. The actual annual-mean imbalance is thought to be about 0.5 W m^{-2} more absorbed than emitted (Loeb et al., 2012; Trenberth et al., 2014).

Figure 2.3 shows aspects of the Earth's radiation budget as observed from satellites (Wielicki et al., 1998). The zonally averaged incident (i.e., incoming) solar radiation at the top of the atmosphere varies seasonally in response to the Earth's motion around the sun. The zonally averaged *albedo*, which is the fraction of the zonally averaged incident radiation that is reflected to space, is highest near the poles, owing to cloudiness as well as snow and ice. It tends to have a weak secondary maximum in the tropics, associated with high cloudiness there. The zonally averaged terrestrial radiation at the top of the atmosphere, also called the *outgoing longwave radiation* or OLR, has its maxima in the subtropics. It is relatively small over the cold poles, but it also has a minimum in the warm tropics, owing to the trapping of terrestrial radiation by the cold, high tropical clouds and by water vapor.

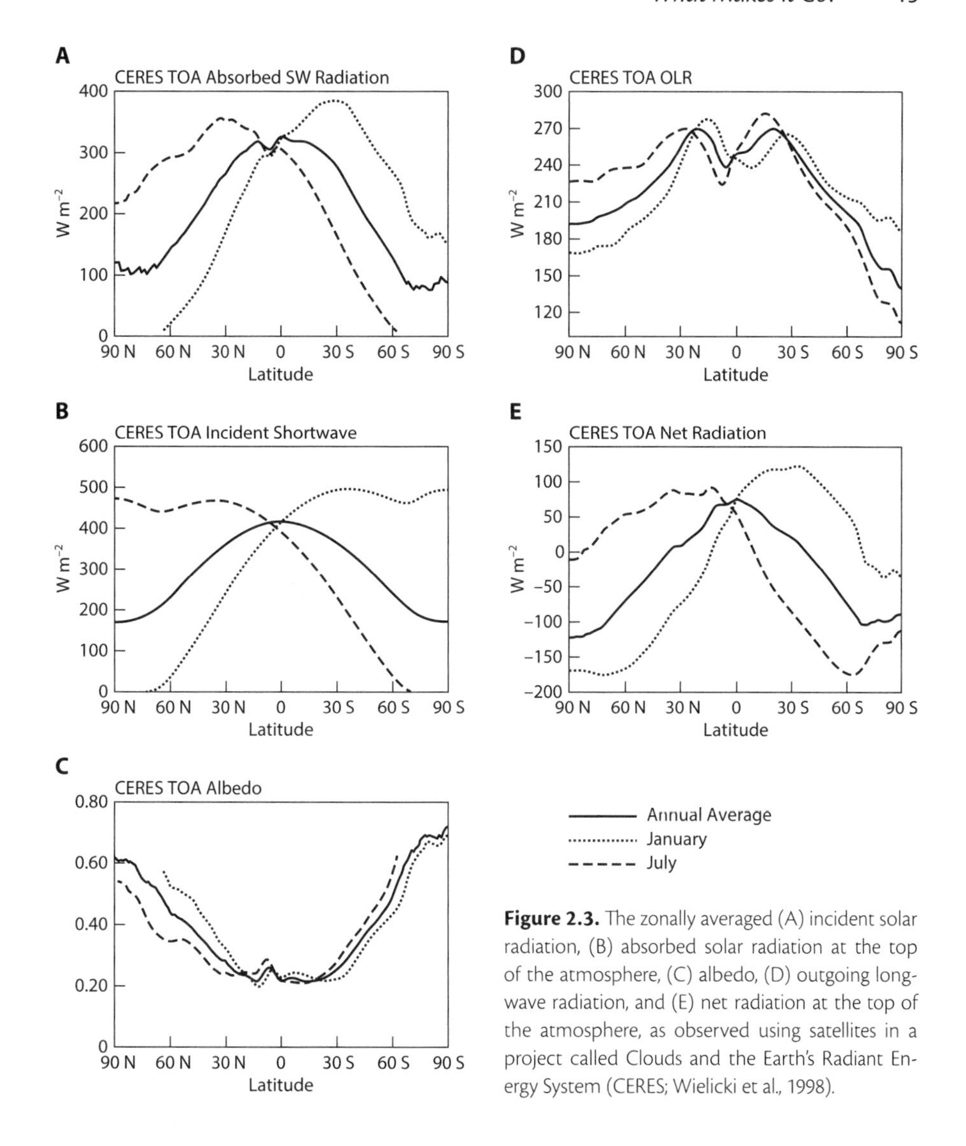

Figure 2.3. The zonally averaged (A) incident solar radiation, (B) absorbed solar radiation at the top of the atmosphere, (C) albedo, (D) outgoing long-wave radiation, and (E) net radiation at the top of the atmosphere, as observed using satellites in a project called Clouds and the Earth's Radiant Energy System (CERES; Wielicki et al., 1998).

The net radiation at the "top of the atmosphere" is the difference between the absorbed solar radiation and the OLR. It is positive in the tropics and negative in higher latitudes. This variation with latitude implies that energy is transported poleward somehow, inside the system. Some of the energy is transported by the atmosphere, and the rest is carried by the oceans.

Meridional Energy Transports by the Atmosphere-Ocean System

Energy transports are produced by the circulations of both the atmosphere and the oceans, and we can regard the global circulations of the atmosphere and oceans as a "response" to this pattern of net radiation. An important point, however, is that the distributions of the albedo and the outgoing longwave radiation are determined in part by the motion field. It is thus an oversimplification to regard these fields as simple forcing functions; they are strongly influenced by the circulation itself.

Consider the energy budget of a column that extends from the center of the Earth to the top of the atmosphere:

$$\frac{\partial E}{\partial t} = N_\infty - \nabla \cdot \mathbf{G}_\infty. \tag{2}$$

Here E is the energy per unit area stored in the column; t is time; N_∞ is the net downward flux of energy at the top of the atmosphere, which is entirely due to radiation. N_∞ has dimensions of energy per unit time per unit area (e.g., W m^{-2}). The energy transport, $-\nabla \cdot \mathbf{G}_\infty$, represents the movement of energy, in the zonal and meridional directions, due to both the winds and the ocean currents, and $\mathbf{G}_\infty$ is a vector with both zonal and meridional components, and dimensions of energy per unit length per unit time (e.g., W m^{-1}). The subscript ∞ on $\mathbf{G}_\infty$ means that it includes all parts of the Earth system, from the center of the Earth out to space.

Suppose that we average (2) over a time interval Δt:

$$\frac{E(t+\Delta t) - E(t)}{\Delta t} = \overline{N_\infty}^t - \nabla \cdot \overline{\mathbf{G}_\infty}^t. \tag{3}$$

Here $\overline{(\ )}^t$ represents a time average; this is a notation used throughout the book. Because the Earth is close to energy balance, $E(t+\Delta t)$ and $E(t)$ cannot be wildly different from each other; this means that the numerator on the left-hand side of (3) is bounded within a finite range regardless of how large Δt is. Therefore, as Δt increases, the left-hand side of (3) decreases in absolute value and eventually becomes negligible compared with the individual terms on the right-hand side. The physical meaning is that energy storage inside the Earth system at particular locations can be neglected if the time-averaging interval is long enough; the minimum time required for such an average would be one year, but ideally the average should be taken over many years. When we apply such a time average, the net radiation across the top of the column must be balanced by transports inside; this condition can be written as

$$\nabla \cdot \overline{\mathbf{G}_\infty}^t = \overline{N_\infty}^t. \tag{4}$$

The global mean of $-\nabla \cdot \mathbf{G}_\infty$ must be exactly zero, not just in a time average but at each instant. The reason is purely mathematical rather than physical: the global mean of the divergence of any vector is zero; you are asked to prove this in problem 1 at the end of this chapter. Equation (4) can be true only if the global mean of $\overline{N_\infty}^t$ is equal to zero, that is, if the Earth is in energy balance. To the extent that this is *not* true, the left-hand side of (3) is *not* negligible.

We now break $\mathbf{G}_\infty$ into its zonal and meridional components, that is,

$$\mathbf{G}_\infty = (G_\infty)_\lambda \mathbf{e}_\lambda + (G_\infty)_\varphi \mathbf{e}_\varphi. \tag{5}$$

Here $\mathbf{e}_\lambda$ and $\mathbf{e}_\varphi$ are unit vectors pointing toward the east and north, respectively. The symbols λ and φ represent longitude and latitude, respectively. We expand the divergence operator in spherical coordinates (see appendix A) as follows:

$$\nabla \cdot \mathbf{G}_\infty = \frac{1}{a\cos\varphi}\frac{\partial (G_\infty)_\lambda}{\partial \lambda} + \frac{1}{a\cos\varphi}\frac{\partial}{\partial \varphi}\left[(G_\infty)_\varphi \cos\varphi\right]. \tag{6}$$

Here a is the radius of the Earth. We multiply both sides of (6) by $a\cos\varphi$ and integrate over all longitudes to obtain

$$\int_0^{2\pi} (\nabla \cdot \overline{\mathbf{G}_\infty}^t) a \cos\varphi \, d\lambda = 2\pi \frac{\partial}{\partial \varphi}\left[\overline{(\mathbf{G}_\infty)_\varphi}^{\lambda,t} \cos\varphi\right] \\ = 2\pi a \overline{N_\infty}^{\lambda,t} \cos\varphi. \tag{7}$$

Here we use the notation

$$\overline{(\ )}^\lambda \equiv \frac{1}{2\pi} \int_0^{2\pi} (\ ) d\lambda \tag{8}$$

to denote a zonal mean. (There is a long but inconvenient tradition of using square brackets to denote zonal means. I have decided not to follow the tradition and hope that others will do the same.) The *combination* of a time average and a zonal mean can be denoted by either $\overline{(\ )}^{\lambda,t}$ or $\overline{(\ )}^{t,\lambda}$. In (7), the zonal derivative has dropped out as a result of the integration with respect to longitude. The second line of (7) results from comparison with (4).

Equation (7) gives us a way to compute the meridional derivative of the poleward energy transport, that is, $(\partial/\partial\varphi)\int_0^{2\pi} \overline{(G_\infty)_\varphi}^t \cos\varphi d\lambda$, in terms of $\overline{N_\infty}^{\lambda,t}$. What we require, however, is a formula for the poleward energy transport itself, rather than its divergence. Hence, we multiply (7) by a, and integrate with respect to latitude, from the South Pole ($\varphi = -\pi/2$) to an arbitrary latitude φ, to obtain

$$\overline{\Theta}^t(\varphi) = 2\pi a^2 \int_{\pi/2}^{\varphi} \overline{N_\infty}^{\lambda,t} \cos\varphi' d\varphi', \tag{9}$$

where we define

$$\Theta(\varphi) \equiv 2\pi a \cos\varphi \, \overline{(G_\infty)_\varphi}^{\lambda,t}, \tag{10}$$

and we use the boundary condition

$$\Theta(-\pi/2) = 0. \tag{11}$$

The dimensions of $\Theta(\varphi)$ are energy per unit time (e.g., W). The boundary condition (10) is *exact*; if it were not true, a finite amount of energy per unit time would be (impossibly) flowing into or out of the South Pole, which is a "point" of zero mass. A similar condition must apply at the North Pole. The right-hand side of (9) is the *area integral* of $\overline{N_\infty}^t$ over the "south polar cap" that extends from the South Pole up to latitude φ. When the upper limit of meridional integration in (8) is set to $\pi/2$, the right-hand side of (9) reduces to the global mean of $\overline{N_\infty}^t$.

Figure 2.4 gives a plot of $\overline{\Theta}^t(\varphi)$, computed by using (9), with values of $\overline{N_\infty}^t$ based on satellite data from the CERES project (Wielicki et al., 1996, 1998). A poleward energy transport is clearly apparent in both hemispheres. The curve of the transport has a pleasingly simple shape, roughly like $\sin(2\varphi)$. Vonder Haar and Oort (1973) were the first to determine $\overline{\Theta}^t(\varphi)$ from satellite measurements of the Earth's radiation budget. The maximum absolute values in middle latitudes of both hemispheres are on the order of 6 PW (a petawatt is 10^{15} J s^{-1}). Because this total energy transport by the climate system can be inferred directly from satellite measurements of the Earth's radiation budget, it is now known with relatively good accuracy.

Figure 2.4 shows that $\overline{\Theta}^t(\varphi)$ is *exactly* zero at both poles. We built that result in for the South Pole, by using (11), but what makes $\overline{\Theta}^t(\varphi)$ miraculously return to zero at the North Pole? The answer is that we have forced it to be zero at the North Pole by "correcting" the data. The observations show that the global mean of $\overline{N_\infty}^t$ is small compared with the local values. It is zero *within the uncertainty of the measurements*, but of course it is not exactly zero. Before computing $\overline{\Theta}^t(\varphi)$ from (9), we apply a

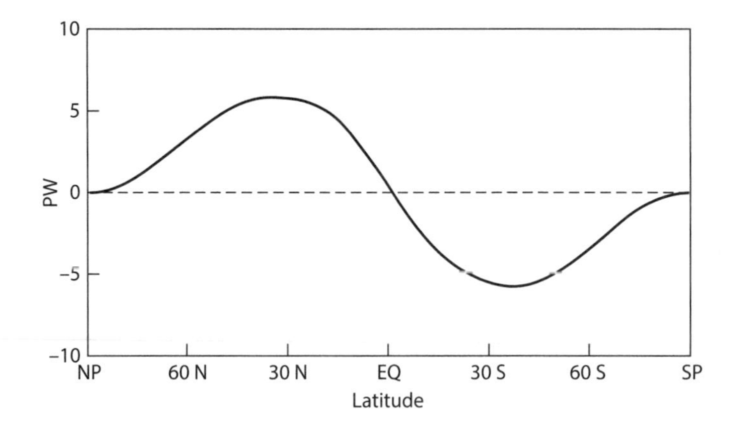

Figure 2.4. The poleward energy transport by the atmosphere and ocean combined, $\overline{\Theta}^t(\varphi)$, as inferred from the observed annually averaged net radiation at the top of the atmosphere using (9). A petawatt (PW) is 10^{15} W. The plot was created using CERES data.

small, globally uniform correction to $\overline{N_\infty}^t$, such that *after the correction* the global mean of $\overline{N_\infty}^t$ is exactly zero. As can be seen from (9), this is sufficient to ensure that $\overline{\Theta}^t(\varphi)$ is zero at the North Pole. The correction to $\overline{N_\infty}^t$ can be interpreted as compensating for the fact that the time rate of change term of (3) is not completely negligible, because the global mean of $\overline{N_\infty}^t$ is small but not zero. The correction also compensates for the inaccuracies of the data used to compute $\overline{N_\infty}^t$.

We can say that the "job" of the global circulations of the atmosphere and oceans is to carry out the meridional energy transport shown in figure 2.4. If the transport of energy from place to place by the atmosphere and oceans could somehow be prevented, then each part of the Earth would have to come into *local* energy balance, by adjusting its temperature, water vapor, and cloudiness so that the outgoing longwave radiation locally balanced the absorbed solar radiation. Such a hypothetical state is referred to as "radiative-convective equilibrium"; modeling studies of radiative-convective equilibrium are discussed in chapter 6. Radiative-convective equilibrium would presumably entail much warmer temperatures in the tropics and much colder temperatures at the poles. The global circulation of the atmosphere and oceans has a moderating effect on the global distribution of temperature, tending to warm the higher latitudes and cool the tropics. As discussed in chapter 7, these same thermal contrasts between the tropics and the poles represent a source of energy (called *available potential energy*) that makes the global circulations of the atmosphere and oceans possible.

If we consider that the global circulation of the atmosphere exists to produce the energy transports shown in figure 2.4, then we can imagine that the "strength of the circulation," as measured for example by the total kinetic energy of the atmosphere, is determined by the magnitude of the required energy transports.

Observations and theory of energy transports by the atmosphere and oceans are discussed further in later chapters.

Surface Boundary Conditions

The global atmospheric circulation is strongly affected by the properties of the Earth's surface and their geographic variations. The most important properties of the Earth's surface are temperature, wetness, topography, heat capacity, albedo, roughness, vegetation, sea ice, land ice, and the energy and moisture budgets of the surface and atmosphere.

Temperature

The temperature of the Earth's surface varies strongly and rapidly over land, and considerably less so over the oceans. The reason for this difference between land and sea temperatures will be discussed in the section on the surface heat capacity.

The oceans cover about two-thirds of the Earth's surface, and their average depth is about 4 km. Water is heavy stuff; the mass of 1 m^3 of water is 10^3 kg, so the mass of the oceans is about 1.3×10^{21} kg. In comparison, the mass of the atmosphere is about 250 times less, roughly 5×10^{18} kg.

Not only is water dense, it has a very high specific heat: about 4200 J kg^{-1} K^{-1}. In contrast, the specific heat of air (at constant pressure) is a little less than a quarter of that, that is, 1000 J kg^{-1} K^{-1}. The total heat capacity of the oceans is thus about 1000 times greater (250×4) than the total heat capacity of the atmosphere. When the oceans say, "Jump," the atmosphere says, "How high?"

The density of the atmosphere decreases exponentially with altitude and can change by 10% or so in the course of a year at a given location, owing to changes in temperature and pressure. In contrast, the density of seawater varies by only a few percent throughout the entire ocean; it is a complex but fairly weak function of temperature, salinity, and pressure. Because of the near incompressibility of water, pressure effects (called *thermobaric* effects) are relatively unimportant; variations in density are mainly due to changes in temperature and salinity. Warmer and fresher water is less dense and tends to float on top; colder and saltier water is more dense and tends to sink. Surface cooling and evaporation create dense water; surface heating and precipitation create light water. Note that the properties of the water are altered mainly near the surface; below the top hundred meters or so, the properties of water parcels remain nearly invariant, even over decades or centuries.

In the study of the atmosphere, we often treat the sea-surface temperature (SST) as a seasonally varying lower boundary condition. Figure 2.5 shows the observed distributions of the SST for January and July. Note the warm currents off the east coasts of North America and Asia, and the cold currents off all west coasts. The warm SSTs at a particular latitude are generally associated with poleward currents; the two best known of these are the Gulf Stream and the Kuroshio. The colder SSTs are generally associated with either equatorward flow (as for example in the case of the California Current) or with upwelling (again, in the region of the California Current, and also along the equator, most noticeably in the eastern Pacific).

As discussed later, the pattern of upwelling is very closely related to the low-level winds; at the same time the low-level winds are strongly tied to the spatial distribution of the SST. The seasonal change of the SST is largest in the Northern Hemisphere, particularly on the western sides of the ocean basins. Note that the seasonal forcing is capable of changing the SSTs by tens of degrees in some mid- and high-latitude locations. The depth to which this seasonal change penetrates is naturally variable but is typically on the order of 100 m. Of course, the temperature of the water at great depth undergoes virtually no seasonal change. In the study of the global atmospheric circulation we often consider the spatial and seasonal distribution of the SST to be "given,"

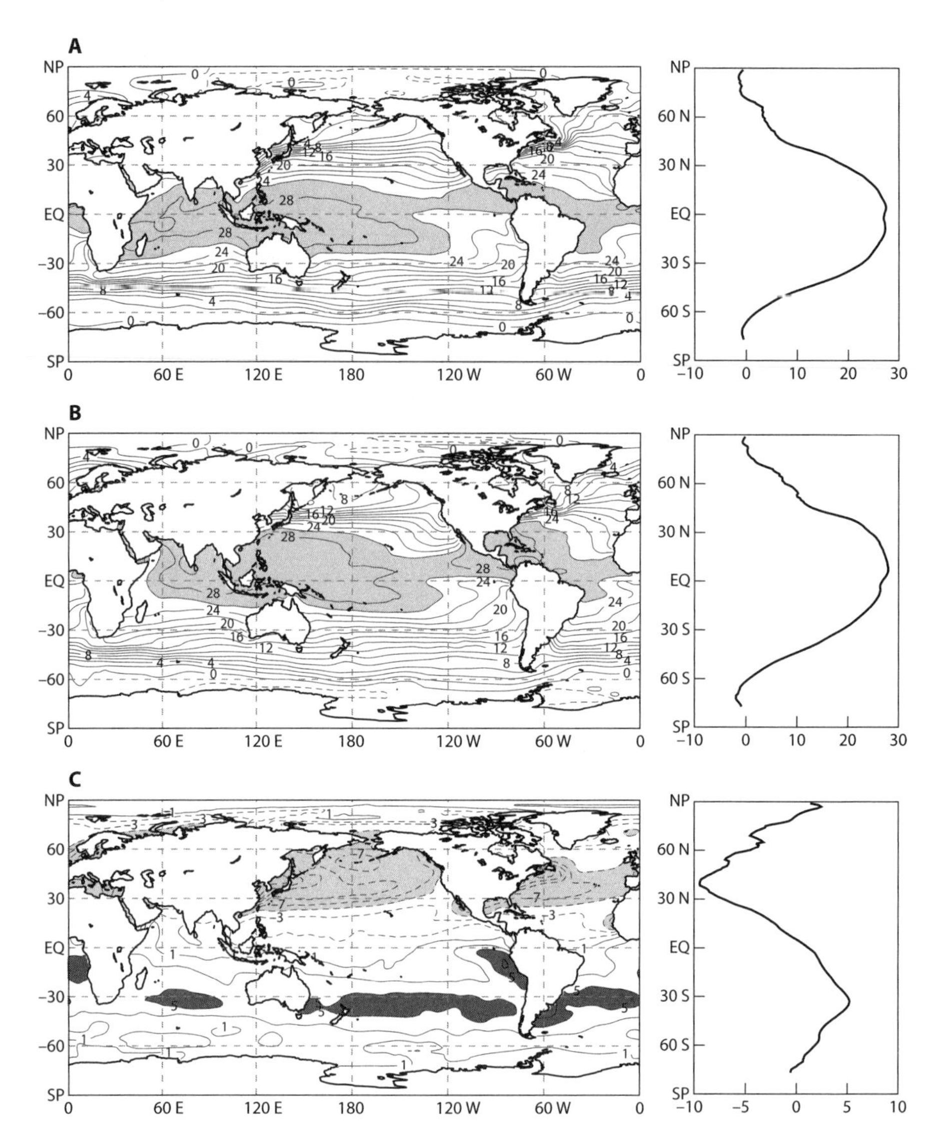

Figure 2.5. (A) SST distribution for January. The contour interval is 2 °C, with shading for values higher than 28 °C. (B) The same distribution for July. (C) SST difference (in Kelvin) between March and September, with light shading for values below −5 K and dark shading for values above 5 K. In each panel, zonal means are shown on the right.

but in reality it is determined in part by what the atmosphere is doing or, rather, what the atmosphere has been doing over time. For example, the distribution of cloudiness strongly affects the flow of solar radiation into the upper ocean, and over time this tends to reduce the SST where clouds are prevalent and the solar insolation at the top of the atmosphere is strong, relative to what the SST would be if the cloudiness were somehow prevented from occurring. The role of clouds in determining the distribution of the SST is a major complication hindering our understanding of the atmosphere and ocean as a coupled system.

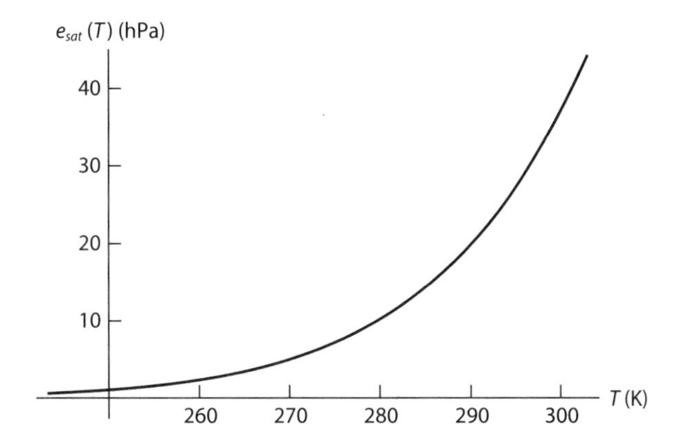

Figure 2.6. The saturation vapor pressure, e_{sat}, as a function of temperature.

Wetness

One of the most important properties of the Earth's surface is that roughly 70% of it is permanently wet and so represents a huge source of moisture. The vapor pressure immediately above a wet surface, called the *saturation vapor pressure*, is a strong function of temperature only. It is approximately given by

$$e_{sat}(T) \cong 6.11 \exp\left[\frac{L}{R_v}\left(\frac{1}{273} - \frac{1}{T}\right)\right] \text{hPa}, \tag{12}$$

where T is the temperature in Kelvin, $L = 2.52 \times 10^6$ J K^{-1} is the latent heat of water vapor, and $R_v = 461$ J K^{-1} kg^{-1} is the gas constant for water vapor. The enormous value of the latent heat of water vapor is one of the reasons why moisture strongly influences the Earth's climate and the global circulation of the atmosphere. The strong temperature dependence of $e_{sat}(T)$ is shown in figure 2.6. Held and Soden (2006) pointed out that for typical surface temperatures the rate of increase is a spectacular 7% per kelvin. Figure 2.7 shows the geographic variation of $e_{sat}(T)$ based on the SSTs plotted in figure 2.5. As discussed later, near the surface over the oceans the actual vapor pressure of the air, that is, the partial pressure of water vapor, "tries" to be $e_{sat}(T)$ but usually falls short by 20% or so. At any rate, the "effective wetness" of the ocean increases with the SST. The largest values of $e_{sat}(T)$ occur in the tropics, of course, and are close to 40 hPa. In that very humid region, about 4% of the near-surface air is water vapor.

The availability of land-surface moisture to the atmosphere is much more complicated and is discussed separately later.

Topography

You should not be surprised to learn that mountains have a strong effect on the global atmospheric circulation. Figure 2.8 shows the locations of the Earth's mountain ranges. Mountains block the wind; this process can be called a *mechanical forcing*. The

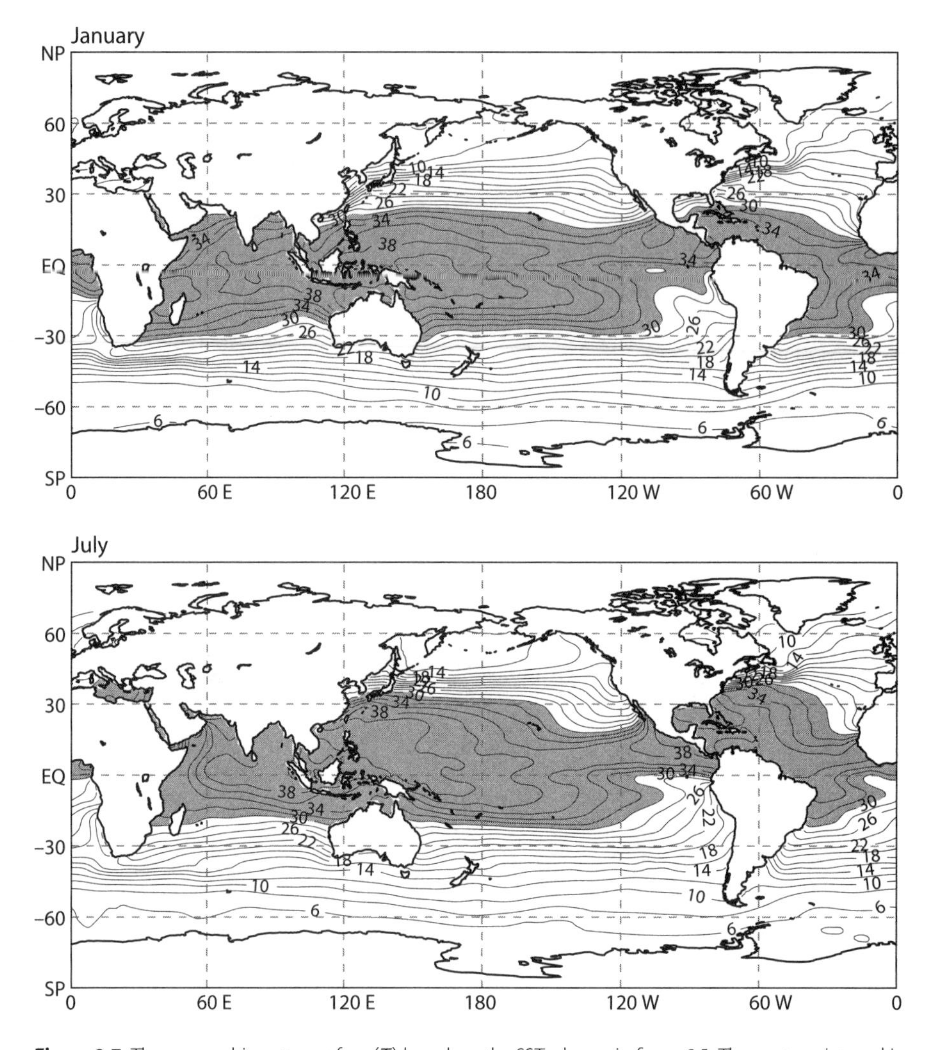

Figure 2.7. The geographic pattern of $e_{sat}(T)$ based on the SSTs shown in figure 2.5. The contour interval is 2 hPa. Values larger than 30 hPa are shaded.

air can flow around or over a mountainous obstacle. What actually happens depends in part on the scale of motion. The distribution of surface pressure across a mountain range can exert a net force on the solid Earth and an equal and opposite net force on the atmosphere. Chapter 5 explains this mechanism.

Mountains can also exert a *thermal forcing* on the atmosphere, because the temperature of a mountain surface can be quite different from that of the surrounding air at the same height. For example, during the northern summer the Tibetan Plateau produces a "warm spot" in the middle troposphere that represents an important aspect of the thermal forcing associated with the Indian summer monsoon. This topic is discussed further in chapter 8.

Finally, mountains strongly influence the geographic distribution of precipitation. Rain and snow are enhanced wherever topography forces the air to flow uphill and also wherever surface heating on sunlit slopes promotes moist convection. In

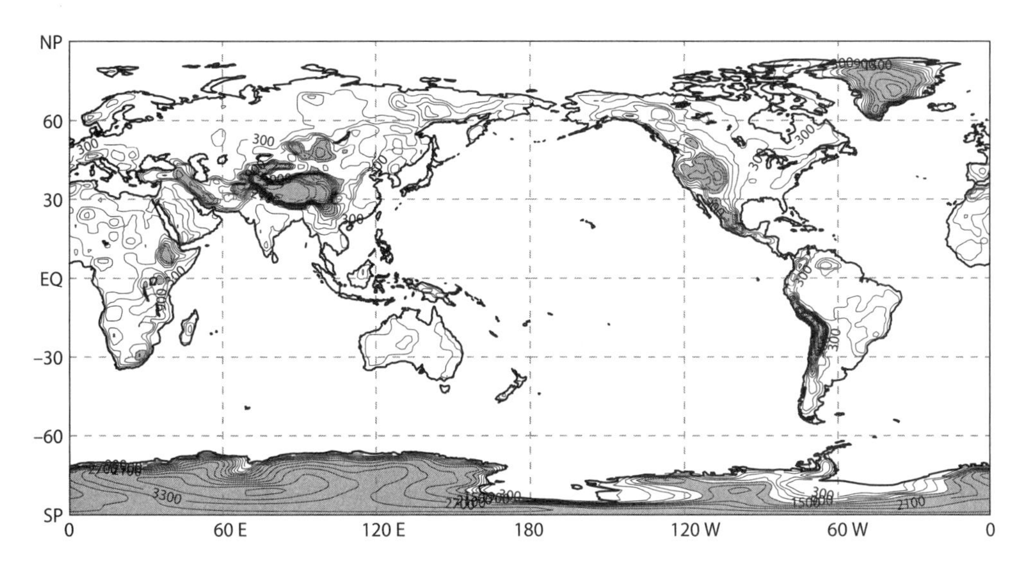

Figure 2.8. The Earth's orography, averaged onto a mesh with a grid spacing of 1° of longitude by 1° of latitude. The contour interval is 300 m. Values higher than 1500 m are shaded.

contrast, precipitation is often reduced on the downstream side of mountainous regions, because the air tends to sink there and also because the moisture content of the air can be depleted by upstream precipitation maxima.

Heat Capacity

The heat capacity of the Earth's surface is the amount of energy needed to change the "skin temperature" of the surface by a given amount. Here the skin temperature, T_S, is defined as the temperature of an equivalent blackbody that would emit infrared radiation at a rate equal to the actual infrared emission by the Earth's surface. The heat capacity is highly variable in space and also varies somewhat less dramatically with time. The concept of heat capacity sounds simple, but it is actually somewhat subtle. The time change of the skin temperature satisfies an equation very similar to (2), that is,

$$C\frac{\partial T_S}{\partial t} = N_S - \nabla \cdot \mathbf{G}_S. \tag{13}$$

Here C is the heat capacity of the surface, N_S is the net downward energy flux at the Earth's surface (due to radiation and other processes discussed later), and $-\nabla \cdot \mathbf{G}_S$ is the horizontal energy transport "inside" the Earth's surface. For the land, we can assume that $\mathbf{G}_S \cong 0$, but for the oceans we expect that energy transport by currents will lead to $\mathbf{G}_S \neq 0$. The heat capacity, C, depends on the composition of the material both at and below the surface, because energy flowing into the surface can be stored through a finite depth. It can also depend on the rate at which the surface energy flux changes with time. For example, a diurnally varying surface energy flux can affect only a shallow layer of soil, with its correspondingly small heat capacity, but a seasonally varying flux can add or remove energy through a much deeper layer of soil, which has a greater heat capacity.

In general, the oceans have very high heat capacity. The geographic distribution of SST does fluctuate seasonally, however, and varies considerably with both longitude and latitude, as shown in figure 2.5.

The land surface has a much smaller heat capacity. This implies that the net surface energy flux averages to nearly zero over land, even for a single day. To understand why, note that with $C \rightarrow 0$ and $\mathbf{G}_S \cong 0$ (appropriate for land), (13) implies that $N_S \cong 0$. For the ocean, with large values of C, daily mean values of N_S can be much larger. The large heat capacity of the ocean means that it is relatively hard to change the SST, because

$$\frac{\partial T_S}{\partial t} = \frac{N_S - \nabla \cdot \mathbf{G}_S}{C}; \tag{14}$$

that is, a large value of C in the denominator on the right-hand side of (14) reduces $\partial T_S / \partial t$ for a given value of $N_S - \nabla \cdot \mathbf{G}_S$. Consequently, the SST can for some purposes be considered a "fixed" lower boundary condition on the atmosphere.

Albedo

The degree to which the surface reflects solar radiation obviously affects its response to the Sun. The surface albedo depends on surface composition and sun angle, among other factors. The ocean has an albedo close to 0.06 when the Sun is high in the sky; that is, it is quite dark. At low sun angles, however, the ocean can reflect considerably more of the incident solar radiation. The albedo of the land surface varies widely, owing to differing compositions of the soil or rock at the surface, differing types and amounts of vegetation cover (discussed further below), and of course the presence or absence of snow.

Roughness

"Rough" surfaces exert a drag on the wind more readily than smooth ones. The surface roughness is another example of a lower boundary condition that is at least partially mechanical in nature. The ocean is relatively smooth, depending on the wind speed, and presents little "roughness" to stimulate momentum exchange with the atmosphere. The land surface is much rougher than the ocean.

Vegetation

The vegetation on the land surface regulates the flow of moisture from the soil, as discussed further below. It also affects both the roughness and albedo of the surface. The pattern of vegetation on the land surface affects the atmosphere in very complicated ways. I have not included a map of vegetation types here, because it is virtually impossible to depict the global distribution of vegetation types without using color, which is not possible in this book. Many good vegetation maps are available online. Obviously, the type, density, and even the health of the land-surface vegetation can affect the surface albedo and surface roughness. These characteristics of the vegetation vary with season, especially in middle latitudes. They can also vary interannually.

The degree to which plants allow moisture to transpire from their leaves into the atmosphere strongly regulates the surface fluxes of sensible and latent heat; strong transpiration cools the surface and reduces the sensible heat flux. Moisture enters plants through their root systems and is transpired through pores in their leaves called *stomata*. The purpose of stomata is gas exchange: carbon dioxide enters and is used in photosynthesis; oxygen is released. Water escapes through the stomata as an accidental side effect. Without vegetation, moisture would move up through the soil and into the atmosphere only through molecular diffusion, which is a very slow process. In effect, the plants are solar-powered water pumps. Sellers et al. (1997) provide an introductory overview.

Sea Ice

The distribution of sea ice (see fig. 2.9) also acts as a thermal lower boundary condition. There are strong seasonal changes in ice cover in the Southern Hemisphere, but not in the Northern Hemisphere. In addition to the obvious strong effect of sea ice on the surface albedo, the ice also acts as an insulator that separates the relatively warm ocean water from the air. Because sea ice is a good insulator, its upper boundary can be much colder than the water beneath. Sea ice is also very smooth, so that little surface drag occurs for a given wind speed. Until recently, the Arctic Ocean was ice-covered all year, while the North Atlantic and the Southern Oceans have long experienced seasonal melting. Of course, the *thickness* of the ice also varies both geographically and seasonally, and the thickness strongly determines the insulating power of the ice. In addition, there typically is a small percentage of open water, especially when the ice is thin. This open water often takes the form of cracks called *leads*. The water in the leads can be much warmer than the ice nearby, especially in winter. Under such conditions, the large-scale average sensible and latent heat fluxes can be dominated by the contributions from the leads, even though leads may cover only a small percentage of the area. Snow that falls on the sea ice insulates it and protects it from the effects of the sun, helping to prevent the ice from melting.

Land Ice

The land-sea distribution and the locations of "permanent" (or, more accurately, nonseasonal) land ice (e.g., the ice sheets that cover Antarctica and Greenland) strongly affect the surface albedo. Over land, the geographic and seasonal variations of the surface albedo are largely determined by the distribution of vegetation, but of course they also depend on snow cover. Permanent land ice is confined mainly to Antarctica and Greenland, in the present climate, although there are many smaller glaciers throughout the world. The Greenland and Antarctic ice sheets are thousands of meters thick in places and so increase the effective topographic height of the Earth's surface. The distribution of land ice can vary dramatically on timescales of thousands of years and longer (e.g., Imbrie and Imbrie, 1979; Crowley and North, 1991).

Energy and Moisture Budgets of the Surface and Atmosphere

Some aspects of the global atmospheric circulation can be regarded as more or less direct responses to the various boundary conditions already mentioned. Examples include the equator-to-pole energy flux by the atmosphere, planetary waves produced by flow over mountains, and monsoons that are strongly tied to the land-sea

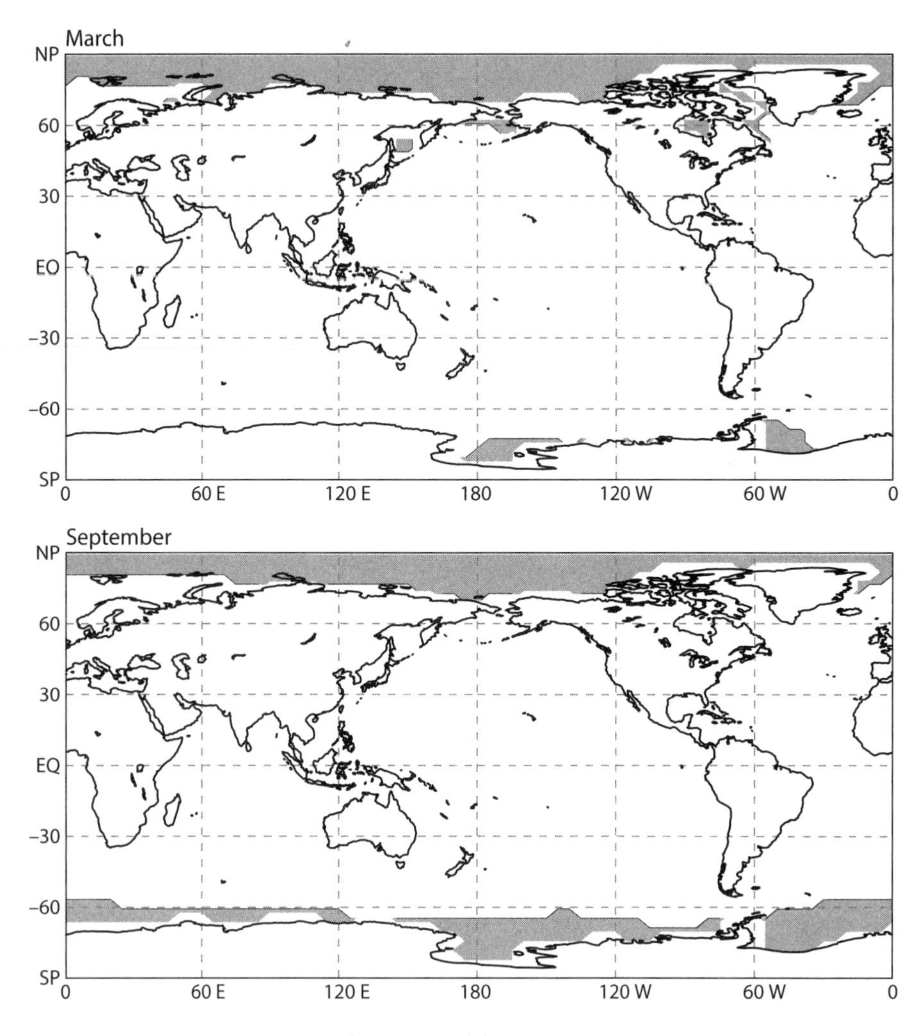

Figure 2.9. The distributions of sea ice (shown in gray) for March and September. The data represent averages for grid cells 5° of longitude wide and 4° of latitude high. The data shown here are representative of the mid-twentieth century. In more recent years, the September Arctic sea ice has been less extensive than shown here.

distribution and the seasonally varying insolation. Of course, many additional time-dependent features of the circulation are less directly tied to the boundary conditions but instead arise from the internal dynamics of the atmosphere. These include winter storms, tropical cyclones, and many other phenomena.

The planetary radiation budget has already been briefly discussed. We now consider the energy and moisture budgets of the Earth's surface and the atmosphere, as shown in table 2.2. The values in this table are known to two significant digits. Of the 239 W m^{-2} that is absorbed by the Earth-atmosphere system, 161 W m^{-2} is absorbed by the Earth's surface. Thus only about $239 - 161 = 78$ W m^{-2} of solar radiation is absorbed by the atmosphere, which is about one-third of the total solar radiation absorbed by the Earth-atmosphere system. The surface receives a total (LW↓ + SW; see the notation defined in table 2.2) of 494 W m^{-2} of incoming radiation. Note that

Table 2.2. Components of the Globally and Annually Averaged Surface Energy Budget

Absorbed solar (SW)	161 W m^{-2}
Downward infrared (LW↓)	333 W m^{-2}
Upward infrared (LW↑)	−396 W m^{-2}
Net longwave (LW)	−63 W m^{-2}
Net radiation (SW + LW)	98 W m^{-2}
Latent heat (LH)	−80 W m^{-2}
Sensible heat (SH)	−17 W m^{-2}

Source: After Trenberth et al. (2009).
Note: A positive sign means that the surface is warmed.

LW↓ is about twice as large as SW! This is the "greenhouse" effect on the surface energy budget. The incoming energy due to the longwave and solar radiation absorbed by the surface is returned to the atmosphere in the form of LW↑, LH, and SH. By far the largest of these is LW↑. The oceans can transport energy from one place to another, so the energy absorbed by the oceans is not necessarily given back in the same place where it is absorbed. Also, the large heat capacity of the upper ocean allows energy storage on seasonal timescales. In contrast, the continents cannot transport energy laterally at any significant rate, and their limited heat capacity forces near energy balance, everywhere, on timescales of a few days at most. Table 2.2 shows that the net radiative heating of the surface, which amounts to 98 W m^{-2}, is balanced primarily by evaporative cooling of the surface at the rate of 80 W m^{-2}. In other words, the surface cools itself by evaporating water.

The globally averaged energy budget of the atmosphere is shown in table 2.3. An interpretation of table 2.3 is that the atmosphere sheds energy through infrared radiation at the rate required to balance the various forms of energy input, and the temperature of the atmosphere adjusts to allow the necessary infrared emission. *The net radiative cooling of the atmosphere, at the rate of −98 W* m^{-2}*, is balanced primarily by the latent energy source due to surface evaporation.* Of course, the latent energy is converted into sensible heat when water vapor condenses. A fraction of the condensed water reevaporates inside the atmosphere. The net condensation rate within the atmosphere is closely balanced by the rate of precipitation at the Earth's surface, which means that the amount of condensed water in the atmosphere is neither increasing nor decreasing with time. The rate at which evaporation introduces moisture into the atmosphere has to be balanced by the rate at which precipitation removes it. Keep in mind that these various balances apply in a globally

Table 2.3. The Globally and Annually Averaged Energy Budget of the Atmosphere

Absorbed solar radiation (240 − 161)	78 W m^{-2}
Net infrared cooling (−239 + 63)	−176 W m^{-2}
Net radiative heating	−98 W m^{-2}
Latent heat input	80 W m^{-2}
Sensible heat input	17 W m^{-2}

Note: The values were obtained by combining the numbers in tables 2.1 and 2.2; a positive sign means that the atmosphere is warmed.

averaged sense rather than locally in space, and in a time-averaged sense rather than instantaneously.

The globally averaged rates of precipitation and evaporation are measures of the "speed" or intensity of the hydrologic cycle. The preceding discussion suggests a second interpretation of the atmospheric energy budget: to a first approximation, *the speed of the hydrologic cycle is "determined by" the rate at which the atmosphere is cooling radiatively.* Of course, this does not mean that the geographic and temporal distributions of precipitation are determined by the corresponding distribution of radiative cooling; in fact, the *local* rate of precipitation tends to be *negatively* correlated with the local atmospheric radiative cooling, because precipitation systems produce high, cold clouds (see below) that reduce the infrared emission to space. This topic is discussed further in chapter 5.

The local rate of precipitation is controlled mainly by dynamical processes, and the rate of evaporation from the Earth's surface is influenced by the surface wind speed. To some extent, the overall strength of the global circulation of the atmosphere is determined by, or at least must be consistent with, the speed of the hydrologic cycle that is required to balance the globally averaged rate of atmospheric radiative cooling.

The net radiative cooling of the atmosphere is strongly affected by high, cold cirrus clouds, many of which are formed within precipitating cloud systems. The cirrus clouds absorb the infrared radiation emitted by the warm atmosphere and surface below; the cirrus themselves emit much more weakly because they are very cold. This means that the cirrus effectively trap infrared radiation inside the atmosphere. For this reason, as the amount of cirrus clouds increases, the radiative cooling of the atmosphere decreases.

Consider together the following points that have been made in the last few paragraphs:

- The radiative cooling of the atmosphere is balanced primarily by latent heat release in precipitating cloud systems.
- Precipitating weather systems produce cirrus clouds.
- Cirrus clouds tend to reduce the radiative cooling of the atmosphere.

In combination, these factors suggest a negative feedback loop that tends to regulate the strength of the hydrologic cycle. To see how this process works, consider an equilibrium in which atmospheric radiative cooling and latent heat release are in balance. Suppose that we perturb the equilibrium by increasing the speed of the hydrologic cycle, including the rate of latent heat release. The same perturbation will increase the rate of cirrus cloud production, which will reduce the rate at which the atmosphere is radiatively cooled. The radiative cooling acts to promote cloud formation through moist convection, so when the radiative cooling rate decreases, cloud activity slows down. In this way, the initial perturbation is damped. This topic is discussed further in chapter 5.

The "effective altitude" for infrared emission by the Earth-atmosphere system is near 5 km above sea level. This simply means that the outgoing longwave radiation at the top of the atmosphere is equivalent to that from a blackbody whose temperature is that of the atmosphere near the 5 km level. Roughly speaking, then, atmospheric motions must carry energy upward from the surface through the first 5 km of the atmosphere, and infrared emission carries the energy the rest of the way out to space.

This upward energy transport by circulating air occurs on both small scales, notably in boundary-layer turbulence and cumulus convection, and also on large scales, notably through midlatitude baroclinic eddies and the tropical Hadley circulation, which are discussed in later chapters. In short, *the atmospheric circulation carries energy upward as well as poleward.*

We now examine in more detail the fluxes of various quantities at the Earth's surface. In addition to the surface solar and terrestrial radiation, we must also consider the turbulent fluxes of momentum, sensible heat, and latent heat. In principle, we should also consider the fluxes of various chemical species, but this important aspect of the climate system is neglected here.

An example of the seasonal variations of the surface shortwave and longwave radiation at a particular station is given in figure 2.10, which shows the variations in the upward and downward shortwave (SW) and longwave (LW) near-surface

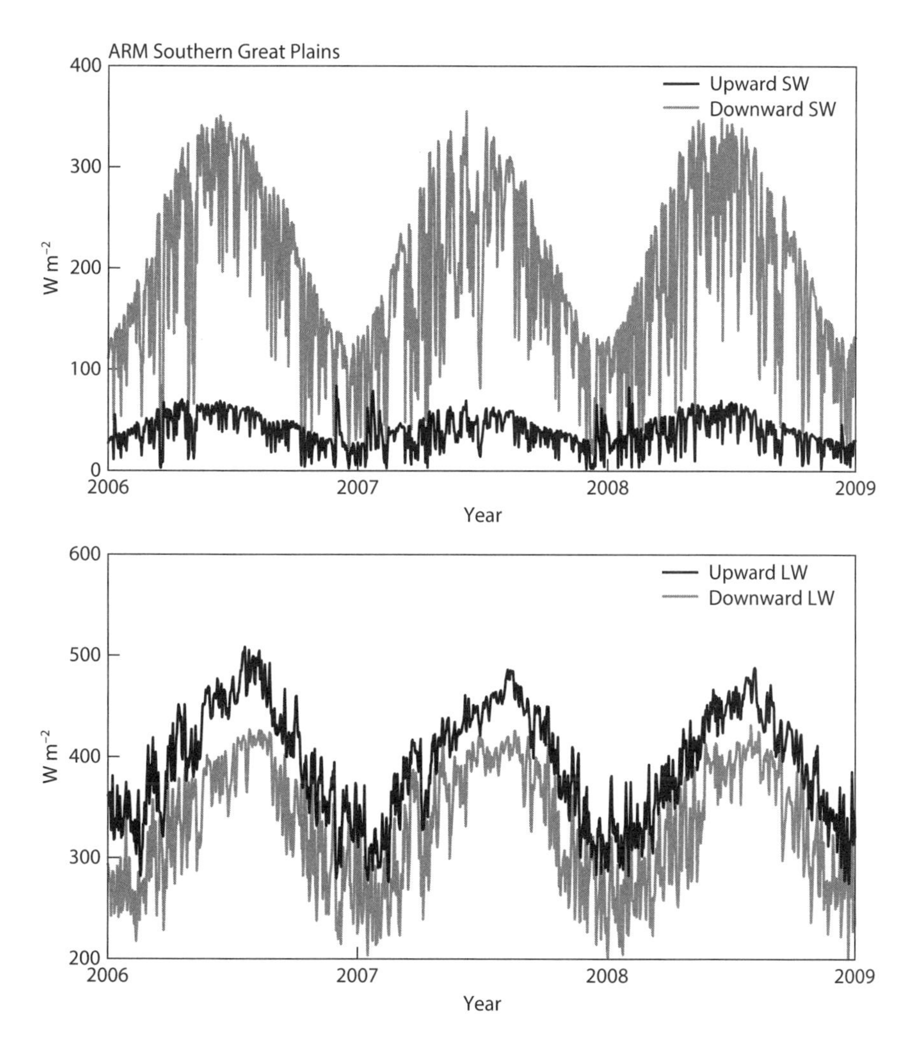

Figure 2.10. Observations of the upward and downward solar (SW) and terrestrial (LW) radiation at the Southern Great Plains (SGP) observing facility of the U.S. Department of Energy's Atmospheric Radiation Measurement Program, in Oklahoma, for the years 2006–2008.

radiation at a field site in Oklahoma. The data cover the three years 2006–2008. The seasonal cycle is clearly evident. High-frequency fluctuations are primarily due to cloudiness. Note that the upward solar radiation has occasional maxima during the winter. These are associated with the increased albedo of the ground following snowstorms.

Data like those shown in figure 2.10 are available for only a few stations around the world. Most of our ideas about the global pattern of surface radiation are based on various estimates that have been carefully worked out but are subject to significant errors (e.g., Wielicki et al., 1996). Figure 2.11A shows the meridional distribution of the zonally averaged solar radiation absorbed by the Earth's surface as a function of latitude for January and July, and the annual mean. Seasonal changes are clearly visible and easy to interpret. Near 50° N in July there is a slight dip or shoulder in the meridional profile of the surface absorbed solar radiation. This dip is associated with cloudiness and indicates that the clouds have a major impact on the energy budget of the ocean in those latitudes. Cloudiness also leads to a weak tropical minimum just north of the equator. The annual mean curve is fairly symmetrical about the equator but shows a minimum near 10° N associated with tropical rain systems. Note also that the annual mean absorbed radiation is lower in the southern high latitudes than in the northern high latitudes.

The zonally averaged net surface longwave energy flux is shown in figure 2.11B. In January, the strongest cooling occurs over Antarctica and in the subtropics of the winter hemisphere. The weakest cooling occurs in cloudy regions, for example, over the Southern Ocean and in the storm tracks of both hemispheres. Although the surface temperature is warmer in summer than in winter, at some latitudes the net longwave cooling of the surface is stronger in winter than in summer! The explanation is that the downward radiation from the atmosphere to the surface increases from winter to summer owing to both the warming of the air and the increase in the atmospheric emissivity due to seasonally increased water vapor content, as well as seasonal changes in cloudiness. This increase in the downward component is so strong that it sometimes overwhelms the increase in the upward component, resulting in a net decrease in surface infrared cooling from winter to summer.

Figure 2.11C shows the zonally averaged net surface radiation (solar and terrestrial combined). High latitudes experience net radiative cooling of the surface in winter, as would be expected. The annual mean net radiation into the surface is positive at all latitudes. It follows that the surface must cool by nonradiative means.

Figure 2.11D shows the zonally averaged latent heat flux. Positive values represent a moistening of the atmosphere and a cooling of the surface. The latent heat flux compensates, to a large extent, for the net radiative heating of the surface shown in the previous figure. Note that the maxima of the latent heat flux occur in the subtropics. Recall that the precipitation maxima occur in the tropics and middle latitudes. This implies that moisture is transported from the subtropics into the tropics, and from the subtropics into middle latitudes. This topic is discussed further in chapter 5.

Figure 2.11E shows the corresponding curves for the surface sensible heat flux. Note that the surface sensible heat flux is generally smaller than the surface latent heat flux. Maxima occur in the winter hemisphere, especially in the northern winter in association with cold-air outbreaks over warm ocean currents at the east coasts of North America and Asia. Local heat flux maxima associated with such cold outbreaks can be on the order of 1000 W m^{-2} on individual days.

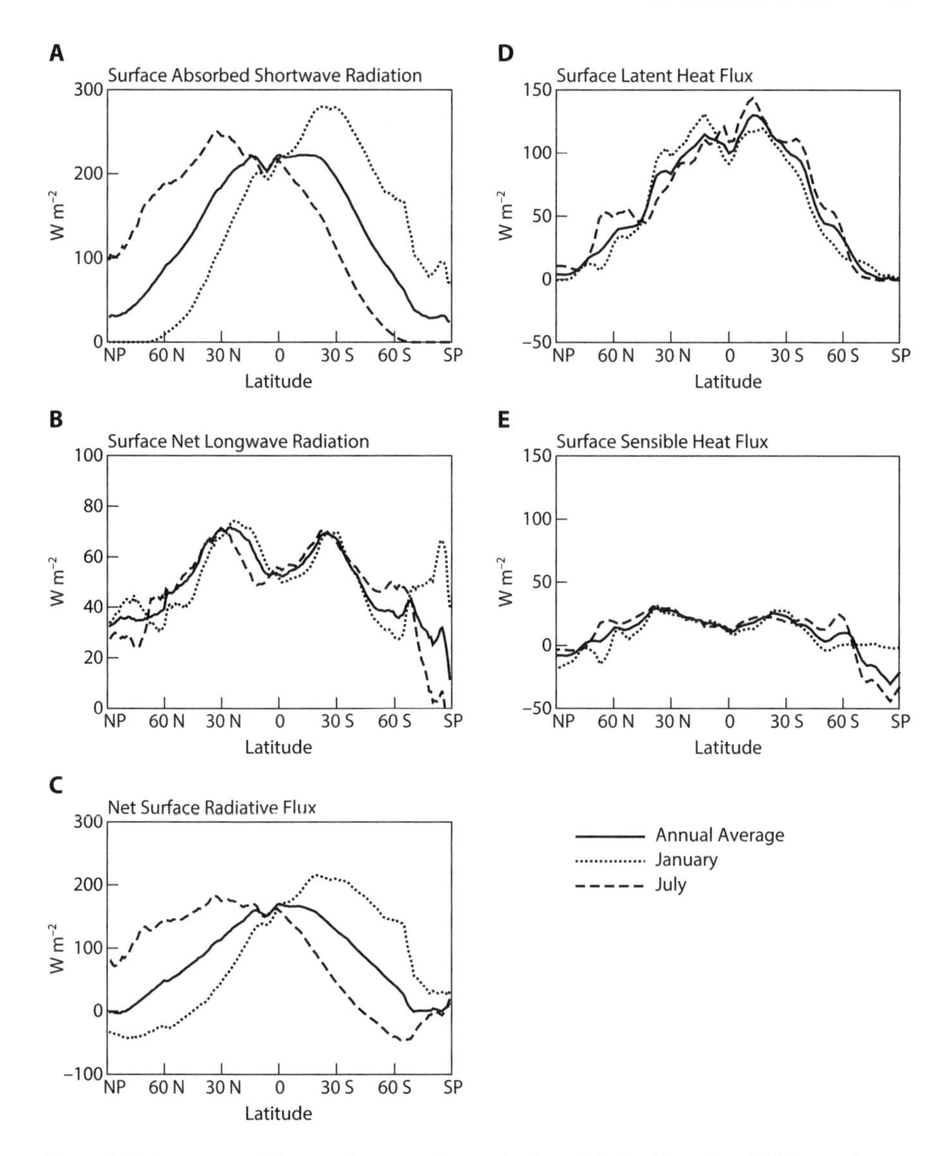

Figure 2.11. A summary of the zonally averaged energy budget of the Earth's surface. (A) The zonally averaged (land and ocean) net solar radiation absorbed by the Earth's surface. (B) The zonally averaged (land and ocean) net infrared cooling of the Earth's surface. (C) The zonally averaged net surface radiation, obtained by combining the data from panels A and B. (D) The zonally averaged surface latent heat flux. Upward fluxes are positive. (E) The zonally averaged surface sensible heat flux. Upward fluxes are positive. Panels A–C are based on Wielicki et al. (1996). Panels D–E are based on the ECMWF reanalysis. Note: These data are ***not*** true observations, although they are based on observations.

Segue

This chapter provides a brief overview of the energy fluxes at the top of the atmosphere, at the Earth's surface, and across the atmosphere. The meridional structure of the net radiation at the top of the atmosphere implies transports by the ocean-atmosphere system. The meridional structure of the net surface energy flux implies energy transports by the ocean. The meridional structure of the net flow of energy into the atmosphere, across its upper and lower boundaries, implies a net transport of energy by the atmosphere. In later chapters we will discuss the nature of these energy transports in more detail, as well as the meridional transports of angular momentum and moisture.

Among the most important points in this chapter are that the net radiative heating of the Earth's surface is balanced mainly by evaporative cooling, and the net radiative cooling of the atmosphere is balanced mainly by latent heat release. Water vapor and clouds play important roles in the Earth's radiation budget. Finally, the lower boundary conditions on the global atmospheric circulation are associated with the distributions of continents and oceans, sea-surface temperature, topography, vegetation, and ice and snow.

With this preparation, we are now ready to take a look at some of the observed features of the global circulation.

Problems

1. a) Prove that for any vector $\mathbf{Q}$,

$$\int_S \nabla \cdot \mathbf{Q} dS = 0,$$

where the integral is taken over a closed surface, for example, the surface of a sphere. We assume that $\mathbf{Q}$ is everywhere tangent to the surface, that is, it "lies in" the surface and so can be described as a "horizontal" vector. Equation (15) shows that the globally averaged divergence of any horizontal vector is zero.

b) Also prove that

$$\int_S \mathbf{k} \cdot (\nabla \times \mathbf{Q}) dS = 0,$$

where the integral is taken over a closed surface. Here $\mathbf{k}$ is a unit vector everywhere perpendicular to the surface. Equation (16) implies that the global mean of the vertical component of the vorticity is zero.

2. a) Suppose that 1 W m^{-2} is supplied to a column of water 100 m deep. Assume that the temperature changes uniformly with depth throughout the column. How much time is needed to increase the temperature of the water by 1 K?

b) Estimate the heat capacity of the entire global ocean in J K^{-1}. If all the solar radiation incident at the top of the atmosphere warmed the ocean uniformly, how long would it take for the temperature of the entire ocean to increase by 1 K?

CHAPTER 3

First Impressions

Introduction

This chapter is intended to provide a quick look at some of the important phenomena of the observed seasonally varying global circulation of the atmosphere. A few additional comments are made about nonseasonal variability. Selected fields are shown and described, with an emphasis on mass, the winds, temperature, and moisture. Among the most important phenomena that can be glimpsed are the tropical Hadley and Walker circulations, the monsoons, planetary waves, and some aspects of the hydrologic cycle. All these subjects will be discussed in more detail in later chapters. Many questions will be raised, but for the most part the answers will be deferred until later chapters.

Many of the plots shown in this and later chapters are based on analyses (actually, *reanalyses*) created at the European Centre for Medium Range Weather Forecasts (ECMWF; Uppala et al., 2005). The existence of such analyses, and their ready availability online, makes a book like this one enormously easier to create now than in decades past.

The Global Distribution of Atmospheric Mass

Mass is arguably the fundamental quantity in any physical description of the atmosphere. The density, ρ, is defined as the mass of air per unit volume. With high accuracy, the surface pressure is equal to the weight of the air above, per unit area, which is given by

$$p_S = \int_{z_S}^{\infty} \rho g dz. \tag{1}$$

Here p is pressure, where the subscript S denotes a surface value; g is the acceleration of gravity; and z is height. The surface pressure varies from place to place, owing partly to the effects of topography and partly to the circulation of the atmosphere. An example is given in figure 3.1, which is a scatter plot of surface pressure against surface

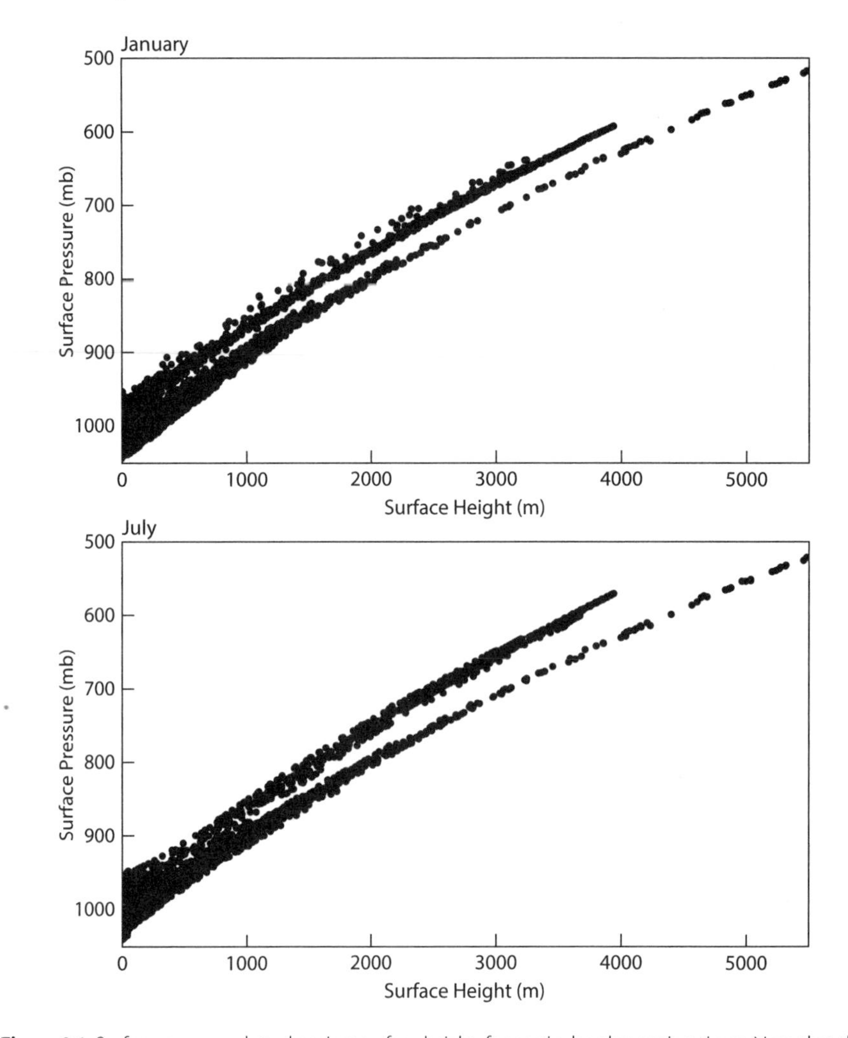

Figure 3.1. Surface pressure plotted against surface height, for particular observation times. Note that the vertical scale decreases upward. For January, the dots along the upper curve come from Greenland. For July, the dots along the upper curve correspond to the region south of 60 °S, so they include Antarctica. The data were taken from a latitude-longitude grid, so the points do not represent equal areas.

elevation, for 00Z (i.e., midnight in Greenwich, England) on January 1 and July 1, 2000. The main point of the figure is that a large fraction of the spatial variability of surface pressure is due to topography; you already knew this, but perhaps you have never seen the data plotted this way before. For each hemisphere, the data fall roughly onto two curves. For the Northern Hemisphere, the upper curve consists mainly of points on Greenland, and for the Southern Hemisphere the upper curve consists mainly of points on Antarctica. The scatter of the data about their mean for each surface elevation indicates the dynamical variability of the surface pressure. The range of variability is particularly large when the surface height is zero, which means that it is large for the oceans. An interpretation is that weather systems are particularly vigorous over the oceans.

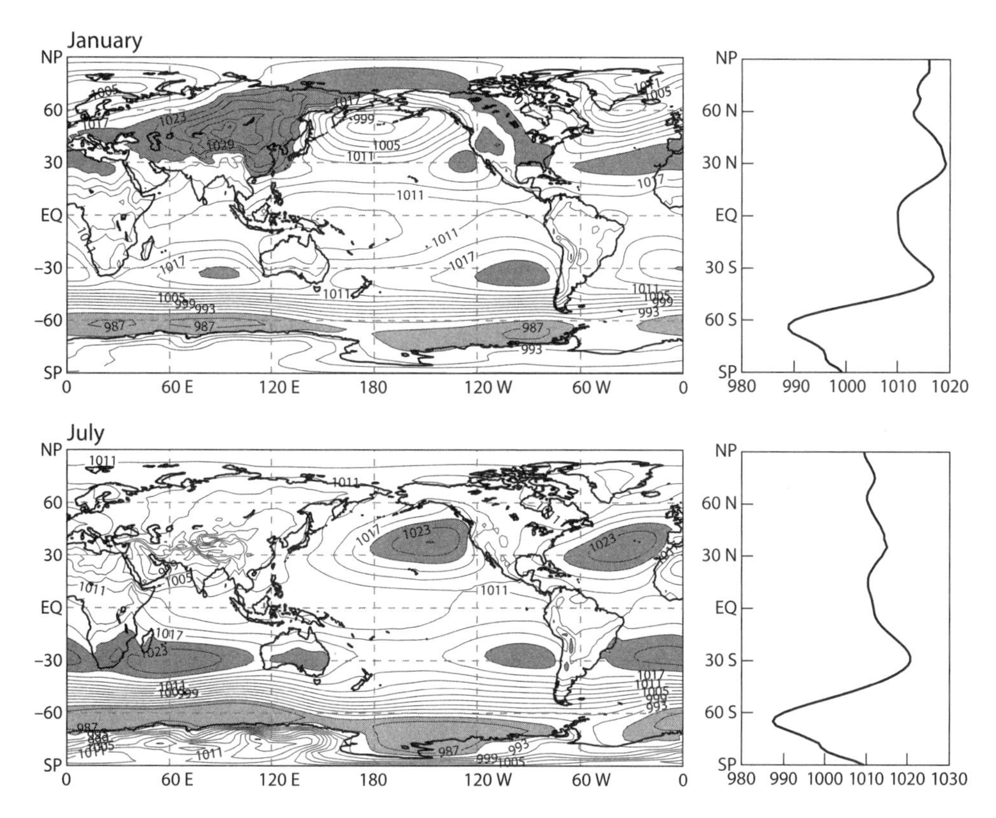

Figure 3.2. Sea-level pressure maps. The contour interval is 3 hPa. Values higher than 1020 hPa have dark shading, and those lower than 990 hPa have light shading. Averages with respect to longitude are shown on the right.

Maps of surface pressure are dominated by minima associated with mountain ranges. Consequently, maps of surface pressure do not clearly show how the horizontal pressure-gradient force varies with changes in the weather. To get around this problem, it is conventional to define *sea-level pressure*, which is computed by adding to the surface pressure a correction designed to represent the additional pressure that would exist if the mountains were not present. Figure 3.2 shows monthly-mean maps of sea-level pressure for January and July. These maps are much smoother than weather maps plotted for particular observation times, because the moving highs and lows that represent individual weather systems have been smoothed out by the time averaging. The plots on the right side of figure 3.2 show the corresponding zonally averaged distributions of the sea-level pressure for January and July.

Throughout this book, we will distinguish between the zonally averaged circulation, one aspect of which is the zonally averaged sea-level pressure shown in the right-side panels of figure 3.2, and departures from the zonal average, which we will call *eddies*. The highs and lows of the sea-level pressure that can be seen in the maps of figure 3.2 are associated with eddies.

Especially in the Northern Hemisphere, there is a pronounced tendency toward low pressure over the oceans and high pressure over the continents in winter, and vice versa in summer. This seasonal shift of air mass between the oceans and continents is

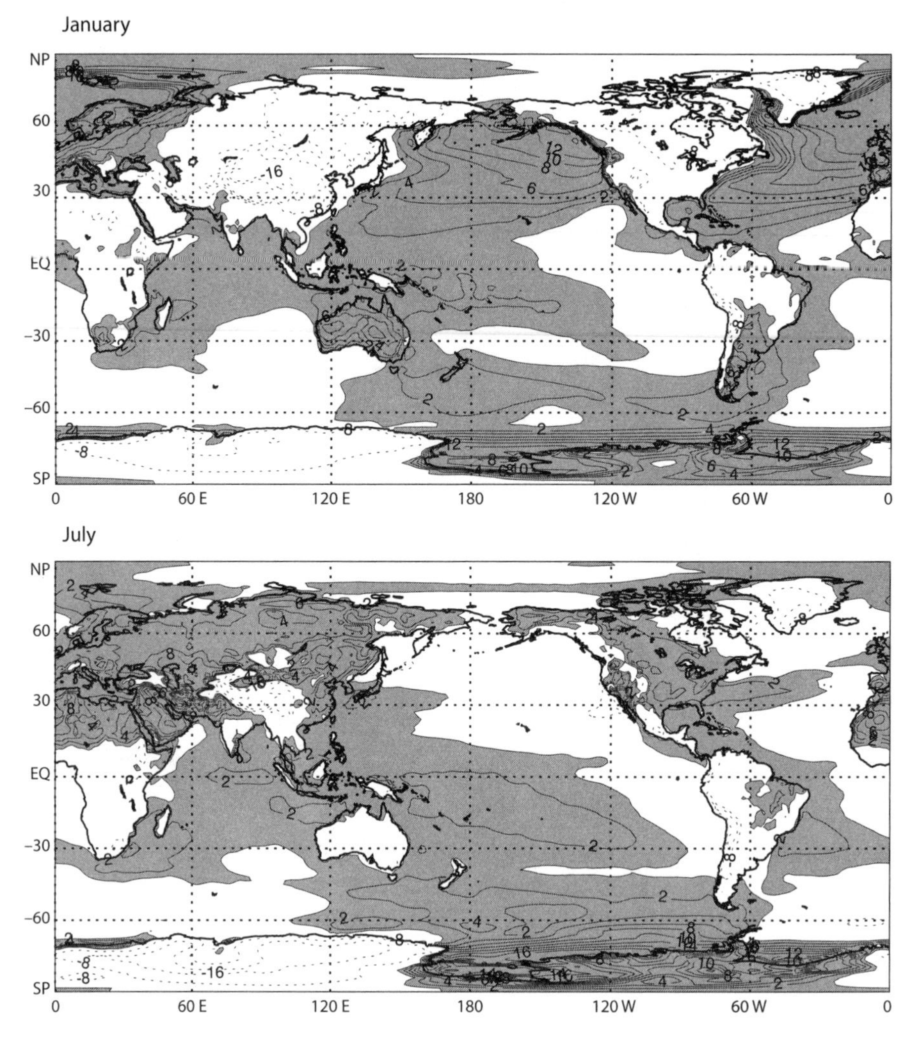

Figure 3.3. The departure of the 2 m temperature from its zonal mean at each latitude, for January and July. Positive values are shaded. The contour interval is 2 K.

associated with a seasonal shift in surface temperature. Figure 3.3 shows the departure of the surface temperature from its zonal mean at each latitude, for January and July. We can say that these are plots of the *eddy surface temperature*. The figure shows that in middle latitudes, especially in the Northern Hemisphere, the surface air temperature is generally colder over the continents than over the oceans in winter, and vice versa in summer. Particularly cold temperatures occur on the eastern side of the Northern Hemisphere continents in January. This is partly because the flow is generally from west to east, so that the air on the east side of the continents has traveled all the way across the continent, gradually cooling along the way. The strong temperature contrast on the east coast of the Northern Hemisphere continents in winter leads to the frequent formation of winter storms, which are often called *baroclinic eddies*. These are discussed later.

Comparison of figures 3.2 and 3.3 shows that there is a tendency for high sea-level pressures to be associated with cold surface temperatures, and vice versa. To understand this, recall that cold air has a higher density than warm air, so that a "pile" of cold air of a given geometric thickness will contain more mass, and therefore weigh more, than a pile of warm air of the same geometric thickness. This can be seen mathematically by combining the ideal gas law with (1) to obtain

$$p_S = \int_{z_S}^{\infty} \left(\frac{pg}{RT} \right) dz. \tag{2}$$

Here R is the gas constant for dry air, and T is temperature.

Figure 3.2 shows that for both January and July there is a tendency for high sea-level pressure to occur in the subtropics. The highs typically appear as "cells," for example, in the North Atlantic and North Pacific in July or off the west coast of South America in January. In many cases, the subtropical highs are found over the eastern parts of the oceans. In the Northern Hemisphere, they are particularly strong in the northern summer (Hoskins, 1996). Strong highs are also apparent in middle latitudes during winter, for example, in Siberia and western North America. Both regions are quite mountainous.

In spherical coordinates, and with pressure as the vertical coordinate, geostrophic balance is expressed by

$$0 = fv_g - \frac{1}{a \cos\varphi} \left(\frac{\partial \phi}{\partial \lambda} \right)_p \quad \text{and} \quad 0 = -fu_g - \frac{1}{a} \left(\frac{\partial \phi}{\partial \varphi} \right)_p, \tag{3}$$

where u_g and v_g are the zonal and meridional components of the geostrophic wind, respectively; f is the Coriolis parameter; and ϕ is the geopotential. With appropriate caution, we can use (3) to relate the pressure gradients to the winds. For example, throughout the year, there is a zonally oriented belt of low pressure across the tropics. Geostrophy tells us that the decrease of sea-level pressure from the subtropics toward the equator implies tropical easterlies near the surface, although we have to be careful about invoking geostrophy near the equator, where the Coriolis parameter passes through zero. Because the sea-level pressure generally decreases from the subtropics to middle latitudes, the geostrophic relation leads us to expect surface westerlies on the poleward side of the subtropical highs. The relatively high pressure over the poles suggests surface easterlies there, although the "sea-level" pressures plotted for Antarctica must be taken with a grain of salt, because the surface of Antarctica is far above sea level.

In the Northern Hemisphere during northern winter, prominent low-pressure cells appear, most notably near the Aleutian Islands and Iceland. These are regions where storm systems are often found on individual days. There is a tendency toward a minimum of the sea-level pressure near 60° N, especially in January but also to some extent in July. A very pronounced belt of low pressure is found over the Southern Ocean north of Antarctica throughout the year, although it is more intense in July (winter) than January (summer).

Generally, there is less seasonal change in the Southern Hemisphere than in the Northern Hemisphere. Also, the departures from the zonal means are much stronger in the Northern Hemisphere than in the Southern Hemisphere.

Figure 3.4. Jule G. Charney, whose many achievements include scale analyses of both extratropical and tropical motions, development of the quasi-geostrophic theory, development (in his Ph.D. thesis at the University of California, Los Angeles) of a classical theory of baroclinic instability, pioneering work on numerical weather prediction, analysis of the interactions of cumulus convection with large-scale motions in tropical cyclones, development of a theory of planetary waves propagating through shear, analysis of blocking, and a theory of desertification. Figure used with permission of the MIT Museum.

The tropical sea-level pressure distribution is generally very smooth and featureless compared with that of middle latitudes. A simple explanation for this was given by Jule Charney (1963) in terms of the differences in dynamical balance between the tropics. Charney (see fig. 3.4) was one of the giants of twentieth-century meteorology. His work is discussed here for the first time, and his name will come up repeatedly throughout the remainder of this book. Charney's explanation of the flatness of tropical sea-level pressures is based on the fact that in middle latitudes the effects of the Earth's rotation are much more important than particle accelerations, while the opposite is true in the tropics. He began his scale analysis (see appendix B) with the equation of horizontal motion, which can be written in simplified form as

$$\frac{D\mathbf{V}_h}{Dt} + f\mathbf{k} \times \mathbf{V}_h = -\nabla_p \phi. \tag{4}$$

Here $\mathbf{V}_h$ is the horizontal wind vector, f is the Coriolis parameter, $\mathbf{k}$ is a unit vector pointing upward, and ϕ is the geopotential. The three terms shown in (4) represent most of the "action" throughout most of the atmosphere. Their orders of magnitude can be estimated as follows:

$$\frac{D\mathbf{V}_h}{Dt} \sim \frac{V^2}{L}, \tag{5}$$

$$f\mathbf{k} \times \mathbf{V}_h \sim f_{\text{midlat}} V, \tag{6}$$

$$\nabla_p \phi \sim \frac{\delta\phi}{L}. \tag{7}$$

Here V is a "velocity scale," which might be on the order of 10 m s^{-1}, L is a length scale, which might be on the order of 10^6 m, and $\delta\phi$ is a typical fluctuation of the geopotential height. Note that $\delta V \sim V$, but $\delta\phi$ is generally much less than ϕ. The numerical values of these scales have been chosen to be representative of *large-scale* motions on the Earth; if we wanted to analyze small-scale motions, we would choose different numerical values. The same numerical values of V and L can be used for both the tropics and middle latitudes because the term *large-scale* is used in the same way for both regions.

In middle latitudes, the acceleration following a particle, $D\mathbf{V}_h/Dt$, is typically negligible in (15) compared with the rotation term. A typical value of $D\mathbf{V}_h/Dt$ can be estimated as $|D\mathbf{V}_h/Dt|(V^2/L) = 10^2/10^6 = 10^{-4}\,\text{m s}^{-2}$. A representative midlatitude value of the Coriolis parameter is $f_{\text{midlat}} \sim 10^{-4}\,\text{s}^{-1}$, so that $f_{\text{midlat}}V \sim 10^{-3}\,\text{m s}^{-2}$, about one order of magnitude larger than $D\mathbf{V}_h/Dt$. Geostrophic balance is therefore approximately satisfied in middle latitudes; that is,

$$f_{\text{midlat}}V \sim \frac{(\delta\phi)_{\text{midlat}}}{L}, \tag{8}$$

or

$$(\delta\phi)_{\text{midlat}} \sim f_{\text{midlat}}VL. \tag{9}$$

Here we added the "midlat" subscript to $\delta\phi$ just for clarity. According to (9), rotation can balance pressure gradients in middle latitudes.

The Coriolis parameter vanishes on the equator, so it is reasonable to expect that sufficiently close to the equator geostrophic balance breaks down (a point discussed further in chapter 8), and particle accelerations tend to balance the pressure gradient force, much as they do on small scales almost everywhere in the atmosphere, and in many engineering contexts (e.g., the flow of water in a pipe):

$$\frac{V^2}{L} \sim \frac{(\delta\phi)_{\text{tropics}}}{L}, \tag{10}$$

or

$$(\delta\phi)_{\text{tropics}} \sim V^2. \tag{11}$$

Comparing (9) and (11), we see that

$$\frac{(\delta\phi)_{\text{tropics}}}{(\delta\phi)_{\text{midlat}}} \sim \frac{V}{f_{\text{midlat}}L} \equiv Ro_{\text{midlat}}, \tag{12}$$

where Ro_{midlat} is a midlatitude Rossby number. By substituting the numerical values given previously, we find that $Ro_{\text{midlat}} \cong 0.1$. Equation (12) therefore tells us that geopotential height fluctuations on pressure surfaces are much smaller in the tropics than in middle latitudes. It follows that pressure fluctuations on height surfaces are much smaller in the tropics than in middle latitudes. The real message is that *the horizontal pressure gradient force is much smaller in the tropics than in middle latitudes. This conclusion holds at any level, so it holds at all levels.*

At this point we can bring in the hydrostatic equation to show that large-scale fluctuations of temperature and surface pressure are also much smaller in the tropics than

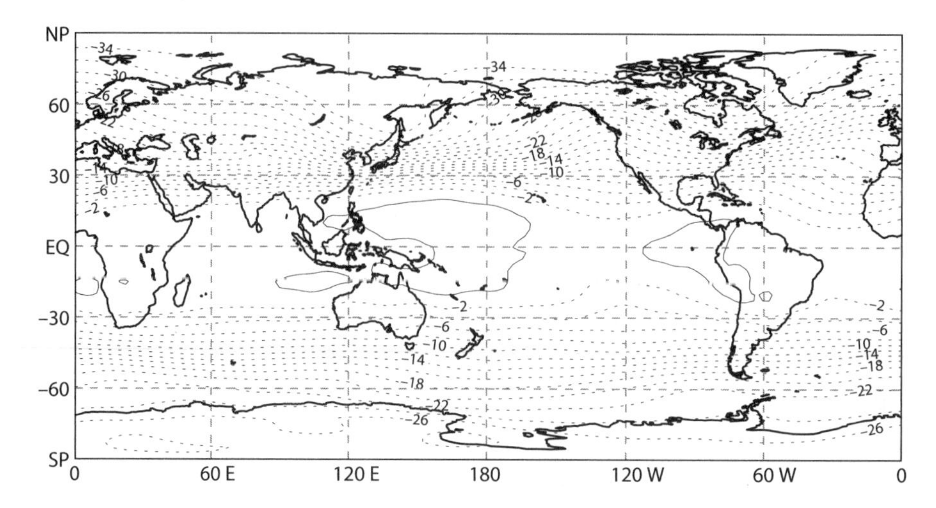

Figure 3.5. The departure of the 500 hPa temperature from its value at Darwin, Australia, for January. The contour interval is 2 K, and the zero contour is solid, while negative contours are dashed. Note the amazing uniformity of the tropical temperatures. This is a consequence of the smallness of the Coriolis parameter in the tropics, as explained by Charney (1963).

in middle latitudes. Suppose that the pressure-gradient force is small at some particular height. Rapid temperature changes in the horizontal direction would imply large horizontal pressure gradients at other heights. The implication is that if the horizontal pressure gradient is small at all levels, then the horizontal temperature gradients must also be small. Figure 3.5 illustrates this situation.

The smallness of horizontal temperature gradients in the tropics is now widely invoked to justify what is called the "weak temperature gradient approximation" (e.g., Sobel et al., 2001). It is important to remember, however, that Charney's primary conclusion is that horizontal pressure gradients are weak in the tropics; the weakness of the horizontal temperature gradient is a secondary conclusion (Romps, 2012).

Zonal Wind

Figure 3.6 shows the latitude-height distribution of the zonally averaged zonal wind for January and July, respectively. The plot extends from the surface to the middle stratosphere. We plot the wind components and other variables against height, rather than pressure, because the pressure coordinate tends to "squash" the stratosphere into a thin region at the top of the plot, obscuring its structure. Although the stratosphere contains only a small fraction of the mass of the whole atmosphere, its dynamical influence extends downward into the troposphere, and so the stratosphere should be of interest even to readers who are mainly focused on the global circulation of the troposphere.

In both January and July, easterlies extend throughout the depth of the tropical troposphere. They are somewhat stronger in July and are concentrated in the Northern Hemisphere in July, and the Southern Hemisphere in January. The near-surface easterlies are stronger in the winter hemisphere.

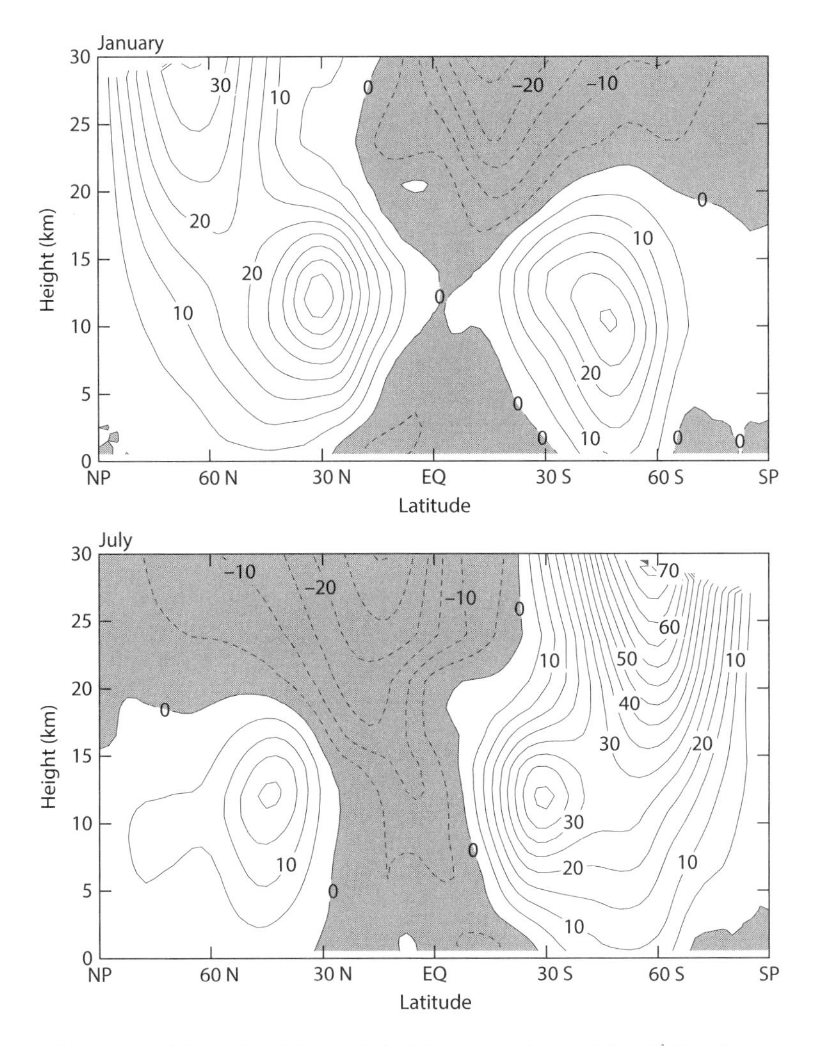

Figure 3.6. Latitude-height section of the zonal wind. The contour interval is 5 m s^{-1}. Easterlies are shaded.

Westerly jets (*jet streams*) are quite prominent in the upper troposphere, especially in the winter hemisphere. These are called the subtropical jets, and they reach their maximum strength during winter at about 30° latitude. Directly below the winter subtropical jet, the surface zonal wind is very weak. The jet extends poleward, however, with a hint of a second maximum in the troposphere near 50° or 60° from the equator, where the zonally averaged surface wind is definitely westerly. The subtropical jets are weaker in summer and are shifted to about 45° latitude. The subtropical jet maxima are consistently found near the 200 hPa level.

Strong westerly jets also occur in the winter stratosphere. In the Southern Hemisphere in July, there is a clear minimum in the westerlies near 150 hPa, at about 40° S. Above and poleward of this minimum is a very powerful stratospheric westerly jet, called the *polar night jet*. A similar but weaker polar night jet occurs in the Northern Hemisphere winter. The stratospheric jets are separated from the troposphere jets by a (weak) minimum of the westerlies.

Easterlies fill the summer hemisphere stratosphere. As discussed later, the summer stratosphere is radiatively controlled, while the winter stratosphere is strongly influenced by dynamics, including upward wave propagation from the troposphere.

The zonally averaged surface winds are quite weak near the poles.

Suppose that the zonal surface wind is nearly geostrophic. Then, in the absence of mountains (e.g., over the oceans) the surface pressure must have a meridional maximum at a latitude where u passes through zero, that is, where surface easterlies meet surface westerlies. Comparison of figures 3.2 and 3.6 shows that, in fact, the subtropical surface-pressure maxima occur at about the same latitudes where the zonal component of the surface wind passes through zero.

Figure 3.7 shows maps of the 850 hPa zonal wind for January and July, respectively. Again, keep in mind that many intense small-scale (~1000 km) features would appear in daily maps but have been smoothed here by time averaging. The monthly-mean

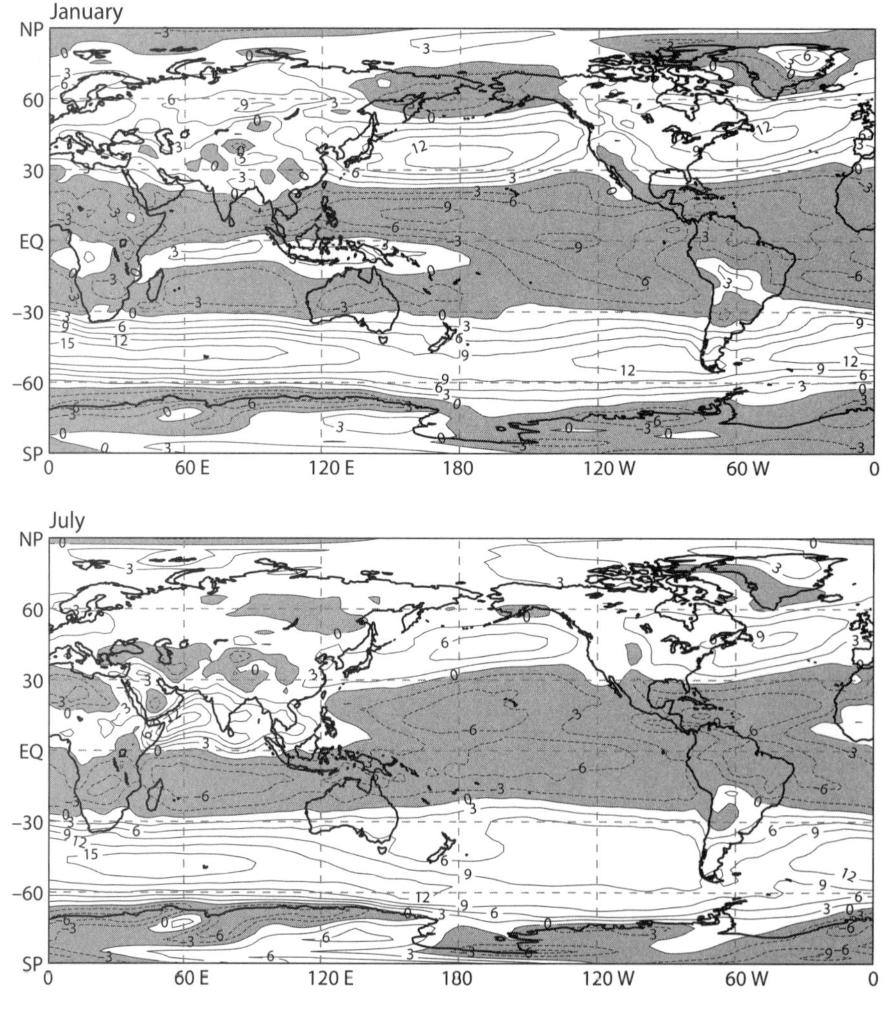

Figure 3.7. Maps of the 850 hPa zonal wind, for January and July. The contour interval is 3 m s^{-1}. Easterlies are shaded.

Figure 3.8. An image of Jupiter obtained from the Cassini probe on December 7, 2000. Note the zonal bands, and the Great Red Spot in the southern hemisphere. The small dark spot at the lower left is the shadow of Jupiter's moon Europa. From http://www.nasa.gov/images/content/414971main_pia02873.jpg.

maps show very obvious alternating bands of easterlies and westerlies, which are qualitatively reminiscent of Jupiter (see fig. 3.8), although Jupiter has more bands; generally speaking, the Earth's atmosphere features easterlies in the tropics, westerlies in middle latitudes, and easterlies again near the poles. Features associated with the strong cells in the sea-level pressure maps are also apparent—for example, the easterlies in the extreme North Pacific in January, associated with the Aleutian Low. Again, variations with longitude are much more evident in the Northern Hemisphere than in the Southern Hemisphere. In general, however, the features seen in the maps have a very zonal orientation, with strong north-south gradients and relatively weak east-west gradients. Note the intensification of both the zonal-mean flow and the eddies of the midlatitude westerlies in winter, in each hemisphere. In winter, the Northern Hemisphere westerlies are particularly strong over the oceans.

The strong positive maximum in the Arabian Sea in July is associated with the Indian monsoon; this is discussed in chapter 8. The westerlies north of Australia (but south of the equator) in January are indicative of the Australian monsoon. In both regions, the sign of the zonal wind reverses seasonally.

Figure 3.9 shows the corresponding maps for 200 hPa. The winds are generally stronger aloft than near the surface; this is true of both the zonal-mean flow and the eddies. Note the very prominent January westerly jet maxima off the east coasts of North America and, especially, Asia. There is also a westerly jet maximum at about 30° S, near the date line. In the Northern winter, there are "dipoles" consisting of easterly-westerly pairs, straddling the equator, near the Americas and also near the longitude of Australia. At some longitudes, the westerlies extend to the equator in January. In July, there are equatorial easterlies at all longitudes, and intensified westerlies in the middle latitudes of the Southern Hemisphere, and weakened ones in the middle latitudes of the Northern Hemisphere. Two intrusions of westerlies are seen in the Northern Hemisphere tropics: one just east of the date line, and another over

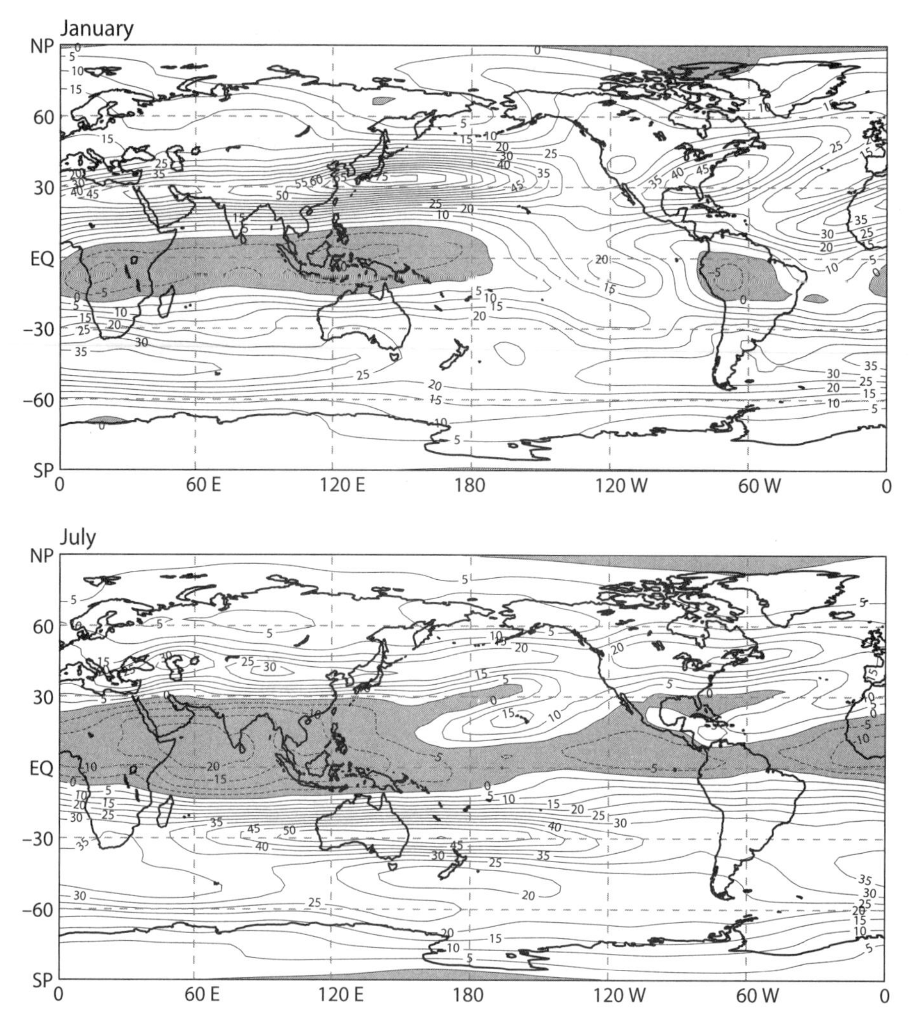

Figure 3.9. Maps of the 200 hPa zonal wind for January and July. The contour interval is 5 m s^{-1}. Easterlies are shaded.

the Atlantic Ocean. As discussed later, these regions of mean westerlies allow waves to propagate from middle latitudes into the tropics.

Meridional Wind

Figure 3.10 shows the latitude-height distribution of the zonally averaged meridional wind for January and July, respectively. The zonal means reach about 2 m s^{-1} in absolute value; the strongest values occur in the tropics. In the winter-hemisphere tropics, in both months, there is a conspicuous dipole structure, with poleward flow aloft and equatorward flow near the surface. Evidently there is convergence near the equator at low levels, and divergence aloft. The convergence zone near the surface shifts from the Southern Hemisphere in January to the Northern Hemisphere in July.

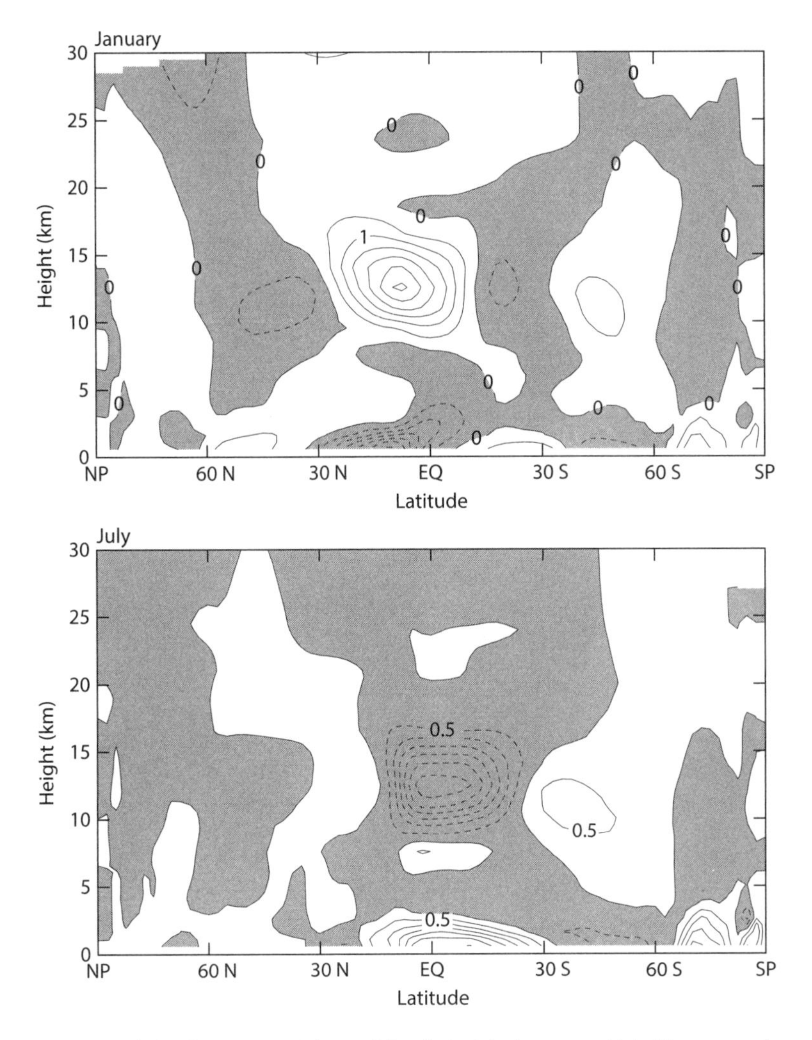

Figure 3.10. Latitude-height sections of the meridional wind, for January and July. The contour interval is 0.5 m s^{-1}. Negative values are shaded.

These features are associated with the Hadley circulation, which is discussed later in this chapter and again in chapter 5. Poleward flow is also found near the surface in middle latitudes, with weak equatorward flow above. There are also weak regions of low-level convergence near 60° S and 60° N.

As with most other fields, the seasonal changes of the meridional wind in the middle latitudes of the Southern Hemisphere are quite weak relative to those in the Northern Hemisphere.

The mass-weighted vertical mean of the time- and zonally averaged meridional wind must be very close to zero at all latitudes, as can be understood from the surface pressure-tendency equation,

$$\frac{\partial p_S}{\partial t} + \nabla \cdot \left(\int_0^{p_S} \mathbf{V}_h dp \right) = 0. \tag{13}$$

which is derived in chapter 4. In (13), $\mathbf{V}_H$ is the horizontal wind vector. The physical meaning of (13) is that the surface pressure changes in time owing to the convergence or divergence of mass in the column above the surface. Averaging (13) around a latitude circle gives

$$\frac{\partial \overline{p_S}^{\lambda}}{\partial t} = \frac{-1}{a\cos\varphi}\frac{\partial}{\partial\varphi}\left(\overline{\int_0^{p_S} v dp}^{\lambda}\cos\varphi\right). \tag{14}$$

In an average over a sufficiently long time, $\partial \overline{p_S}^{\lambda}/\partial t$ must become negligible at each latitude, because $\overline{p_S}^{\lambda}$ is bounded within a fairly narrow range, so that (14) reduces to

$$\frac{\partial}{\partial\varphi}\left(\overline{\int_0^{p_S} v dp}^{\lambda}\cos\varphi\right) = 0. \tag{15}$$

Equation (15) means that $\overline{\int_0^{p_s} v dp}^{\lambda,t}\cos\varphi$ is independent of latitude. Since $\cos\varphi = 0$ at both poles, we conclude that

$$\boxed{\overline{\int_0^{p_S} v dp}^{\lambda,t} = 0 \text{ at all latitudes}}. \tag{16}$$

The physical interpretation of (16) is very simple. Suppose, for example, that $\overline{\int_0^{p_s} v dp}^{\lambda}$ was positive at the equator, so that air was systematically flowing from the Southern Hemisphere into the Northern Hemisphere. That situation could occur at a given instant, but if it continued over time, the surface pressure in the Southern Hemisphere would eventually decrease to zero, and the surface pressure in the Northern Hemisphere would increase to roughly double its normally observed average value. The pressure-gradient force would of course resist such a scenario. The implications of (16) for angular momentum transport are discussed in chapter 5.

It follows from (3) that the zonally averaged meridional component of the geostrophic wind is exactly zero:

$$\boxed{\overline{v_g}^{\lambda} = 0}. \tag{17}$$

In (17), the zonal average is taken along an isobaric surface, and (17) can be violated when an isobaric surface intersects the Earth's surface. Equation (17) implies that all zonally averaged meridional circulations are *completely* ageostrophic. An implication is that important large-scale circulations are not necessarily close to geostrophic balance. We note, however, that the strongest features in figure 3.10 are found in the tropics, where geostrophy would be expected to lose its grip.

Equation (14) shows that the mass-weighted vertical integral of the meridional wind (i.e., the vertically integrated meridional mass flux) can lead to small systematic changes with time in the meridional distribution of atmospheric mass. Figure 3.11 shows the variation with season of the vertically and zonally averaged meridional velocity, *as inferred from the seasonal changes of the distribution of mass.* The top panel shows variations with latitude, and the bottom panel shows the seasonal cycle at the equator. The values are on the order of 1 mm s^{-1}, much too small to determine directly from observations of the wind.

The globally averaged surface pressure is very nearly invariant with time, apart from small changes associated with the seasonal cycle of atmospheric water vapor.

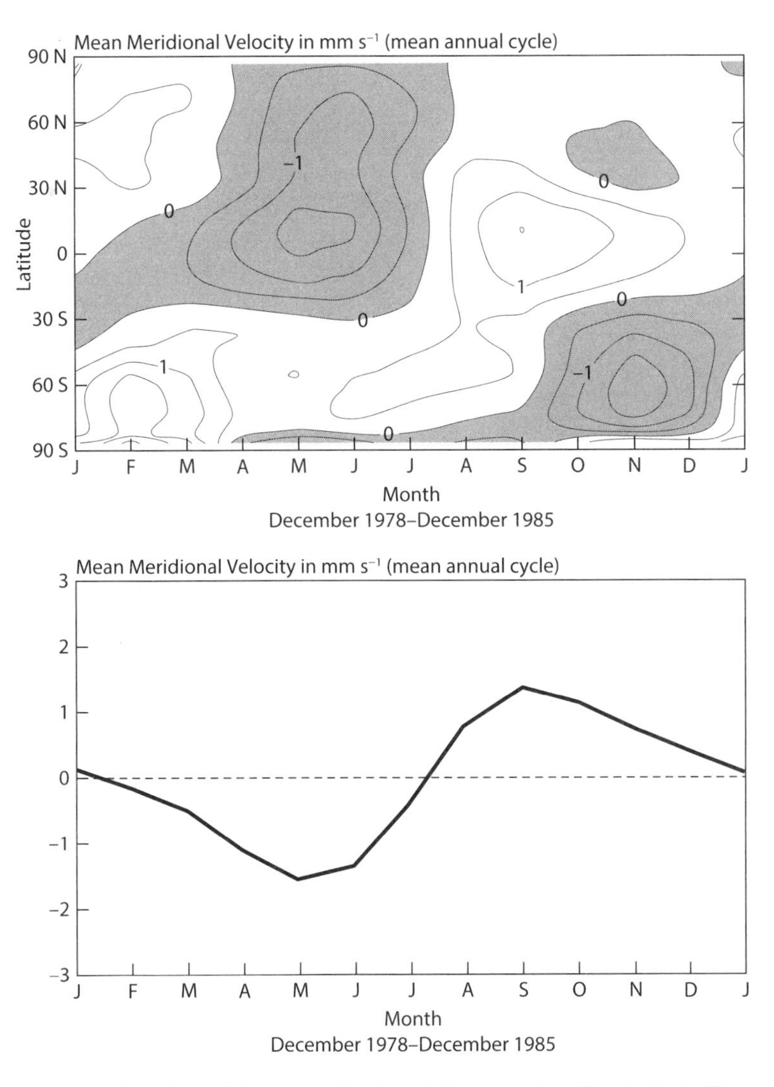

Figure 3.11. Seasonal variations of the mass-weighted vertically averaged meridional velocity. The top panel shows variations with latitude, and the bottom panel shows the climatological seasonal cycle at the equator. From Trenberth et al. (1987). Copyright © 1988 by the American Geophysical Union.

Trenberth et al. (1987) discuss observations of the seasonal changes of the hemispherically averaged and globally averaged surface pressures associated with dry air and with water vapor. These are shown in figure 3.12. The fluctuations are on the order of 0.1% of the total mass of the atmosphere. The observed distribution of water vapor is discussed later in this chapter.

Figure 3.13 shows maps of the 850 hPa meridional wind for January and July, respectively. Figure 3.14 shows the corresponding maps for the 200 hPa surface. Unlike the zonal wind, the time-averaged meridional wind does not show a banded east-west structure; the east-west gradients are at least as strong as the north-south gradients. In the Northern Hemisphere, southerly and northerly flows tend to alternate, with a structure that resembles zonal wavenumber 3 or 4, that is, three or four maxima and minima

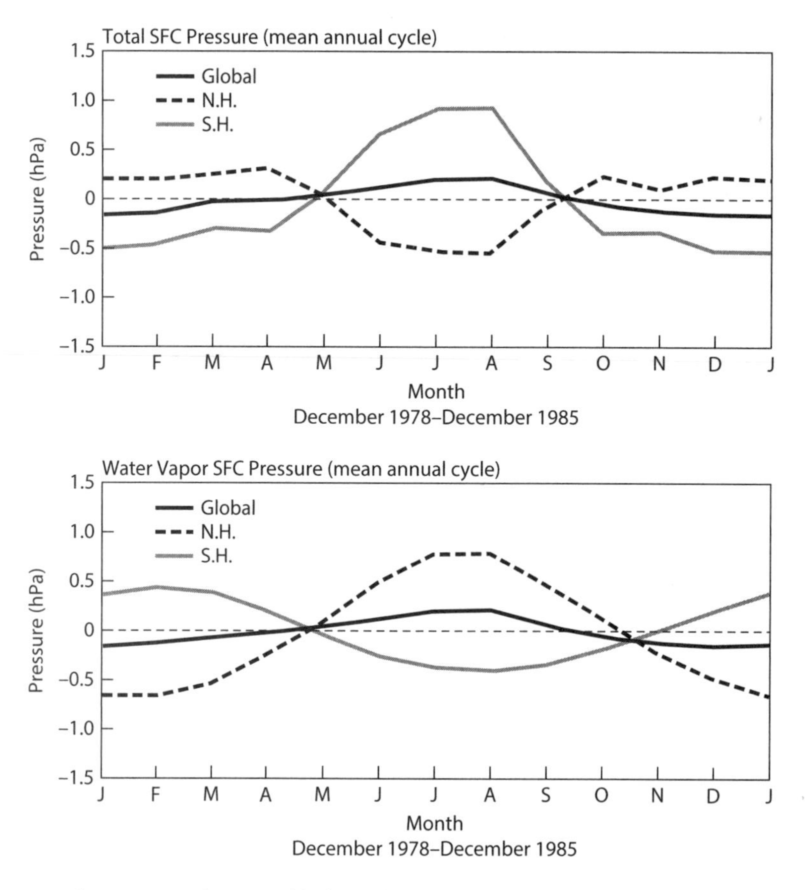

Figure 3.12. The variations with season of the hemispherically averaged and globally averaged surface pressures associated with dry air and with water vapor. From Trenberth et al. (1987). Copyright © 1988 by the American Geophysical Union.

around a latitude circle. The time-averaged meridional currents in the Southern Hemisphere are generally weaker than those in the Northern Hemisphere. The intensities of the meridional currents are stronger at 200 hPa than at 850 hPa. In the Northern Hemisphere especially, stronger features tend to predominate in winter than in summer.

As can be seen in figure 3.10, the zonally averaged meridional flow in the tropics of the winter hemisphere is generally toward the summer pole at low levels, and toward the winter pole aloft. This is not very apparent in figures 3.13 or 3.14, however. Figure 3.13 shows a strong southerly flow at 850 hPa just north of the equator in July, associated with the Indian summer monsoon. The northerly flow near 120° E in January is associated with the winter monsoon but is relatively inconspicuous.

In many parts of the world, the mean meridional wind reverses seasonally. Examples include the Arabian Sea, most of the North Pacific, and the southern Great Plains of North America.

Figure 3.15 shows the streamlines at 850 hPa for January and July, and figure 3.16 shows the corresponding streamlines at 200 hPa. Streamlines indicate direction but not magnitude. Such features as the subtropical highs and midlatitude lows of sea-level pressure are clearly evident in the 850 hPa winds. There are strong cross-equatorial flows at 850 hPa in both the Pacific and Indian Oceans in July. The 200 hPa streamlines

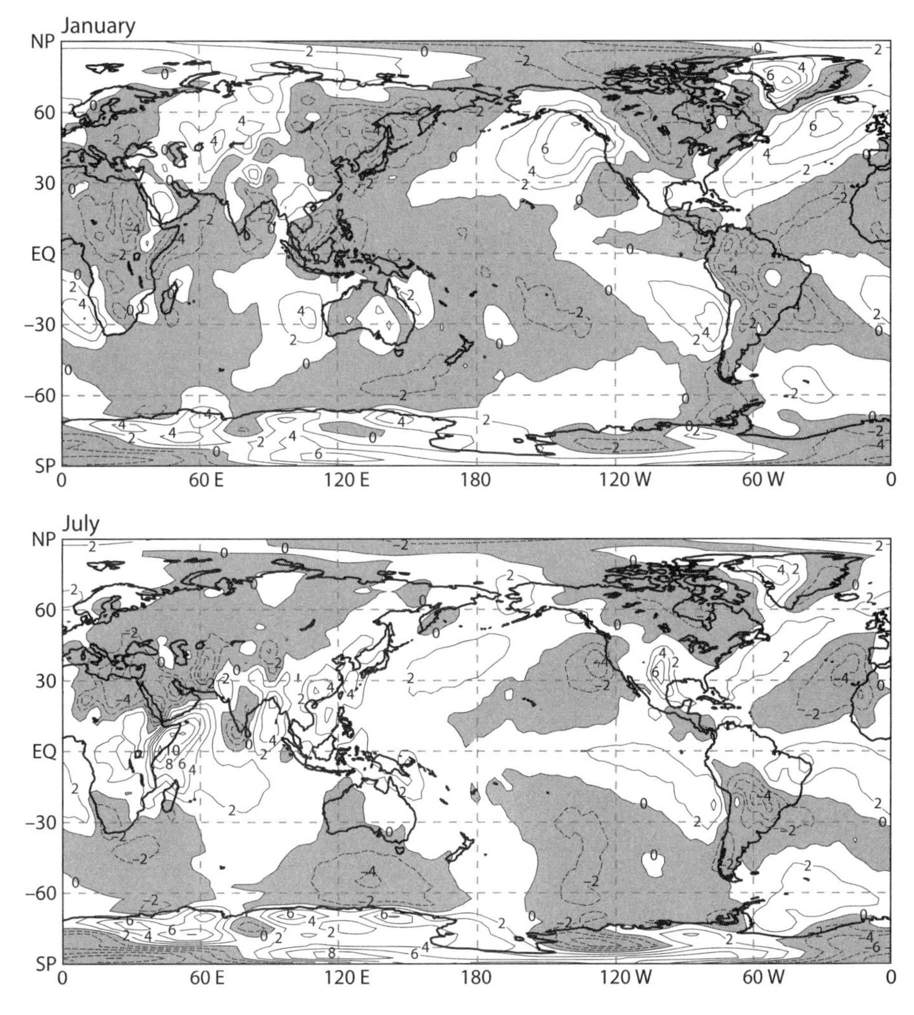

Figure 3.13. Maps of the 850 hPa meridional wind, for January and July. The contour interval is 2 m s^{-1}. Northerlies are shaded.

show wavy patterns in the midlatitude winter, and also tropical phenomena, including a strong monsoon-induced anticyclone over the Indian subcontinent in July.

Vertical Velocity and the Mean Meridional Circulation

The *mean meridional circulation* (MMC) is the name given to the circulation of mass in the latitude-height plane. It can be analyzed by starting from the continuity equation in pressure coordinates, which is

$$\nabla_p \cdot \mathbf{V}_h + \frac{\partial \omega}{\partial p} = 0, \tag{18}$$

where ω is the vertical velocity as seen in pressure coordinates, that is, the time rate of change of pressure following a particle of air. Taking the zonal average of (18), and

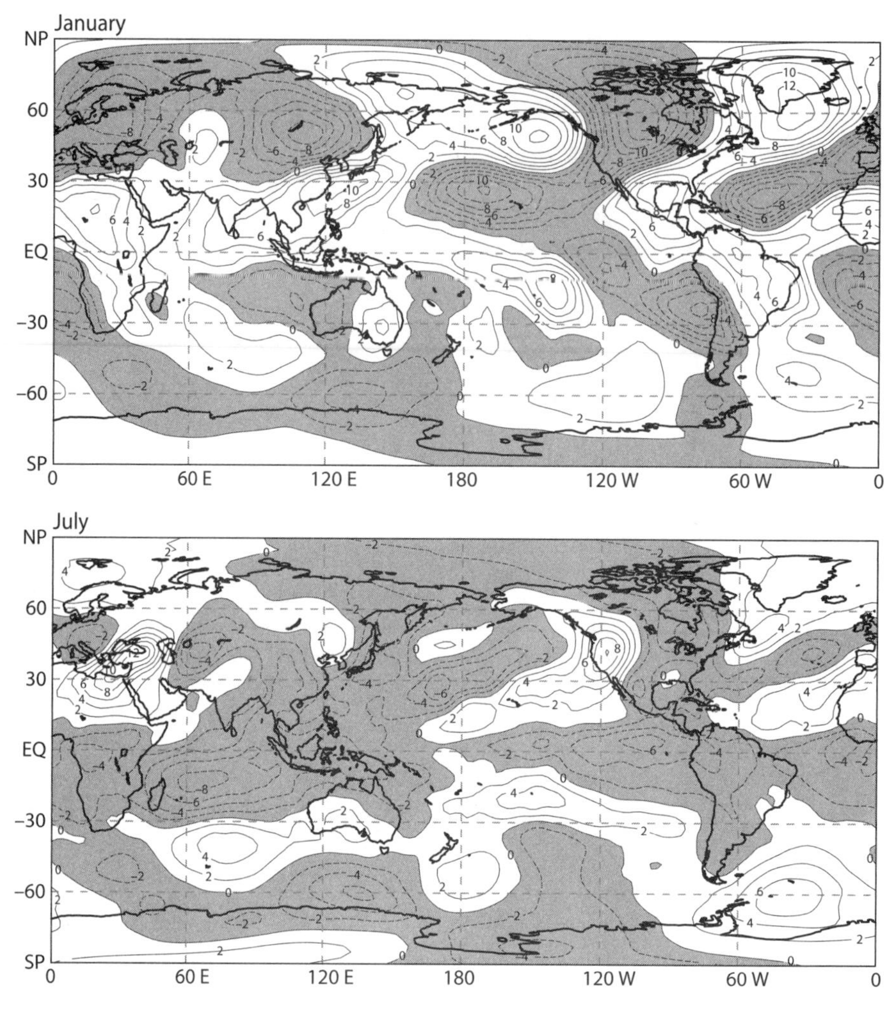

Figure 3.14. Maps of the 200 hPa meridional wind, for January and July. The contour interval is 2 m s^{-1}. Northerlies are shaded.

using the form of the divergence operator in spherical coordinates as discussed in chapter 2, we find that

$$\frac{1}{a\cos\varphi}\frac{\partial}{\partial\varphi}(\overline{v}^{\lambda}\cos\varphi)+\frac{\partial\overline{\omega}^{\lambda}}{\partial p}=0. \tag{19}$$

It is useful to discuss the MMC in terms of a *streamfunction*, ψ. Plots of ψ conveniently depict the zonally averaged vertical velocity and the zonally averaged meridional velocity together in one diagram. The definition of ψ is embodied in the two equations

$$\overline{v}^{\lambda}2\pi a\cos\varphi\equiv g\frac{\partial\psi}{\partial p}, \tag{20}$$

$$\overline{\omega}^{\lambda}2\pi a^{2}\cos\varphi\equiv -g\frac{\partial\psi}{\partial\varphi}. \tag{21}$$

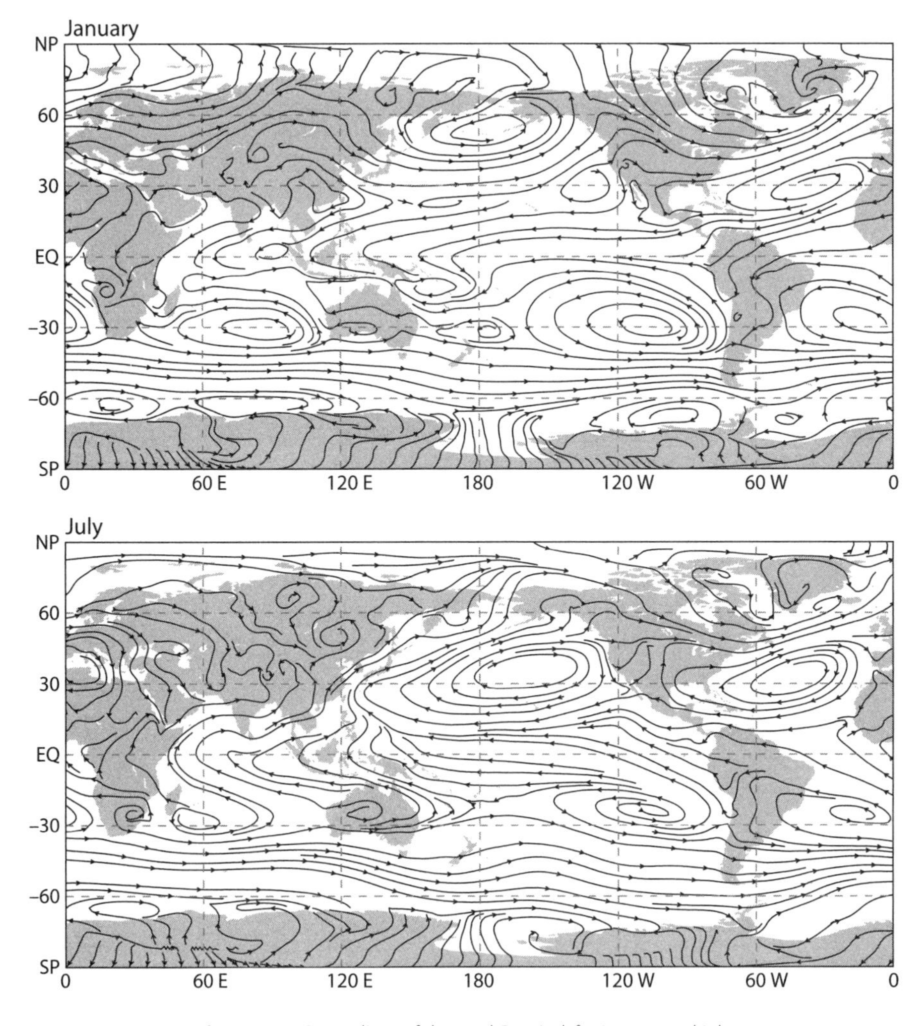

Figure 3.15. Streamlines of the 850 hPa wind, for January and July.

The motivation for this definition is that for any given distribution of ψ the zonally averaged continuity equation, (19), is automatically satisfied; this is easily verified by substitution. Because ψ is defined in terms of its derivatives, it can be determined only within an arbitrary constant. In other words, an arbitrary constant can be added to ψ without changing $\overline{v}^{\lambda}$ or $\overline{\omega}^{\lambda}$. Note that ψ is independent of longitude, because it is defined in terms of longitudinal averages.

A minor technical difficulty is that the zonal averages used to define ψ cannot be carried out for pressure surfaces that intersect the ground. We are going to ignore that issue. It can be skirted, for example by using a terrain-following vertical coordinate system, but the details are too technical to consider here.

In view of (20) and (21), we can compute ψ from either $\overline{v}^{\lambda}$ or $\overline{\omega}^{\lambda}$, but observations of $\overline{v}^{\lambda}$ are generally considered more reliable, so $\overline{v}^{\lambda}$ is preferable. To perform the vertical integration, we need a boundary condition. At $p = 0$, we have $\omega = 0$; this means that the atmosphere is neither gaining nor losing mass by exchanges with space. It follows that

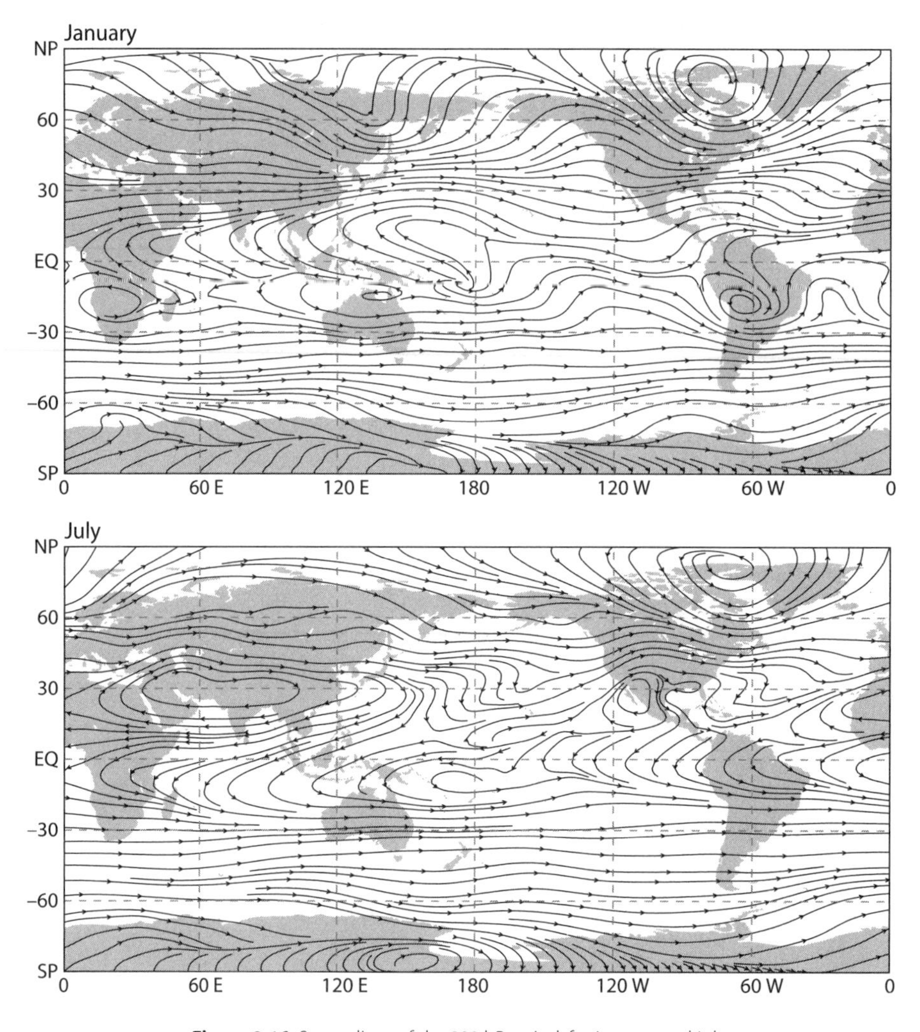

Figure 3.16. Streamlines of the 200 hPa wind, for January and July.

$$\frac{\partial \psi}{\partial \varphi} = 0 \text{ at } p = 0; \tag{22}$$

that is, the top of the atmosphere is a line of constant streamfunction. Similarly, it is easy to see that lines of constant ψ cannot intersect the Earth's surface; if they did, that would imply a flow of air across the Earth's surface. Therefore,

$$\frac{\partial \psi}{\partial \varphi} = 0 \text{ at } p = p_S. \tag{23}$$

In fact, since the zonal mean of the time-averaged vertically integrated meridional wind is approximately zero, ψ must take the same latitude-independent value at $p = 0$ and $p = p_S$. It is conventional to choose this value to be zero; that is, we use the upper boundary condition

$$\psi = 0 \text{ at the top of the atmosphere}; \tag{24}$$

this choice determines the arbitrary constant mentioned previously.

Note from (20) and (21) that

$$\psi \sim \left(\frac{\delta p}{g}\right)\cdot\frac{L}{t}\cdot L \sim \frac{M}{L^2}\cdot\frac{L^2}{t} = \frac{M}{t}; \tag{25}$$

that is, ψ has dimensions of mass per unit time and is typically expressed in units of 10^{12} g s^{-1}, which is the same as 10^9 kg s^{-1}. This unit is sometimes called the *sverdrup* (abbreviated Sv) in the oceanographic literature.

Figure 3.17 shows the latitude-height distribution of the streamfunction of the MMC for January and July, respectively. The observed meridional wind was used to create these plots by vertical integration of (20) and incorporation of (24). The figure shows that deep rising motion occurs in the summer-hemisphere tropics, with sinking motion on either side. The strongest tropical rising motion is near 300 hPa, but notice that weak rising motion continues into the tropical stratosphere. The strongest subsidence is in the winter-hemisphere subtropics, again near 300 hPa. Rising motion occurs

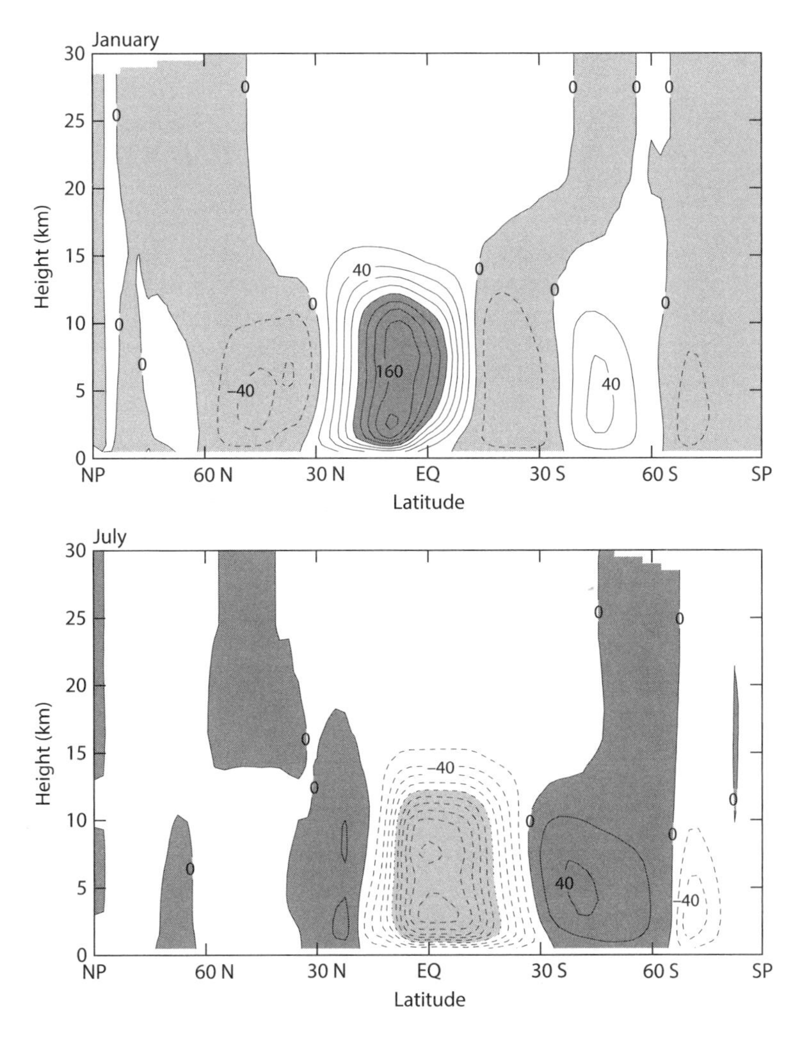

Figure 3.17. The streamfunction of the mean meridional circulation. Positive values represent counterclockwise circulations; negative values represent clockwise circulations. The units are 10^{12} g s^{-1}.

in middle latitudes and is strongest in the winter. Maximum values tend to occur near 500 hPa. Sinking motion is found near the poles, mainly in the lower troposphere.

The dominant cellular structures in the tropics are called *Hadley cells*. They are very important components of the global circulation. There is a "large" Hadley cell at each solstice, with its rising branch in the summer-hemisphere tropics and its body extending into the winter-hemisphere subtropics. Its peak magnitude is about 160 x 10^{12} g s^{-1}. A weaker Hadley cell is found in the summer hemisphere. Both Hadley cells are *direct* circulations, which means that their rising branches are warmer than their sinking branches. As discussed later, direct circulations convert potential energy into kinetic energy.

The cover of this book shows a portion of a Hadley cell over Africa in August, when the large Hadley cell has its rising branch north of the equator and its sinking branch in the southern subtropics. The rising branch of the cell is marked by bright thunderstorms across Africa that extend over the Atlantic Ocean, slightly north of the equator. The sinking branch is associated with clear air over southern Africa and shallow stratocumulus clouds over the South Atlantic.

Because of the seasonal growth and decay of the Hadley cells in the two hemispheres, the zonally averaged meridional wind at the equator reverses seasonally. Near the solstices its direction is from the winter hemisphere into the summer hemisphere at low levels, and from the summer hemisphere into the winter hemisphere in the upper troposphere. Bowman and Cohen (1997) discuss the interhemispheric transports associated with the seasonally changing Hadley circulations.

There are also *indirect* circulations in the middle latitudes, which are most pronounced in the Southern Hemisphere in both seasons. These are called *Ferrel cells*. The sinking branches of the Ferrel cells are adjacent to the sinking branches of the Hadley cells; both are found in the subtropics, near 30° of latitude both north and south of the equator. In these latitude belts, the sinking air diverges horizontally, both toward the pole and toward the equator. The poleward branch is the inflow to the rising motion of the Ferrel cell, and the equatorward branch is the inflow to the rising branch of the Hadley cell. Recall that the zonally averaged sea-level pressure has its maximum in the subtropics; we can think of the diverging subtropical meridional flows as being pushed by the meridional pressure-gradient force, from the sea-level pressure maximum toward lower pressure on both sides. Further physical interpretation of the Ferrel cells is discussed later.

Finally, the polar regions play host to weak direct circulations.

There is an important but very slow meridional circulation in the stratosphere, called the *Brewer-Dobson circulation* (Butchart, 2014). It is too weak to show up in figure 3.7, but it carries mass upward across the tropopause in the tropics and then toward the winter pole. The Brewer-Dobson circulation plays an important role in the stratospheric transports of ozone and angular momentum, and is discussed further in chapter 9.

It is interesting to examine the seasonal change of the MMC, as shown in figure 3.18. Rough symmetry between the hemispheres is seen approximately one month after the equinoxes. Near the solstices, the Hadley cells are best developed on the winter side and least developed on the summer side. Explanations for this behavior have been proposed by Lindzen and Hou (1988), Hack et al. (1989), and Dima and Wallace (2003). The rising motion on the summer side of the equator, near the solstice, is associated with the summer monsoons (Newell et al., 1974; Schulman, 1973; Dima and Wallace, 2003), which are discussed in chapter 8.

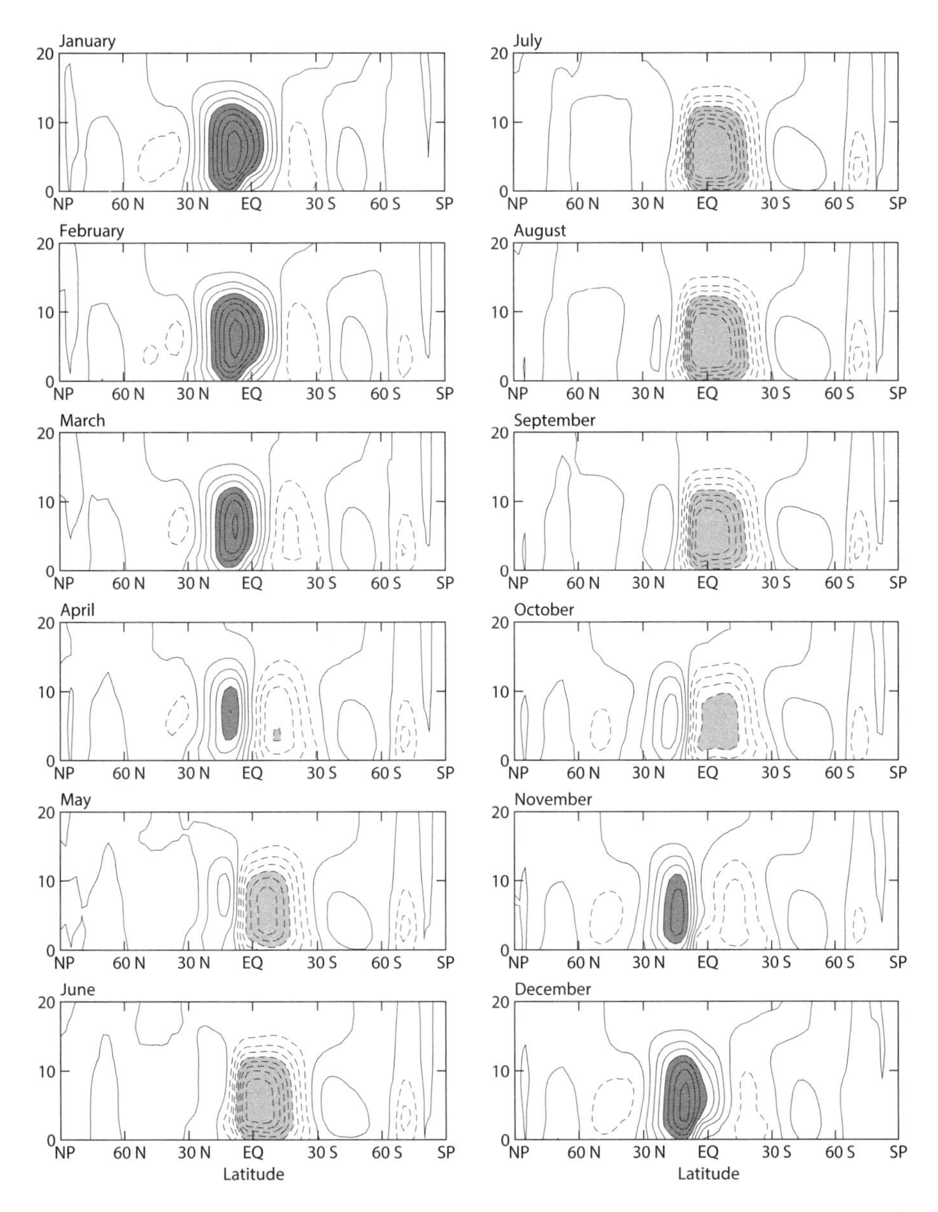

Figure 3.18. Seasonal change of the mean meridional circulation. The contour interval is 25×10^{12} g s^{-1}. Shading is used for values larger than 100×10^{12} g s^{-1} and smaller than -100×10^{12} g s^{-1}. A similar figure appeared in Dima and Wallace (2003).

The correspondence between the zonally averaged vertical motion and the zonally averaged meridional motion is fairly evident. The meridional currents can be interpreted as outflows from or inflows to the vertical currents. Figure 3.19 shows maps of the 500 hPa vertical velocity for January and July, respectively. The units are 10^{-4} Pa s^{-1}. The strongest maxima and minima have absolute values of roughly 10^{-1} Pa s^{-1}, which is about 100 hPa per day. Figure 3.17 shows that regular bands of rising and sinking motion are arranged along latitude circles. The bands are apparent in figure

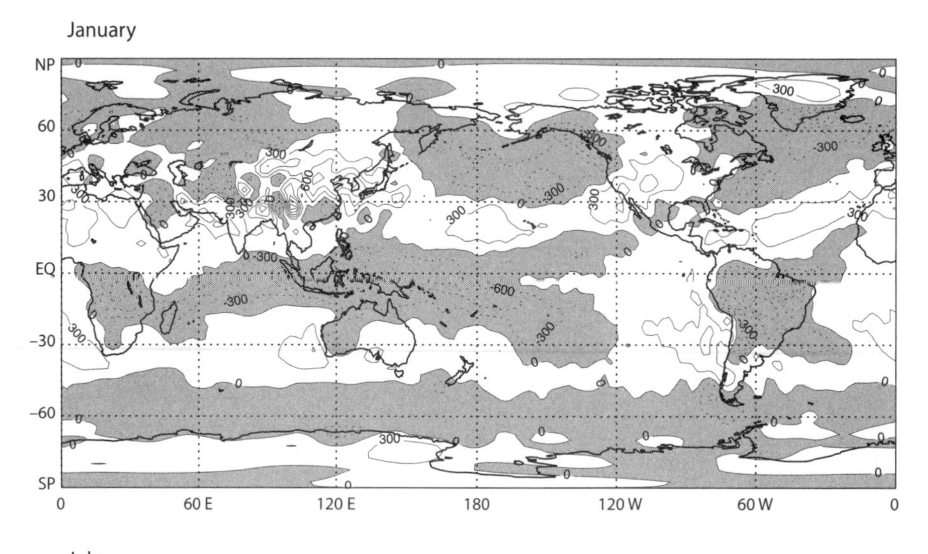

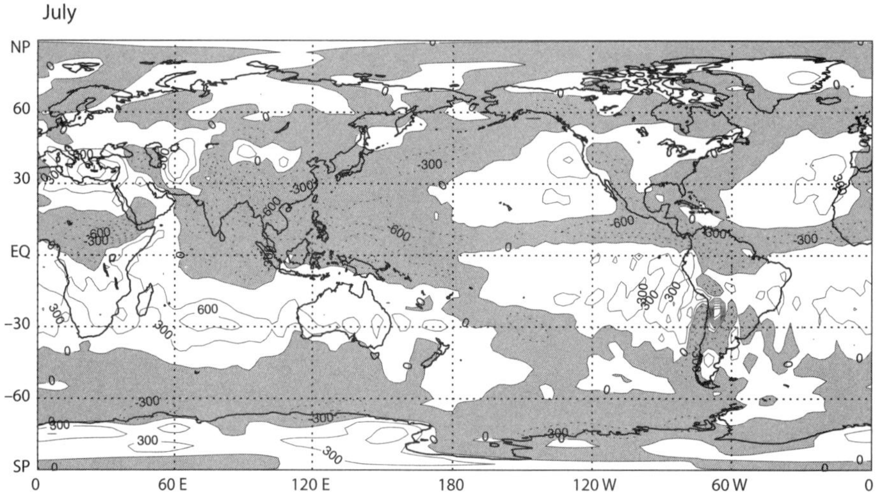

Figure 3.19. Maps of ω, the vertical "pressure velocity" (in mb s^{-1}), for the 500 hPa surface. Negative values (corresponding to rising motion) are shaded.

3.19, especially in the July data. There is some tendency for rising motion in the tropics; sinking motion in the subtropics, especially in the winter hemisphere; rising motion in middle latitudes; and sinking motion over the poles. Sinking motion tends to be associated with surface-pressure maxima, and rising motion with surface-pressure minima. For example, the subtropical highs are clearly associated with large-scale sinking motion in the middle troposphere. The seasonal change of the large-scale vertical motion is very spectacular in the region of the Tibetan Plateau. Rising motion occurs in summer, and sinking motion in winter. These changes are associated with the Indian monsoon, as discussed in chapter 8. There are some cases of rising motion upstream of mountain ranges, and sinking motion downstream; examples are the Rocky Mountains and the Himalayas, in winter. It is not easy to see a clear pattern of orographically forced vertical motions associated with the other major mountain ranges of the world, however. They do exist, but a more refined analysis is needed to detect them.

Note the rising motion over southern Africa and tropical South America in January, in the same regions where there are water vapor maxima at 850 hPa in January, as we will see later. Large water vapor mixing ratios tend to occur in regions of rising motion, and small water vapor mixing ratios in regions of sinking motion. In particular, deserts, such as the Sahara, are regions of tropospheric sinking motion.

Temperature

Figure 3.20 shows the latitude-height distributions of the zonally averaged temperature for January and July, respectively. At low levels, the warmest air is near the equator, but near 100 hPa the coldest air is over the equator. In fact, some of the lowest temperatures in the atmosphere are found near the tropical tropopause. It is also true

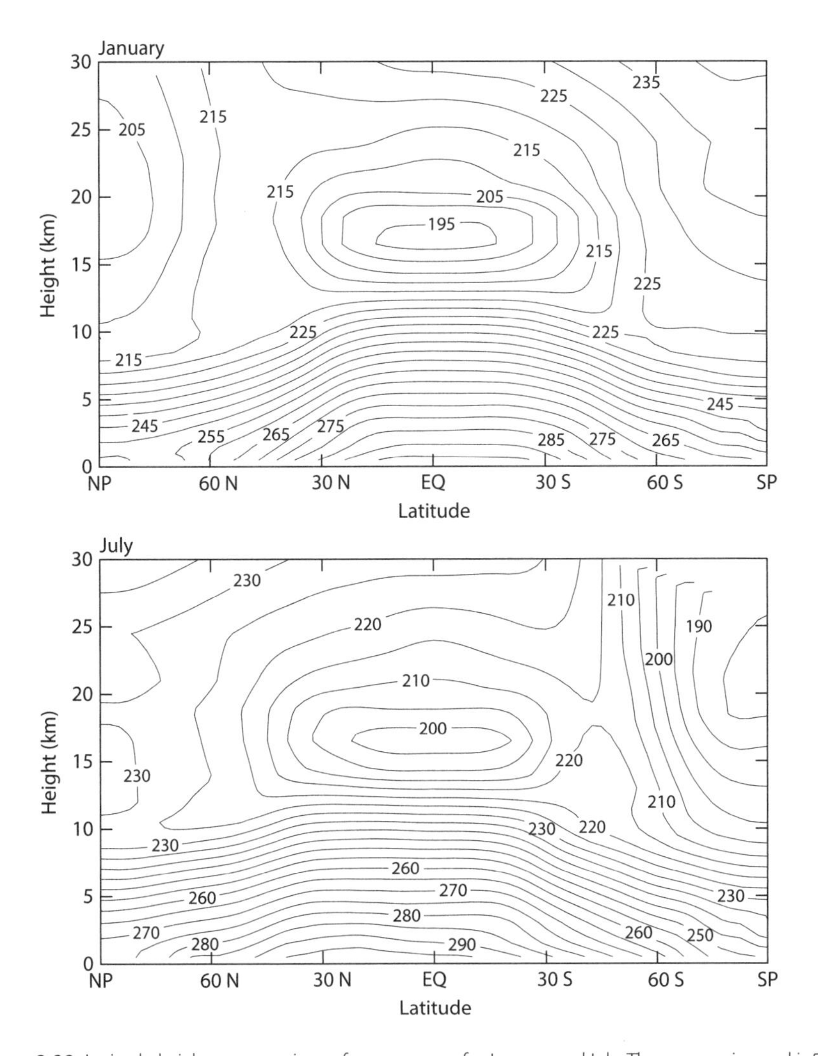

Figure 3.20. Latitude-height cross sections of temperature, for January and July. The contour interval is 5 K.

that the tropopause is highest in the tropics and lowest near the poles. The tropopause height changes almost discontinuously with latitude in the subtropics, particularly in the winter hemisphere. In the stratosphere, extremely cold temperatures are found above the winter pole, especially in the Southern Hemisphere. The summer pole is much warmer owing to the absorption of solar radiation by ozone.

It has been suggested that for a wide class of sufficiently massive atmospheres, the tropopause must occur near the 100 hPa level, because at higher pressures the atmosphere is relatively opaque to the thermal infrared radiation, while at lower pressures it is relatively transparent (Robinson and Catling, 2013).

The midlatitude low-level temperature gradients are quite strong in the winter hemisphere. Above 200 hPa, the summer pole is considerably warmer than the winter pole; in fact, the zonally averaged temperature increases monotonically from the equator to the summer pole near 100 hPa, which suggests easterlies in the summer stratosphere, as are indeed observed. In the winter hemisphere, the warmest 100 hPa temperatures occur in middle latitudes. The strong decrease of temperature between midlatitudes and the poles is consistent with the polar night jets mentioned earlier.

Generally speaking, the lapse rate, $-\partial T/\partial z$, is largest in the tropics and smaller (or even negative) near the poles. In January, a temperature "inversion" (i.e., increase in temperature with altitude) appears over the North Pole.

The thermal wind equations relate the vertical shear of the geostrophic wind to the horizontal temperature gradient, namely,

$$\frac{\partial u_g}{\partial p} = \frac{R}{fpa}\left(\frac{\partial T}{\partial \varphi}\right)_p, \text{ and } \frac{\partial v_g}{\partial p} = \frac{-R}{fpa\cos\varphi}\left(\frac{\partial T}{\partial \lambda}\right)_p. \tag{26}$$

Thermal wind balance between the meridional temperature gradient and the vertical shear of the zonal wind is well satisfied, as can be qualitatively confirmed by comparison of figures 3.6 and 3.20.

Figure 3.21 shows maps of the 850 hPa temperature for January and July, and figure 3.22 shows the corresponding results for 200 hPa. The expected winter-to-summer warming at 850 hPa is clearly evident in the Northern Hemisphere, but not in the Southern Hemisphere, except over land. Monthly-mean temperatures over the high Antarctic terrain reach about −50 °C in July, while those over the Arctic Ocean in January do not fall below −35 °C. In the tropics there is very little seasonal change, and the temperature distribution is very uniform.

Naturally, the temperature gradient points mainly from the poles toward the tropics at 850 hPa, but wavy patterns are plainly visible in the Northern Hemisphere in January. There are some regions in which the mean temperature actually increases poleward, at 850 hPa, for example, from northern Africa eastward to India in July, and north of Australia in January. From thermal wind considerations, we might expect easterlies aloft in these regions. This expectation is borne out in figure 3.9. In winter, there is a tendency for the eastern side of the Northern Hemisphere continents to be colder than the western side, which leads to particularly strong meridional temperature gradients on the east coasts. We know that such strong temperature gradients favor a rapid upward increase of the westerlies; also, such highly baroclinic regions are preferred centers of cyclogenesis.

The strongest temperature gradients at 200 hPa are found in high latitudes, especially in winter. The wavelike eddy pattern is much stronger at 200 hPa than at 850 hPa. Particularly noticeable are the maxima over the North Pacific in January, over

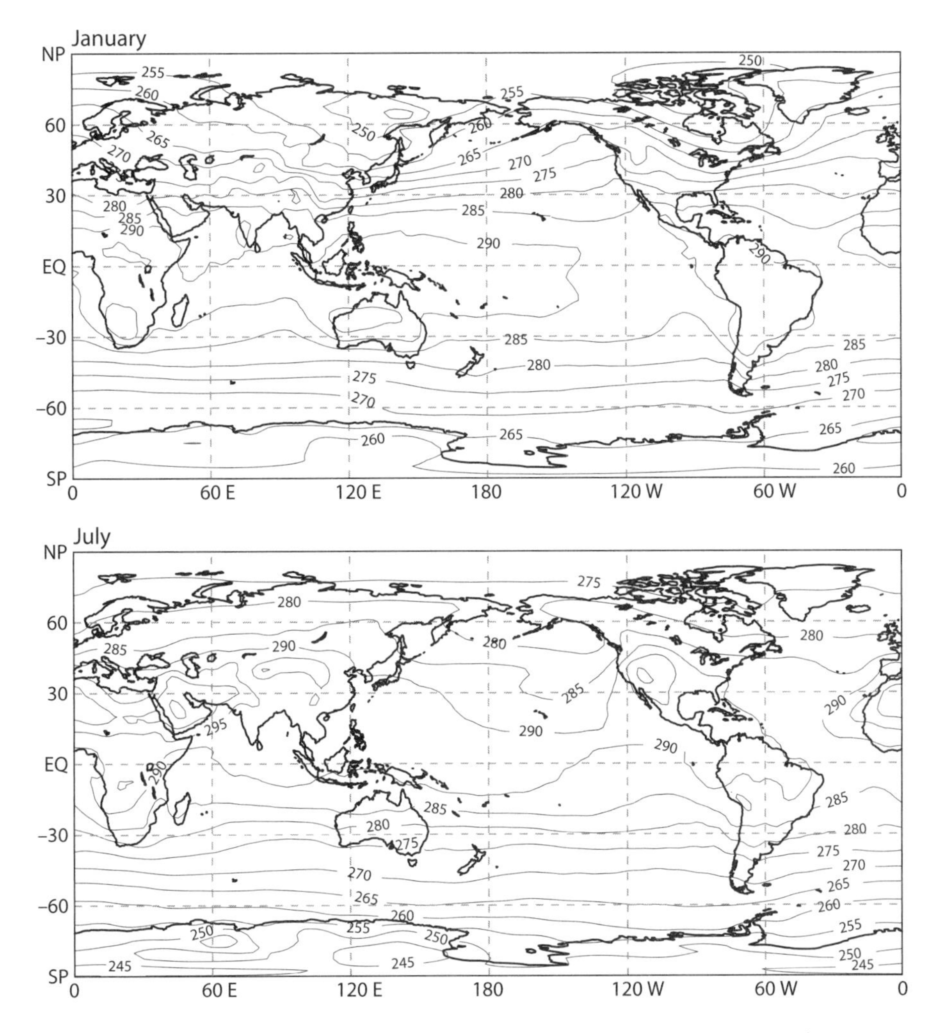

Figure 3.21. Maps of the 850 hPa temperature, for January and July. The contour interval is 5 K.

eastern North America in January, and over southern Asia in July. Note than in each of these regions the 200 hPa temperature increases from the tropics toward middle latitudes, which implies a tendency for the westerlies to weaken above this level.

Recall that the potential temperature is defined by

$$\theta \equiv T\left(\frac{p_0}{p}\right)^{\kappa}, \tag{27}$$

where p_0 is a constant reference pressure, usually chosen to be 1000 hPa, and $\kappa \equiv R/c_p$, where R is the gas constant, and c_p is the specific heat of air at constant pressure. As will be discussed in chapter 4, important facts about the potential temperature are that it is conserved under dry adiabatic processes and that (with minor exceptions) it increases with altitude throughout the atmosphere, as is evident in figure 3.23. In the troposphere, potential temperature surfaces slope downward from the polar regions to the tropics, while in the stratosphere they do the opposite. The stratosphere is

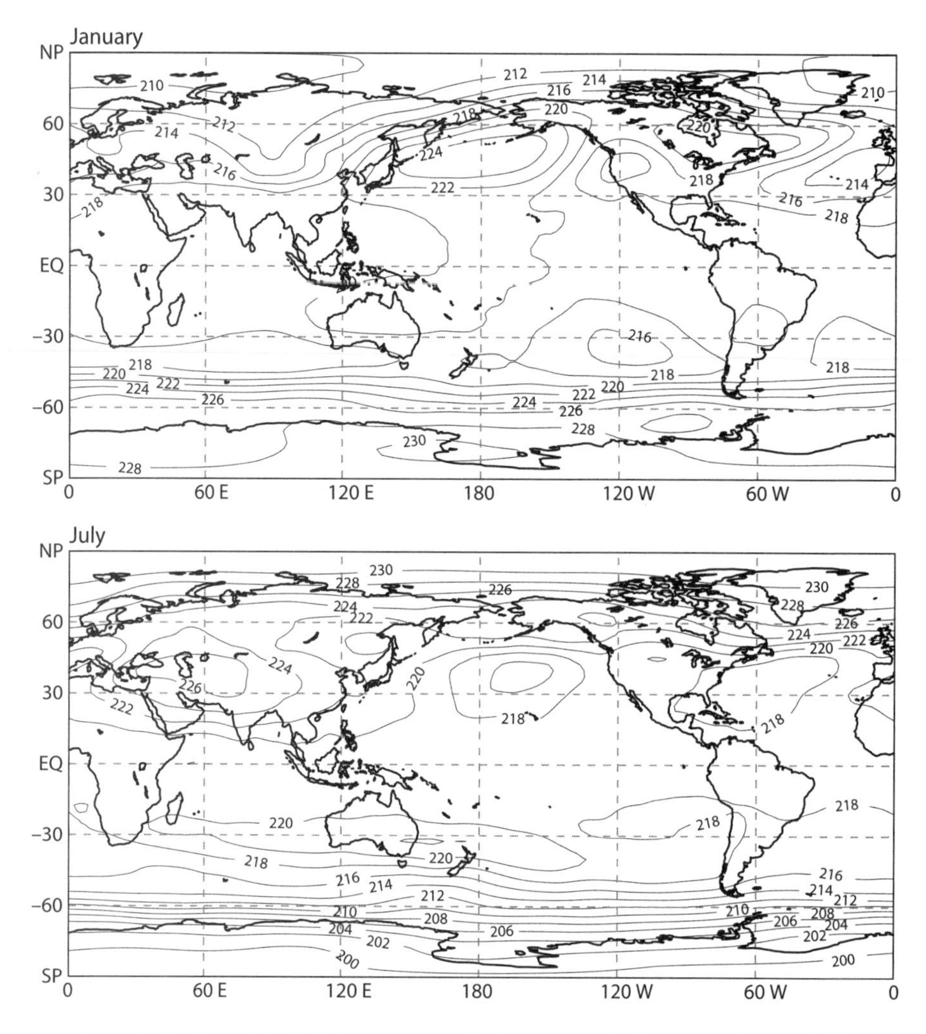

Figure 3.22. Maps of the 200 hPa temperature, for January and July. The contour interval is 5 K.

easily identified because the potential temperature increases very sharply there with height, indicating that the static stability is very strong. In contrast, the static stability is particularly weak in the tropical upper troposphere.

Hoskins (1991) distinguishes the following three regimes, which can be identified in figure 3.23: There is an "Overworld," in which by definition the potential temperature surfaces are everywhere above the tropopause. From the data, we see that such air has potential temperatures of about 390 K or greater; that is, the tropical tropopause roughly coincides with the surface on which the potential temperature is 390 K. The "Middleworld" is defined to have potential temperature surfaces that cross the tropopause, which means that air moving along isentropic surfaces in the Middleworld can move between the troposphere and the stratosphere. The Middleworld is found outside the tropics, in middle latitudes. Finally, the "Underworld" has potential temperature surfaces that intersect the Earth's surface, so that air moving isentropically in the Underworld can "sample" the properties of the Earth's surface and communicate them to the atmosphere. Much of the Underworld is found in the

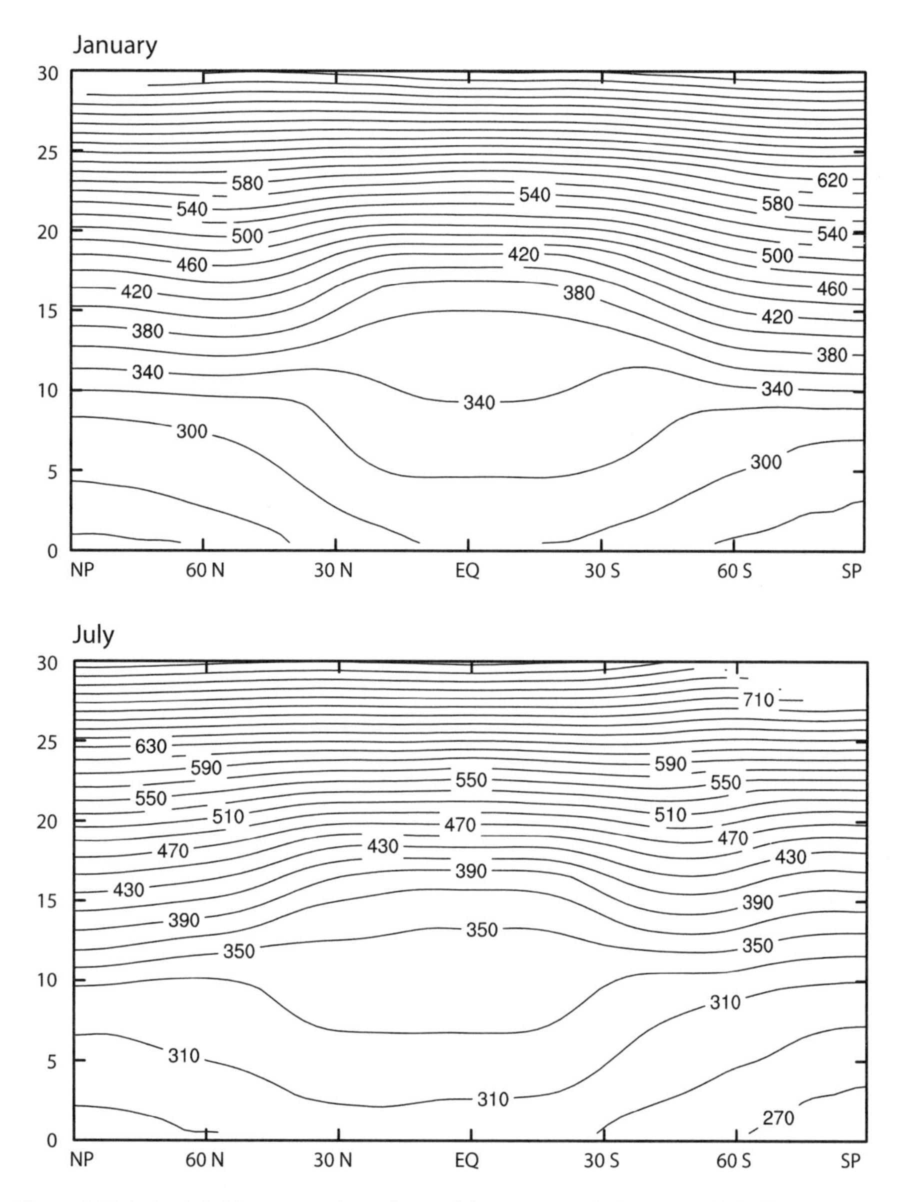

Figure 3.23. Latitude-height cross sections of potential temperature, for January and July. The contour interval is 20 K.

high-latitude troposphere. From figure 3.23, it appears that the largest potential temperatures in the Underworld are on the order of 300 K, although in reality much larger values of the potential temperature do occur locally near the Earth's surface, for example, on a summer afternoon in the Sahara desert.

As you may know, potential temperature can be used as a vertical coordinate, and such an approach has many advantages in theoretical analyses, numerical modeling, and the interpretation of observations (e.g., Hoskins et al., 1985). We will frequently use potential temperature as a vertical coordinate in this book. Chapter 4 introduces the relevant equations.

Moisture

The globally averaged evaporation and precipitation rates, which must balance in a time average, are not accurately known but are a little less than 3 kg m^{-2} day^{-1}. It follows that an average water molecule has a *residence time* of about eight days in the atmosphere, between its introduction by surface evaporation and its removal by precipitation.

The *water vapor mixing ratio* is the density of water vapor divided by the density of dry air. Figure 3.24 shows the latitude-height distributions of the zonally averaged water vapor mixing ratio for January and July, respectively. The figure should be viewed with caution, however, because the data presented here have been fed through the analysis/forecast system of a numerical weather prediction center, which can easily distort the distribution of water vapor. The panels show the most humid air near the equator, and the driest air near the winter pole. The seasonal change in the Northern Hemisphere is quite dramatic. There is an extremely rapid decrease of the mixing ratio with height at all latitudes. The largest zonally and temporally averaged values, near the surface in the tropics, are close to 18 g kg^{-1}, which means that about 2% of the air is water vapor. Even minute quantities of water vapor in the upper troposphere and stratosphere can be very important radiatively.

Because the mixing ratio is greatest near the surface, regions of low-level mass convergence, such as those apparent in the zonally averaged meridional wind, tend to be regions of vertically integrated moisture convergence as well. Although as much mass diverges at upper levels as converges at lower levels, the diverging air aloft is dry, while the converging air near the surface is humid. A net convergence of water vapor occurs in such regions. The situation is reversed in desert regions, where the dry upper tropospheric air converges and descends.

Figure 3.25 shows maps of the 850 hPa water vapor mixing ratio for January and July, respectively. As would be expected, the largest values occur in the tropics and the

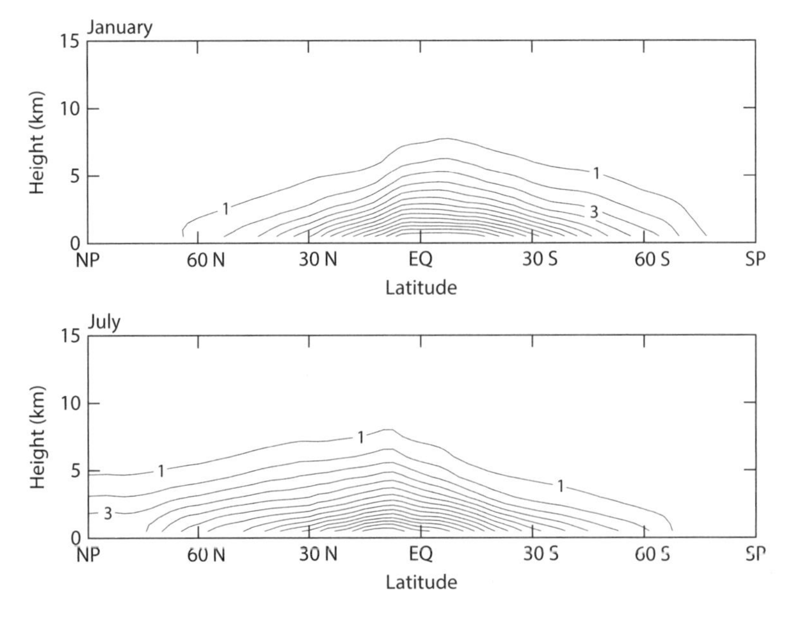

Figure 3.24. Latitude-height section of the water vapor mixing ratio. The contour interval is 1 g kg^{-1}.

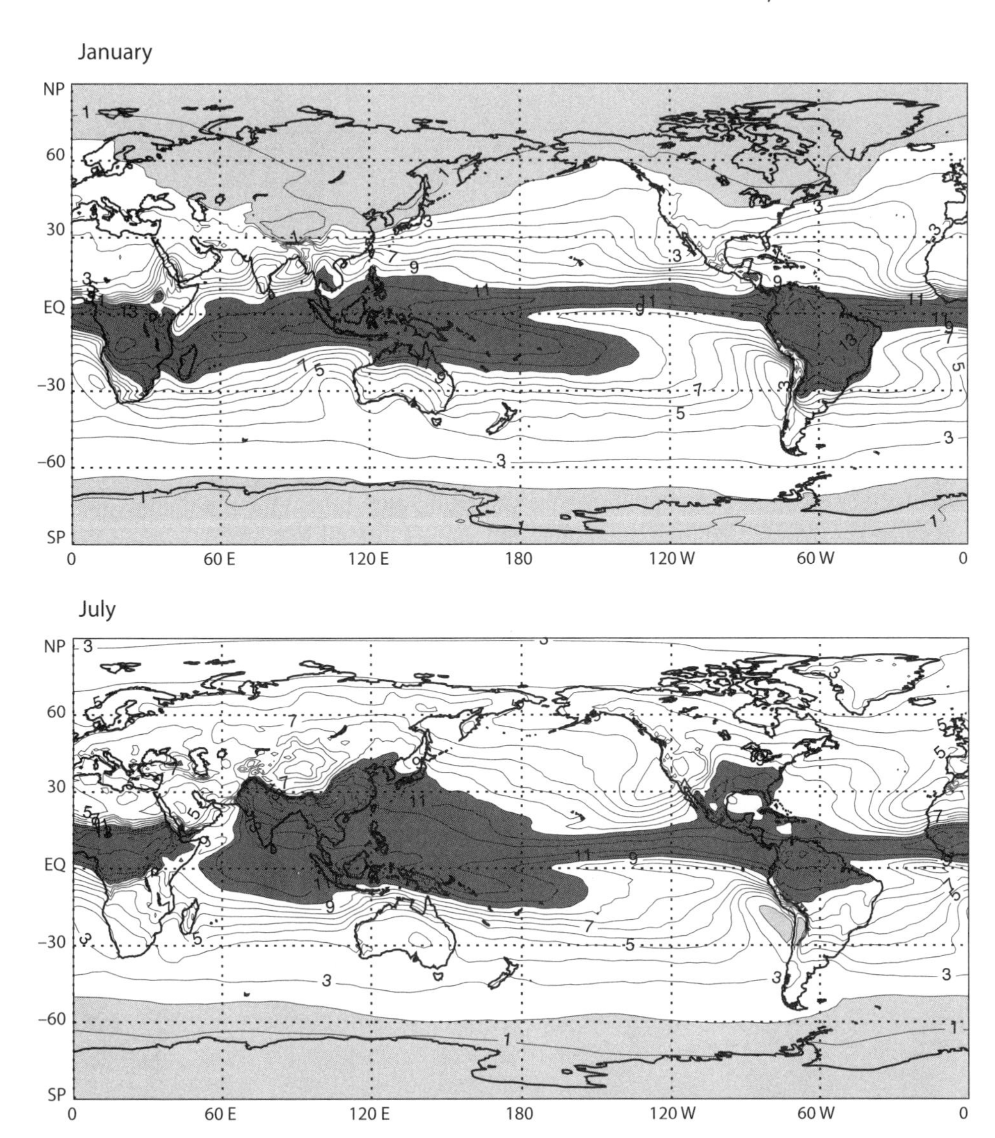

Figure 3.25. Maps of the 850 hPa water vapor mixing ratio. The contour interval is 1 g kg^{-1}. Values greater than 10 g kg^{-1} are darkly shaded.

summer hemisphere. A very clear maximum extends around the circumference of the Earth in the tropics, mainly somewhat north of the equator. Meridional moisture gradients are often quite sharp, and there are also strong east-west variations. For example, in January there are strong maxima over southern Africa and in the Amazon basin. Minima are found in the subtropical highs. There are very dramatic seasonal changes over the midlatitude continents, with larger values in summer. Major desert regions like the Sahara and western North America are clearly associated with water vapor minima.

The *relative humidity* can be defined as the ratio of the actual water vapor pressure to the saturation vapor pressure; the latter is an increasing function of temperature

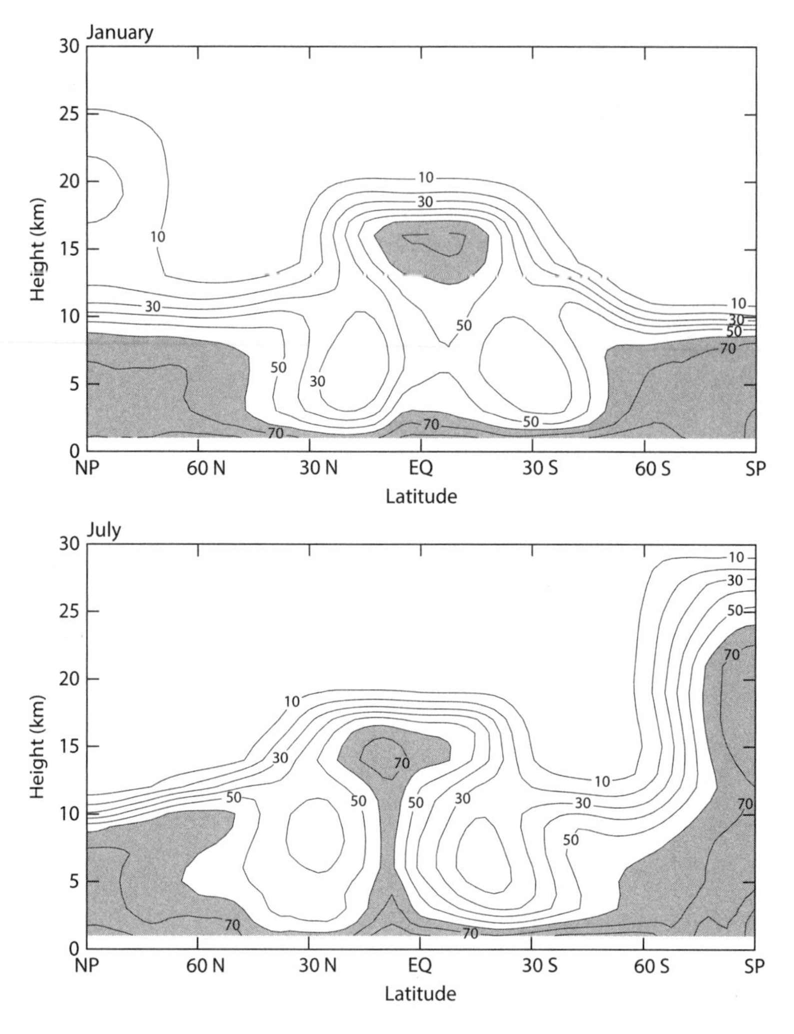

Figure 3.26. The observed latitude-height distribution of the zonally averaged relative humidity, in percent, as analyzed by ECMWF. Values greater than 60% are shaded.

only, and was introduced in chapter 2. Locally, a relative humidity of 1 or more, corresponding to water vapor "saturation," normally indicates the presence of cloud. The zonally averaged relative humidity is shown in figure 3.26. The troposphere is very humid in rainy regions of the tropics; high relative humidities are also found in middle latitudes, especially in winter, where strong low-pressure systems are frequent.

Lots of Questions

This chapter was designed to impart some familiarity with basic features of the observed global circulation but without much attempt to explain why the circulation appears as it does. The observations presented raise many questions, such as the following:

- What determines the magnitude of the pole-to-equator gradient of the surface temperature?
- Why are seasonal changes generally weaker in the Southern Hemisphere than in the Northern Hemisphere?
- What determines the observed lapse rate of temperature? Why does the lapse rate change in moving from the tropics to the middle latitudes to the poles?
- What determines the intensities and latitudes of the jet streams? Why do the winter jet maxima occur at particular longitudes?
- Why are the subtropical highs typically found on the eastern side of the ocean basins?
- Why is the intertropical convergence zone mainly north of the equator?
- What accounts for the observed patterns of large-scale rising and sinking motion?
- Why is there such a strong belt of low pressure around Antarctica? What determines the locations and intensities of the Aleutian and Icelandic Lows?
- What determines the vertical distribution of water vapor?
- Why is the upper-level circulation "wavy"? Why is the lower-level circulation "lumpy?"
- What determines the height of the tropopause as a function of latitude? Why is the tropical tropopause so cold? Why does the tropopause height have a discontinuity near the subtropical jets?
- What mechanisms generate the observed variations with longitude of the monthly-mean fields?
- What accounts for the observed large-scale pattern of winds in a monsoon?
- How much do individual Januaries and Julys, for particular years, differ from the "average" January and July conditions shown here? What causes such year-to-year variations?
- What are the geographic patterns of the day-to-day weather fluctuations that accompany the monthly-mean maps shown here, and how do these fluctuations affect the time means?
- Why does the global circulation appear "smooth" rather than "noisy?"

These and many other questions will be discussed in the remainder of this book. To address them, we will need the ideas presented in chapter 4.

Problems

1. Estimate the total water vapor content of the atmosphere, in kilograms. Explain how you arrived at your answer.
2. Make a rough estimate of the total kinetic energy of the atmosphere, in joules. If all the solar radiation absorbed by the Earth were used to supply this kinetic energy, how long would it take to accumulate the observed amount? *Note*: In reality the rate at which kinetic energy is generated in the atmosphere is much less than the rate at which the Earth absorbs solar radiation.
3. Derive the surface pressure tendency equation,

$$\frac{\partial p_S}{\partial t} + \nabla \cdot \left(\int_0^{p_S} \mathbf{V}_h dp \right) = 0,$$

by starting from the continuity equation in pressure coordinates,

$$\nabla_p \cdot \mathbf{V} + \frac{\partial \omega}{\partial p} = 0.$$

CHAPTER 4

The Rules of the Game

Introduction

This chapter gives a quick review of concepts to be used later. Topics covered include the conservation principles for momentum, moisture, energy in its various forms, and potential vorticity. Some important approximations, including the quasi-hydrostatic and quasi-geostrophic approximations, are briefly introduced.

Conservation of the Mass of Dry Air

Let $\mathbf{V}$ denote the three-dimensional wind vector. The vector $\rho\mathbf{V}$ is the flux of mass across a unit area. Consider a "control volume" CV, fixed in space, through which fluid can flow (see fig. 4.1). Conservation of mass is expressed by

$$\frac{d}{dt}\int_{CV} \rho dV = \int_{CV} \frac{\partial \rho}{\partial t} dV = -\int_{\sigma} \rho \mathbf{V} \cdot \mathbf{d\sigma}, \tag{1}$$

where the first two integrals are taken over the control volume, and $\mathbf{d\sigma}$ is an outward normal vector whose magnitude is that of a (differential) element of the bounding surface of the control volume. Using Gauss's theorem, we can rewrite (1) as

$$\int_{CV} \left[\frac{\partial \rho}{\partial t} + \nabla \cdot (\rho \mathbf{V})\right] dV = 0. \tag{2}$$

Since the control volume is arbitrary, we conclude that

$$\boxed{\frac{\partial \rho}{\partial t} + \nabla \cdot (\rho \mathbf{V}) = 0}. \tag{3}$$

Mass conservation can also be expressed using a Lagrangian framework, in which we consider time rates of change following particles. For this purpose, we use the Lagrangian (particle-following) time derivative, D/Dt. The Lagrangian time derivative can be expanded as

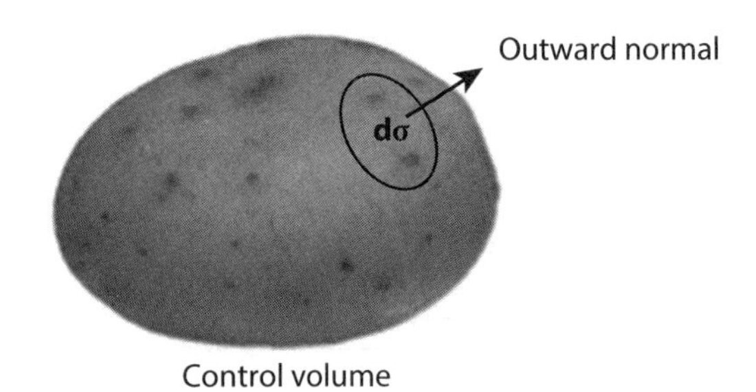

Figure 4.1. Sketch illustrating conservation of mass. Here $\mathbf{d\sigma}$ is a small element of area on the boundary of a control volume. Control volumes are always shaped like potatoes—hence the expression "great geophysical potato."

$$\frac{D}{Dt} = \frac{\partial}{\partial t} + \mathbf{V} \cdot \nabla. \tag{4}$$

To demonstrate this, let $\mathbf{r}$ be the position vector that corresponds to the particle's location, and write the Lagrangian time derivative of an arbitrary variable ψ as

$$\frac{D\psi}{Dt} = \frac{\psi(\mathbf{r} + \Delta\mathbf{r}, t + \Delta t) - \psi(\mathbf{r}, t)}{\Delta t}. \tag{5}$$

Think of the two values of ψ on the right-hand side of (5) as "measured values" for a particular particle. The measurements are made at times $t + \Delta t$ and t, when the particle's positions are $\mathbf{r} + \Delta\mathbf{r}$ and $\mathbf{r}$, respectively. We can write

$$\psi(\mathbf{r} + \Delta\mathbf{r}, t + \Delta t) - \psi(\mathbf{r}, t) = \frac{\partial\psi}{\partial t}\Delta t + \nabla\psi \cdot \Delta\mathbf{r} + \text{higher-order terms}. \tag{6}$$

Dividing both sides of (6) by Δt, and using $\Delta\mathbf{r}/\Delta t \equiv \mathbf{V}$, we obtain

$$\frac{D\psi}{Dt} = \frac{\partial\psi}{\partial t} + \mathbf{V} \cdot \nabla\psi. \tag{7}$$

The left-hand side of (7), that is, $D\psi/Dt$, is a Lagrangian description of the change of ψ experienced by a moving particle, and the right-hand side, that is, $(\partial\psi/\partial t) + \mathbf{V} \cdot \nabla\psi$, is the sum of the time rate of change of ψ as seen in a fixed Eulerian coordinate system, and a term representing the effects of advection as seen in the Eulerian framework. Equation (7) is valid in any frame of reference; that is, both the left-hand side of (7) and the *sum* of the terms on the right-hand side are independent of coordinate frame. The individual terms on the right-hand side of (7) do depend on the Eulerian coordinate system used, however.

By using (7), we can rewrite (3) as

$$\frac{D\rho}{Dt} + \rho\nabla \cdot \mathbf{V} = 0 \tag{8}$$

or as

$$\nabla \cdot \mathbf{V} = -\frac{1}{\rho}\frac{D\rho}{Dt} = \frac{1}{\alpha}\frac{D\alpha}{Dt}, \tag{9}$$

where $\alpha \equiv 1/\rho$ is the specific volume. For an *incompressible* fluid of fixed density, with $(1/\rho)\,D\rho/Dt = (1/\alpha)\,D\alpha/Dt = 0$, (9) implies that $\nabla \cdot \mathbf{V} = 0$; that is, the three-dimensional wind field must be nondivergent. Air is, of course, compressible. Liquid water is an example of a fluid that is much less compressible than air.

The mass budget of a column of air can be derived from (3) as follows. We integrate with respect to height, through the entire column, to obtain

$$\int_{z_S}^{\infty} \frac{\partial \rho}{\partial t} dz + \int_{z_S}^{\infty} \nabla \cdot (\rho \mathbf{V})\, dz = 0. \tag{10}$$

Here z_S is the surface height, which varies in space and could conceivably vary with time. In deriving (10), we used the fact that $\lim_{z \to \infty} \rho = 0$. Moving the integrals inside the derivatives, and taking into account the possible spatial and temporal variability of z_S, we obtain

$$\frac{\partial}{\partial t}\left(\int_{z_S}^{\infty} \rho dz\right) + \nabla \cdot \left(\int_{z_S}^{\infty} \rho \mathbf{V}_h dz\right) + \rho_S \left[\frac{\partial z_S}{\partial t} + (\mathbf{V}_h)_S \cdot \nabla z_S - w_S\right] = 0, \tag{11}$$

where $\mathbf{V}_h$ is the horizontal velocity vector, subscript S denotes the Earth's surface, and w is the vertical velocity, so that

$$\mathbf{V} = \mathbf{V}_h + w\mathbf{e}_z, \text{ where } \mathbf{e}_z \text{ is a unit vector that points upward.} \tag{12}$$

The expression $\rho_S[(\partial z_S/\partial t) + (\mathbf{V}_h)_S \cdot \nabla z_S - w_s]$ represents the flux of mass across the Earth's surface. As a physical lower boundary condition, we declare that in fact no mass crosses the Earth's surface, which implies that

$$\boxed{\frac{\partial z_S}{\partial t} + (V_h)_S \cdot \nabla z_S - w_s = 0}. \tag{13}$$

Equation (13) will be used repeatedly later in the book. We would need to keep the $\partial z_S/\partial t$ term of (13) if, for example, we planned to study the effects of earthquakes. In that case, we would specify $\partial z_S/\partial t$ to describe how the lower boundary moved. Normally, of course, the height of the Earth's surface can be assumed to be independent of time (and we will assume so in this book), in which case (13) reduces to

$$w_S = (\mathbf{V}_h)_S \cdot \nabla z_S. \tag{14}$$

Equation (14) says that the three-dimensional surface wind vector is tangent to the Earth's surface.

It is interesting to note that there is a nonzero flux of mass across the Martian surface, near the poles. The Martian atmosphere consists almost entirely of carbon dioxide (CO_2). In the winter, polar temperatures are low enough so that the atmosphere (partially) condenses onto the surface; during the summer, the atmosphere gains mass near the summer pole as the frozen CO_2 returns to gaseous form (e.g., Lewis and Prinn, 1984; Zent, 1996). This results in seasonal changes of the Martian surface pressure on the order of 30% (Hess et al., 1977).

The condensation of water in the Earth's atmosphere is analogous to the condensation of CO_2 in the Martian atmosphere, but of course water makes up only a tiny fraction of the Earth's atmosphere, so the resulting changes in the surface pressure are small. If the Earth's atmosphere became cold enough to condense nitrogen (that's not going to happen!), large changes in the surface pressure could result.

With the use of (13), (11) reduces to

$$\frac{\partial}{\partial t}\left(\int_{z_S}^{\infty} \rho dz\right) + \nabla \cdot \left(\int_{z_S}^{\infty} \rho \mathbf{V}_h dz\right) = 0, \tag{15}$$

which expresses conservation of mass for the air column. According to (15), the total mass of air in the column, per unit horizontal area, changes in time only as a result of fluxes of air across the sides of the column.

Recall that to an excellent degree of approximation a vertical increment of pressure is proportional to the corresponding vertical increment of mass of air per unit area; that is,

$$dp = -\rho g\, dz. \tag{16}$$

Equation (16), which will be discussed again later in this chapter, implies that the surface pressure is (approximately) given by

$$p_S = \int_{z_S}^{\infty} \rho g\, dz. \tag{17}$$

Use of (17) in (15) gives

$$\frac{\partial p_S}{\partial t} + \nabla \cdot \left(\int_0^{p_S} \mathbf{V}_h\, dp \right) = 0. \tag{18}$$

This is the surface pressure tendency equation, which was introduced in chapter 3.

Conservation of Atmospheric Moisture

While we are on the subject of mass, let's discuss atmospheric moisture. About 99% of it is water vapor. The remainder is divided between liquid and ice.

Recall that the mixing ratio of water vapor, q_v, is the density of water vapor divided by the density of dry air. In the troposphere, q_v ranges from around 20 g kg^{-1} in warm tropical places like Hong Kong in summer, to a minimum of less than 1 kg^{-1} in cold places like the upper troposphere or near the surface in the Antarctic winter. We can also define mixing ratios for liquid water and ice. The total water mixing ratio, q_T, can be separated into various subspecies by phase and/or particle size, as follows:

$$q_T = q_v + q_c + q_i + q_r + q_s. \tag{19}$$

Here q_c is the mixing ratio of "cloud water," which consists of droplets small enough to have negligible fall speeds; q_i is the mixing ratio of "cloud ice," which consists of crystals small enough to have negligible fall speeds; q_r is the mixing ratio of rainwater; and q_s is the mixing ratio of snow. In fact, much more elaborate breakdowns are possible based on fine-grain distinctions among particles of various sizes and shapes. There is evidence that such complexity does matter for detailed analyses of the global circulation of the atmosphere, but it is beyond the scope of this book, and in fact we will even ignore the ice phase for simplicity. We therefore replace (19) by

$$q_T = q_v + l, \tag{20}$$

where l represents the mixing ratio of liquid water regardless of drop size, including both q_c and q_r.

Conservation of total water is expressed by

$$\frac{\partial}{\partial t}(\rho q_T) + \nabla \cdot (\rho \mathbf{V} q_T + \mathbf{F}_{q_T} - \mathbf{P}) = 0, \tag{21}$$

where $\mathbf{F}_{qv}$ is the vector flux of total water vapor due to molecular diffusion, and $\mathbf{P}$ is the vector flux of liquid water (or ice) due to precipitation, which involves motion relative to the moving air. We include a minus sign in the definition of $\mathbf{P}$ so that a positive value of the vertical component of $\mathbf{P}$ will correspond to *downward*-moving rain or snow. We can write separate conservation equations for q_v and l, as follows:

$$\frac{\partial}{\partial t}(\rho q_v) + \nabla \cdot (\rho \mathbf{V} q_v + \mathbf{F}_{q_v}) = -C, \tag{22}$$

$$\frac{\partial}{\partial t}(\rho l) + \nabla \cdot (\rho \mathbf{V} l - \mathbf{P}) = C. \tag{23}$$

Here C is the net rate of condensation per unit mass; condensation converts vapor into liquid. When we add (22) and (23), the condensation terms cancel, and we recover (21).

As discussed in detail in chapter 6, vertical moisture transport in the atmosphere is often dominated by fluxes due to small-scale motions. In this book the term *small scale* means smaller than a few hundred kilometers in the horizontal direction. Small-scale motions specifically include turbulent eddies with scales from centimeters to a few kilometers, convective and wave motions with scales from a few kilometers to a few tens of kilometers, and mesoscale weather systems (such as squall lines) with scales ranging from a few tens of kilometers up to a few hundred kilometers. Fluxes due to these small-scale motions are typically many orders of magnitude larger than the molecular fluxes, so in the study of the global circulation we can often (but not always) ignore the molecular fluxes.

The small-scale fluxes are vectors that include both horizontal and vertical components, but they mainly affect the large-scale circulation through *vertical* exchanges; that is, it is the vertical fluxes that matter most. For example, when considering the effects of small-scale moisture fluxes on large-scale weather systems, we can assume that

$$\boxed{\nabla \cdot \mathbf{F}_{q_T} \cong \frac{\partial}{\partial z}(F_{q_T})_z}, \tag{24}$$

where we now interpret $\mathbf{F}_{q_T}$ as a flux due to small-scale motions rather than molecular diffusion, and (F_{q_T}) is the vertical component of $\mathbf{F}_{q_T}$. Although (24) is written for the water vapor flux, similar approximations also apply to fluxes of other quantities, including radiative fluxes. The justification for (24) is *not* that the horizontal fluxes themselves are smaller than the vertical fluxes. In fact, for small-scale motions, the horizontal and vertical fluxes are often about equally strong. It is the flux *divergence*, however, rather than the flux itself, that influences the time rate of change. The justification for (24) is that the depth of the atmosphere, over which the vertical flux converges or diverges, is shallow compared with the horizontal extent of the large-scale motions, over which a horizontal flux converges or diverges.

The approximation (24) is *not* valid for the fluxes associated with large-scale eddies. The horizontal fluxes due to large-scale eddies, such as winter storms, are considerably stronger than the vertical fluxes. For large-scale eddies, both horizontal and vertical flux divergences are comparable in magnitude; both matter.

When considering large-scale motion, we can also approximate the precipitation term of (21) by

$$\nabla \cdot \mathbf{P} \cong \frac{\partial P}{\partial z}. \tag{25}$$

Substituting (24) and (25) into (21), we obtain

$$\boxed{\frac{\partial}{\partial t}(\rho q_T) + \nabla \cdot (\rho \mathbf{V} q_T) + \frac{\partial}{\partial z}\left[(F_{q_T}) - P\right] = 0}, \tag{26}$$

which applies to large-scale motions and will be used later.

Conservation of Momentum on a Rotating Sphere

The length of a sidereal day is 86,164 s, so the Earth rotates about its axis with an angular velocity of $2\pi/(86{,}164 \text{ s}) \cong 7.292 \times 10^{-5} \text{ s}^{-1}$. This angular velocity can be represented by a vector, $\mathbf{\Omega}$, pointing toward the celestial north pole, as shown in figure 4.2. Consider a coordinate system that is rotating with the Earth. As discussed by Holton (2004), Newton's statement of momentum conservation, as applied in the rotating coordinate system, is

$$\frac{D\mathbf{V}}{Dt} = -2\mathbf{\Omega} \times \mathbf{V} - \mathbf{\Omega} \times (\mathbf{\Omega} \times \mathbf{r}) - \nabla \phi_a - \alpha \nabla p - \alpha \nabla \cdot \mathbf{F}. \tag{27}$$

Here $\mathbf{r}$ is a position vector extending from the center of the Earth to a particle of air whose position is generally changing with time. The gravitational potential is ϕ_a. The pressure-gradient term is $-\alpha\nabla p$, where $\alpha \equiv 1/\rho$ is the specific volume. The quantity $\mathbf{F}$ is the stress tensor (see appendix A) associated with molecular viscosity. The dimensions of $\mathbf{F}$ are density times velocity squared; for example, the units could be $(\text{kg m}^{-3})\,(\text{m s}^{-1})^2 = \text{kg m}^{-1}\,\text{s}^{-2}$. Note that $\nabla \cdot \mathbf{F}$ is a vector.

The term $-2\mathbf{\Omega} \times \mathbf{V}$ represents the Coriolis acceleration, whose direction is perpendicular to $\mathbf{V}$, and the term $-\mathbf{\Omega} \times (\mathbf{\Omega} \times \mathbf{r})$ represents the centrifugal acceleration (see fig. 4.3). You should be able to show that

$$\mathbf{V}_e = \mathbf{\Omega} \times \mathbf{r} = (\Omega r \cos\varphi)\,\mathbf{e}_\lambda, \tag{28}$$

where $\mathbf{e}_\lambda$ is a unit vector pointing east, φ is latitude, and $\mathbf{V}_e$ is the velocity (as seen in the inertial frame) that a particle at radius r and latitude φ experiences owing to the

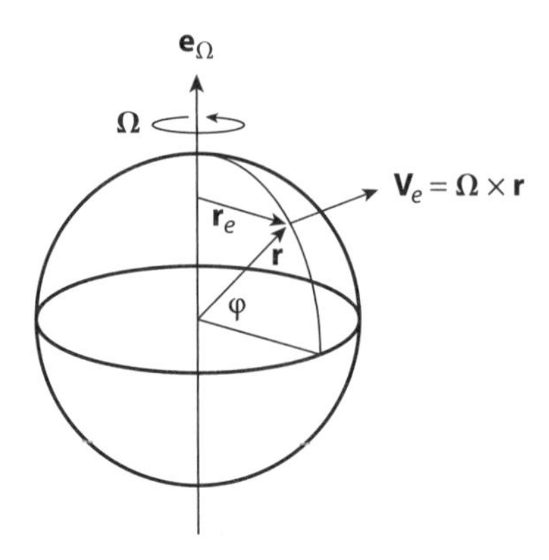

Figure 4.2. The sketch defines notation used in the text.

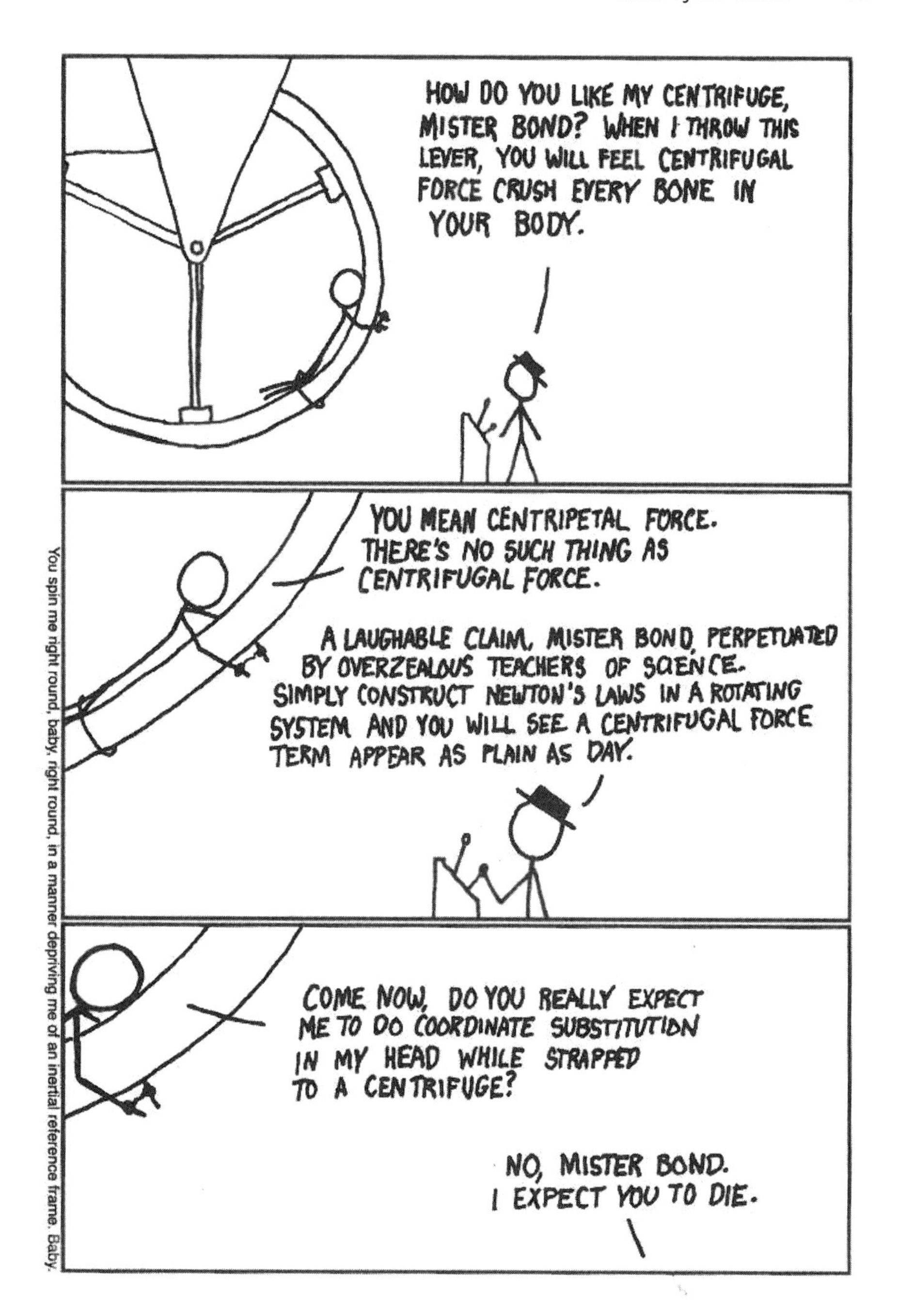

Figure 4.3. A cartoon from the strip *xkcd* by Randall Munroe.

Earth's rotation (see fig. 4.2). The total velocity of the particle is $\mathbf{V} + \mathbf{V}_e$. With this notation, we find that

$$\begin{aligned} -\boldsymbol{\Omega} \times (\boldsymbol{\Omega} \times \mathbf{r}) &= (\Omega^2 r \cos\varphi)\, \mathbf{e}_\lambda \times \mathbf{e}_\Omega \\ &= \Omega^2 \mathbf{r}_e, \end{aligned} \tag{29}$$

where $\mathbf{r}_e$ is the vector shown in figure 4.2, and $\mathbf{e}_\Omega$ is a unit vector pointing toward the celestial north pole. The centrifugal acceleration points outward, in the direction of $\mathbf{r}_e$, which is perpendicular to the axis of the Earth's rotation. It can be shown that

$$\Omega^2 \mathbf{r}_e = \nabla\left[\frac{1}{2}|\boldsymbol{\Omega}\times\mathbf{r}|^2\right]. \tag{30}$$

According to (30), the centrifugal acceleration can be regarded as the gradient of a potential, called the *centrifugal potential.* The "apparent" gravity, $\mathbf{g}$, due to the combined effects of true gravity and the centrifugal acceleration, can be defined as

$$\mathbf{g} = \mathbf{g}_a + |\boldsymbol{\Omega}|^2 \mathbf{r}_e, \tag{31}$$

where $\mathbf{g}_a \equiv -\nabla\phi_a$, and using (30) we see that the potential of $\mathbf{g}$ is

$$\begin{aligned} \phi &= \phi_a - \frac{1}{2}|\boldsymbol{\Omega}\times\mathbf{r}|^2 \\ &= \phi_a - \frac{1}{2}(\Omega r\cos\varphi)^2, \end{aligned} \tag{32}$$

so that $\mathbf{g} \equiv -\nabla\phi$. We refer to ϕ as the *geopotential.* The horizontal pressure-gradient force vanishes when the pressure is uniform along isosurfaces of the geopotential. Because of the centrifugal acceleration (and also because of spatial inhomogeneities of true gravity due to inhomogeneities of the Earth's mass), these isosurfaces are not quite spherical. The centrifugal acceleration causes them to bulge outward in low latitudes and pull inward near the poles; the resulting shape of a geopotential isosurface is an *oblate spheroid.* For most purposes, however,

$$\mathbf{g} \cong \mathbf{g}_a = -g\mathbf{e}_r, \tag{33}$$

because the centrifugal acceleration is small compared with $\mathbf{g}_a$. In (33), $\mathbf{k}$ is a unit vector pointing upward, away from the center of the Earth. When we use (33) with a spatially constant value of g, as is conventional, we must also approximate the shape of the Earth as a sphere, with topographic bumps.

We can now write the three-dimensional equation of motion as

$$\boxed{\frac{D\mathbf{V}}{Dt} = -2\boldsymbol{\Omega}\times\mathbf{V} - \nabla\phi - \alpha\nabla p - \alpha\nabla\cdot\mathbf{F}}. \tag{34}$$

Another useful form of this equation is

$$\boxed{\frac{\partial\mathbf{V}}{\partial t} + (\nabla\times\mathbf{V} + 2\boldsymbol{\Omega})\times\mathbf{V} + \nabla\left(\frac{1}{2}\mathbf{V}\cdot\mathbf{V} + \phi\right) = -\alpha\nabla p - \alpha\nabla\cdot\mathbf{F}}. \tag{35}$$

Here $\nabla\times\mathbf{V}$ is the three-dimensional vorticity vector, and $(1/2)\mathbf{V}\cdot\mathbf{V}$ is the kinetic energy per unit mass. To obtain (35) from (34), we used the vector identity

$$(\mathbf{V}\cdot\nabla)\mathbf{V} = (\nabla\times\mathbf{V})\times\mathbf{V} + \nabla\left(\frac{1}{2}\mathbf{V}\cdot\mathbf{V}\right). \tag{36}$$

We will have occasion to use both (34) and (35).

We have already used spherical coordinates (λ, φ, r). The unit vectors in spherical coordinates are $\mathbf{e}_\lambda$, $\mathbf{e}_\varphi$, and $\mathbf{e}_r$, respectively. For future reference, the gradient, divergence, curl, and Laplacian can be expressed in spherical coordinates as follows:

$$\nabla A = \left(\frac{1}{r\cos\varphi}\frac{\partial A}{\partial\lambda}, \frac{1}{r}\frac{\partial A}{\partial\varphi}, \frac{\partial A}{\partial r}\right), \tag{37}$$

$$\nabla \cdot \mathbf{H} = \frac{1}{r\cos\varphi}\frac{\partial H_\lambda}{\partial \lambda} + \frac{1}{r\cos\varphi}\frac{\partial}{\partial \varphi}(H_\varphi \cos\varphi) + \frac{1}{r^2}\frac{\partial}{\partial r}(H_r r^2), \tag{38}$$

$$\nabla \times \mathbf{H} = \left\{ \frac{1}{r}\left[\frac{\partial H_r}{\partial \varphi} - \frac{\partial}{\partial r}(rH_\varphi)\right], \frac{1}{r}\frac{\partial}{\partial r}(rH_\lambda) - \frac{1}{r\cos\varphi}\frac{\partial H_r}{\partial \lambda}, \frac{1}{r\cos\varphi}\left[\frac{\partial H_\varphi}{\partial \lambda} - \frac{\partial}{\partial \varphi}(H_\lambda \cos\phi)\right]\right\}, \tag{39}$$

$$\nabla^2 A = \frac{1}{r^2\cos\varphi}\left[\frac{\partial}{\partial \lambda}\left(\frac{1}{\cos\varphi}\frac{\partial A}{\partial \lambda}\right) + \frac{\partial}{\partial \varphi}\left(\cos\varphi\frac{\partial A}{\partial \varphi}\right) + \frac{\partial}{\partial r}\left(r^2\cos\varphi\frac{\partial A}{\partial r}\right)\right]. \tag{40}$$

Here A is an arbitrary scalar, and $\mathbf{H} = (H_\lambda, H_\varphi, H_r)$ is an arbitrary vector. For further discussion, see appendix A.

The divergence operator can be expanded as

$$\nabla \cdot \mathbf{H} = \frac{1}{r\cos\varphi}\frac{\partial H_\lambda}{\partial \lambda} + \frac{1}{r\cos\varphi}\frac{\partial}{\partial \varphi}(H_\varphi \cos\varphi) + \frac{\partial H_r}{\partial r} + \frac{2H_r}{r}. \tag{41}$$

Because the Earth's atmosphere is very thin compared with the radius of the Earth, the last term of (41) is negligible, and the divergence operator can be approximated by

$$\nabla \cdot \mathbf{H} \cong \frac{1}{a\cos\varphi}\frac{\partial H_\lambda}{\partial \lambda} + \frac{1}{a\cos\varphi}\frac{\partial}{\partial \varphi}(H_\varphi \cos\varphi) + \frac{\partial H_r}{\partial r}. \tag{42}$$

Note that r has been replaced by a in the first two terms. In this book we normally use (42) rather than (41), largely because it is traditional to do so; it is not clear that the approximation (42) actually makes our work simpler. The approximation would not be applicable to a deep atmosphere, such as that of a star, or Jupiter.

The velocity vector can be expressed in terms of zonal, meridional, and radial components; that is,

$$\mathbf{V} \equiv u\mathbf{e}_\lambda + v\mathbf{e}_\varphi + w\mathbf{e}_r, \tag{43}$$

where the velocity components are defined by

$$u \equiv r\cos\varphi\frac{D\lambda}{Dt}, v \equiv r\frac{D\varphi}{Dt}, w \equiv \frac{Dr}{Dt}, \tag{44}$$

and

$$\begin{aligned} \frac{D}{Dt} &\equiv \frac{\partial}{\partial t} + \frac{D\lambda}{Dt}\frac{\partial}{\partial \lambda} + \frac{D\varphi}{Dt}\frac{\partial}{\partial \varphi} + \frac{Dr}{Dt}\frac{\partial}{\partial r} \\ &= \frac{\partial}{\partial t} + \frac{u}{r\cos\varphi}\frac{\partial}{\partial \lambda} + \frac{v}{r}\frac{\partial}{\partial \varphi} + w\frac{\partial}{\partial r}. \end{aligned} \tag{45}$$

The directions in which the unit vectors $\mathbf{e}_\lambda$, $\mathbf{e}_\varphi$, and $\mathbf{e}_r$ actually point depend on longitude and latitude. Therefore, as an air particle moves from place to place, the directions of the unit vectors change. Consequently, an expanded version of the acceleration term of (34) contains a total of six terms:

$$\frac{D\mathbf{V}}{Dt} = \left(\frac{Du}{Dt}\mathbf{e}_\lambda + u\frac{D\mathbf{e}_\lambda}{Dt}\right) + \left(\frac{Dv}{Dt}\mathbf{e}_\varphi + v\frac{D\mathbf{e}_\varphi}{Dt}\right) + \left(\frac{Dw}{Dt}\mathbf{e}_r + w\frac{D\mathbf{e}_r}{Dt}\right). \tag{46}$$

Geometric reasoning leads to the following formulas:

$$\begin{aligned} \frac{D\mathbf{e}_\lambda}{Dt} &= \frac{D\lambda}{Dt}\sin\varphi\mathbf{e}_\varphi - \cos\varphi\frac{D\lambda}{Dt}\mathbf{e}_r \\ &= \left(\frac{u\tan\varphi}{r}\right)\mathbf{e}_\varphi - \frac{u}{r}\mathbf{e}_r, \end{aligned} \tag{47}$$

$$\begin{aligned}\frac{D\mathbf{e}_\varphi}{Dt} &= \frac{D\lambda}{Dt}\sin\varphi\mathbf{e}_\lambda - \frac{D\varphi}{Dt}\mathbf{e}_r \\ &= \left(\frac{u\tan\varphi}{r}\right)\mathbf{e}_\lambda - \frac{v}{r}\mathbf{e}_r,\end{aligned} \tag{48}$$

$$\begin{aligned}\frac{D\mathbf{e}_r}{Dt} &= \cos\varphi\frac{D\lambda}{Dt}\mathbf{e}_\lambda + \frac{D\varphi}{Dt}\mathbf{e}_\varphi \\ &= \frac{u}{r}\mathbf{e}_\lambda - \frac{v}{r}\mathbf{e}_\varphi.\end{aligned} \tag{49}$$

You should try to understand where (47)–(49) come from; one approach is to visualize the movement and rotation of the unit vectors using a pencil and a globe. When (47)–(49) are taken into account, (34) can be expanded into component form as

$$\begin{aligned}\frac{Du}{Dt} + \frac{uw}{r} - \frac{uv\tan\varphi}{r} &= fv - \bar{f}w - \frac{\alpha}{r\cos\varphi}\frac{\partial p}{\partial\lambda} - \alpha(\nabla\cdot\mathbf{F})_\lambda, \\ \frac{Dv}{Dt} + \frac{vw}{r} - \frac{u^2\tan\varphi}{r} &= fu - \frac{\alpha}{r}\frac{\partial p}{\partial\varphi} - \alpha(\nabla\cdot\mathbf{F})_\varphi, \\ \frac{Dw}{Dt} + \left(\frac{u^2+v^2}{r}\right) &= \bar{f}u - \alpha\frac{\partial p}{\partial r} - \alpha(\nabla\cdot\mathbf{F})_r - g.\end{aligned} \tag{50}$$

Here we define

$$f \equiv 2\Omega\sin\varphi \quad \text{and} \quad \bar{f} \equiv 2\Omega\cos\varphi. \tag{51}$$

The set of equations in (50) are the components of the equation of motion in spherical coordinates. On the left-hand side of (50), the terms $uw/r, uv\tan\varphi/r, vw/r, u^2\tan\varphi/r$, and $(u^2+v^2)/r$ arise from the spatial variations of the unit vectors in spherical coordinates. They are called *metric terms*. Finally, by using the continuity equation, we can rewrite (50) in the flux form:

$$\begin{aligned}\frac{\partial}{\partial t}(\rho u) + \nabla\cdot(\rho\mathbf{V}u) + \rho\frac{uw}{r} - \rho uv\frac{\tan\varphi}{r} &= \rho fv - \rho\bar{f}w - \frac{1}{r\cos\varphi}\frac{\partial p}{\partial\lambda} - (\nabla\cdot\mathbf{F})_\lambda, \\ \frac{\partial}{\partial t}(\rho v) + \nabla\cdot(\rho\mathbf{V}v) + \rho\frac{vw}{r} + \rho u^2\frac{\tan\varphi}{r} &= -\rho fu - \frac{1}{r}\frac{\partial p}{\partial\varphi} - (\nabla\cdot\mathbf{F})_\varphi, \\ \frac{\partial}{\partial t}(\rho w) + \nabla\cdot(\rho\mathbf{V}w) - \rho\left(\frac{u^2+v^2}{r}\right) &= \rho\bar{f}u - \frac{\partial p}{\partial r} - \rho g - (\nabla\cdot\mathbf{F})_r.\end{aligned} \tag{52}$$

Conservation of Angular Momentum

As discussed in introductory physics textbooks (e.g., Feynman et al., 1963), the angular momentum per unit volume of a particle, **L**, with respect to some origin, is a vector, given by the cross product of the particle's three-dimensional linear momentum vector (per unit volume), $\rho\mathbf{V}$, and the displacement vector separating the particle from the origin, **r**:

$$\mathbf{L} = \mathbf{r}\times\rho\mathbf{V}. \tag{53}$$

The angular momentum vector of the Earth's atmosphere, with respect to an origin at the center of the Earth, could be computed by applying (53) to each air parcel and integrating over the entire atmosphere. The result would be a vector.

Such a computation would reveal that the angular momentum of the air is due mostly to the rotation of the Earth rather than to the motion of the air relative to the Earth. In addition, as we have already discussed, the motion of the air relative to the Earth includes strong, mostly zonal jets. For these reasons, the angular momentum vector for the atmosphere as a whole points roughly at the "North Star," that is, along the Earth's axis of rotation. In practice, when we discuss the angular momentum of the atmosphere we are almost always concerned with the *component* of the angular momentum vector parallel to the axis of the Earth's rotation. This component (per unit mass) is given by

$$\boxed{M \equiv (\Omega r\cos\varphi + u)\,r\cos\varphi}. \tag{54}$$

Taking the Lagrangian time derivative of (54), and using (44), we find that

$$\frac{DM}{Dt} = r\cos\varphi\frac{Du}{Dt} - (2\Omega r\cos\varphi + u)(v\sin\varphi + w\cos\varphi). \tag{55}$$

Both the meridional and vertical velocity appear on the right-hand side of (55). Eliminating Du/Dt using (50), we find that

$$\begin{aligned}\frac{DM}{Dt} = r\cos\varphi\left[\frac{uv\tan\varphi}{r} - \frac{uw}{r} + fv - \tilde{f}w - \frac{\alpha}{r\cos\varphi}\frac{\partial p}{\partial\lambda} - \alpha(\nabla\cdot\mathbf{F})_\lambda\right] \\ -(2\Omega r\cos\varphi + u)(v\sin\varphi + w\cos\varphi).\end{aligned} \tag{56}$$

All the Coriolis and metric terms cancel, so that (56) simplifies drastically to

$$\frac{DM}{Dt} = -\alpha\frac{\partial p}{\partial\lambda} - \alpha r\cos\varphi(\nabla\cdot\mathbf{F})_\lambda. \tag{57}$$

The flux form is

$$\boxed{\frac{\partial}{\partial t}(\rho M) + \nabla\cdot(\rho\mathbf{V}M) = -\frac{\partial p}{\partial\lambda} - r\cos\varphi(\nabla\cdot\mathbf{F})_\lambda}. \tag{58}$$

Conservation of Kinetic Energy and Potential Energy

The kinetic energy equation can be derived from the equation of motion. We begin by forming the dot product of the momentum equation in the form of (35) with **V**, which gives

$$\frac{\partial K}{\partial t} + \mathbf{V}\cdot\nabla(K+\phi) = -\alpha\mathbf{V}\cdot\nabla p - \alpha\mathbf{V}\cdot\nabla\cdot\mathbf{F}, \tag{59}$$

where

$$K \equiv \frac{1}{2}\mathbf{V}\cdot\mathbf{V} \tag{60}$$

is the kinetic energy per unit mass. Note that K, and therefore the total energy, depend on the choice of coordinate system. For example, the value of the kinetic energy at a given place in the atmosphere will differ greatly between a coordinate system that is rotating with the Earth and an inertial coordinate system. This is not a problem for (59) because it involves $\partial K/\partial t$ and ∇K, but not K itself. Since ϕ is independent of time in height coordinates, we can modify (59) to

$$\frac{\partial}{\partial t}(K+\phi)+\mathbf{V}\cdot\nabla(K+\phi)=-\alpha\mathbf{V}\cdot\nabla p-\alpha\mathbf{V}\cdot\nabla\cdot\mathbf{F}, \tag{61}$$

or

$$\frac{D(K+\phi)}{Dt}=-\alpha\mathbf{V}\cdot\nabla p-\alpha\mathbf{V}\cdot(\nabla\cdot\mathbf{F}). \tag{62}$$

The sum $K+\phi$ is called the *mechanical energy per unit mass*, and (62) is called the *mechanical energy equation*.

On the right-hand side of (62), the rate at which work is done by the pressure force, per unit mass, is represented by $-\alpha\mathbf{V}\cdot\nabla p$. This expression can be manipulated as follows:

$$\begin{aligned}-\alpha\mathbf{V}\cdot\nabla p&=-\alpha\nabla\cdot(p\mathbf{V})+\alpha(p\nabla\cdot\mathbf{V})\\&=-\alpha\nabla\cdot(p\mathbf{V})+p\frac{D\alpha}{Dt}.\end{aligned} \tag{63}$$

On the second line of (63), we used the continuity equation to eliminate $\alpha\nabla\cdot\mathbf{V}$. The $\nabla\cdot(p\mathbf{V})$ term on the second line of (63) has the form of a flux divergence and so represents a spatial redistribution of energy by the pressure force. The $p(D\alpha/Dt)$ term represents the work done by volume expansion (analogous to the work done in inflating a balloon). We refer to $p(D\alpha/Dt)$ as the *expansion-work* term.

Because α is a constant for a fluid of constant density (e.g., "shallow water," which is discussed near the end of this chapter), the internal energy of a constant-density fluid is nonconvertible, although it is not zero. Because it is nonconvertible, the internal energy of a constant-density fluid plays no role in the energy cycle; it can be ignored.

The friction term of (62) can also be expanded to reveal two physically distinct parts, as follows:

$$-\alpha\mathbf{V}\cdot(\nabla\cdot\mathbf{F})=-\alpha\nabla\cdot(\mathbf{F}\cdot\mathbf{V})-\delta, \tag{64}$$

where

$$\delta\equiv-\alpha(\mathbf{F}\cdot\nabla)\cdot\mathbf{V} \tag{65}$$

is the rate of kinetic energy dissipation per unit mass. The quantity $\nabla\cdot(\mathbf{F}\cdot\mathbf{V})$ in (64) is a divergence, so it represents a spatial redistribution of kinetic energy (and momentum) as the frictional stress (represented by $\mathbf{F}$) causes neighboring air parcels to do work on each other. Because this is merely a spatial redistribution of energy, it does not change the total amount of kinetic energy in the atmosphere, except where friction does work on the lower boundary. In contrast, *the dissipation rate, δ, is a true sink of kinetic energy*. It is shown in appendix C that

$$\boxed{\delta\geq 0}. \tag{66}$$

Kinetic energy dissipation converts macroscopic kinetic energy into microscopic kinetic energy. This is why, as discussed later in this chapter, δ appears as a source of thermodynamic energy, that is, as "frictional heating." It is a weak but persistent source of internal energy for the atmosphere. The vertical integral of $\rho\delta$ is believed to be about 5 W m^{-2}, averaged over the globe. Kinetic energy dissipation is ultimately the result of molecular viscosity. Similarly, molecular thermal conductivity is ultimately responsible for the dissipation of thermal fluctuations. It is an amazing fact that even though the molecular processes act on scales of a few millimeters, they have profound effects on the global-scale circulation of the atmosphere!

Substitution of (63) and (64) into (62) gives the mechanical energy equation in the form

$$\boxed{\frac{D(K+\phi)}{Dt} = -\alpha\nabla\cdot(p\mathbf{V} + \mathbf{F}\cdot\mathbf{V}) + p\frac{D\alpha}{Dt} - \delta}. \tag{67}$$

Energy Transports and Dissipation Due to Small-Scale Motions

When **F** is dominated by small-scale turbulent momentum flux, we can use the approximation

$$\nabla\cdot(\overline{\mathbf{F}\cdot\mathbf{V}}) \cong \frac{\partial}{\partial z}(\overline{\mathbf{V}}_h\cdot\mathbf{F}_V). \tag{68}$$

Here $\mathbf{F}_V$ is the upward flux of horizontal momentum. We introduced an approximation similar to (68) in our discussion of atmospheric moisture. The frictional work term shown in (68) is small throughout most of the atmosphere; it matters most of all in the turbulent planetary boundary layer (PBL) near the surface. In particular, energy exchange occurs when the surface wind stress pushes on the ocean, giving rise to the *wind-driven* ocean circulation. The rate at which the atmosphere does work on the oceans can be roughly estimated as follows. As discussed in chapter 5, the surface frictional stress is typically less than or on the order of 0.1 Pa. With a few exceptions, the ocean currents have speeds on the order of 0.1 m s^{-1} or slower. The rate at which the ocean gains energy owing to the stress applied by the atmosphere is given by the product of the stress with the speed of the current, which, using the values just mentioned, is on the order of 10^{-2} W m^{-2}—which is tiny compared with other energy fluxes, such as the net surface radiation. The work that the surface wind stress does on the ocean is obviously quite important for the ocean and for the climate system as a whole. Nevertheless, from the point of view of the atmospheric energy budget, the rate at which energy is lost through work done on the ocean is utterly negligible.

In a similar way, we can approximate the dissipation rate by

$$\delta \cong -\frac{\mathbf{F}_V}{\rho}\cdot\frac{\partial\overline{\mathbf{V}}_H}{\partial z}. \tag{69}$$

As discussed in chapter 7, this is an example of what is called a *gradient-production term*. Specifically, it represents the rate of production of *turbulence kinetic energy* (TKE) by conversion from the kinetic energy of the mean flow. The physical picture is that the kinetic energy of the mean flow is converted to TKE; that is, it shifts to smaller scales. Then, the TKE is dissipated in the true molecular sense. We can apply (69) to estimate the rate of dissipation in the PBL, as follows: Most of the vertical shear of the horizontal wind typically occurs in the lower part of the PBL, where the momentum flux is fairly close to its surface value. We can therefore approximate the integral of (69) through the depth of the PBL by

$$\int_{PBL} \rho\delta dz \cong -(\mathbf{F}_V)_S\cdot\mathbf{V}_M, \tag{70}$$

where $\mathbf{V}_M$ is the mean horizontal wind in the middle of the PBL, near the top of the shear layer. The bulk aerodynamic formula tells us that

$$(\mathbf{F}_V)_S = -\rho_S C_D|\mathbf{V}_M|\mathbf{V}_M. \tag{71}$$

Substitution of (71) into (70) gives

$$\int_{PBL} \rho\delta dz \cong \rho_S C_D |\mathbf{V}_M|^3, \tag{72}$$

which shows that the rate of dissipation in the PBL increases very strongly as the wind speed increases. For $|\mathbf{V}_M| = 10$ m s^{-1}, we find that $\int_{PBL} \rho\delta dz \cong 1$ W m^{-2}. The rate at which kinetic energy is dissipated in the PBL is thus considerably larger than the rate at which the atmosphere does frictional work on the ocean.

Bister and Emanuel (1998) point out that very large dissipation rates must occur in hurricanes, because of the large wind speeds in such storms and the cubic dependence of the dissipation rate on wind speed, as shown in (72). They show that dissipative heating acts to intensify the storm, leading to an increase in the maximum wind speed by as much as 25%. Businger and Businger (2001) elaborate on the importance of the dissipation in regions of strong winds, showing that the vertically integrated dissipation rate in a moderately strong storm can be in excess of 1000 W m^{-2}.

Where Does Mechanical Energy Come From?

Using continuity, we can convert the mechanical energy equation (67) to flux form:

$$\frac{\partial}{\partial t}[\rho(K+\phi)] + \nabla\cdot[\rho\mathbf{V}(K+\phi) + p\mathbf{V} + \mathbf{F}\cdot\mathbf{V}] = \rho p\frac{D\alpha}{Dt} - \rho\delta. \tag{73}$$

All contributions to $(\partial/\partial t)\,[\rho(K+\phi)]$, on the left-hand side of (73), represent transport processes, which merely redistribute energy in space. In contrast, the expansion-work term, $p(D\alpha/Dt)$, does not have to integrate to zero, although it can be either positive or negative at a given place and time. Recall, however, that the dissipation term is always a sink. It follows that, in an average over the whole atmosphere and over time, *the $p(D\alpha/Dt)$ term must be positive*; that is, it must act as a source of mechanical energy:

$$\boxed{\int_V \overline{\rho p\frac{D\alpha}{Dt}}^t dV = \int_V \overline{\delta\rho}^t dV \geq 0}. \tag{74}$$

Here we integrated over the entire mass of the atmosphere. Equation (74) means that, on the average, the pressure force must do positive expansion work to compensate for the dissipation of kinetic energy. To make $\int_V \rho p(D\alpha/Dt)dV$ positive, expansion must take place, on the average, at a higher pressure than compression. For example, there can be expansion in the lower troposphere and compression in the upper troposphere.

Given that in an average sense the expansion work term of (73) must act as a source of mechanical energy, we should ask where this energy comes from. We show below that it comes from the thermodynamic energy of the atmosphere. Expansion work represents an energy conversion process, which can have either sign locally but is positive when averaged over the whole atmosphere and over time.

Similarly, given that the dissipation term of (73) represents a sink of mechanical energy, we should ask where the energy goes. The answer is that it appears as a source of thermodynamic energy. Dissipation is therefore another energy conversion process—a conversion that runs in only one direction.

Equation (74) simply means that the rate of kinetic energy dissipation must be equal, on the average, to the rate of mechanical energy generation. As mentioned earlier, this rate has been estimated to be about 5 W m^{-2}. For comparison, recall that the solar radiation absorbed by the Earth-atmosphere system is about 240 W m^{-2}.

Evidently the climate system is not very efficient at converting the absorbed solar energy into atmospheric mechanical energy.

Mechanical energy generation can also be expressed in another way. To see this, note that

$$\begin{aligned} p\frac{D\alpha}{Dt} &= \frac{D}{Dt}(p\alpha) - \alpha\frac{Dp}{Dt} \\ &= \frac{D}{Dt}(p\alpha) - \omega\alpha. \end{aligned} \tag{75}$$

Equation (75) shows that the expansion-work term is closely related to the product $\omega\alpha$. Substituting (75) into the mechanical energy equation (73), we find that

$$\frac{\partial}{\partial t}[\rho(K+\phi)] + \nabla\cdot[\rho\mathbf{V}(K+\phi) + \mathbf{F}\cdot\mathbf{V}] = -\rho(\omega\alpha) - \rho\delta + \frac{\partial p}{\partial t}. \tag{76}$$

In comparing (76) with (73) we find that the pressure-work term of (73), involving $\nabla\cdot(p\mathbf{V})$, has disappeared via a cancellation, but "in its place" we have picked up a new term involving the local time rate of change of the pressure, $\partial p/\partial t$, on the right-hand side of (76).

In an average over the whole atmosphere, and over time, the $-\omega\alpha$ term of (76) must be positive; that is, it must act as a source of mechanical energy:

$$\boxed{-\int_V \overline{\omega\alpha\rho}^{\,t}\, dV = \int_V \overline{\delta\rho}^{\,t}\, dV \geq 0}. \tag{77}$$

Comparison of (77) and (74) shows that

$$-\int_V \overline{\omega\alpha\rho}^{\,t}\, dV = \int_V \overline{\rho p\frac{D\alpha}{Dt}}^{\,t}\, dV. \tag{78}$$

We can therefore use $\omega\alpha$ and $p(D\alpha/Dt)$ as alternative measures of the rate of conversion between mechanical and thermodynamic energy. Most of the time we will use $\omega\alpha$.

Conservation of Thermodynamic Energy

The internal energy of a perfect gas is given by

$$e = c_v T, \tag{79}$$

where c_v, the specific heat of air at constant volume, is a constant. For dry air, $c_v = 5R/2 \cong 713\ \mathrm{J\,kg^{-1}\,K^{-1}}$. More generally, the internal energy also includes the latent heat associated with the potential condensation of water vapor, and we find that for moist air

$$e \cong c_v T + Lq_v. \tag{80}$$

We could also add the latent heats of other atmospheric constituents, such as nitrogen, oxygen, and carbon dioxide, to represent the effects of their potential condensation. We do not bother to do so because those constituents do not condense under conditions encountered in the Earth's atmosphere. Equation (80) is approximate, because we have neglected the specific heat of the water vapor, as well as the specific heat of any liquid (or ice) that might be present. In atmospheric science we frequently define the internal energy as the internal energy of dry air and treat the latent heat as an "external" source or sink of internal energy.

When thermodynamic energy is added to a system, the energy input equals the sum of the work done and the change in the internal energy:

$$\frac{D}{Dt}(c_v T) + p\frac{D\alpha}{Dt} = -\alpha\nabla\cdot(\mathbf{R} + \mathbf{F}_S) + LC + \delta. \tag{81}$$

Here $\mathbf{F}_S$ is the vector flux of internal energy due to molecular diffusion; $\mathbf{R}$ is the vector flux of energy due to radiation (note the notational conflict with the gas constant, which should not be confusing), L is the latent heat of water vapor, and C is the rate of condensation per unit mass. The dissipation rate appears in (81) as a source of internal energy, that is, "frictional heating." Equation (81) is a statement of the conservation of thermodynamic energy, applied to a moving particle of air. It is sometimes called the *First Law of Thermodynamics*, although that terminology seems rather medieval.

An alternative statement of the conservation of thermodynamic energy, obtained using (75) in (81), is

$$\boxed{\frac{D}{Dt}(c_p T) = \omega\alpha - \alpha\nabla\cdot(\mathbf{R} + \mathbf{F}_S) + LC + \delta}, \tag{82}$$

where

$$c_p = R + c_v \cong 1004\ \mathrm{J\ kg^{-1}\ K^{-1}}, \tag{83}$$

a nice round, easily remembered number. Equation (82) can be called the "enthalpy form" of the thermodynamic energy equation, where the enthalpy is defined by

$$\eta \equiv c_p T. \tag{84}$$

More generally, the enthalpy can be written as

$$\eta = e + p\alpha. \tag{85}$$

For saturated air containing liquid, it turns out that (84) must be replaced by

$$\eta \cong c_p T - Ll, \tag{86}$$

where L is the latent heat of condensation (e.g., Lorenz 1979; Emanuel, 1994), and l is the liquid water mixing ratio.

A third form of the thermodynamic equation is

$$\frac{Dp}{Dt} - \left(\frac{c_p}{c_v}RT\right)\frac{D\rho}{Dt} = -\alpha\nabla\cdot(\mathbf{R} + \mathbf{F}_S) + LC + \delta, \tag{87}$$

which can be derived from (81) or (82) by using the equation of state and the continuity equation. The quantity $(c_p/c_v)RT$ turns out to be the square of the speed of sound. We will not use (87) in this book, but we include it here for completeness.

Finally, the conservation of thermodynamic energy can also be expressed in terms of the potential temperature, which was defined in chapter 3. We can show that

$$\boxed{c_p\left(\frac{T}{\theta}\right)\frac{D\theta}{Dt} = -\alpha\nabla\cdot(\mathbf{R} + \mathbf{F}_S) + LC + \delta}. \tag{88}$$

In the absence of heating and dissipation, $D\theta/Dt = 0$; that is, θ is conserved following a particle. This is one of the reasons that θ is a particularly useful quantity.

In summary, the thermodynamic energy equation can be expressed in the four equivalent forms (81), (82), (87), and (88).

The Vertically Integrated Enthalpy

Lorenz (1955) pointed out that in a hydrostatically balanced atmosphere, the mass-weighted vertical integral of the enthalpy is equal to the sum of the mass-weighted vertical integrals of the internal and potential energies. To demonstrate this result, we begin from hydrostatics with the equation

$$\frac{\partial p}{\partial z} = -\rho g. \tag{89}$$

The vertically integrated potential energy, P, satisfies

$$\begin{aligned} P &\equiv \int_0^\infty gz\rho dz \\ &= -\int_0^\infty \left(\frac{\partial p}{\partial z} z\right) dz \\ &= -\int_0^\infty \left[\frac{\partial}{\partial z}(pz) - p\right] dz \\ &= -\left[(pz)\Big|_{z=0}^{z=\infty} - \int_0^\infty \rho RT\, dz\right] \\ &= \int_0^\infty \rho RT\, dz. \end{aligned} \tag{90}$$

Equation (90) says that the total potential energy of the column is proportional to the mass-weighted average temperature of the column. The explanation is that warmer air occupies a larger volume, so that a warmer column is "taller." It follows from (90) that

$$\boxed{\begin{aligned} P + I &= \int_0^\infty \rho RT\, dz + \int_0^\infty c_v T \rho dz \\ &= \int_0^\infty (c_v + R) T \rho dz \\ &= \int_0^\infty c_p T \rho dz. \end{aligned}} \tag{91}$$

Here $I \equiv \int_0^\infty c_v T \rho dz$ is the vertically integrated internal energy. Note that $\phi + c_v T = c_p T$ is *not true*; this would imply that $\phi = RT$, which is obviously nonsense. Equation (91) is used in chapter 7, where we discuss available potential energy.

Conservation of Total Energy

We now use the water vapor conservation equation in the form

$$\frac{D}{Dt}(Lq_v) = -\alpha \nabla \cdot (L\mathbf{F}_{q_v}) - LC, \tag{92}$$

where L is the latent heat of condensation. When we add (81) and (92), the condensation terms cancel, and we get

$$\frac{D}{Dt}(c_v T + Lq_v) = -p\frac{D\alpha}{Dt} - \alpha \nabla \cdot (\mathbf{R} + \mathbf{F}_h) + \delta. \tag{93}$$

Here $\mathbf{F}_h \equiv \mathbf{F}_S + L\mathbf{F}_{q_v}$, is the molecular flux of "moist static energy." The reason for this terminology will be explained later. The left-hand side of (93) is the Lagrangian time rate of change of the total internal energy given by (80).

Adding (67) and (93), we get

$$\boxed{\frac{D}{Dt}(K + \phi + c_v T + Lq_v) = -\alpha\nabla \cdot (p\mathbf{V} + \mathbf{F}\cdot\mathbf{V} + \mathbf{R} + \mathbf{F}_h)}. \tag{94}$$

The $p(D\alpha/Dt)$ ("expansion work") terms of (67) and (93) have canceled, as have the dissipation terms, because these terms represent conversions between thermodynamic and mechanical energy. From (94) we see that *the total energy is given by the sum of the kinetic, potential, internal, and latent energies.*

The continuity equation can be used to convert (94) to flux form:

$$\frac{\partial}{\partial t}[\rho(K+\phi+c_v T+Lq_v)]+\nabla\cdot[\rho\mathbf{V}(K+\phi+c_v T+Lq_v)+p\mathbf{V}+\mathbf{F}\cdot\mathbf{V}+\mathbf{R}+\mathbf{F}_h]=0. \tag{95}$$

For the atmosphere as a whole, energy fluxes across the upper and lower boundaries are very important. For this reason, it is useful to distinguish between horizontal and vertical fluxes of energy, as follows:

$$\begin{aligned}&\frac{\partial}{\partial t}(\rho e_T) + \nabla_h\cdot\left[\rho\mathbf{V}_h e_T + p\mathbf{V}_h + (\mathbf{F}\cdot\mathbf{V})_h + \mathbf{R}_h(\mathbf{F}_h)_h\right]\\ &\qquad + \frac{\partial}{\partial z}\left[\rho w e_T + pw + (\mathbf{F}\cdot\mathbf{V})_z + R_z + (F_h)_z\right] = 0.\end{aligned} \tag{96}$$

Here we use the shorthand notation

$$e_T \equiv K + \phi + c_v T + Lq_v, \tag{97}$$

and the subscripts h and z denote the "horizontal part" of a vector (i.e., a vector in the horizontal plane) and the (positive upward) vertical component of a vector, respectively. Note that the vector flux of moist static energy, $\mathbf{F}_h$, is an exception to this rule. In fact, we are using the symbol $(\mathbf{F}_h)_h$ to denote the horizontal part of the vector flux of moist static energy. We vertically integrate (96) through the entire atmospheric column, using Leibniz's rule to take the integrals inside the derivatives, and write the result as

$$\begin{aligned}&\frac{\partial}{\partial t}\left(\int_{z_S}^{\infty}\rho e_T dz\right) + \nabla_h\cdot\left(\int_{z_S}^{\infty}\rho\mathbf{V}_h e_T dz\right) + (\rho e_T)_S\left(\frac{\partial z_S}{\partial t} + \mathbf{V}_H\cdot\nabla z_S - w_S\right)\\ &\qquad + \nabla_h\cdot\left[\int_{z_S}^{\infty}(p\mathbf{V}_h + (\mathbf{F}\cdot\mathbf{V})_h + \mathbf{R}_h + (\mathbf{F}_h)_h)\,dz\right]\\ &\qquad = -p_S\left[(\mathbf{V}_h)_S\cdot\nabla z_S - w_S\right] - \{[(\mathbf{F}\cdot\mathbf{V})_h]_S\cdot\nabla z_S - [(\mathbf{F}\cdot\mathbf{V})_z]_S\}\\ &\qquad - \{[(\mathbf{F}_h)_h]_S\cdot\nabla z_S - [(F_h)_z]_S\} - (\mathbf{R}_z)_\infty - \{[\mathbf{R}_h]_S\cdot\nabla z_S - (\mathbf{R}_z)_S\}.\end{aligned} \tag{98}$$

Using the boundary condition that no mass crosses the Earth's surface, we can simplify (99) to

$$\begin{aligned}&\frac{\partial}{\partial t}\left(\int_{z_S}^{\infty}\rho e_T dz\right) + \nabla_H\cdot\left[\int_{z_S}^{\infty}(\rho\mathbf{V}_h e_T + p\mathbf{V}_h + (\mathbf{F}\cdot\mathbf{V})_h + \mathbf{R}_h + (\mathbf{F}_h)_h)\,dz)\right]\\ &\qquad = p_S\frac{\partial z_S}{\partial t} - \{[(\mathbf{F}\cdot\mathbf{V})_h]_S\cdot\nabla z_S - [(\mathbf{F}\cdot\mathbf{V})_z]_S\}\\ &\qquad - \{[(\mathbf{F}_h)_h]_S\cdot\nabla z_S - [(F_h)_z]_S\} - (\mathbf{R}_z)_\infty - \{[\mathbf{R}_h]_S\cdot\nabla z_S - (\mathbf{R}_z)_S\}.\end{aligned} \tag{99}$$

The terms on the right-hand side of (99) represent the effects of fluxes at the upper and lower boundaries. As pointed out earlier, the only flux of energy at the upper boundary is that due to radiation, denoted by $-(\mathbf{R}_z)_\infty$. For the lower boundary there are terms representing pressure work, frictional work, the flux of moist static energy, and radiation.

When $\partial z_S/\partial t = 0$, the pressure-work term vanishes. Over the ocean, however, the surface height fluctuates owing to the passage of waves; this enables the atmosphere and ocean to exchange energy via pressure work. Thus, energy is added to the waves that make the energy exchange possible. Even over land, the vegetation moves as the wind blows through it, so the pressure-work term can be nonzero. In addition, an earthquake can impart energy to the atmosphere through the pressure-work term; this is, of course, a minuscule effect in terms of the global circulation. The friction terms on the right-hand side of (99) represent the work done by surface drag. As already examined in chapter 2, the surface flux of moist static energy, $(\mathbf{F}_h)_S$ is a very important energy source for the atmosphere; it is considered further in chapter 5. Finally, again as discussed in chapter 2, the radiative energy flux is a major mode of energy exchange at both the upper and lower boundaries of the atmosphere.

When we integrate (99) horizontally over the entire sphere, the horizontal flux divergence term vanishes, and we get

$$\begin{aligned}\frac{d}{dt}\left[\int_A\left(\int_{z_S}^{\infty}\rho e_T dz\right)dA\right] = & \int_A p_S\frac{\partial z_S}{\partial t}dA - \int_A\{[(\mathbf{F}\cdot\mathbf{V})_h]_S\cdot\nabla z_S - [(\mathbf{F}\cdot\mathbf{V})_z]_S\}dA \\ & - \int_A\{[(\mathbf{F}_h)_h]_S\cdot\nabla z_S - [(F_h)_z]_S\}dA - \int_A(\mathbf{R}_z)_\infty dA \\ & - \int_A\{[\mathbf{R}_h]_S\cdot\nabla z_S - (R_z)_S\}dA.\end{aligned} \tag{100}$$

Here $\int_A (\ )dA$ denotes the integral over the sphere. Equation (100) shows that in the absence of heating and friction and when the height of the Earth's surface is independent of time,

$$\frac{d}{dt}\left[\int_A\left(\int_{z_S}^{\infty}\rho e_T dz\right)dA\right] = 0; \tag{101}$$

that is, the total energy of the atmosphere is invariant.

The pressure-work and frictional-work terms of (100) remove energy from the global atmosphere, which means that in an overall sense the atmosphere does work on the lower boundary, rather than vice versa. It follows that, in a time average, the radiative and molecular flux terms of (100) must act as an energy source. As discussed earlier in this chapter, the rate at which the atmosphere does frictional work on the lower boundary is on the order of tenths of a watt per square meter. The pressure-work term is typically even smaller. As discussed in chapter 2, the remaining individual terms on the right-hand side of (100) are typically larger by several orders of magnitude. It follows that these remaining terms must very nearly cancel in a time average; that is,

$$0 \cong -\int_A\left\{\overline{[(F_h)_h]_S\cdot\nabla z_S - [(F_h)_z]_S}^t\right\}dA - \int_A\overline{(R_z)_\infty}^t dA - \int_A\left\{\overline{[R_h]_S\cdot\nabla z_S - (R_z)_S}^t\right\}dA. \tag{102}$$

This means that the total "heating" of the atmosphere, given by the right-hand side of (102), is very nearly zero. Equation (102) can be written more simply as

$$\int_V\left[-\nabla\cdot\overline{(R+F_h)}^t\right]dV \cong 0, \tag{103}$$

where $\int_V (\ )\rho dV$ denotes a volume integral over the entire atmosphere.

Static Energies

Another useful equation, closely related to the total energy equation, can be obtained by adding $\partial p/\partial t$ to both sides of (95) and then using the equation of state and (83). The result is

$$\frac{\partial}{\partial t}[\rho(K+\phi+c_pT+Lq_v)]+\nabla\cdot[\rho\mathbf{V}(K+\phi+c_pT+Lq_v)+\mathbf{R}+\mathbf{F}_h+\mathbf{F}\cdot\mathbf{V}]=\frac{\partial p}{\partial t}. \quad (104)$$

Take a moment to compare (95) and (104). The advected quantity in (95) is the total energy $K+\phi+c_vT+Lq_v$, while the advected quantity in (104) is $K+\phi+c_pT+Lq_v$. They are not the same, because $c_p \neq c_v$. The transport term associated with $\mathbf{V}$ in (95) is $\nabla\cdot[\rho\mathbf{V}+(K+\phi+c_vT+Lq_v)+p\mathbf{V}]$, while in (104) it is $\nabla\cdot[\rho\mathbf{V}+(K+\phi+c_pT+Lq_v)]$. In view of the equation of state, *these two transport terms are actually equal to each other.* A nice property of (104) is that the quantity under the time rate of change is the same as the quantity that is advected by the wind. The disadvantage of (104) is the $\partial p/\partial t$ term on the right-hand side. However, $\partial p/\partial t$ becomes negligible in a sufficiently long time average and may be negligible in other contexts, too.

We now make some approximations. An air parcel zipping along at a rather extreme 100 m s^{-1} has a kinetic energy per unit mass of 5×10^3 J kg^{-1}, and of course, the kinetic energy of a parcel moving at a more typical 10 m s^{-1} is 100 times smaller. For comparison, a parcel on the 200 hPa surface has a potential energy per unit mass (relative to sea level) of about 1.2×10^5 J kg^{-1}, and the internal energy per unit mass of a parcel with a temperature of only 200 K is about 1.5×10^5 J kg^{-1}. These examples illustrate that the contribution of the kinetic energy to the total energy is typically negligible. In addition, the friction and pressure-tendency terms of (104) can often be neglected. With these simplifying approximations, (104) reduces to

$$\boxed{\frac{\partial}{\partial t}(\rho h)+\nabla\cdot(\rho\mathbf{V}h+\mathbf{R}+\mathbf{F}_h)\cong 0}, \quad (105)$$

where

$$\boxed{h\equiv c_pT+\phi+Lq_v} \quad (106)$$

is the moist static energy. According to (105), the moist static energy is approximately conserved under both moist adiabatic and dry adiabatic processes. Since precipitation does not affect the water vapor mixing ratio, the moist static energy is approximately conserved even when precipitation is occurring. Our earlier comparison of (104) and (95) shows that although the moist static energy is not equal to the total energy, the *transport* of moist static energy is a good approximation to the *transport* of total energy.

The preceding discussion shows that conservation of moist static energy is an approximation to the total energy equation, rather than the thermodynamic energy equation. That is why there is no dissipation term in (105); such a term would of course appear in any version of the thermodynamic energy equation, although we might justify neglecting it under some conditions.

Because the water vapor mixing ratio, q_v, is conserved under dry adiabatic processes, conservation of moist static energy implies that the dry static energy,

$$\boxed{s\equiv c_pT+\phi}, \quad (107)$$

is approximately conserved under dry adiabatic processes. The dry static energy normally increases with altitude in the atmosphere, because, as you will show when you work problem 15 at the end of this chapter, the rate of change of the dry static energy with height has the same sign as the rate of change of the potential temperature with height.

The moist and dry static energies are used extensively later in this book and are also very widely used in the research literature. The vertical profile of the moist static energy and its geographic variations are discussed in detail in chapters 5 and 6.

Entropy

For any gas or liquid, the entropy per unit mass, ε, satisfies

$$T\frac{D\varepsilon}{Dt} = \frac{De}{Dt} + p\frac{D\alpha}{Dt}. \tag{108}$$

Comparing (108) with the thermodynamic energy equation, (81), we see that

$$T\frac{D\varepsilon}{Dt} = -\alpha\nabla\cdot(\mathbf{R} + \mathbf{F}_s) + LC + \delta. \tag{109}$$

Using the equation of state with (108), we can show that

$$\varepsilon = c_p \ln\left(\frac{T}{T_0}\right) - R\ln\left(\frac{p}{p_0}\right). \tag{110}$$

Here T_0 and p_0 are suitable reference values that arise as "constants of integration." The reference values are needed because the argument of the logarithm must always be nondimensional, as discussed in appendix B. There is a simple relationship between the entropy and the potential temperature, namely,

$$\varepsilon = c_p \ln\left(\frac{\theta}{\theta_0}\right). \tag{111}$$

Here θ_0 is the potential temperature corresponding to T_0 and p_0. For saturated air containing liquid water, (111) must be replaced by

$$\varepsilon \cong c_p \ln\left(\frac{\theta}{\theta_0}\right) - \frac{Ll}{T} \tag{112}$$

(e.g., Lorenz, 1979; Emanuel, 1994).

Equation (109) can be rearranged as

$$\frac{D\varepsilon}{Dt} = \frac{[-\alpha\nabla\cdot(\mathbf{R} + \mathbf{F}_s) + LC + \delta]}{T}. \tag{113}$$

We use continuity to rewrite (113) in flux form, then integrate over the entire mass of the atmosphere to obtain

$$\frac{d}{dt}\int_V \rho\varepsilon dV = \int_V \left[\frac{-\nabla\cdot(\mathbf{R} + \mathbf{F}_s) + \rho LC + \rho\delta}{T}\right] dv. \tag{114}$$

In an average over a sufficiently long time, (114) reduces to

$$\int_V \overline{\left[\frac{-\alpha\nabla\cdot(\mathbf{R} + \mathbf{F}_s) + LC + \delta}{T}\right]\rho}^{\,t} dV = 0. \tag{115}$$

Because $\delta \geq 0$, dissipation never decreases and normally increases the entropy. This allows us to convert (115) into an inequality by dropping the dissipation rate:

$$\boxed{\int_V \overline{\left[\frac{-\alpha\nabla\cdot(\mathbf{R}+\mathbf{F}_s)+LC}{T}\right]}^t \rho\, dV \leq 0}. \tag{116}$$

This important result means that *for the atmosphere as a whole, heating must act to decrease the entropy.*

For (116) to be satisfied, heating must occur, on the average, where the temperature is high, and cooling must occur, on the average, where the temperature is low. This implies that heating and cooling processes must act to increase temperature contrasts with time. One way for this to happen is for heating to occur in the tropics and near the surface, where the air is warm, and cooling near the poles and up high, where the air is cold. To see that this conclusion can be drawn from (116), let

$$Q \equiv -\alpha\nabla\cdot(\mathbf{R}+\mathbf{F}_s)+LC \tag{117}$$

denote the heating rate, and divide Q and T into averages representing the volume integral in (117) (denoted by overbars), and departures from the averages (denoted by primes). Then we can write

$$Q = \overline{Q} + Q', \text{ and } T = \overline{T} + T'. \tag{118}$$

It follows that

$$\begin{aligned}
\overline{\left(\frac{Q}{T}\right)} &= \overline{\left(\frac{\overline{Q}+Q'}{\overline{T}+T'}\right)} \\
&= \overline{\frac{\overline{Q}(1+Q'/\overline{Q})}{\overline{T}(1+T'/\overline{T})}} \\
&\cong \frac{\overline{Q}}{\overline{T}}\overline{(1+Q'/\overline{Q})(1+T'/\overline{T})} \\
&= \frac{\overline{Q}}{\overline{T}}\overline{\left[1+\frac{Q'}{\overline{Q}}-\frac{T'}{\overline{T}}-\frac{Q'T'}{\overline{Q}\,\overline{T}}\right]} \\
&= \frac{\overline{Q}}{\overline{T}}\left(1-\frac{\overline{Q'T'}}{\overline{Q}\,\overline{T}}\right) \\
&= \frac{\overline{Q}}{\overline{T}}-\frac{\overline{Q'T'}}{\overline{T}^2}.
\end{aligned} \tag{119}$$

For $\overline{Q}=0$ [see (103)], we get

$$\overline{\left(\frac{Q}{T}\right)} = -\frac{\overline{Q'T'}}{\overline{T}^2}. \tag{120}$$

If we add energy where the temperature is already warm, and remove energy where the temperature is already cool, then $\overline{Q'T'}>0$, so $\overline{(Q/T)}$ will be negative. A process that tends to "heat where it's hot, and cool where it's cold" acts to increase the temperature variance, that is, the square of the departure of the temperature from its mean value, and it tends to decrease the entropy. We show in chapter 7 that such a process also generates available potential energy.

We have concluded that heating tends to increase the temperature contrasts within the atmosphere. For the system to achieve a steady state, some other process must oppose this tendency. That process is energy transport, which on the average carries energy from warm regions to cool regions, for example, from the tropics toward the polar regions, and from the warm surface towards the cold upper atmosphere. The energy transports by the global circulation tend to cool where the temperature is high (e.g., the tropical lower troposphere) and tend to warm where the temperature is low (e.g., the polar troposphere).

A final important point is that the global entropy budget is fundamentally different from the global energy budget. Energy is conserved, which means that in a time average the fluxes of energy into the system must be balanced by fluxes out. In contrast, *the Earth makes entropy* via a wide variety of dissipative processes. As a result, *there is a net entropy flux out of the Earth system*, via radiation, and in a time average this outward flux of entropy is equal to the Earth's entropy production rate. It is therefore possible to use satellite measurements of the entropy flow across the top of the atmosphere to infer the Earth's entropy production rate. This has been done. For further discussion, see Stephens and O'Brien (1993).

The Primitive Equations

A number of important approximations to the equation of motion are commonly used in the analysis of the large-scale circulation systems:

- Replace r by a everywhere, where a is the radius of the Earth. An approximation of this form can be justified for an atmosphere that is thin compared with the radius of the planet, and so it is called the *thin atmosphere approximation*. It is a good approximation for Earth but would not apply, for example, to Jupiter.
- Drop the terms of (50) that contain $\overline{f}$. This means that the horizontal component of Ω disappears from the equations. There is an ongoing discussion concerning the circumstances under which this approximation breaks down.
- Neglect uw/r and vw/r, the curvature terms involving w, in the equations for u and v, respectively, and neglect $u^2 + v^2/r$ in the equation of vertical motion.
- Replace the equation of vertical motion by the *hydrostatic equation*:

$$\frac{\partial p}{\partial z} = -\rho g. \tag{121}$$

Justifications of these approximations can be found in standard textbooks on atmospheric dynamics (e.g., Holton, 2004). The resulting system is often called the *primitive equations*.

The hydrostatic equation, (121), deserves some further discussion. With an appropriate boundary condition, (121) allows us to compute $p(z)$ from $\rho(z)$. Even when the air is moving, (121) gives a good approximation to $p(z)$, simply because Dw/Dt and the vertical component of the friction force are almost always small compared with g. Equation (121) as applied to moving air is called the *hydrostatic approximation*, and it is applicable to virtually all meteorological phenomena, even including violent thunderstorms.

For large-scale circulations, the approximate $p(z)$ determined through the use of (121) can be used to compute the pressure-gradient force in the equation of horizontal motion. Such use is known as the *quasi-static approximation*. This approximation applies very accurately to large-scale motions, but it introduces unacceptable errors for small-scale motions, such as thunderstorms. When the quasi-static approximation is made, the effective kinetic energy is due to the horizontal wind only; the contribution of the vertical component, w, is neglected. For large-scale motions, $w \ll (u, v)$, so that this quasi-static kinetic energy is very close to the true kinetic energy. Holton (2004) discusses this topic further.

Replacing the equation of vertical motion with (121) introduces two fundamental changes. First, *the equation of vertical motion can no longer be used to determine the vertical velocity.* The alternative method used to obtain the vertical velocity depends on the vertical coordinate system used, which is expected, because the actual meaning of the vertical velocity also depends on the vertical coordinate system. For example, in the case of pressure coordinates, the vertical velocity is $\omega \equiv Dp/Dt$, and ω can be determined using the continuity equation with the upper boundary condition $\omega = 0$ at $p = 0$. In this book we will determine the vertical velocity as needed, on a case-by-case basis.

The second fundamental change that is introduced with use of (121) is that *it allows only one (three-dimensional) thermodynamic variable to be predicted*, whereas the full equation set allows two. For example, if the density is predicted, then (121) can be used to determine the pressure by vertical integration from $p = 0$ at the top of the atmosphere, and the ideal gas law can then be used to determine the temperature. As a result of this change, the quasi-static approximation filters vertically propagating sound waves. That is good, because vertically propagating sound waves have no perceptible effects on weather or climate and can cause practical difficulties in models; for most purposes, it is better to eliminate them (e.g., Arakawa and Konor, 2009).

When we consider large-scale motions, the tendencies due to molecular fluxes are overwhelmed by those associated with turbulence, convection, and gravity waves. In addition, for large-scale motions the flux divergences associated with turbulence, convection, and small-scale waves can be accurately approximated by the vertical derivative of the vertical component of the flux.

With these approximations, we can replace (50) by

$$\begin{aligned} \frac{Du}{Dt} - \frac{uv\tan\varphi}{a} &= fv - \frac{\alpha}{a\cos\varphi}\frac{\partial p}{\partial\lambda} - \alpha\frac{\partial F_u}{\partial z}, \\ \frac{Dv}{Dt} + \frac{u^2\tan\varphi}{a} &= -fu - \frac{\alpha}{a}\frac{\partial p}{\partial\varphi} - \alpha\frac{\partial F_v}{\partial z}, \\ 0 &= -g - \alpha\frac{\partial p}{\partial z}. \end{aligned} \tag{122}$$

Also, the vector equation of horizontal motion is given by

$$\frac{\partial \mathbf{V}_h}{\partial t} + (\zeta_z + f)\mathbf{k}\times\mathbf{V}_h + \nabla_z K + w\frac{\partial \mathbf{V}_h}{\partial z} = -\alpha\nabla_z p - \alpha\frac{\partial \mathbf{F}_V}{\partial z}, \tag{123}$$

where

$$\zeta_z \equiv \mathbf{k}\cdot(\nabla_z \times \mathbf{V}_h) \tag{124}$$

is the vertical component of the relative vorticity as seen in height coordinates,

$$K \equiv \frac{1}{2}\mathbf{V}_h\cdot\mathbf{V}_h \tag{125}$$

is the kinetic energy per unit mass associated with the *horizontal velocity only*, and the vector $\mathbf{F}_V$ is the vertical flux of horizontal momentum due to small-scale motions.

Potential Temperature as a Vertical Coordinate

The most basic requirement for a variable to be used as a vertical coordinate is that it vary monotonically with height. Even this requirement can be relaxed, however; for example, a vertical coordinate can be independent of height over some layer of the atmosphere, provided that the layer is not too deep. We have already used both height coordinates and pressure coordinates in this book. One of the most useful vertical coordinates is potential temperature. With only minor exceptions, θ increases with height throughout the atmosphere. We now derive the basic equations of atmospheric motion as expressed in θ coordinates, and without using the hydrostatic approximation. For strong reasons that will become clear, potential temperature will be used as a vertical coordinate in numerous places throughout the rest of this book.

The starting point is the basic equations in height coordinates, showing the horizontal and vertical velocities and derivatives separately. The equations of horizontal and vertical motion, with rotation and friction omitted for simplicity, are

$$\frac{D\mathbf{V}_h}{Dt} = -\frac{1}{\rho}\nabla_z p, \tag{126}$$

$$\frac{Dw}{Dt} = -\frac{1}{\rho}\frac{\partial p}{\partial z} - g. \tag{127}$$

The continuity equation is

$$\left(\frac{\partial \rho}{\partial t}\right)_z + \nabla_z \cdot (\rho \mathbf{V}_h) + \frac{\partial}{\partial z}(\rho w) = 0. \tag{128}$$

We can write the thermodynamic energy equation as

$$\dot{\theta} \equiv \frac{D\theta}{Dt} = \frac{Q}{\Pi}, \tag{129}$$

where Q is the heating rate per unit mass, and Π is the Exner function, which satisfies both

$$c_p T = \Pi\theta, \tag{130}$$

and

$$\Pi = c_p\left(\frac{p}{p_0}\right)^{\kappa}. \tag{131}$$

Finally, we include the prognostic equation for an arbitrary scalar Λ, which is

$$\left[\frac{\partial}{\partial t}(\rho\Lambda)\right]_z + \nabla_z \cdot (\rho \mathbf{V}_h \Lambda) + \frac{\partial}{\partial z}(\rho w \Lambda) = \rho S_\Lambda, \tag{132}$$

where S_Λ is the source of Λ per unit mass.

We now transform these equations to θ-coordinates. Using methods described in appendix D, we can write the horizontal pressure gradient as

$$\frac{1}{\rho}\nabla_z p = \frac{1}{\rho}\nabla_\theta p - \frac{1}{\rho}\frac{\partial p}{\partial z}\nabla_\theta z. \tag{133}$$

We can rewrite the first term on the right-hand side of (133), using (130), (131), and the ideal gas law:

$$\begin{aligned}\frac{1}{\rho}\nabla_\theta p &= RT\frac{\nabla_\theta p}{p}\\ &= \frac{RT}{\kappa}\frac{\nabla_\theta \Pi}{\Pi}\\ &= \frac{c_p T}{\Pi}\nabla_\theta \Pi\\ &= \theta\nabla_\theta \Pi\\ &= \nabla_\theta(\Pi\theta)\\ &= \nabla_\theta(c_p T)\\ &= \nabla_\theta s - g\nabla_\theta z.\end{aligned} \tag{134}$$

Here s is the dry static energy. Similarly, we can write the vertical pressure-gradient force as

$$\begin{aligned}\frac{1}{\rho}\frac{\partial p}{\partial z} &= \frac{RT}{p}\frac{\partial p}{\partial z}\\ &= \frac{RT}{\kappa\Pi}\frac{\partial \Pi}{\partial z}\\ &= \frac{c_p T}{\Pi}\frac{\partial \Pi}{\partial z}\\ &= \theta\frac{\partial \Pi}{\partial z}\\ &= \frac{\partial}{\partial z}(\Pi\theta) - \Pi\frac{\partial \theta}{\partial z}\\ &= \frac{\partial}{\partial z}(c_p T) - \Pi\frac{\partial \theta}{\partial z}\\ &= \frac{\partial s}{\partial z} - g - \Pi\frac{\partial \theta}{\partial z}\\ &= \frac{\partial \theta}{\partial z}\left(\frac{\partial s}{\partial \theta} - \Pi\right) - g.\end{aligned} \tag{135}$$

Substituting (134) and (135) into (133), we find that

$$\begin{aligned}\frac{1}{\rho}\nabla_z p &= (\nabla_\theta s - g\nabla_\theta z) - \left[\frac{\partial \theta}{\partial z}\left(\frac{\partial s}{\partial \theta} - \Pi\right) - g\right]\nabla_\theta z\\ &= \nabla_\theta s - \frac{\partial \theta}{\partial z}\left(\frac{\partial s}{\partial \theta} - \Pi\right)\nabla_\theta z.\end{aligned} \tag{136}$$

Using (135) and (136), we can now rewrite the equations of horizontal and vertical motion as

$$\frac{D\mathbf{V}_h}{Dt} = -\left[\nabla_\theta s - \frac{\partial \theta}{\partial z}\left(\frac{\partial s}{\partial \theta} - \Pi\right)\nabla_\theta z\right], \tag{137}$$

$$\frac{Dw}{Dt} = -\frac{\partial \theta}{\partial z}\left(\frac{\partial s}{\partial \theta} - \Pi\right). \tag{138}$$

When the dry static energy, s, was introduced earlier in this chapter, we pointed out that it is approximately conserved under dry adiabatic processes, and it tends to increase with height in a statically stable atmosphere. It may be confusing to see the same variable in the expressions for the horizontal and vertical pressure-gradient forces as expressed in isentropic coordinates, but it really is the same variable, playing two very different roles in the equations. In the context of the horizontal and vertical pressure-gradient forces, the dry static energy is sometimes called the *Montgomery potential*, but we don't use that term in this book.

Next, we transform the continuity equation, (128). Using methods discussed in appendix D, we can rewrite (128) as

$$\left(\frac{\partial \rho}{\partial t}\right)_\theta - \frac{\partial \theta}{\partial z}\frac{\partial \rho}{\partial \theta}\left(\frac{\partial z}{\partial t}\right)_\theta + \nabla_\theta \cdot (\rho \mathbf{V}_h) - \frac{\partial \theta}{\partial z}\left[\frac{\partial}{\partial \theta}(\rho \mathbf{V}_h)\right]\cdot \nabla_\theta z + \frac{\partial \theta}{\partial z}\frac{\partial}{\partial \theta}(\rho w) = 0, \quad (139)$$

or

$$\frac{\partial z}{\partial \theta}\left(\frac{\partial \rho}{\partial t}\right)_\theta - \frac{\partial \rho}{\partial \theta}\left(\frac{\partial z}{\partial t}\right)_\theta + \nabla_\theta \cdot (\rho \mathbf{V}_h) - \left[\frac{\partial}{\partial \theta}(\rho \mathbf{V}_h)\right]\cdot \nabla_\theta z + \frac{\partial}{\partial \theta}(\rho w) = 0. \quad (140)$$

Note that

$$\left[\frac{\partial}{\partial t}\left(\rho \frac{\partial z}{\partial \theta}\right)\right]_\theta = \frac{\partial z}{\partial \theta}\left(\frac{\partial \rho}{\partial t}\right)_\theta + \rho \frac{\partial}{\partial \theta}\left(\frac{\partial z}{\partial t}\right)_\theta, \quad (141)$$

And

$$\nabla_\theta \cdot \left(\rho \frac{\partial z}{\partial \theta}\mathbf{V}_h\right) = \frac{\partial z}{\partial \theta}\nabla_\theta \cdot (\rho \mathbf{V}_h) + \rho \mathbf{V}_h \cdot \nabla_\theta \left(\frac{\partial z}{\partial \theta}\right). \quad (142)$$

Substituting (141) and (142) into (140), we obtain

$$\left(\frac{\partial \rho_\theta}{\partial t}\right)_\theta + \nabla_\theta \cdot (\rho_\theta \mathbf{V}_h) - \frac{\partial}{\partial \theta}\left\{\rho\left[\left(\frac{\partial z}{\partial t}\right)_\theta + \mathbf{V}_h \cdot \nabla_\theta z - w\right]\right\} = 0, \quad (143)$$

where

$$\rho_\theta \equiv \rho \frac{\partial z}{\partial \theta}, \quad (144)$$

is called the *pseudodensity* and has dimensions of mass per unit area per unit temperature (e.g., kg m^{-2} K^{-1}). It follows from (144) that

$$\frac{1}{\rho}\frac{\partial}{\partial z} = \frac{1}{\rho_\theta}\frac{\partial}{\partial \theta}. \quad (145)$$

Referring to equations (11) and (13) in our discussion of mass conservation at the beginning of this chapter, we recognize

$$\mu \equiv -\rho\left[\left(\frac{\partial z}{\partial t}\right)_\theta + \mathbf{V}_h \cdot \nabla_\theta z - w\right] \quad (146)$$

as the upward mass flux across an isentropic surface. The minus sign appears in (146) because an increase in the height of an isentropic surface favors a downward "flow" of mass across that surface. Definition (146) allows us to simplify the continuity equation, (143), to

$$\left(\frac{\partial \rho_\theta}{\partial t}\right)_\theta + \nabla_\theta \cdot (\rho_\theta \mathbf{V}_h) + \frac{\partial \mu}{\partial \theta} = 0. \quad (147)$$

In a similar way, the conservation equation for an intensive scalar, (132), becomes

$$\left[\frac{\partial}{\partial t}(\rho_\theta \Lambda)\right]_\theta + \nabla_\theta \cdot (\rho_\theta \mathbf{V}_h \Lambda) + \frac{\partial}{\partial \theta}(\mu \Lambda) = \rho_\theta S_\Lambda. \quad (148)$$

The advective form can be obtained by combining (147) with (148):

$$\left(\frac{\partial \Lambda}{\partial t}\right)_\theta + \mathbf{V}_h \cdot \nabla_\theta \Lambda + \frac{\mu}{\rho_\theta}\frac{\partial \Lambda}{\partial \theta} = S_A. \tag{149}$$

As a special case of (149), the advective form of the potential temperature equation is

$$\mu = \rho_\theta \dot{\theta}, \tag{150}$$

where we utilized $\dot{\theta} \equiv S_\theta$. Using (150), we can rewrite (146), (147), and (148) as

$$\left(\frac{\partial z}{\partial t}\right)_\theta = -\mathbf{V}_h \cdot \nabla_\theta z + w - \frac{\partial z}{\partial \theta}\dot{\theta}, \tag{151}$$

$$\left(\frac{\partial \rho_\theta}{\partial t}\right)_\theta + \nabla_\theta \cdot (\rho_\theta \mathbf{V}_h) + \frac{\partial}{\partial \theta}(\rho_\theta \dot{\theta}) = 0, \tag{152}$$

$$\left[\frac{\partial}{\partial t}(\rho_\theta \Lambda)\right]_\theta + \nabla_\theta \cdot (\rho_\theta \mathbf{V}_h \Lambda) + \frac{\partial}{\partial \theta}(\rho_\theta \dot{\theta} \Lambda) = \rho_\theta S_\Lambda. \tag{153}$$

The Lagrangian time derivative is expressed by

$$\frac{D}{Dt}(\ ) = \left(\frac{\partial}{\partial t}\right)_\theta (\ ) + \mathbf{V}_h \cdot \nabla_\theta (\ ) + \dot{\theta}\frac{\partial}{\partial \theta}(\ ). \tag{154}$$

In summary, the prognostic equations needed to describe nonhydrostatic motions in θ-coordinates are (137), (138), (151), (152), and (153). Not counting the scalar Λ, the prognostic variables of the θ-coordinate model are $\mathbf{V}_h$, w, ρ_θ, and z. The corresponding prognostic variables of the z-coordinate model are $\mathbf{V}_h$, w, ρ, and θ.

Finally, we have to determine Π and s. We can use ρ_θ and $z(\theta)$ in (144) to find the density, ρ. Knowing ρ and θ, we can determine Π from the equation of state in the form

$$\Pi = c_p \left(\frac{\rho R \theta}{p_0}\right)^{\frac{\kappa}{1-\kappa}}. \tag{155}$$

We can then obtain s using $s = \Pi\theta + gz$. The pressure and temperature can easily be diagnosed from Π and θ, although they are not needed in the equations under discussion in this section.

In the quasi-static limit, the pseudodensity is still predicted using (152). The pressure can be computed from the pseudodensity by vertically integrating

$$\rho_\theta = -\frac{1}{g}\frac{\partial p}{\partial \theta}. \tag{156}$$

To obtain (156) we used (144) and the hydrostatic equation in the form

$$\frac{\partial p}{\partial z} = -\rho g. \tag{157}$$

Once the pressure is known, Π can be obtained from its definition, and the temperature can then be calculated from θ. The equation of vertical motion, (139), reduces to the hydrostatic equation in the form

$$\frac{\partial s}{\partial \theta} - \Pi = 0. \tag{158}$$

The dry static energy can be obtained by vertically integrating (158), starting from the lower boundary condition

$$s = gz, \text{ where } \theta = 0. \tag{159}$$

From s and the temperature the height of each θ-surface can be found, so that (151) is not needed and should not be used. With the use of (158), the horizontal momentum equation, (137), simplifies to

$$\frac{D\mathbf{V}_h}{Dt} = -\nabla_\theta s. \tag{160}$$

As can be seen in (160), in the hydrostatic limit the horizontal pressure-gradient force in θ-coordinates is a gradient. This fact is used later in the derivation of the potential vorticity equation.

Figure 4.4 shows the observed zonally averaged isentropic pseudodensity for January. The vertical axis in the figure is potential temperature, and the lower part of the axis includes low values of θ that can occur near the poles but are never found in the tropics. As a result, no contours are plotted in the lower portion of the figure, for the tropical regions, and we can consider that the pseudodensity is actually zero there, which explains why the pseudodensity rapidly decreases downward near the surface in the tropics. This topic is discussed further in chapter 7. Away from the surface, the pseudodensity generally decreases with height, like ρ.

Vorticity and Potential Vorticity

The concept of potential vorticity (PV) is necessary to understand the role of planetary waves in the global circulation. To derive the most general form of the potential vorticity equation, we start from the quasi-static equation of horizontal motion in isentropic coordinates, now restoring the Coriolis and friction terms:

$$\left(\frac{\partial \mathbf{V}_h}{\partial t}\right)_\theta + (\zeta_\theta + f)\mathbf{k} \times \mathbf{V}_h + \nabla_\theta (K + s) + \dot{\theta}\frac{\partial \mathbf{V}_h}{\partial \theta} = -\frac{1}{\rho_\theta}\frac{\partial \mathbf{F}_V}{\partial \theta}. \tag{161}$$

Here

$$\zeta_\theta \equiv \mathbf{k} \cdot (\nabla_\theta \times \mathbf{V}_h) \tag{162}$$

can be interpreted as the projection of the vertical component of the vorticity onto the unit vector normal to the isentropic surface, and $\mathbf{F}_V$ is the vertical flux of horizontal momentum due to small-scale eddies.

The vorticity equation in isentropic coordinates can be derived by applying $\mathbf{k} \cdot \nabla \times$ to (161). The gradient term of (161), that is, $\nabla_\theta (K + s)$, drops out, because the curl of any gradient is zero. The remaining terms can be simplified and combined using the vector identities

$$\mathbf{k} \cdot \nabla_\theta \times (\mathbf{k} \times \mathbf{H}) = \nabla_\theta \cdot \mathbf{H}, \tag{163}$$

and

$$\mathbf{k} \cdot \nabla_\theta \times \mathbf{H} = -\nabla_\theta \cdot (\mathbf{k} \times \mathbf{H}), \tag{164}$$

where $\mathbf{H}$ is an arbitrary horizontal vector. Using (163)–(164) and the fact that the Coriolis parameter is independent of time, we find that

$$\left(\frac{\partial \eta}{\partial t}\right)_\theta + \nabla_\theta \cdot \left[\mathbf{V}_h \eta - \mathbf{k} \times \left(\dot{\theta}\frac{\partial \mathbf{V}}{\partial \theta} + \frac{1}{\rho_\theta}\frac{\partial \mathbf{F}_V}{\partial \theta}\right)\right] = 0, \tag{165}$$

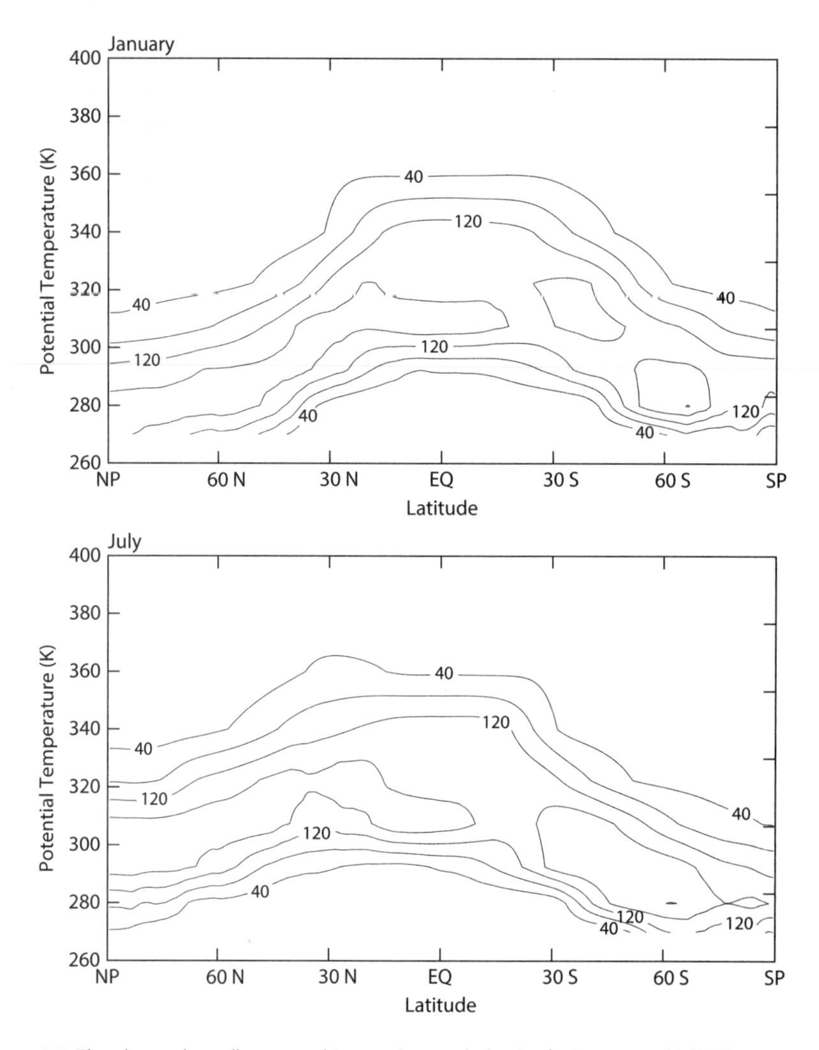

Figure 4.4. The observed zonally averaged isentropic pseudodensity for January and July. The contour interval is 40 kg m^{-2} K^{-1}.

where

$$\eta \equiv \zeta_\theta + f \tag{166}$$

is the absolute vorticity. Notice that in (165) the contribution from the "vertical advection term" of (161) appears inside a horizontal divergence operator! It thus appears to represent a *horizontal redistribution of PV along θ-surfaces*. It does not look like vertical advection, and it does not look like a source or sink. This counterintuitive result, due to Haynes and McIntyre (1987), comes from the use of (164).

We now define

$$\boxed{Z \equiv \frac{\eta}{\rho_\theta}} \tag{167}$$

as the *Ertel potential vorticity*. Then, we can rewrite (165) as

$$\boxed{\left(\frac{\partial}{\partial t}\right)_\theta(\rho_\theta Z) + \nabla_\theta \cdot \left[\rho_\theta \mathbf{V}_h Z - \mathbf{k} \times \left(\dot{\theta}\frac{\partial \mathbf{V}_h}{\partial \theta} + \frac{1}{\rho_\theta}\frac{\partial \mathbf{F}_\mathrm{V}}{\partial \theta}\right)\right] = 0}. \tag{168}$$

This amazing result was derived and discussed by Haynes and McIntyre (1987) and is called the *impermeability theorem*. Equation (168) implies that for isentropic surfaces that do not intersect the Earth's surface, *the area average of the mass-weighted PV on an isentropic surface cannot change, even in the presence of heating and friction.* It's like Las Vegas: what happens on the isentropic surface stays on the isentropic surface. An exception occurs for the Underworld of Hoskins (1991), which by definition has isentropic surfaces that intersect the Earth's surface. The PV of the Underworld can change owing to boundary effects on such surfaces. For further discussion, see Bretherton and Schär (1993). We will return to (168) in chapter 9.

The PV is of interest in part because it obeys a very simple conservation equation, that is, (168). As emphasized by Hoskins et al. (1985), however, a more important reason to be interested in the PV is that, with appropriate boundary conditions, it essentially determines the wind and temperature distributions for a large class of balanced circulations. Examples are given later.

Figure 4.5 shows the zonally averaged PV, which is (almost) a monotonic function of latitude and passes through zero very close to the equator. The stratosphere is a region of high PV because the pseudodensity is small there. The winter polar stratosphere has particularly large values. Intrusions of stratospheric air into the troposphere are characterized by anomalously large values of the PV. The troposphere is a region of nearly uniform PV, relative to the stratosphere, which suggests that PV is mixed or homogenized by the tropospheric weather systems. In middle and high latitudes of each hemisphere the tropopause roughly coincides with a constant-PV surface, namely, the ±2 PVU surface, where a PVU or *potential vorticity unit* is 10^{-6} m^2 K s^{-1} kg^{-1}. Kunz et al. (2011) provide some cautionary comments on this PV-based definition of the tropopause height. As mentioned earlier, the tropopause height is discontinuous in the subtropics, and one possible definition of the tropics is that it is the region on the equatorward side of these discontinuities.

The Quasi-Geostrophic System

Charney (1948) gave a systematic justification for what is now called the *quasi-geostrophic* (QG) system of equations. At that time, the QG system was of interest because it provided a way to do numerical weather prediction on the primitive digital computers that were just becoming available. From a modern perspective, the QG system is useful mainly because it makes it possible to give simple (but approximate) explanations for synoptic-scale weather phenomena in middle latitudes. We will use the QG system in chapter 8.

Holton (2004) and Vallis (2006) provide thorough introductions to QG dynamics. The QG system can be derived via an expansion in powers of the Rossby number, which looks very formal, although I've never found it particularly convincing. Here we briefly review the basic idea, which involves various approximations in the momentum equation and the thermodynamic equation, without trying to justify the approximations in detail.

Using pressure as a vertical coordinate, we can rewrite the equation of horizontal motion as

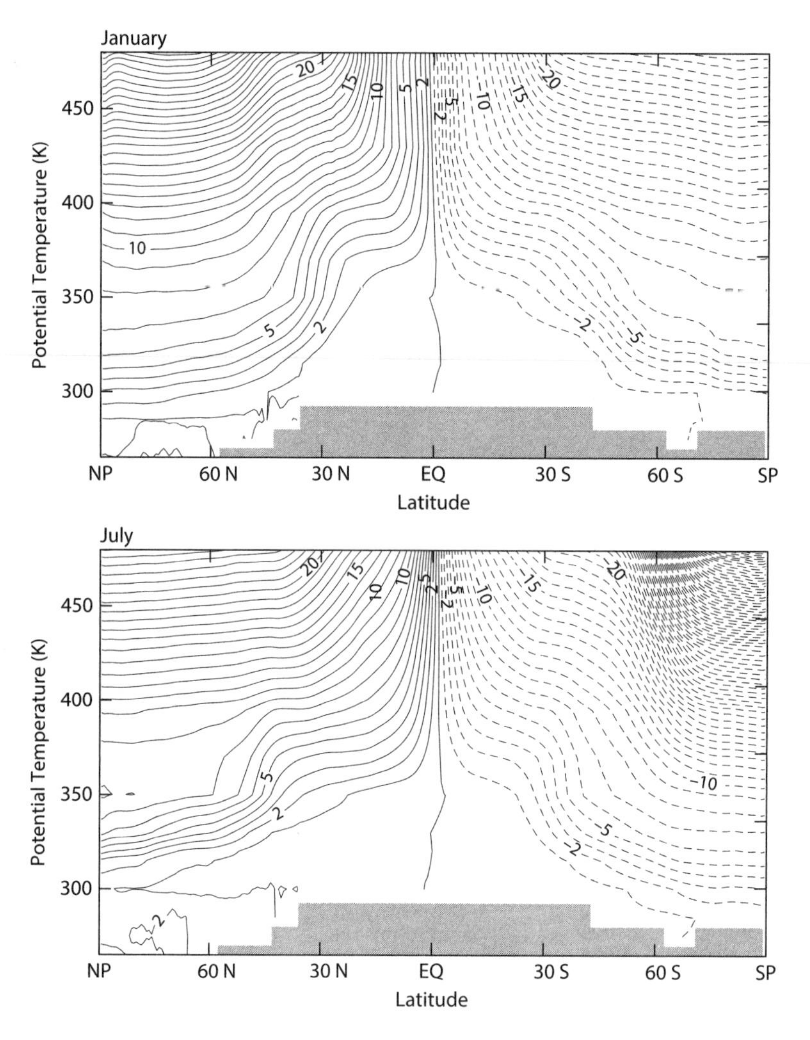

Figure 4.5. The zonally averaged Ertel potential vorticity plotted as a function of latitude and potential temperature. The contour interval is 1 PVU (*potential vorticity unit*), which is defined as 10^{-6} m^2 K s^{-1} kg^{-1}. The shaded area at the bottom represents isentropic surfaces that are on the "ground." A similar figure appears in Edouard et al. (1997).

$$\frac{D\mathbf{V}}{Dt} + f\mathbf{k}\times\mathbf{V} = -\nabla\phi. \tag{169}$$

Here we neglect friction, for simplicity, and omit the subscript h on $\mathbf{V}$. By definition, the geostrophic wind vector, $\mathbf{V}_g$, satisfies

$$f\mathbf{k}\times\mathbf{V}_g \equiv -\nabla\phi, \tag{170}$$

where ϕ is the geopotential. Using (170), we can rewrite (169) as

$$\frac{D\mathbf{V}}{Dt} + f\mathbf{k}\times(\mathbf{V}-\mathbf{V}_g) = 0, \tag{171}$$

or

$$\frac{\partial \mathbf{V}}{\partial t} + (\mathbf{V} \cdot \nabla)\mathbf{V} + \omega \frac{\partial \mathbf{V}}{\partial p} + f\mathbf{k} \times (\mathbf{V} - \mathbf{V}_g) = 0. \tag{172}$$

As was discussed in chapter 3, the nondimensional Rossby number is defined by

$$Ro \equiv \frac{V}{fL}, \tag{173}$$

where V and L are, respectively, a wind speed and a horizontal length scale that are typical of the motion system under consideration. Recall from chapter 3 that $Ro \ll 1$ for synoptic-scale motions in middle latitudes. Let

$$\beta \equiv \frac{1}{a} \frac{df}{d\varphi} \tag{174}$$

denote the rate of change of the Coriolis parameter with latitude. Charney showed that when the conditions

$$Ro \ll 1 \tag{175}$$

and

$$\frac{\beta L}{f} \leq Ro \tag{176}$$

are satisfied, (172) can be approximated by

$$\boxed{\left(\frac{\partial}{\partial t} + \mathbf{V}_g \cdot \nabla\right)\mathbf{V}_g + f\mathbf{k} \times (\mathbf{V} - \mathbf{V}_g) = 0}. \tag{177}$$

Equation (177) is the QG form of the horizontal momentum equation. Condition (175) means that momentum advection is weak compared with the Coriolis term, and (176) means that the Coriolis parameter varies within a fairly small range across the meridional scale of the weather system under consideration; that is, the weather system is not too "wide" in the meridional direction. In progressing from (172) to (177), the local tendency of the wind was approximated by the local tendency of the geostrophic wind, horizontal advection of the horizontal wind was approximated by geostrophic advection of the geostrophic wind, and vertical advection of the horizontal wind was neglected.

The "potential temperature" form of the thermodynamic energy equation is (88). For the case of no heating, it can be written using pressure coordinates as

$$\frac{\partial \theta}{\partial t} + (\mathbf{V} \cdot \nabla)\theta + \omega \frac{\partial \theta}{\partial p} = 0. \tag{178}$$

In the QG system, (178) is linearized about a basic-state potential temperature that varies only in the vertical. We write

$$\theta(\lambda, \varphi, p, t) = \theta_{bs}(p) + \theta'(\lambda, \varphi, p, t), \tag{179}$$

where $\theta_{bs}(p)$ is the basic-state potential temperature, and $\theta'(\lambda, \varphi, p, t)$ is the departure from the basic state. We approximate (178) by

$$\left(\frac{\partial}{\partial t} + \mathbf{V}_g \cdot \nabla\right)\theta' + \omega \frac{\partial \theta_{bs}}{\partial p} = 0. \tag{180}$$

Here horizontal advection is by the geostrophic wind, and vertical advection acts only on the basic-state potential temperature. Even though vertical advection was neglected in the momentum equation, it has to be retained in the thermodynamic energy equation because the Earth's atmosphere is strongly stratified.

Using the hydrostatic equation in the form

$$\frac{\partial \phi}{\partial p} = -\frac{R}{p}\left(\frac{p}{p_0}\right)^{\kappa}\theta \tag{181}$$

we can show that

$$\boxed{\theta\left(\frac{\partial \theta_{bs}}{\partial p}\right)^{-1} = \frac{1}{S}\frac{\partial \phi}{\partial p}}, \tag{182}$$

where

$$S(p) \equiv -\frac{\alpha_{bs}}{\theta_{bs}}\frac{\partial \theta_{bs}}{\partial p} \tag{183}$$

is the static stability, and α_{bs} is the basic-state specific volume. By substitution from (182), we can rewrite the thermodynamic energy equation (180) in terms of the geopotential:

$$\boxed{\left(\frac{\partial}{\partial t} + \mathbf{V}_g \cdot \nabla\right)\left(\frac{1}{S}\frac{\partial \phi}{\partial p}\right) + \omega = 0}. \tag{184}$$

This is the QG form of the thermodynamic energy equation.

Next, we form the vorticity equation corresponding to (177). It is

$$\frac{\partial \zeta_g}{\partial t} + (\mathbf{V}_g \cdot \nabla)\zeta_g + \beta_0 v_g + f_0 \nabla \cdot \mathbf{V} = 0, \tag{185}$$

where

$$\begin{aligned}\zeta_g &\equiv \mathbf{k} \cdot (\nabla \times \mathbf{V}_g) \\ &\cong \frac{\nabla^2 \phi}{f_0}\end{aligned} \tag{186}$$

is the geostrophic vorticity. In the divergence term of (185) we replaced f by f_0, and β by β_0. These are constant values appropriate for the "middle" of the latitude band under consideration. To obtain the second line of (186), we neglected a term involving the variation of the Coriolis parameter with latitude. Using (186) and the continuity equation,

$$\nabla \cdot \mathbf{V} + \frac{\partial \omega}{\partial p} = 0, \tag{187}$$

we can rewrite (185) as

$$\boxed{\left(\frac{\partial}{\partial t} + \mathbf{V}_g \cdot \nabla\right)\left(\frac{\nabla^2 \phi}{f_0} + \beta_0 y\right) - f_0 \frac{\partial \omega}{\partial p} = 0}, \tag{188}$$

where we used

$$\left(\frac{\partial}{\partial t} + \mathbf{V}_g \cdot \nabla\right) y = v_g. \tag{189}$$

Equation (188) is the QG vorticity equation.

The final step in the derivation of the QG system is to eliminate ω between (184) and (188). Using (170), we can show that

$$\frac{\partial}{\partial p}\left[\left(\frac{\partial}{\partial t}+\mathbf{V}_g\cdot\nabla\right)\left(\frac{1}{S}\frac{\partial\phi}{\partial p}\right)\right]=\left(\frac{\partial}{\partial t}+\mathbf{V}_g\cdot\nabla\right)\left[\frac{\partial}{\partial p}\left(\frac{1}{S}\frac{\partial\phi}{\partial p}\right)\right]. \tag{190}$$

This allows us to rewrite (184) as

$$\left(\frac{\partial}{\partial t}+\mathbf{V}_g\cdot\nabla\right)\left[\frac{\partial}{\partial p}\left(\frac{1}{S}\frac{\partial\phi}{\partial p}\right)\right]+\frac{\partial\omega}{\partial p}=0. \tag{191}$$

Elimination of $\partial\omega/\partial p$ between (188) and (191) gives

$$\boxed{\left(\frac{\partial}{\partial t}+\mathbf{V}_g\cdot\nabla\right)Z_{QG}=0}, \tag{192}$$

where

$$\boxed{Z_{QG}=\frac{\nabla^2\phi}{f_0}+\beta_0 y+\frac{\partial}{\partial p}\left(\frac{f_0}{S}\frac{\partial\phi}{\partial p}\right)} \tag{193}$$

is called the *QG pseudopotential vorticity* (QGPPV), and (192) is called the *QGPPV equation*. If Z_{QG} is known, and appropriate boundary conditions are provided, (193) can be solved as a boundary-value problem for ϕ. Once ϕ has been determined, the geostrophic wind can be obtained from (170), and the potential temperature from (182). We say that the QGPPV can be "inverted" to solve for the winds and temperature (e.g., Hoskins et al., 1985). The QGPPV is thus the key to QG dynamics.
We will use these ideas in chapter 8.

The Shallow Water Equations

Finally, we briefly discuss the shallow water equations, which will also be used, several times, in chapter 8. The shallow water equations are a two-dimensional system based on three assumptions or idealizations, as follows:

1. The fluid is incompressible, and the density is uniform everywhere and for all time.

Water is much less compressible than air, so this assumption partially explains the name of the shallow water system. The assumption of incompressibility has two important consequences. First, the continuity equation simplifies to

$$\nabla\cdot\mathbf{V}=0; \tag{194}$$

see (5). Because the three-dimensional divergence vanishes, convergence (divergence) in the horizontal must be accompanied by divergence (convergence) in the vertical. We assume that the fluid is bounded below by an impermeable surface (like the Earth's surface) of possibly variable height h_T and bounded above by a *free surface* (like the surface of a lake) at height h (see fig. 4.6). With these assumptions, (194) leads to a continuity equation of the form

$$\boxed{\frac{\partial h}{\partial t}+\nabla\cdot(h\mathbf{V})=0}, \tag{195}$$

where h is the depth of the fluid.

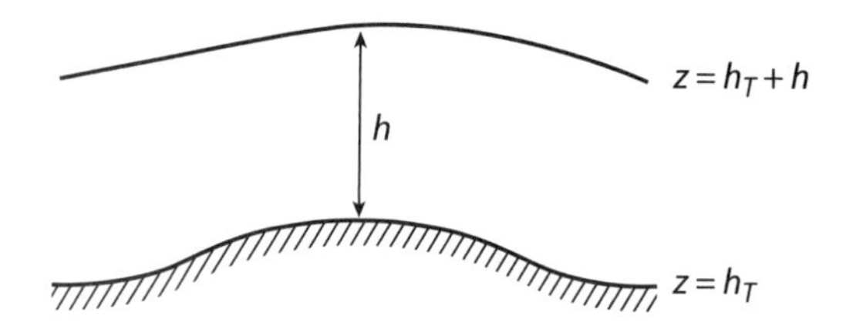

Figure 4.6. Schematic illustrating shallow water flowing over a "mountain."

The second consequence of incompressibility is that the expansion-work term vanishes in the mechanical energy and thermodynamic energy equations [see (67) and (81)], so that there is no conversion between internal energy and mechanical energy. The internal energy can therefore be ignored, and there is no need for a thermodynamic energy equation or a temperature. Any kinetic energy that is dissipated is just "gone," so when dissipative friction is included, the shallow water equations do not conserve total energy.

2. There is no vertical shear of the horizontal wind. This is where "shallowness" comes in.
3. The motion is quasi-hydrostatic, so that the pressure at a given level is proportional to the depth of the water above that level, which varies owing only to changes in the height of the free surface.

Using the first and third assumptions, we can write the horizontal pressure-gradient force (HPGF) as

$$\begin{aligned} \mathbf{HPGF} &= -\frac{1}{\rho}\nabla[g\rho(h_T + h)] \\ &= -g\nabla(h_T + h), \end{aligned} \tag{196}$$

where h_T is the height of the bottom topography relative to an arbitrary reference level, so that $h + h_T$ is the height of the free surface. We can then write the shallow water form of the horizontal momentum equation as

$$\boxed{\frac{D\mathbf{V}}{Dt} + f\mathbf{k}\times\mathbf{V} = -g\nabla(h + h_T)}. \tag{197}$$

A friction term can be added if desired.

Segue

This chapter outlined the conservation principles for momentum, mechanical and thermodynamic energy, and potential vorticity. Energy conversion processes were discussed in detail, and widely used approximations were introduced. The equations of motion using potential temperature as a vertical coordinate were derived. Finally, important approximations, including the quasi-geostrophic system of equations, as well as the idealized shallow water system were introduced.

These ideas will be used throughout the following chapters. Next, in chapter 5, we discuss the balance requirements for moisture, energy, and angular momentum, in terms of the zonally averaged circulation.

Problems

1. The eastward velocity associated with the Earth's rotation is $\mathbf{V}_e = \Omega r \cos\varphi \mathbf{e}_\lambda$. Show that

$$\mathbf{e}_r \cdot (\nabla \times \mathbf{V}_e) = 2\Omega \sin\varphi.$$

2. Prove the following about the unit vectors in spherical coordinates:

$$\begin{aligned} \nabla \cdot \mathbf{e}_\lambda &= 0, \\ \nabla \cdot \mathbf{e}_\varphi &= -\frac{\tan\varphi}{r}, \\ \nabla \cdot \mathbf{e}_r &= \frac{2}{r}, \\ \nabla \times \mathbf{e}_\lambda &= \frac{\mathbf{e}_\varphi}{r} + \frac{\tan\varphi}{r}\mathbf{e}_r, \\ \nabla \times \mathbf{e}_\varphi &= -\frac{\mathbf{e}_\lambda}{r}. \end{aligned}$$

3. Starting from the equation of motion, calculate the radius of a geostationary orbit. You will have to take into account the variation of the Earth's gravity with distance from the center of the Earth.
4. Without using a coordinate system, prove that

$$\Omega^2 \mathbf{r}_e = \nabla\left(\frac{1}{2}|\Omega \times \mathbf{r}|^2\right).$$

5. a) Suppose that you can bench-press 100 kg at the North Pole. How much can you lift at the equator? Assume that the Earth is a perfect sphere and that the acceleration due to the Earth's gravity is horizontally uniform.
   b) What length of day would be required to make the centrifugal acceleration cancel the gravitational acceleration at the Earth's surface, at 40° N?
6. Using a two-dimensional spherical coordinate system, (λ, φ), prove by direct calculation (i.e., "the long way") that

$$\frac{D\mathbf{V}}{Dt} + 2\Omega \times \mathbf{V} + \Omega \times (\Omega \times \mathbf{r}) = \frac{\partial \mathbf{V}}{\partial t} + (\nabla \times \mathbf{V} + 2\Omega) \times \mathbf{V} + \nabla\left(\frac{1}{2}\mathbf{V} \cdot \mathbf{V}\right) - \nabla\left[\frac{1}{2}|\Omega \times \mathbf{r}|^2\right].$$

7. Prove that for a closed volume

$$\int_V \frac{DA}{Dt} \rho dV = \frac{d}{dt} \int_V A \rho dV.$$

8. Show that for an isentropic process

$$\frac{p}{p_0} = \left(\frac{\rho}{\rho_0}\right)^\gamma = \left(\frac{T}{T_0}\right)^{\frac{\gamma}{\gamma-1}},$$

   where the subscript 0 denotes a reference state, and $\gamma \equiv c_p / c_v$.
9. Show that for an arbitrary process,

$$\alpha dp = \theta d\Pi,$$
$$Td\varepsilon = \Pi d\theta,$$
$$\eta = \Pi\theta + \text{constant}.$$

10. A process that mixes (i.e., homogenizes) the potential temperature increases the entropy. The following two exercises illustrate this principle, in slightly different ways:
   a) Consider two parcels of equal mass. Parcel one has potential temperature

   $$\theta_1 = \theta_0 + \Delta\theta,$$

   and parcel two has potential temperature

   $$\theta_2 = \theta_0 - \Delta\theta,$$

   where

   $$0 \leq \Delta\theta < \theta_0.$$

   Show that for a given value of θ_0, the combined entropy of the two parcels is maximized if the two parcels have the same potential temperature.
   b) Consider a process that mixes potential temperature; that is,

   $$\frac{\partial\theta}{\partial t} = \frac{\partial}{\partial x}\left(K\frac{\partial\theta}{\partial x}\right),$$

   where $K \geq 0$, acting over a domain that is either periodic or else closed in the sense that $K(\partial\theta/\partial x) = 0$ on the boundaries. Prove that this process causes the domain-averaged entropy to increase with time.
14. Starting from

$$\int_V p\frac{D\alpha}{Dt}\rho dV > 0,$$

show that the entropy, ε, satisfies

$$\int_V T\frac{D\varepsilon}{Dt}\rho dV > 0,$$

and that

$$\int_V \Pi\frac{D\theta}{Dt}\rho dV > 0.$$

Discuss the physical meaning of these inequalities.
15. Prove that in a hydrostatic atmosphere the dry static energy, s, satisfies

$$\frac{\partial s}{\partial z} \cong \Pi\frac{\partial\theta}{\partial z}.$$

16. Derive an expression for the isentropic pseudodensity $\rho_0(\theta)$ for the special case of an isothermal (uniform temperature) and hydrostatic atmosphere. Assuming a surface pressure of 1000 hPa and a temperature of 270 K, plot $\rho_0(\theta)$ for θ in the range zero to infinity.
17. Derive the mechanical energy equation in θ-coordinates, in the form

$$\left[\frac{\partial}{\partial t}(\rho_\theta K)\right]_\theta + \nabla_\theta \cdot [\rho_\theta \mathbf{V}(K+\phi)] + \frac{\partial}{\partial \theta}\left[\rho_\theta \dot{\theta}(K+\phi) - z\left(\frac{\partial p}{\partial t}\right)_\theta + \mathbf{V} \cdot \mathbf{F}_\mathrm{V}\right]$$
$$= -\rho_\theta \omega \alpha - \rho_\theta \delta$$

by starting from the equation of motion in θ-coordinates.

18. Consider an air parcel at rest at the sea surface on the equator. If the parcel rises from the surface to an altitude of 15 km, conserving its angular momentum, what is its zonal velocity? For purposes of this problem, define the axial component of the angular momentum by

$$M \equiv r\cos\varphi\,(\Omega r\cos\varphi + u),$$

where r is the radial distance of the parcel from the center of the Earth. To answer the question, first consider a parcel at rest at the surface, where $r = a$. Show that if the parcel is moved to radius r, without changing its angular momentum, its zonal velocity is approximately given by

$$u \cong -2\Omega(r-a).$$

19. Show that the *three-dimensional* pressure gradient force can be written as $-\theta\nabla\Pi$. Do not use the hydrostatic approximation.
20. Demonstrate that $\Pi = c_p(\rho R\theta/p_0)^{\kappa/1-\kappa}$ is a form of the equation of state.
21. As discussed in the text, the centrifugal acceleration causes the isosurfaces of the geopotential to bulge at the equator, relative to the poles. How large is the equatorial bulge, in kilometers? Assume for simplicity that g is uniform.
22. Using

$$f\mathbf{k} \times \mathbf{V}_g = -\nabla\phi,$$

prove that

$$\frac{\partial}{\partial p}\left[\left(\frac{\partial}{\partial t} + \mathbf{V}_g \cdot \nabla\right)\left(\frac{1}{S}\frac{\partial \phi}{\partial p}\right)\right] = \left(\frac{\partial}{\partial t} + \mathbf{V}_g \cdot \nabla\right)\left[\frac{\partial}{\partial p}\left(\frac{1}{S}\frac{\partial \phi}{\partial p}\right)\right].$$

CHAPTER 5

Go with the Flow

Overview

The atmosphere is subject to spatially and temporally varying sources and sinks of energy, moisture, and angular momentum at its upper and lower boundaries. The circulation transports energy, moisture, and angular momentum so as to maintain globally averaged balances over time. The sources and sinks themselves are affected by the circulation. In this chapter we show how these balance requirements are satisfied. We focus on the zonally averaged sources, sinks, and vertical and meridional transports but with an emphasis on the systematic and powerful effects of the zonally varying component of the circulation on the zonally averaged circulation.

As recounted in the beautiful book by Lorenz (1967), early theories of the global circulation were aimed at explaining the (then very inadequately observed) zonally averaged circulation. Many of the relevant processes depend in an essential way on variations of the winds, temperature, and the like, along latitude circles. We cannot understand the zonally averaged circulation without accounting for the effects of these departures from the zonal mean. We define an *eddy* as a circulation feature that varies with longitude. The important role of eddies in the maintenance of the zonally averaged circulation was only gradually appreciated, by such early scholars as Dove (1837), Defant (1921), Jeffreys (1926), Rossby (1941, 1947), Bjerknes (1948), and Starr (1948), among others. The importance of deep cumulus convection for the global circulation was first understood by Riehl and Malkus (1958), as will be discussed in chapter 6. Vertical momentum fluxes due to small-scale gravity waves are also important, as discussed in chapter 9.

During recent decades, the zonally varying structure of the global circulation has been studied intensively for its own sake, rather than just in terms of its effects on the zonally averaged circulation. This more modern perspective is adopted in chapter 8 and beyond.

The processes that maintain the observed zonally averaged circulation act on many scales. They involve energy conversions and transports due to radiation and small-scale convection, and transports of angular momentum, energy, moisture, and potential vorticity by a variety of large-scale weather systems. In this chapter we glimpse those various processes only indirectly, through their effects on the zonally averaged circulation. Later chapters discuss the nature of small-scale convection and large-scale eddies in detail and explain how they interact with the zonally averaged circulation.

This chapter follows in the footsteps of the heroic analysis of Peixoto and Oort (1992). Advancing technology and the rise of global forecast centers have made the task much, much easier. In our analysis:

- We used ECMWF reanalysis products and other modern data sources that did not exist 20 years ago.
- We analyzed the data using isentropic coordinates rather than pressure coordinates.
- We had the luxury of accessing the data over the Internet and analyzing it using much improved computer hardware and software.

The main goals of this chapter are to present some basic information about the observed global atmospheric circulations of dry air, water, energy, and angular momentum, and to illustrate some useful analysis concepts.

Tools for Discussing Eddies and Their Effects on the Zonally Averaged Circulation

The rest of this book makes use of notations that allow us to distinguish between eddies and zonal means. We will occasionally (but not systematically) distinguish between *stationary* eddies, which are anchored to features (such as mountain ranges) on the Earth's surface and so appear in time-averaged (e.g., monthly mean) maps, and *transient* eddies that move and thus are smeared out to invisibility in sufficiently long time averages. We adopt the notations shown in table 5.1, some of which have already been used. In the following discussion, we consider various statistics derived from the spatial and temporal distributions of fields called v and T. We use these symbols only for convenience. In principle, v and T can stand for any variables. They could be the same variable.

The time average of the product of v and T can be expanded as follows:

$$\begin{aligned} \overline{vT}^{t} &= \overline{(\overline{v}^{t} + v')(\overline{T}^{t} + T')}^{t} \\ &= \overline{\overline{v}^{t}\,\overline{T}^{t} + v'T' + \overline{v}^{t}T' + v'\,\overline{T}^{t}}^{t} \\ &\cong \overline{v}^{t}\,\overline{T}^{t} + \overline{v'T'}^{t}. \end{aligned} \tag{1}$$

Table 5.1. Notation Used in Describing Eddies and Zonal Means

Notation	Meaning
$\overline{(\ )}^{t}$	Time mean
$(\ )'$	Departure from time mean, or *transient component*
$\overline{(\ )}^{\lambda}$	Zonal mean
$(\ )^{*}$	Departure from zonal mean, or *eddy component*

As indicated, the last line of (1) is only approximate, unless the averaging interval is infinite or unless the time averages are taken over discrete, nonoverlapping blocks of time. The product $\overline{v'T'}^t$ is the temporal covariance of v and T. We can decompose the first term of (1) into its zonal mean and eddy components:

$$\begin{aligned}\overline{v}^t\,\overline{T}^t &= \overline{(\overline{v}^\lambda + v^*)}^t\,\overline{(\overline{T}^\lambda + T^*)}^t\\ &= (\overline{v}^{t,\lambda} + \overline{v^*}^t)(\overline{T}^{t,\lambda} + \overline{T^*}^t)\\ &= \overline{v}^{t,\lambda}\,\overline{T}^{t,\lambda} + \overline{v^*}^t\,\overline{T^*}^t + \overline{v}^{t,\lambda}\,\overline{T}^{*} + \overline{v^*}^t\,\overline{T}^{t,\lambda}.\end{aligned} \tag{2}$$

It follows that

$$\overline{\overline{v}^t\,\overline{T}^t}^\lambda = \overline{v}^{t,\lambda}\,\overline{T}^{t,\lambda} + \overline{\overline{v^*}^t\,\overline{T^*}^t}^\lambda. \tag{3}$$

Similarly,

$$\overline{v'T'}^{\lambda,t} = \overline{\overline{v'}^\lambda\,\overline{T'}^\lambda}^t + \overline{v'^*\,T'^*}^{\lambda,t}. \tag{4}$$

Both (3) and (4) are exact. Finally, substitution into the zonal average of (1) gives

$$\overline{vT}^{t,\lambda} = \overline{v}^{t,\lambda}\,\overline{T}^{t,\lambda} + \overline{\overline{v^*}^t\,\overline{T^*}^t}^\lambda + \overline{\overline{v'}^\lambda\,\overline{T'}^\lambda}^t + \overline{v'^*\,T'^*}^{\lambda,t}. \tag{5}$$

Formulas for some statistics of interest are given in table 5.2.

The following are some examples of quantities in each category. Suppose that v is the meridional wind, and T is temperature. Then, vT can be called the meridional flux of temperature, and is defined at each point in space. Applying a time average,

Table 5.2. Terminology Used in Discussing Zonally Averaged Flow and Eddy Covariances

1	$\overline{v}^t, \overline{T}^t$	Time mean fields
2	$\overline{vT}^t$	Time average of a product
3	$\overline{v}^\lambda\,\overline{T}^\lambda$	Product of zonal means
4	$\overline{v}^t\,\overline{T}^t$	Product of time averages $\overline{(1)}^\lambda \times \overline{(1)}^\lambda$
5	$\overline{v'T'}^t = \overline{vT}^t - \overline{v}^t\,\overline{T}^t$	Total transient (temporal covariance) $(2) - \overline{(4)}^\lambda$
6	$\overline{v}^{t,\lambda}\,\overline{T}^{t,\lambda}$	Stationary symmetric $\overline{(1)}^\lambda \times \overline{(1)}^\lambda$
7	$\overline{\overline{v^*}^t\,\overline{T^*}^t}^\lambda = \overline{\overline{v}^t\,\overline{T}^t}^\lambda - \overline{v}^{t,\lambda}\,\overline{T}^{t,\lambda}$	Stationary eddy (zonal covariance of time averages) $\overline{(4)}^\lambda - (6)$
8	$\overline{\overline{v'}^\lambda\,\overline{T'}^\lambda}^t = \overline{\overline{v}^\lambda\,\overline{T}^\lambda}^t - \overline{v}^{t,\lambda}\,\overline{T}^{t,\lambda}$	Transient symmetric (temporal covariance of zonal means) $\overline{(3)}^\lambda - (6)$
9	$\overline{v'^*\,T'^*}^{t,\lambda} = \overline{v'T'}^{t,\lambda} - \overline{\overline{v'}^\lambda\,\overline{T'}^\lambda}^t$	Transient eddy (combined zonal and temporal covariance) $(5) - (8)$
10	$\overline{v^*T^*}^{t,\lambda} = \overline{\overline{v^*}^t\,\overline{T^*}^t}^\lambda + \overline{v'^*\,T'^*}^{t,\lambda}$	Total eddy $(7) + (9)$

we get $\overline{vT}^t$, which is the time-averaged meridional flux of temperature. A product of time averages, such as $\overline{v}^t\overline{T}^t$, can represent the flux of temperature due to the time-averaged meridional wind and the time-averaged temperature. This quantity does not include additional fluxes that may arise from fluctuations on timescales shorter than the averaging interval.

The product $\overline{v}^\lambda\overline{T}^\lambda$ is the contribution to the meridional flux of temperature from the zonally averaged meridional wind and temperature, such as the Hadley circulation. The total *eddy* temperature flux, listed in row 10 of the table, can be written as the total flux minus $\overline{v}^\lambda\overline{T}^\lambda$. The transient symmetric flux, shown in row 8 of the table, arises from the temporally varying part of the zonally averaged circulation. The transient eddy flux, shown in row 9 of the table, comes from the parts of v and T that vary both zonally and in time.

Figure 5.1 provides an example. The top map shows the eddy part of the 850 hPa meridional wind averaged over the month of January, that is, $\overline{v^*}^t$; the corresponding map of the full field was given earlier. The middle panel shows a similar plot for the eddy temperature field at 850 hPa, that is, $\overline{T^*}^t$; again, a map of the full field was presented earlier. The eddy meridional wind field looks similar to the full meridional wind field simply because the zonal mean of the meridional wind is fairly small at all latitudes. In contrast, the eddy temperature field looks very different from the full temperature field; the eddy field is lumpy, while the full field has a strong tendency toward east-west stripes. The stripes are "removed" when the zonal mean is subtracted from the full field to construct the eddy field. The bottom panel of figure 5.1 shows a map of the product, $\overline{v^*}^t\,\overline{T^*}^t$. To its right is the corresponding zonal mean plot showing $\overline{(\overline{v^*}^t\,\overline{T^*}^t)}^\lambda$, that is, the zonally averaged meridional temperature flux due to stationary eddies. The flux is small except in middle latitudes of the Northern Hemisphere; as discussed later, strong stationary eddies are produced by flow over topography in the Northern Hemisphere in winter.

In this chapter we discuss "total" eddy fluxes as seen in isentropic coordinates. The distinct roles of transient and stationary eddies are discussed in later chapters.

An Isentropic View of the Mass Circulation

The remainder of this chapter makes extensive use of the isentropic vertical coordinate system introduced in chapter 4. The zonally and temporally averaged continuity equation in θ-coordinates is

$$\frac{\partial \overline{\rho_\theta}^\lambda}{\partial t} + \frac{1}{a\cos\varphi}\frac{\partial}{\partial\varphi}\left(\overline{\rho_\theta v}^\lambda\cos\varphi\right) + \frac{\partial}{\partial\theta}\left(\overline{\rho_\theta\dot{\theta}}^\lambda\right) = 0. \tag{6}$$

In (6), it is understood that the zonal averages, time derivative, and meridional derivative are taken along isentropic surfaces. Using the eddy notation introduced earlier, we can separate the zonally averaged meridional mass flux that appears in (6) into two parts:

$$\boxed{\overline{\rho_\theta v}^\lambda = \overline{\rho_\theta}^\lambda\overline{v}^\lambda + \overline{\rho_\theta^* v^*}^\lambda}. \tag{7}$$

The first term on the right-hand side of (7) is the product of the zonally averaged pseudodensity with the zonally averaged meridional wind. The second term arises from *correlated fluctuations of the pseudodensity and the meridional wind.* For example, if ρ_θ^* is large where v^* is large, and vice versa, then $\overline{\rho_\theta^* v^*}^\lambda$ will be positive. The

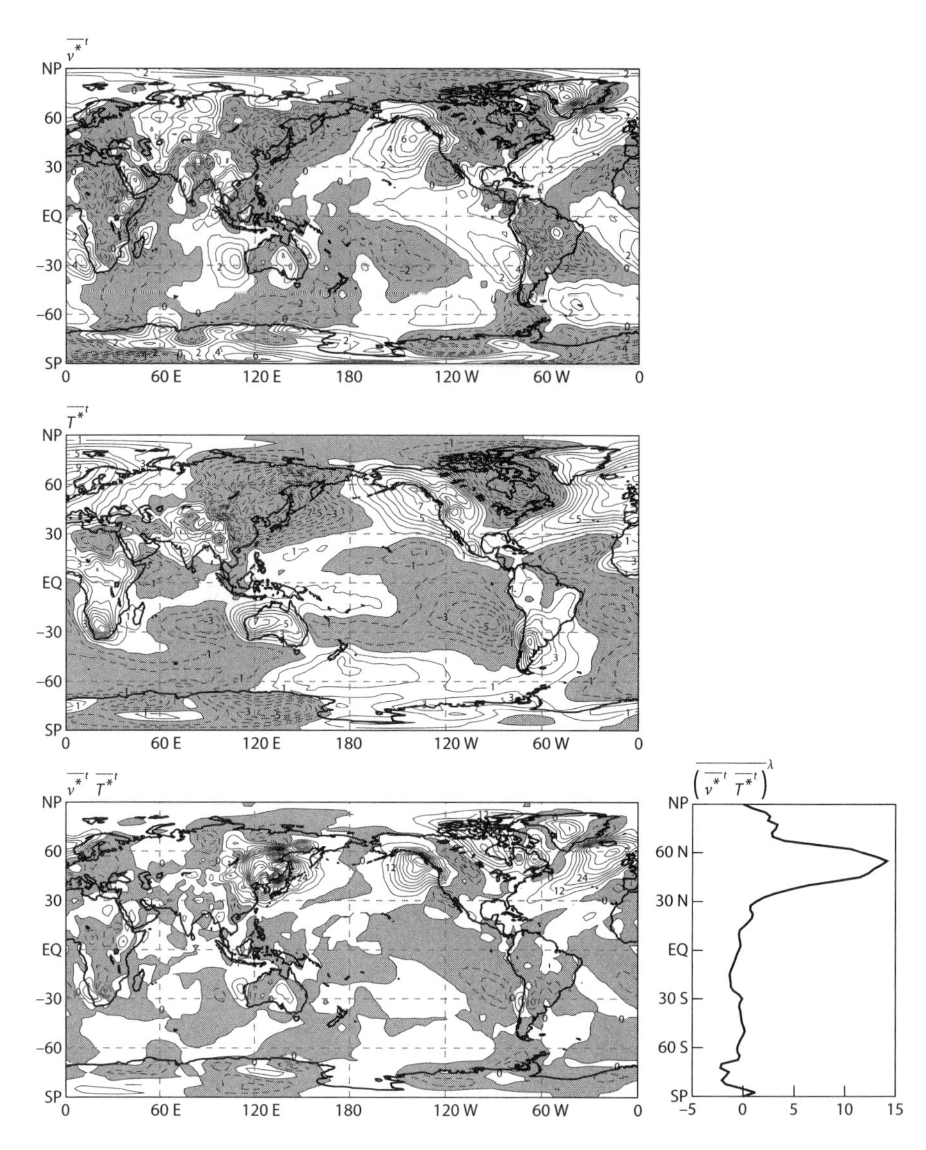

Figure 5.1. An example of the decompositions of fields into their zonal mean and eddy parts, and the zonal mean of an eddy product. The top and center panels show maps of the stationary eddy parts of the January-mean meridional wind and temperature at 850 hPa, respectively. The bottom-left and bottom-right panels show the product and the zonal mean of the product, respectively. See the text for details.

eddy mass flux $\overline{\rho_\theta^* v^*}^\lambda$ is sometimes called the meridional *bolus mass flux*, and $\overline{\rho_\theta^* v^*}^\lambda / \overline{\rho_\theta}^\lambda$ can be called the meridional *bolus velocity*. Here is an interpretation: Think of ρ_θ as the width of a flexible pipe, and v as the rate of flow through the pipe. If the width of the pipe and the flow fluctuate *in a correlated way*, then there can be a net mass transport even if the mean velocity, that is, $\overline{v}^\lambda$, is zero. For example, suppose that the pipe is wide where the flow is toward the north and narrow where the flow is toward the south. There will be a net mass flux toward the north, even if $\overline{v}^\lambda = 0$. The mass transport by the zonally averaged meridional wind is thus supplemented by a kind of "fluid-dynamical peristalsis," in which meridional velocity fluctuations are correlated with variations in the pseudodensity, leading to an eddy mass flux along isentropic surfaces.

The zonally averaged meridional mass flux in height coordinates is $\overline{\rho v}^{\lambda}$, where ρ is the (ordinary) density, and the zonal average is taken at a fixed height. We could decompose $\overline{\rho v}^{\lambda}$ so as to define a bolus mass flux in height coordinates, but we don't bother to do so, because variations of ρ on height surfaces are so small that the implied bolus mass flux is negligible. *In middle latitudes, especially in winter, the large variations of ρ_θ on isentropic surfaces make $\overline{\rho_\theta^* v^*}^{\lambda}$ large enough to be important.* The variations are small in the tropics, because the horizontal temperature gradients are weak there (see chapter 3).

For the reasons explained in chapter 2, we can neglect the tendency term of (6) if we average over a sufficiently long time:

$$\frac{1}{a\cos\varphi}\frac{\partial}{\partial\varphi}\left(\overline{\rho_\theta v}^{\lambda,t}\cos\varphi\right) + \frac{\partial}{\partial\theta}\left(\overline{\rho_\theta\dot{\theta}}^{\lambda,t}\right) = 0. \tag{8}$$

Equation (8) shows that the time average of $\dot{\theta}$ induces a time-averaged meridional circulation; that is, $\overline{\rho_\theta\dot{\theta}}^{\lambda,t} \neq 0$ implies that $\overline{\rho_\theta v}^{\lambda,t} \neq 0$ somewhere. The other way of thinking about (8) is that there cannot be any time and zonally averaged meridional mass flow on θ-surfaces unless there is heating. This conclusion follows from mass conservation in θ-space (Dutton, 1976; Townsend and Johnson, 1985; Hsu and Arakawa, 1990; Edouard et al., 1997).

As discussed in chapter 3, the zonally averaged meridional mass flux is very close to zero when averaged over time and vertically integrated. The same result can be obtained by integrating (8) with respect to θ over the range $\theta = 0$ to $\theta \to \infty$ and using $(\rho_\theta)_{\theta=0} = (\rho_\theta)_{\theta\to\infty} = 0$.

The top two panels of figure 5.2 show the distribution of the meridional mass flux, $\overline{\rho_\theta v}^{\lambda,t}$, with latitude and potential temperature for January (left) and July (right). The middle two panels show the contribution of the "mean" mass flux, $\overline{\overline{\rho_\theta}^{\lambda}\overline{v}^{\lambda}}^{t}$, and the bottom two panels show the contributions from the bolus mass flux, $\overline{\rho_\theta^* v^*}^{\lambda,t}$. The transport by the mean mass flux dominates in the tropics, where ρ_θ^* and the bolus mass flux are small. In middle latitudes, on the other hand, $\overline{v}^{\lambda}$ is small owing to geostrophy (as discussed in chapter 3), but ρ_θ^* and the bolus mass flux are large, especially in winter. In middle latitudes, $\overline{\rho_\theta^* v^*}^{\lambda,t}$ essentially "takes over" from the tropical zonally averaged flow. Here $\overline{\rho_\theta^* v^*}^{\lambda,t}$ includes contributions from both stationary and transient eddies. The top two panels show that the sum of $\overline{\overline{\rho_\theta}^{\lambda}\overline{v}^{\lambda}}^{t}$ and $\overline{\rho_\theta^* v^*}^{\lambda,t}$, that is, the total zonally averaged mass flux, has about the same magnitude in both the tropics and middle latitudes. Equation (8) shows that changes of $\overline{\rho_\theta v}^{\lambda,t}$ with latitude along isentropic surfaces are indicative of heating or cooling. We can see such changes between 20° and 40° N in January, for example. In the upper troposphere, $\overline{\rho_\theta v}^{\lambda,t}$ decreases poleward in January near 20° N, indicating cooling there, and in the lower troposphere it increases poleward near 40° N, indicating heating there. As discussed in chapters 6 and 8, the subtropical cooling is due to longwave radiation, and the midlatitude heating is due to the combined (and coupled) effects of surface energy fluxes and latent heat release.

A streamfunction for the mass circulation as seen in θ-coordinates can be defined by

$$\begin{aligned} \overline{\rho_\theta v}^{\lambda,t} 2\pi a\cos\varphi &\equiv -\frac{\partial\overline{\psi_\theta}^{t}}{\partial\theta}, \\ \overline{\rho_\theta\dot{\theta}}^{\lambda,t} 2\pi a^2\cos\varphi &\equiv \frac{\partial\overline{\psi_\theta}^{t}}{\partial\varphi}. \end{aligned} \tag{9}$$

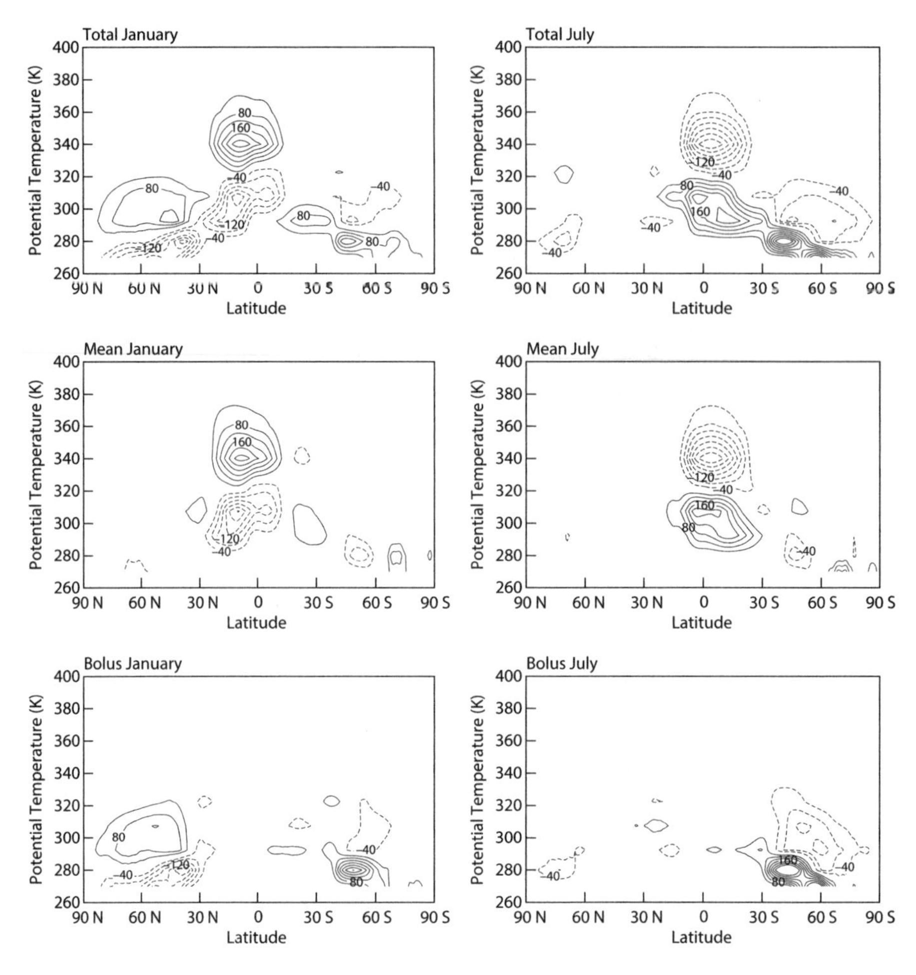

Figure 5.2. Isentropic meridional mass flux for January (left) and July (right). The top panels show the total flux, the middle panels show the mean mass flux, and the bottom panels show the bolus flux. The contour interval is 40 kg m^{-1} K^{-1} s^{-1}. The zero contour is not shown.

Here the subscript θ on $\overline{\psi_\theta}'$ is a reminder that the zonal and temporal averages used to define it were taken along isentropic surfaces. You should confirm that ψ_θ has the same dimensions (i.e., mass per unit time) as the streamfunction defined using pressure coordinates in chapter 3. The sign convention for the streamfunction in (9) is consistent with the one used in chapter 3. Figure 5.3 shows plots of the streamfunction of the seasonally varying mean meridional circulation as seen in p-coordinates (right panels) and θ-coordinates (left panels). *The actual data used are exactly the same with both coordinate systems.* As discussed in chapter 3, the streamfunction as seen in p-coordinates features a "large" Hadley cell at each solstice, with its rising branch in the summer-hemisphere tropics and its body extending into the winter-hemisphere subtropics. The peak magnitude of the streamfunction is about 16×10^{10} kg s^{-1}. A weaker Hadley circulation occurs in the summer hemisphere. Both Hadley cells are "direct" circulations. This jargon means that their rising branches are warm, and their sinking branches are cool. Similar observations were discussed in chapter 3.

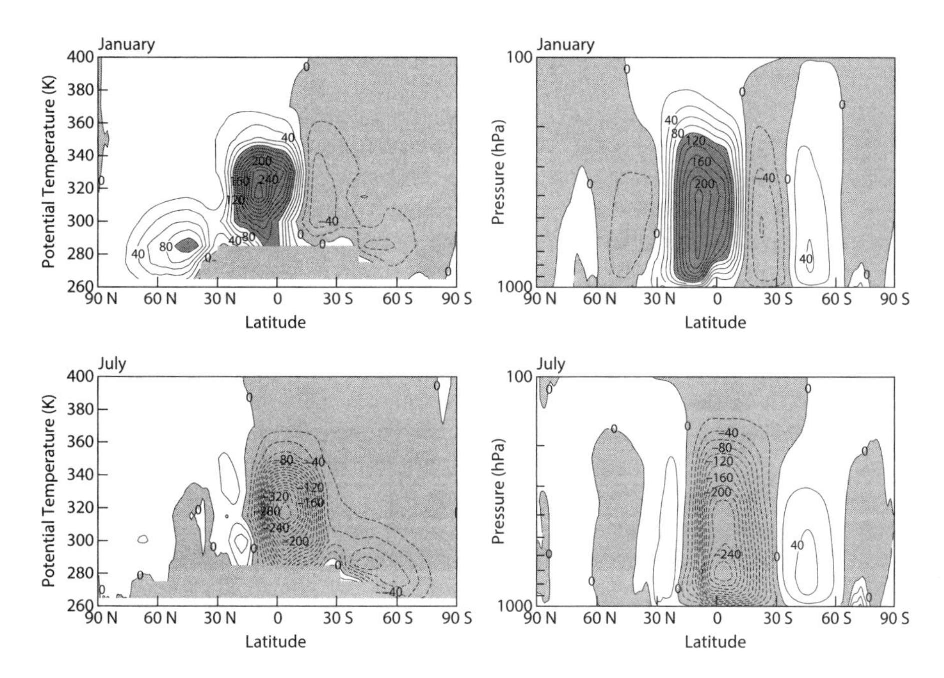

Figure 5.3. The streamfunction of the seasonally varying mean meridional circulation as seen in θ-coordinates (left panels) and in p-coordinates (right panels). Units are 10^9 kg s^{-1}. In the left panels, the upper boundary of the blocked-out region near the bottom represents the surface of the Earth, along which the potential temperature is, of course, a function of latitude. A similar figure was published by Townsend and Johnson (1985).

You may be surprised to see, in figure 5.3, that the streamfunction plots look very different when θ-coordinates are used. Hadley cells still appear, with somewhat stronger mass circulations, but *these direct circulations extend all the way to the poles*, primarily in the winter hemisphere. Ferrel cells do not appear at all. Here is one part of the explanation: when we use p-coordinates, the zonally averaged potential temperature equation is

$$\frac{\partial \overline{\theta}^{\lambda}}{\partial t}+\frac{1}{a\cos\varphi}\frac{\partial}{\partial\varphi}\left[\left(\overline{v}^{\lambda}\overline{\theta}^{\lambda t}+\overline{v^{*}\theta^{*}}^{\lambda}\right)\cos\varphi\right]+\frac{\partial}{\partial p}\left(\overline{\omega}^{\lambda}\overline{\theta}^{\lambda}+\overline{\omega^{*}\theta^{*}}^{\lambda}\right)=\overline{\dot{\theta}}^{\lambda}. \qquad (10)$$

In (10), the zonal averages and the temporal and meridional derivatives are taken along isobaric surfaces. We see that meridional fluxes of potential temperature are associated with both the mean meridional circulation (MMC) and the eddies. In the tropics and in the subtropics of the winter hemisphere, the MMC is very strong, and the fluctuations of potential temperature on isobaric surfaces are quite weak, so the meridional flux of potential temperature along isobaric surfaces is due mainly to the Hadley circulation. In middle latitudes, the Hadley circulation is weak when viewed in pressure coordinates, and the fluctuations of potential temperature on isobaric surfaces are large, especially in the winter hemisphere, owing to baroclinic eddies, so the eddy potential temperature flux along isobaric surfaces can and does become large enough to take over the job of transporting potential temperature poleward.

To derive an equation that corresponds to (10) in θ-coordinates, we simply multiply (6), which is the isentropic-coordinate version of the zonally averaged continuity equation, by θ. Using the fact that the derivatives in (6) are taken along isentropic surfaces, we can rearrange the result to

$$\frac{\partial}{\partial t}\left(\overline{\rho_\theta}^\lambda \theta\right) + \frac{1}{a\cos\varphi}\frac{\partial}{\partial \varphi}\left(\overline{\rho_\theta v}^\lambda \theta \cos\varphi\right) + \frac{\partial}{\partial \theta}\left(\overline{\rho_\theta \dot{\theta}}^\lambda \theta\right) = \overline{\rho_\theta \dot{\theta}}^\lambda, \tag{11}$$

which is closely analogous to (10). While (10) contains "eddy terms," (11) does not. The reason is that, *by definition*, $\theta^* = 0$ on θ-surfaces. The eddy flux of potential temperature along isentropic surfaces must therefore be exactly zero. This is one way of understanding why the Hadley circulations have to extend all the way to the poles when depicted in θ-coordinates. The Hadley circulations are the only game in town; there is no eddy flux of θ to hand off to.

Although the inputs to the calculation of the isentropic streamfunction are the pseudodensity and the meridional wind, figure 5.3 also contains information about atmospheric heating and cooling. Whenever the streamfunction contours cross isentropic surfaces, heating (for upward motion) or cooling (for downward motion) is implied. For example, figure 5.3 implies heating in the tropical rising branches of the Hadley circulations, and cooling in the sinking branches near the poles. As discussed previously, there are also signs of cooling in the subtropics, and heating slightly farther poleward, in middle latitudes.

These ideas and methods have recently been extended by Pauluis et al. (2010) and Pauluis and Mrowiec (2013) to analyze the observed mass circulation in *moist* entropy space. They have found that the mass circulation as seen in moist entropy space is about twice as strong as the mass circulation presented here using θ-coordinates.

Water Vapor Transports

As discussed in chapter 4, the total water mixing ratio satisfies

$$\left(\frac{\partial}{\partial t}\right)_z (\rho q_T) + \nabla_z \cdot (\rho \mathbf{V}_h q_T) + \frac{\partial}{\partial z}\left[\rho w q_T + (F_{q_T})_z - P\right] = 0, \tag{12}$$

where $\mathbf{V}_h$ is the horizontal wind vector, $(F_{q_T})_z > 0$ is the upward flux of total water vapor due to small-scale motions, and P is the downward flux of liquid water due to precipitation. We have written the horizontal and vertical advection terms separately, and, as indicated by the subscripts z on the temporal and horizontal derivatives in (12), we are using height as the vertical coordinate. Now, we translate (12) into θ-coordinates:

$$\left(\frac{\partial}{\partial t}\right)_\theta (\rho_\theta q_T) + \nabla_\theta \cdot (\rho_\theta \mathbf{V}_h q_T) + \frac{\partial}{\partial \theta}\left[\rho_\theta \dot{\theta} q_T + (F_{q_T})_z - P\right] = 0. \tag{13}$$

From this point on, we will omit the subscript θ on the temporal and horizontal derivatives, and also the subscript z on F_{q_T}. Averaging (13) with respect to longitude and time, we find that

$$\frac{1}{a\cos\varphi}\frac{\partial}{\partial \varphi}\left(\overline{\rho_\theta v q_T}^{\lambda,t}\cos\varphi\right) + \frac{\partial}{\partial \theta}\left(\overline{\rho_\theta \dot{\theta} q_T}^{\lambda,t} + \overline{F_{q_T}}^{\lambda,t} - \overline{P}^{\lambda,t}\right) = 0. \tag{14}$$

Finally, we integrate (14) vertically through the entire depth of the atmosphere:

$$\frac{\partial}{\partial \varphi}\left(\int_0^\infty \overline{\rho_\theta v q_T}^{\lambda,t} d\theta \cos\varphi\right) = \left[\overline{(F_{q_T})_S}^{\lambda,t} - \overline{P_S}^{\lambda,t}\right] a\cos\varphi. \tag{15}$$

To obtain (15) we have used $(\rho_\theta)_{\theta=0} = (\rho_\theta)_{\theta\to\infty} = 0$, and $(F_{q_T})_{\theta\to\infty} = 0$. Because of the vertical integration, (15) would convey exactly the same information regardless of which vertical coordinate system was used in (13); the integral on the left-hand side would take slightly different forms with different vertical coordinate systems, but its numerical value would be the same in all cases. Equation (15) shows that the divergence of the vertically integrated meridional moisture flux balances the surface fluxes of moisture due to turbulence and precipitation.

We can use (15) to determine the vertically integrated atmospheric meridional transport of total water as a function of latitude, using the vertical fluxes on the right-hand side as input. In chapter 2, we used a very similar method to determine the meridional transport of energy by the atmosphere and ocean combined, using the observed net radiation at the top of the atmosphere as input. According to (15), moisture must be laterally "imported" into latitude belts where precipitation exceeds evaporation, and it must be exported where evaporation exceeds precipitation. Evaporation and precipitation must balance when averaged over the globe.

We now examine the individual terms on the right-hand side of (15). The surface latent heat flux, $L(F_q)_S$, is the surface evaporation rate multiplied by the latent heat of condensation. It satisfies a so-called bulk aerodynamic formula of the form

$$L(F_q)_s = L\rho_S c_T |\mathbf{V}_S|(q_g - q_a). \tag{16}$$

Over a water surface, q_g is just the saturation mixing ratio evaluated using the surface temperature of the water and the surface air pressure. The "effective" value of q_g over land is much more difficult to determine because it depends on such variables as the soil moisture, the amount of vegetation, and the state of the vegetation (e.g., whether or not photosynthesis is occurring). Maps of the surface latent heat flux are shown in figure 5.4. Note the maxima over the subtropical oceans. There are no strong downward surface moisture fluxes, but weak downward fluxes do occur and lead to the formation of dew or frost.

Figure 5.5 shows maps and zonal means of the precipitation rate for January and July. Because precipitation tends to be very "noisy" in both space and time, average values are difficult to determine accurately. The values shown for such remote regions as the South Pacific Ocean cannot be strongly defended, although this situation should improve over the next few years because of additional data from the Global Precipitation Measurement Mission (Hou et al., 2014). It is clear from the figure that the rainiest regions of the world are in the tropics. The tropical maximum is associated with the rising branch of the Hadley circulation and the monsoons. Precipitation minima occur in the subtropics, where many of the Earth's major deserts are found. The seasonal shifts of tropical precipitation are quite spectacular and are most clearly seen in and near South America, Africa, India, and Southeast Asia. In January, heavy rain falls over the Amazon basin, over southern Africa, the Indian Ocean, the maritime continent north of Australia, in the South Pacific convergence zone that extends southeastward from the intersection of the date line with the equator, and across most of the tropical Pacific and Atlantic Oceans north of the equator. In July, the tropical rains have generally shifted to the north. Heavy rainfall occurs in the extreme northern part of South America, the neighboring Caribbean Sea and tropical North Atlantic Ocean, over northern equatorial tropical Africa, over India and neighboring regions of Southeast Asia, and to the north of the maritime continent off the east coast of tropical Asia.

Minima of the zonally averaged precipitation occur in the subtropics, which is where the major deserts of the world are found. Recall that the subtropics are home

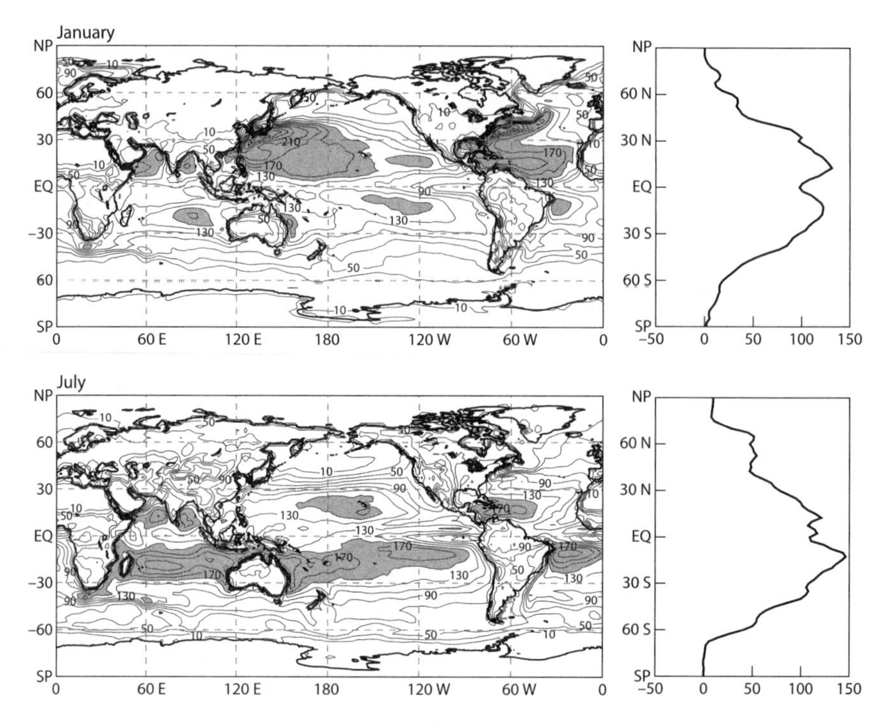

Figure 5.4. Maps of the surface latent heat flux, for January and July, based on ECMWF reanalyses. The contour interval is 20 W m^{-2}. Values higher than 150 W m^{-2} are shaded. The corresponding zonal averages are shown on the right. These are not real observations, although they are influenced by observations.

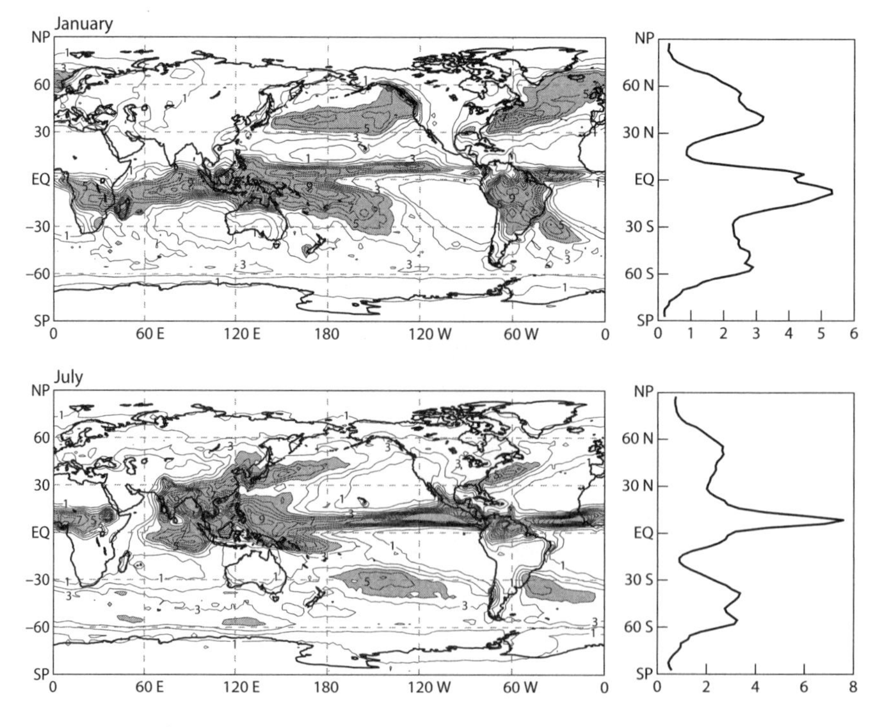

Figure 5.5. Maps of the January and July precipitation rate, from the Global Precipitation Climatology Center (Adler et al., 2003). The contour interval is 1 mm day^{-1}. Values greater than 4 mm day^{-1} are shaded.

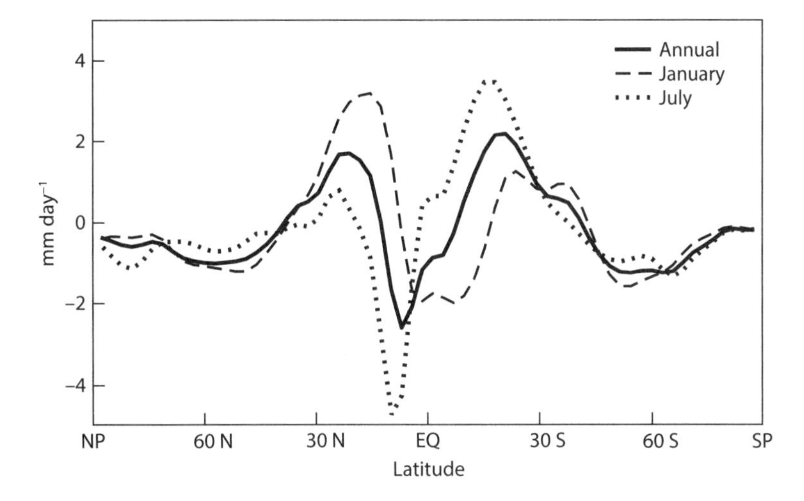

Figure 5.6. The zonally averaged difference between evaporation and precipitation, for January and July, and the annual mean.

to the descending branch of the Hadley circulation. The subsidence carries dry air downward and suppresses precipitation. Secondary precipitation maxima occur in the middle latitudes and are associated with winter cyclones and summer monsoons. The warm currents off the east coast of North America and Asia receive heavy precipitation mainly in January. The northwestern portion of the United States receives heavy precipitation in January but not in July.

Regions that receive plentiful precipitation throughout the year include eastern North America, the extreme southern tip of South America, and England.

Figure 5.6 shows the zonally averaged difference between evaporation and precipitation, for January and July. This is the input needed on the right-hand side of (15). Evaporation greatly exceeds precipitation in the subtropics, so in those regions the Earth's surface is a source of total water. In the tropical rain band and the midlatitude storm tracks, precipitation exceeds evaporation.

We now multiply both sides of (15) by $2\pi a$ and integrate with respect to latitude from the South Pole to an arbitrary latitude φ to obtain

$$\boxed{2\pi a \cos\varphi \int_0^\infty \overline{\rho_\theta v q_T}^{\lambda,t} d\theta = 2\pi a^2 \int_{-\frac{\pi}{2}}^{\varphi} \left[\overline{(F_{q_T})_S}^{\lambda,t} - \overline{P}^{\lambda,t}\right] \cos\varphi' d\varphi'}. \tag{17}$$

The quantity on the left-hand side of (17) is the total northward transport of moisture across latitude φ. To obtain (17), we used the boundary condition that the northward transport is zero at the South Pole. Compare (17) with (2.9).

We used the same method in chapter 2 to determine the northward transport of energy by the atmosphere and ocean combined, based on observations of the Earth's radiation budget. For reasons discussed in chapter 2, we corrected $\overline{(F_{q_T})_S}^t - \overline{P}^t$, that is, the time-averaged difference between evaporation and precipitation, to have a global mean of zero before performing the integration in (17). The results are shown in figure 5.7. The tropical and subtropical latent energy transport is toward the equator; that is, it is the "wrong way" to contribute to the poleward energy transport needed to satisfy planetary energy balance. The explanation is that most of the water vapor is in the lower troposphere, where the Hadley cells carry it toward the equator. The

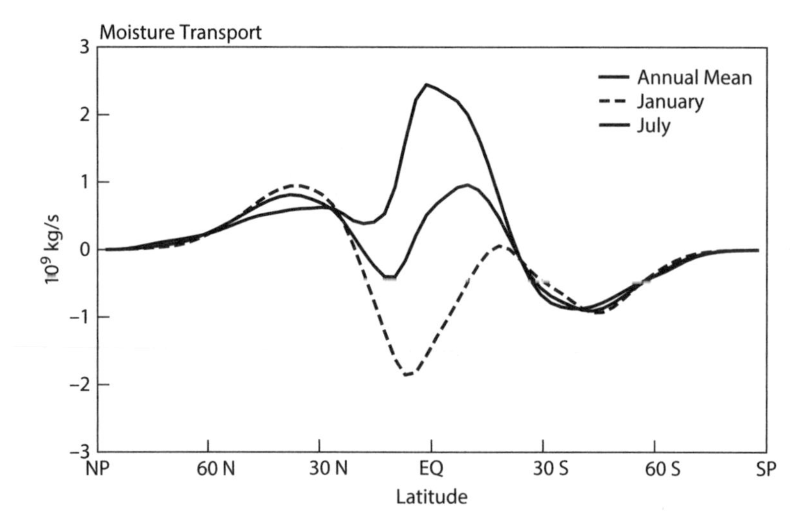

Figure 5.7. The vertically integrated meridional transport of water vapor, for January, July, and the annual mean.

midlatitude transport of water vapor (and latent energy) is poleward, but its magnitude is limited by cold winter temperatures.

The ECMWF reanalysis includes the vertically integrated northward transport of both water vapor and liquid water as products. We could have plotted them in figure 5.7, but we chose to show the results of integrating (17) to illustrate the concept.

The curves shown in figure 5.7 should not be interpreted in terms of cause and effect. Evaporation can "push" moisture from the subtropics toward the precipitation centers, or precipitation can "pull" it, or both can happen. More information is needed to interpret what is going on.

With this preparation, we now examine the vertical distribution of the zonally averaged meridional moisture flux, $\overline{\rho_\theta v q_T}^\lambda$, using isentropic coordinates. The flux is the product of three quantities: the pseudodensity, the meridional wind, and the total water mixing ratio. This product can be expanded to show four contributions, as follows:

$$\begin{aligned}\rho_\theta v q_T &= \left[\overline{\rho_\theta v}^\lambda + (\rho_\theta v)^*\right]\left(\overline{q_T}^\lambda + q_T^*\right) \\ &= \overline{\rho_\theta v}^\lambda\,\overline{q_T}^\lambda + \overline{\rho_\theta v}^\lambda q_T^* + (\rho_\theta v)^*\,\overline{q_T}^\lambda + (\rho_\theta v)^* q_T^*.\end{aligned} \tag{18}$$

Zonal averaging of (18) causes two of the four terms to drop out:

$$\overline{\rho_\theta v q_T}^\lambda = \overline{\rho_\theta v}^\lambda\,\overline{q_T}^\lambda + \overline{(\rho_\theta v)^* q_T^*}^\lambda. \tag{19}$$

Substituting from (7), we obtain

$$\boxed{\overline{\rho_\theta v q_T}^\lambda = \left(\overline{\rho_\theta}^\lambda\,\overline{v}^\lambda + \overline{\rho_\theta^* v^*}^\lambda\right)\overline{q_T}^\lambda + \overline{(\rho_\theta v)^* q_T^*}^\lambda}. \tag{20}$$

Equation (20) shows that the zonally averaged total moisture, $\overline{q_T}^\lambda$, is advected by both the mean mass flux and the bolus mass flux (Gent and McWilliams, 1990). The third term of (20) can be interpreted as the meridional eddy flux of total water, because the mass flux involved, that is, $(\rho_\theta v)^*$, has a mean value of zero, and the term vanishes if

q_T is zonally uniform. We adopt the following shorthand terminology for the ensuing discussion:

$\overline{\rho_\theta v q_T}^\lambda$ is the *total flux*;
$(\overline{\rho_\theta}^\lambda)(\overline{v}^\lambda)(\overline{q_T}^\lambda)$ is the *mean flux*;
$(\overline{\rho_\theta^* v^*}^\lambda)\overline{q_T}^\lambda$ is the *bolus flux*; and
$\overline{(\rho_\theta v)^* q_T^*}^\lambda$ is the *eddy flux*.

The bolus flux of moisture does involve eddies, because it is proportional to the bolus mass flux, $\overline{\rho_\theta^* v^*}^\lambda$, but it does not involve the eddy part of the total water, q_T^*.

Substituting (20) into (14), we find that

$$\frac{1}{a\cos\varphi}\frac{\partial}{\partial\varphi}\left\{\left[\overline{\left(\overline{\rho_\theta}^\lambda\overline{v}^\lambda+\overline{\rho_\theta^* v^*}^\lambda\right)\overline{q_T}^\lambda}^t+\overline{(\rho_\theta v)^* q_T^*}^{\lambda,t}\right]\cos\varphi\right\} + \frac{\partial}{\partial\theta}\left(\overline{\rho_\theta\dot{\theta}q_T}^{\lambda,t}+\overline{F_{q_T}}^{\lambda,t}-\overline{P}^{\lambda,t}\right)=0. \tag{21}$$

Before proceeding, we make one adjustment to (21). Let q_{GM} be the mass-weighted average of q_T over the entire atmosphere. We multiply the temporally and zonally averaged continuity equation, (8), by q_{GM} and subtract the result from (21) to obtain

$$\frac{1}{a\cos\varphi}\frac{\partial}{\partial\varphi}\left\{\left[\overline{\left(\overline{\rho_\theta}^\lambda\overline{v}^\lambda+\overline{\rho_\theta^* v^*}^\lambda\right)\left(\overline{q_T}^\lambda-q_{GM}\right)}^t+\overline{(\rho_\theta v)^* q_T^*}^{\lambda,t}\right]\cos\varphi\right\} + \frac{\partial}{\partial\theta}\left[\overline{\rho_\theta\dot{\theta}(q_T-q_{GM})}^{\lambda,t}+\overline{F_{q_T}}^{\lambda,t}-\overline{P}^{\lambda,t}\right]=0. \tag{22}$$

Equation (22) involves both meridional and vertical fluxes of the zonally averaged difference $\overline{q_T}^\lambda - q_{GM}$, that is, the difference of $\overline{q_T}^\lambda$ from its global mean. Think of $\overline{q_T}^\lambda$ as the sum of q_{GM} and $\overline{q_T}^\lambda - q_{GM}$. The mass circulation carries both q_{GM} and $\overline{q_T}^\lambda - q_{GM}$. Any redistribution of q_{GM} by the global mass circulation has no effect on the distribution of total moisture, simply because q_{GM} is spatially constant by definition. For this reason, the flux of $\overline{q_T}^\lambda - q_{GM}$ is more informative than the flux of $\overline{q_T}^\lambda$. A similar approach will be used with the fluxes of moist static energy and angular momentum, later in this chapter.

Incorporating the terminology introduced earlier, figure 5.8 shows the total meridional moisture flux, the mean flux, the bolus flux, and the eddy flux, for both January and July. The moisture flux is appreciable only in the lower half of the troposphere, because the cold air aloft cannot contain much water vapor. In each hemisphere, the total moisture flux diverges from the subtropics and converges into both the tropical and midlatitude precipitation maxima, as expected. The mean flux dominates in the tropics. It is equatorward in the lower troposphere, and poleward aloft. The bolus flux is weak. The eddy flux carries moisture poleward, that is, from more humid regions to drier regions.

The form of (21) allows us to define a streamfunction for the circulation of total water. We use the observed total flux of vapor to evaluate ψ_{q_T}, the streamfunction associated with the atmospheric circulation of total water. Here ψ_{q_T} is defined by

$$\overline{\rho_\theta v(q_T-q_{GM})}^{\lambda,t}2\pi a\cos\varphi \equiv -\frac{\partial\psi_{q_T}}{\partial\theta}, \tag{23}$$

and

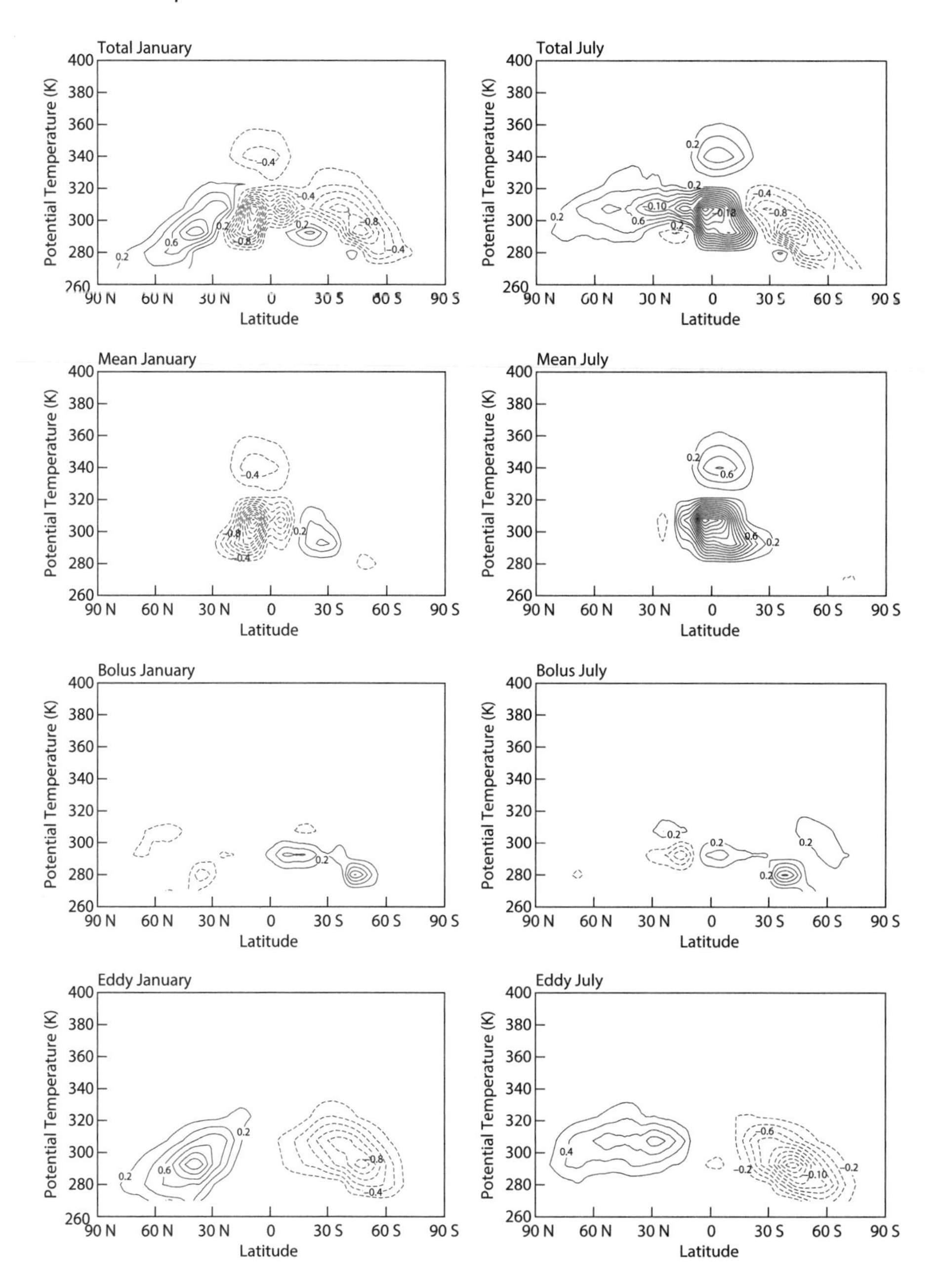

Figure 5.8. Isentropic meridional total water mass fluxes for January (left) and July (right). The top row shows the total flux, the second row shows the flux associated with the zonally averaged meridional wind, the third row shows the bolus flux, and the bottom row shows the eddy flux. The contour interval is 0.2 kg K^{-1} m^{-1} s^{-1}.

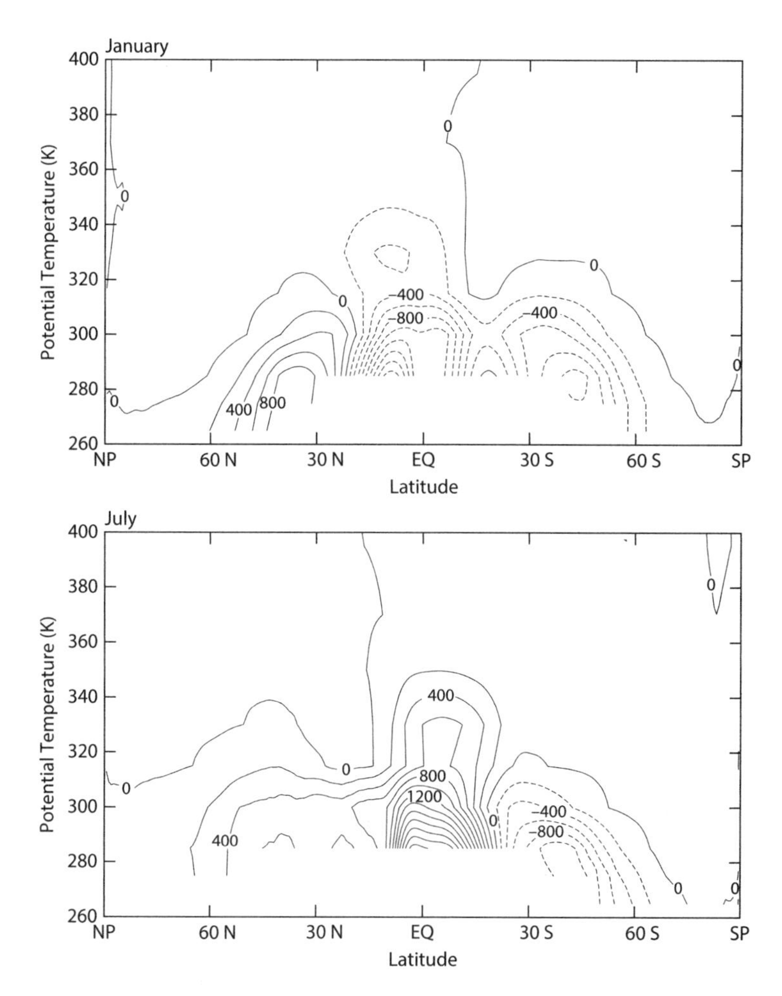

Figure 5.9. The streamfunction associated with the atmospheric circulation of total water, as seen in isentropic coordinates. The contour interval is 10^6 kg s^{-1}. The cut-off region at the bottom of the figure is "underground."

$$\left[\overline{\rho_\theta \dot{\theta}(q_T - q_{GM})}^{\lambda,t} + \overline{F}_{q_T}^{\lambda,t} - \overline{P}^{\lambda,t}\right] 2\pi a^2 \cos\varphi \equiv \frac{\partial \psi_{q_T}}{\partial \varphi}. \tag{24}$$

We can set $\psi_{q_T} = 0$ at the top of the atmosphere, because there is no vertical flux of moisture there.

The vertical transport of total water is very different from that of water vapor, because of the contributions from precipitating liquid water and ice. In contrast, the horizontal transport of total water is almost equal to the horizontal transport of water vapor, simply because the amount of water vapor in the air is about 100 times larger than the amount of liquid water and ice. We can therefore use the observed distribution of water vapor to obtain a good approximation to the meridional flux of total water. That is how we evaluate the left-hand side of (23).

Figure 5.9 shows contour plots of ψ_{q_T} for January and July. There is very little poleward moisture transport in the arid upper branches of the Hadley cells. Consider 10°

N in January or 10° S in July. As we integrate down from the top of the atmosphere to compute the streamfunction using (23), the first strong moisture flux that we encounter is toward the equator, in the lower half of the troposphere. The streamfunction plots tell us that moisture enters the atmosphere via surface evaporation in the subtropics, flows across the equator, and then leaves the atmosphere via precipitation in the tropics of the opposite hemisphere. This is a clockwise circulation for January and a counterclockwise circulation for July, as shown by the contours of ψ_{q_T}.

As discussed in chapter 6, the rising branch of the Hadley circulation contains very strong upward moisture transport by turbulence and cumulus convection, and very strong downward moisture transport by precipitation. These vertical fluxes extend through almost the entire depth of the troposphere. Figure 5.9 does not show closely spaced vertical contours of the moisture streamfunction extending through the troposphere, however. This indicates that the most vigorous upward moisture fluxes due to turbulence and convection are very nearly balanced by the downward flux due to precipitation.

Moist Static Energy Transports

We now present a very similar analysis for the meridional transport of moist static energy. As a reminder, we showed in chapter 4 that the moist static energy is approximately conserved under both moist and dry adiabatic processes, even when precipitation is occurring. Although the moist static energy is not equal to the total energy, the transport of moist static energy is very nearly the equal to the transport of total energy.

Figure 5.10 shows the latitude-height distribution of moist static energy. In the tropics and to some extent in the middle latitudes of the summer hemisphere, the moist static energy decreases with height in the lower troposphere. As mentioned in chapter 4, the dry static energy normally increases with altitude, but in the lower troposphere, where water vapor is plentiful, especially in the tropics, the decrease of the water vapor mixing ratio with height overwhelms the increase of s, so that the moist static energy decreases with altitude. Above the middle troposphere, the water vapor mixing ratio is so small that the moist static energy is nearly equal to the dry static energy; therefore, the moist static energy increases with height in the upper troposphere. For this reason the moist static energy has a minimum in the tropical middle troposphere, as shown in figure 5.10. The midtropospheric minimum is important and interesting, for reasons discussed in chapter 6.

In isentropic coordinates, the conservation of moist static energy is expressed by

$$\frac{\partial}{\partial t}(\rho_\theta h) + \nabla \cdot (\rho_\theta \mathbf{V} h) + \frac{\partial}{\partial \theta}(\rho_\theta \dot{\theta} h + F_h + R) = 0. \tag{25}$$

As a reminder, F_h and R are the (positive upward) fluxes of moist static energy due to small-scale motions and radiation, respectively. Averaging over longitude and time, we obtain

$$\frac{1}{a\cos\varphi}\frac{\partial}{\partial \varphi}\left(\overline{\rho_\theta v h}^{\lambda,t}\cos\varphi\right) + \frac{\partial}{\partial \theta}\left(\overline{\rho_\theta \dot{\theta} h}^{\lambda,t} + \overline{F_h}^{\lambda,t} + \overline{R}^{\lambda,t}\right) = 0. \tag{26}$$

Finally, we vertically integrate (26) through the entire atmospheric column:

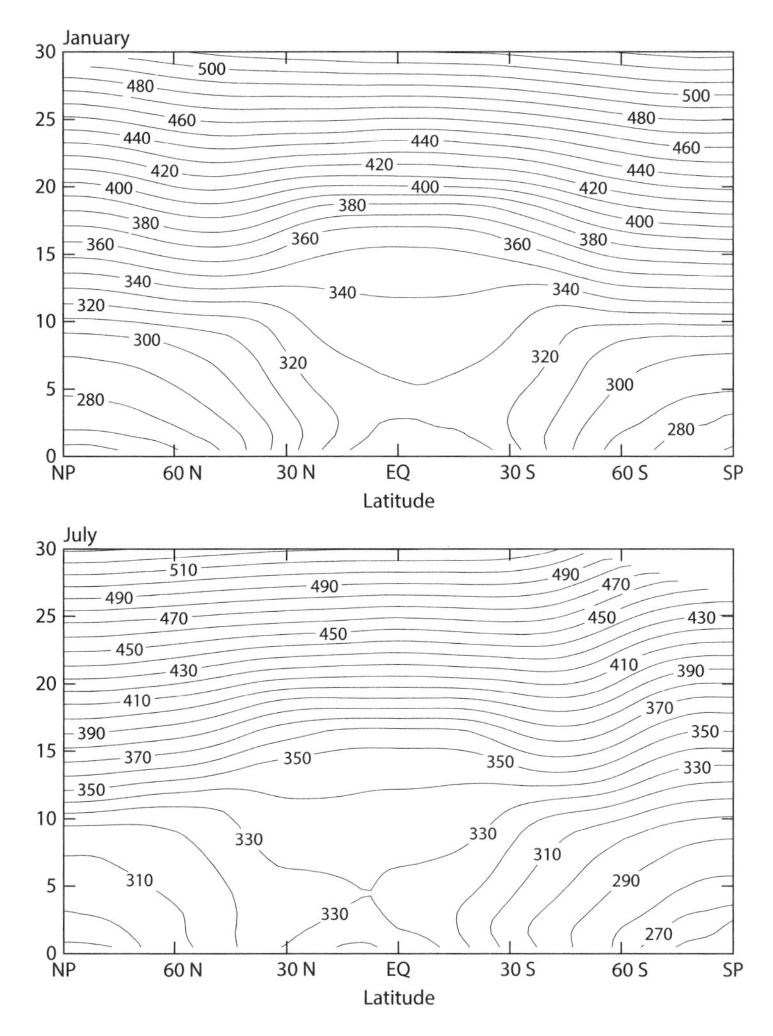

Figure 5.10. The observed latitude-height distribution of the zonally averaged moist static energy (in kJ kg^{-1}), for January and July, as analyzed by ECMWF.

$$\frac{\partial}{\partial\varphi}\left(\int_0^\infty \overline{\rho_\theta v h}^{\lambda,t} d\theta \cos\varphi\right) = a\cos\varphi\left(\overline{F}_{hS}^{\lambda,t} + \overline{R}_S^{\lambda,t} - \overline{R}_\infty^{\lambda,t}\right). \tag{27}$$

Here we used $(\rho_\theta)_{\theta=0} = (\rho_\theta)_{\theta\to\infty} = 0$, and $(F_h)_{\theta\to\infty} = 0$.

The moist static energy flux due to small-scale motions, F_h, is the sum of the *sensible heat flux* and the water vapor flux. The sensible heat flux is the small-scale flux of dry enthalpy; roughly speaking, it is a flux of temperature. Its surface value satisfies a bulk aerodynamic formula of the form

$$(F_s)_S = c_p \rho_S c_T |\mathbf{V}_S| (T_g - T_a). \tag{28}$$

Here c_T is a nondimensional transfer coefficient (e.g., Stull, 1988). The dimensions of $(F_s)_S$ are energy per unit area per unit time. The subscript g denotes a value representative of the lower boundary (ground or ocean), and the subscript a represents a value representative of a level inside the atmosphere but near the surface. As can be

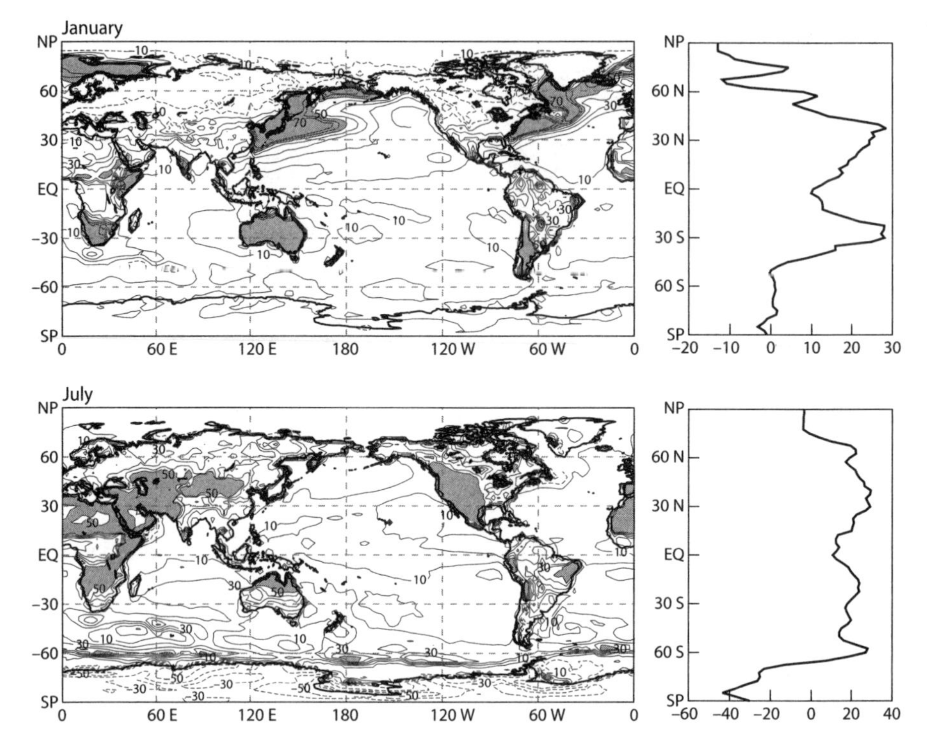

Figure 5.11. Maps of the surface sensible heat flux, for January and July, based on ECMWF reanalyses. The contour interval is 10 W m^{-2}. Values higher than 50 W m^{-2} are shaded. Note that the contour interval and shading convention are different from those of figure 5.4. The corresponding zonal averages are shown on the right. These are not real observations, although they are influenced by observations.

seen from (28), the surface sensible heat flux is upward when the ground or ocean is warmer than the air. It tends to cool the ground and warm the air, thus trying to put itself out of business. The absorption of solar radiation by the ground is an example of a process that can maintain $(F_s)_S$ upward.

Figure 5.11 shows maps and zonal averages of the sensible heat flux for January and July. The largest values occur over the midlatitude oceans in winter, near the eastern coasts of the continents. These strong sensible heat fluxes are associated with fast currents of cold air moving from the cold continents out over warm ocean currents. Equation (28) shows that strong sensible heat fluxes are to be expected under such conditions. Large values also occur over the summer and tropical continents, especially where the surface is dry and lacking in vegetation—for example, over the Sahara desert. There are no large negative values of the surface sensible heat flux, because a downward sensible heat flux tends to damp the turbulence; the buoyancy force does not "want" warm air to move downward or cold air to move upward.

The surface moist static energy flux can be obtained by adding the surface sensible heat flux to the surface latent heat flux; the result is shown in figure 5.12.

Figure 5.13 shows maps of $-[(\overline{R}^{\lambda,t})_S - (\overline{R}^{\lambda,t})_\infty]$, which can be called the net *atmospheric radiative cooling* (ARC). The ARC is strongest in the winter hemisphere. In northern summer, the zonally averaged ARC is almost bimodal, with a near discontinuity close to the equator, owing to the strong upwelling infrared radiation from the Northern Hemisphere continents. The strongest cooling tends to occur in

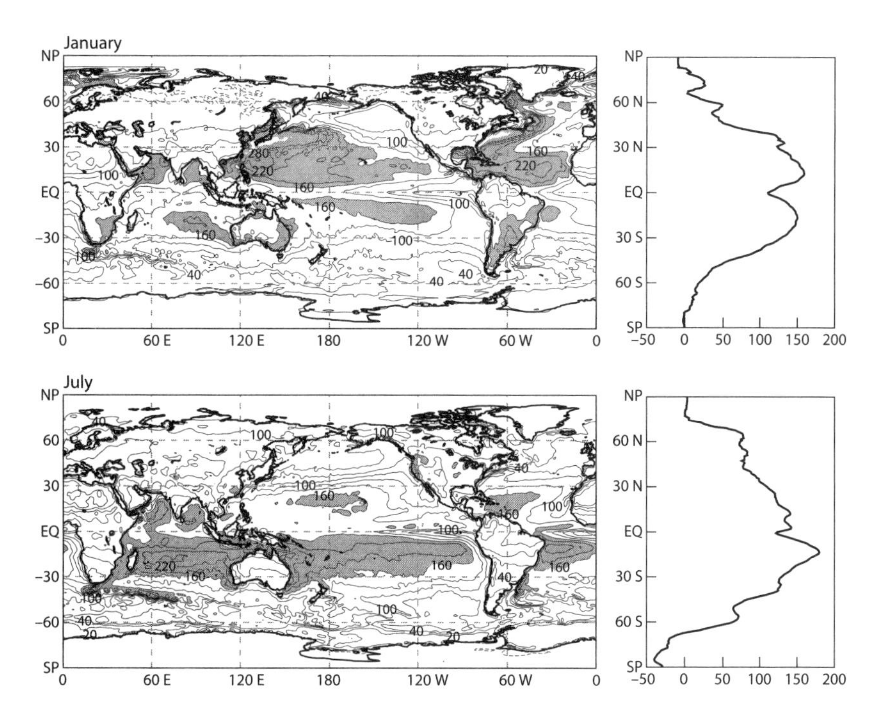

Figure 5.12. Maps of the surface moist static energy flux, for January and July, based on ECMWF reanalyses. The contour interval is 20 W m^{-2}. Values higher than 150 W m^{-2} are shaded. The corresponding zonal averages are shown on the right. These are not real observations, although they are influenced by observations.

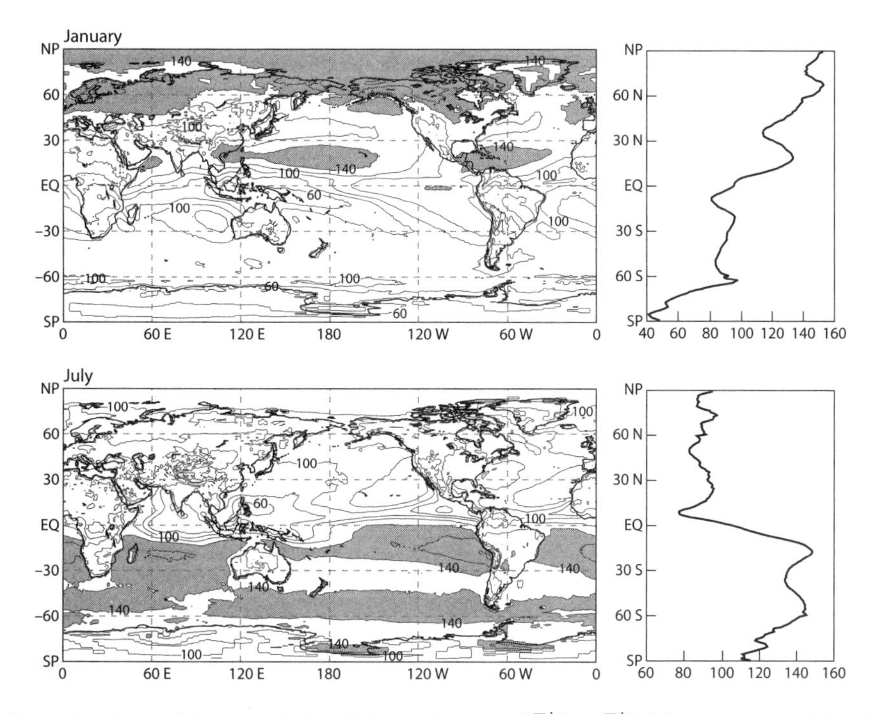

Figure 5.13. Maps of the atmospheric radiative cooling rate, $-[(\overline{R}^{\lambda,t})_S - (\overline{R}^{\lambda,t})_\infty]$, for January and July, based on satellite data (Wielicki et al., 1996, 1998). The corresponding zonal averages are shown on the right. The contour interval is 20 W m^{-2}; values larger than 140 W m^{-2} are shaded.

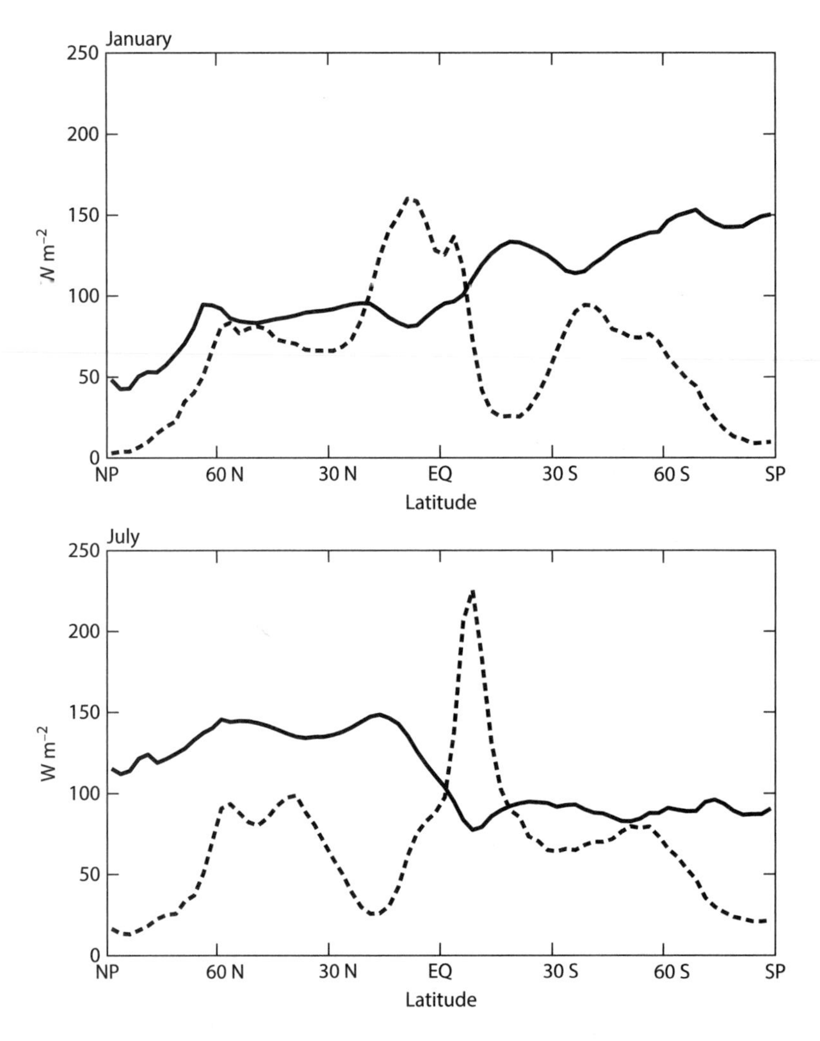

Figure 5.14. The zonally averaged atmospheric radiative cooling rate (solid lines) and the zonally averaged precipitation rate expressed as a vertically integrated latent heating rate (dashed lines), for January and July. They are anticorrelated.

regions of little precipitation, for example, over the subtropical oceans of the winter hemisphere. As discussed in chapter 2, the *globally averaged* net radiative cooling of the atmosphere is nearly proportional to the globally averaged precipitation rate, but *locally* the radiative cooling rate is negatively correlated with the precipitation rate, because the high clouds associated with precipitating weather systems reduce the outgoing longwave radiation at the top of the atmosphere. Figure 5.14 shows that, especially in the tropics, the zonally averaged atmospheric radiative cooling rate tends to be large where the zonally averaged precipitation rate is small, and vice versa.

To compute the vertically integrated northward transport of energy by the atmosphere, we have to sum the terms on the right-hand side of (27); that is, $(\overline{F}_h^{\lambda,t})_S + (\overline{R}^{\lambda,t})_S - (\overline{R}^{\lambda,t})_\infty$. We used CERES data for $(\overline{R}^{\lambda,t})_\infty$ and the ECMWF reanalysis for $(\overline{F}_h^{\lambda,t})_S + (\overline{R}^{\lambda,t})_S$. With this approach, the total (atmosphere plus ocean) transport is

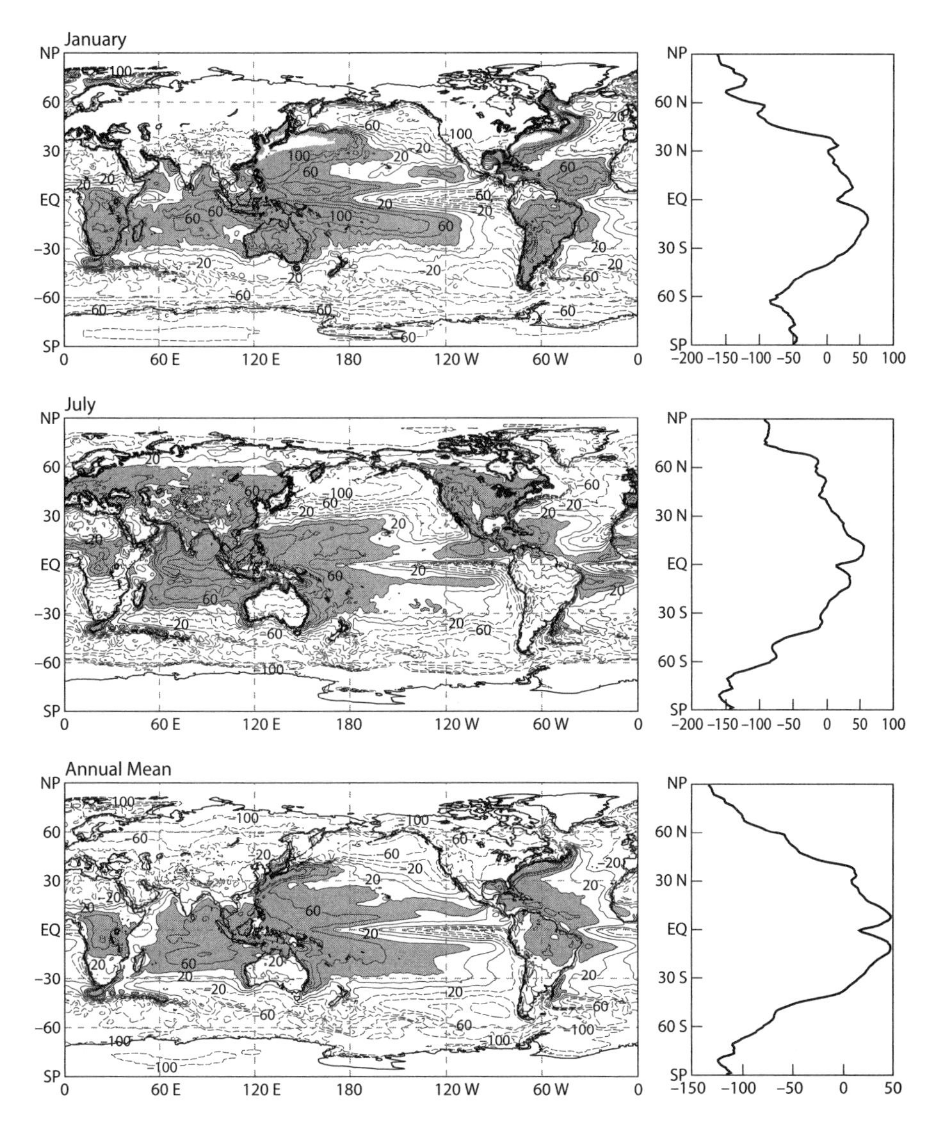

Figure 5.15. Maps of the net source of moist static energy due to fluxes at the surface and the top of the atmosphere, $(\overline{F}_h^{\lambda,t})_S + (\overline{R}^{\lambda,t})_S - (\overline{R}^{\lambda,t})_T$, for January, July, and the annual mean. The contour interval is 20 W m^{-2}; values larger than 40 W m^{-2} are shaded. The corresponding zonal averages are shown on the right. These are not real observations, although they are influenced by observations.

determined entirely from the CERES data, as it was in chapter 2; the ocean transport is determined entirely from the ECMWF reanalysis; and the atmospheric transport is based on data from both sources.

Figure 5.15 shows the inferred net source of moist static energy for the atmosphere. The tropical atmosphere experiences a source of moist static energy, owing primarily to the surface moist static energy flux, and the polar regions experience a sink owing to radiative cooling.

We now multiply both sides of (27) by $2\pi a$ and integrate with respect to latitude from the South Pole to an arbitrary latitude φ to obtain

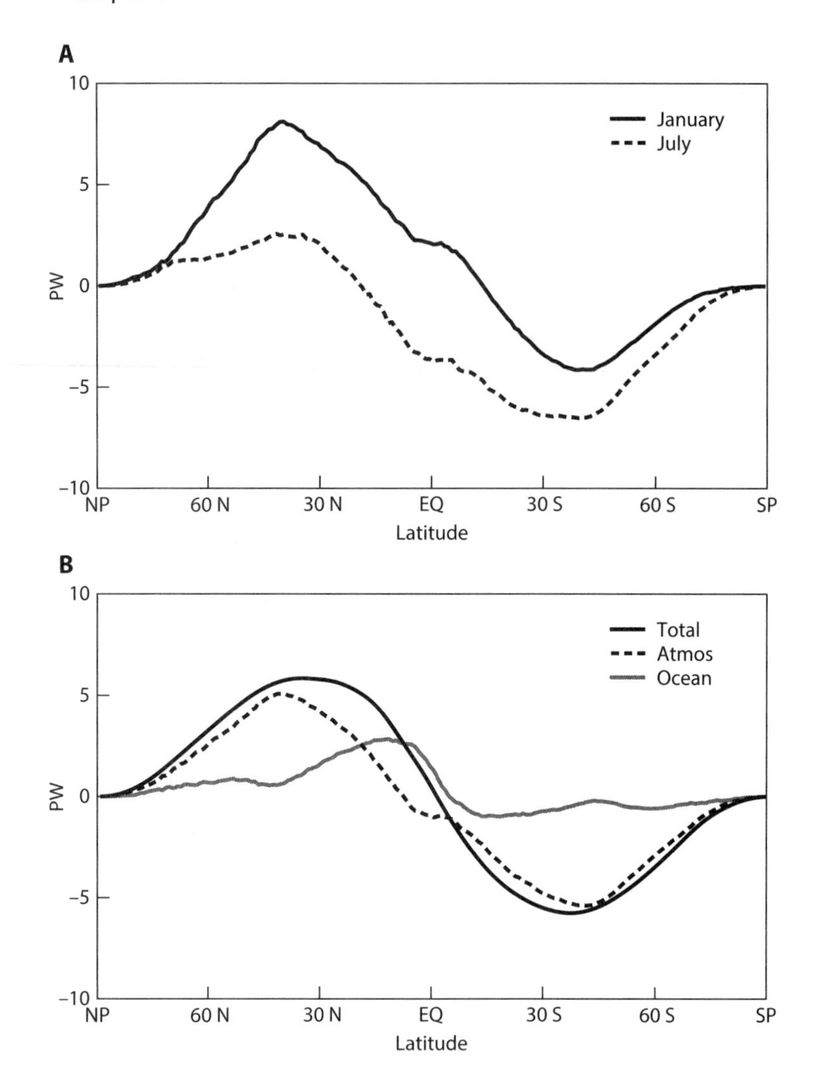

Figure 5.16. (A) The vertically integrated northward transport of moist static energy by the atmosphere, for January and July. (B) The annual-mean vertically integrated northward transport of moist static energy by the atmosphere. Also shown for comparison are the total transport by the atmosphere-ocean system, which was discussed in chapter 2, and the difference between the total transport and the atmospheric transport, which can be interpreted as the transport by the oceans.

$$\boxed{2\pi a \cos\varphi \int_0^\infty \overline{\rho_\theta v h}^{\lambda,t} d\theta = 2\pi a^2 \int_{\frac{\pi}{2}}^{\varphi} \left[(\overline{F}_h^{\lambda,t})_S + (\overline{R}^{\lambda,t})_S - (\overline{R}^{\lambda,t})_\infty \right] \cos\varphi' d\varphi'}. \quad (29)$$

The quantity on the left-hand side of (29) is the total northward transport of moist static energy across latitude φ. Compare (29) with (17). As usual, we correct $\left[(\overline{F}_h^{\,t})_S + (\overline{R}^{\,t})_S - (\overline{R}^{\,t})_\infty\right]$ to produce a global mean of zero, and we set the northward transport to zero at the South Pole.

The total northward transport of moist static energy by the atmosphere is shown in the upper panel of figure 5.16, for January and July. Here we choose to show the ECMWF reanalysis product for the northward transport of total energy by the atmosphere, instead of the results of the integration of (29). The northward transport by the atmosphere increases toward the north from about 40° S to 40° N; that is, it diverges, implying that the atmosphere is exporting moist static energy from this wide band of latitudes. The poleward transport shows a noticeable seasonal cycle in the Northern Hemisphere, with the strongest transport in the northern winter, but the Southern Hemisphere transport shows relatively little seasonal change.

The lower panel of figure 5.16 shows the *annual mean* energy transport by the oceans and atmosphere combined, computed from satellite data exactly as in chapter 2. Also shown is the annual mean northward energy transport by the atmosphere, based on the ECMWF analysis. Finally, the ocean energy transport is shown as the difference between the total transport and the atmospheric transport. The ocean energy transport is smaller than that by the atmosphere, but not negligible, especially in the Northern Hemisphere. The transport is difficult to determine directly from measurements of the ocean's temperature and currents, because there are insufficient data. Results analogous to those of figure 5.16 have also been presented by Vonder Haar and Oort (1973), Oort and Vonder Haar (1976), Carissimo et al. (1985), Savijärvi (1988), and Masuda (1988). Trenberth and Caron (2001) provide a much more detailed discussion.

Figure 5.17 shows the total, mean, bolus, and eddy meridional moist static energy fluxes. By analogy with our earlier discussion of the meridional moisture fluxes, the meridional moist static energy fluxes are expressed in terms of the difference $h - h_{GM}$, where h_{GM} is the mass-weighted average of h over the entire atmosphere. The mean, bolus, and eddy fluxes all make important contributions to the total flux. The eddy flux of moist static energy is almost entirely due to the eddy flux of latent heat (i.e., water vapor), because dry static energy fluctuations on isentropic surfaces are small. The total moist static energy flux converges strongly on the 300 K surface in the middle latitudes of the winter hemisphere.

We can define a streamfunction, ψ_h, for the temporally and zonally averaged transport of moist static energy, using

$$\overline{\rho_\theta v(h - h_{GM})}^{\lambda,t} 2\pi a \cos\varphi \equiv -\frac{\partial \psi_h}{\partial \theta}, \tag{30}$$

and

$$\left[\overline{\rho_\theta \dot{\theta}(h - h_{GM})}^{\lambda,t} + \overline{F_h}^{\lambda,t} + \overline{R}^{\lambda,t}\right] 2\pi a^2 \cos\varphi \equiv \frac{\partial \psi_h}{\partial \varphi}. \tag{31}$$

There is a new twist, however. When we defined streamfunctions for the circulations of dry air and moisture, we set them to zero at the top of the atmosphere, on the grounds that there is no vertical mass flow there. We cannot do this with ψ_h, because radiation carries a flux of energy at the top of the atmosphere. We can write the required upper boundary condition as

$$R_\infty 2\pi a^2 \cos\varphi \equiv \left(\frac{\partial \psi_h}{\partial \varphi}\right)_\infty, \tag{32}$$

where R_∞ is the net radiation at the top of the atmosphere. As discussed in chapter 2, the global mean of R_∞ is expected to be close to zero. When it is zero, (32) implies that $(\psi_h)_\infty$ takes the same value at both poles. An additional condition is needed to determine the constant of integration, however. A simple possibility, used here, is to choose the constant so that the cosine-weighted meridional average of $(\psi_h)_\infty$ is zero.

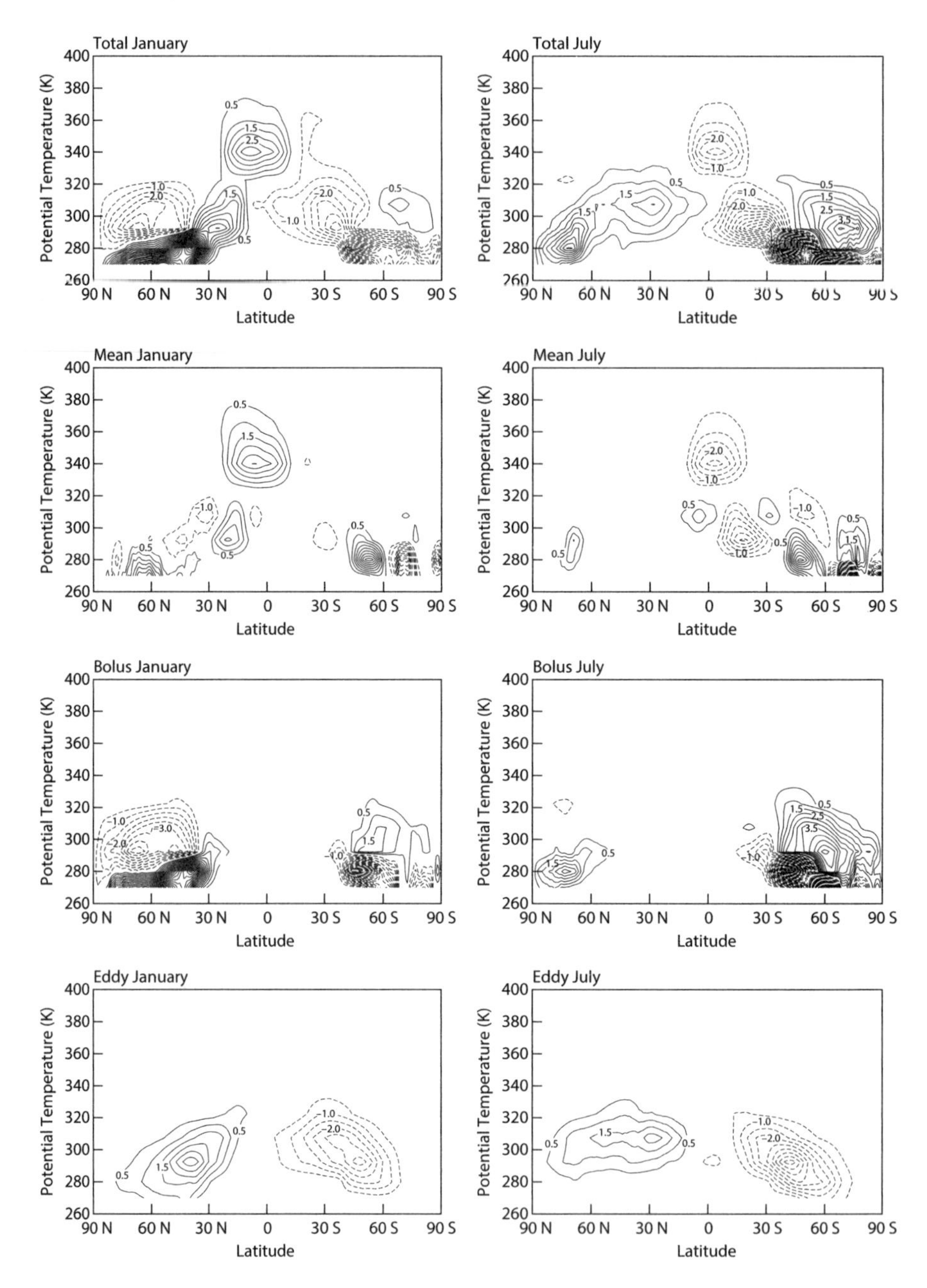

Figure 5.17. Isentropic meridional moist static energy mass fluxes for January (left) and July (right). The top row shows the total flux, the second row shows the flux associated with the zonally averaged meridional wind, the third row shows the bolus flux, and the bottom row shows the eddy flux. The contour interval is 0.5×10^6 W K^{-1} m^{-1}.

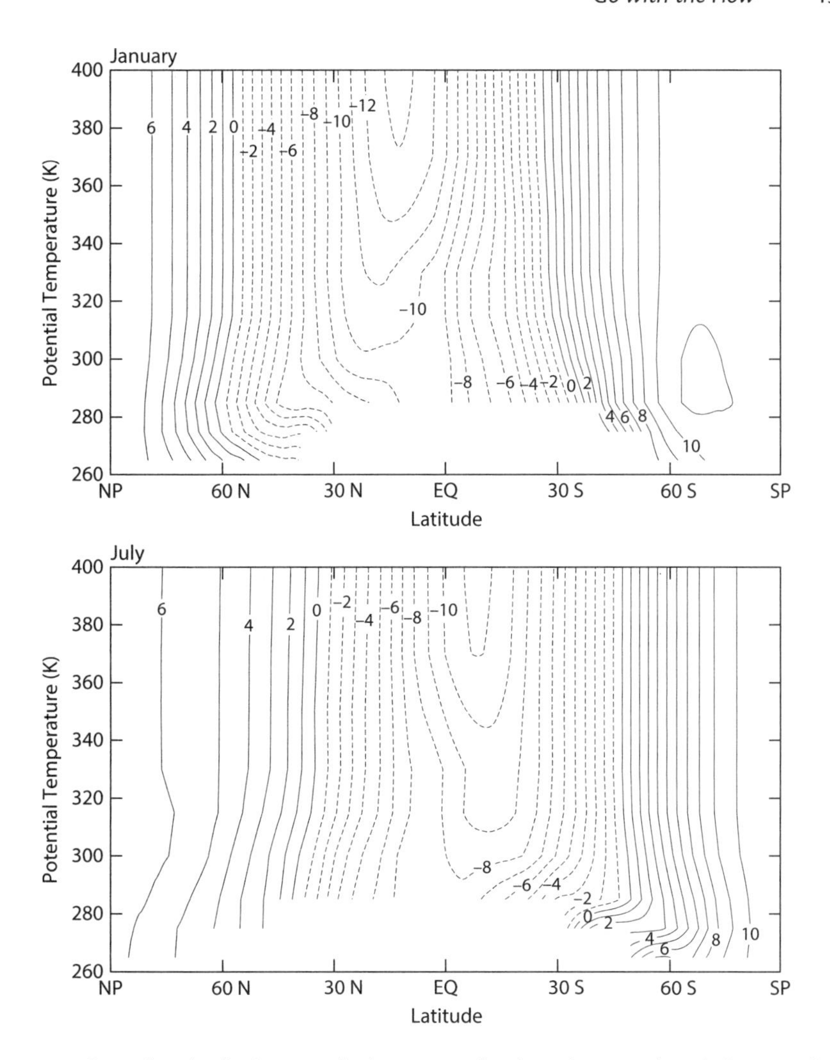

Figure 5.18. Streamfunction for the atmospheric transport of moist static energy through the atmosphere, for January, July, and the annual mean. The contour interval is 1 PW. The cut-off region at the bottom of the figure is "underground."

Figure 5.18 shows plots of ψ_h for January and July, and the annual mean. The upper boundary condition (32) strongly determines the solution. The isolines of the streamfunction are nearly vertical at all latitudes, indicating that the vertical flow of energy due to radiation is dominant.

Angular Momentum

As discussed in chapter 4, the component of the atmosphere's angular momentum vector that points in the direction of the Earth's axis of rotation is given by

$$M \equiv (\Omega a \cos\varphi + u) a \cos\varphi, \tag{33}$$

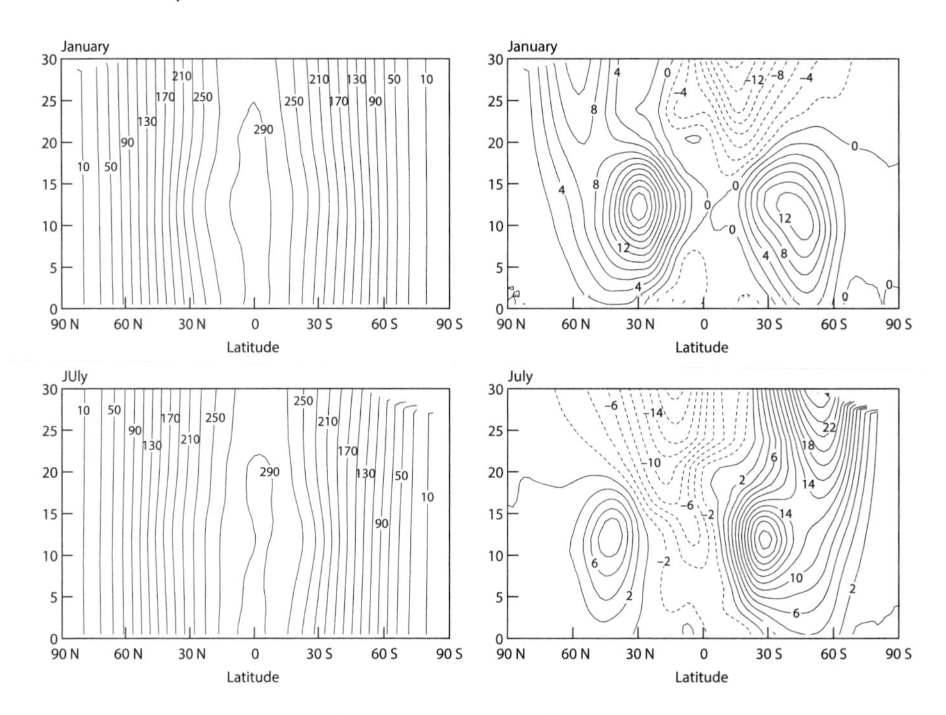

Figure 5.19. The observed absolute (left panels) and relative (right panels) atmospheric angular momentum per unit mass, for January and July. The units are 10^7 m^2 s^{-1}.

per unit mass. Here we used the Earth's radius as an approximation to the radial distance from the center of the Earth to a particle inside the atmosphere. The $\Omega a^2 \cos^2\varphi$ term of (30) represents the angular momentum due to the Earth's rotation, and the $ua\cos\varphi$ term represents the *relative* angular momentum due to the rotation of the atmosphere relative to the Earth's surface. Figure 5.19 shows the observed zonally averaged total angular momentum and relative angular momentum, both per unit mass, for January and July. The absolute angular momentum is generally about an order of magnitude larger than the relative angular momentum; it varies strongly with latitude, and only slightly with height. It is positive everywhere and takes its largest values near the equator. The plots of the zonally averaged *relative* angular momentum naturally resemble those of the zonally averaged zonal wind, which are presented in chapter 3. In the tropical upper troposphere, the absolute angular momentum contours tilt or bulge slightly toward the winter pole, in both seasons, which suggests some tendency for the air flowing poleward in the main Hadley cell to conserve its absolute angular momentum.

Early theories of the zonally averaged circulation (e.g., Held and Hou, 1980; Lindzen, 1990) were based on the assumption of conservation of angular momentum in the upper branch of the Hadley cell. As the air flows poleward in the upper branch of a Hadley cell, a tendency towards conservation of angular momentum implies a poleward increase of the westerlies, which are observed to take their maximum values in the jet streams near the latitudes where the Hadley circulations stop. These *subtropical jets* can thus be interpreted as consequences of the poleward flow of air in the Hadley circulations. They are sometimes called *Hadley-driven jets*. However, they are much weaker than they would be if angular momentum were truly conserved in the

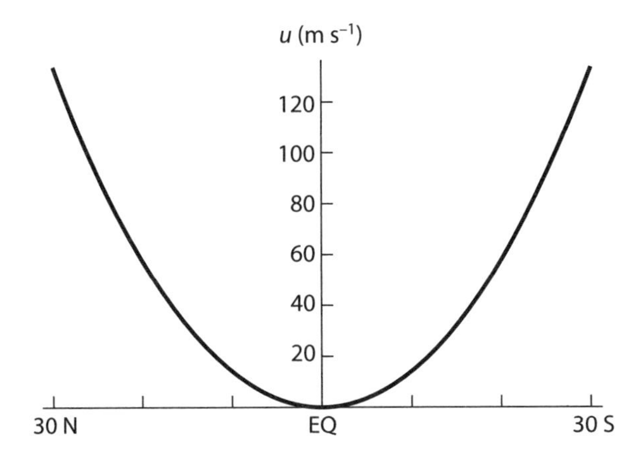

Figure 5.20. The hypothetical variation of the zonal wind with latitude under conservation of angular momentum, for the special case in which the zonal wind vanishes at the equator.

poleward branches of the Hadley cells. Figure 5.20 shows the zonal wind that would occur if a particle started on the equator with $u = 0$ and moved toward either pole while exactly conserving its angular momentum. Starting the particle from 10° off the equator would hardly change the plot. The zonal wind values shown at 30° north and south are much larger than those observed in the subtropical jets, which implies that there are important sinks of angular momentum in the upper branches of the Hadley cells. For further discussion, see the reviews by Held and Hoskins (1985) and Schneider (2006).

In isentropic coordinates, conservation of angular momentum is expressed by

$$\frac{\partial}{\partial t}(\rho_\theta M) + \nabla \cdot (\rho_\theta \mathbf{V} M) + \frac{\partial}{\partial \theta}(\rho_\theta \dot{\theta} M + F_u a \cos\varphi) = -\rho_\theta \frac{\partial s}{\partial \lambda}. \tag{34}$$

Here F_u is the vertical flux of zonal momentum due to small-scale eddies, and

$$\begin{aligned} s &\equiv c_p T + gz \\ &= \Pi\theta + gz \end{aligned} \tag{35}$$

is the dry static energy, where

$$\Pi \equiv c_p \left(\frac{p}{p_0}\right)^\kappa \tag{36}$$

is the Exner function. As a reminder, the term on the right-hand side of (34) represents the effects of the zonal pressure-gradient force. Applying a zonal average and a time average to (34), we obtain

$$\frac{1}{a\cos\varphi}\frac{\partial}{\partial \varphi}\left(\overline{\rho_\theta v M}^{\lambda,t}\cos\varphi\right) + \frac{\partial}{\partial \theta}\left(\overline{\rho_\theta \dot{\theta} M}^{\lambda,t} + \overline{F_u}^{\lambda,t} a\cos\varphi\right) = -\overline{\rho_\theta \frac{\partial s}{\partial \lambda}}^{\lambda,t}. \tag{37}$$

Referring to (33), we can think of $\overline{\rho_\theta v M}^{\lambda,t}$ as the sum of a flux associated with the *Earth angular momentum*, $\Omega(a\cos\varphi)^2$, and a flux associated with the relative angular momentum, $ua\cos\varphi$. The Earth angular momentum is independent of longitude and approximately independent of height and time. Because $\Omega(a\cos\varphi)^2$ is large, the flux of Earth angular momentum is large at individual levels, especially in the tropics.

Nevertheless, the *vertically integrated* flux of Earth angular momentum is negligible at every latitude in a sufficiently long time average, simply because, as discussed in chapter 3, the vertically integrated meridional mass flux is negligible in a sufficiently long time average.

The pressure-gradient term of (37) can be rewritten in a very interesting way, as follows. Using the hydrostatic equation in isentropic coordinates, that is,

$$\frac{\partial s}{\partial \theta} = \Pi, \tag{38}$$

we can write

$$\begin{aligned}
-\rho_\theta \frac{\partial s}{\partial \lambda} &= \frac{1}{g}\frac{\partial p}{\partial \theta}\frac{\partial s}{\partial \lambda} \\
&= \frac{1}{g}\frac{\partial}{\partial \theta}\left(p\frac{\partial s}{\partial \lambda}\right) - \frac{p}{g}\frac{\partial}{\partial \theta}\left(\frac{\partial s}{\partial \lambda}\right) \\
&= \frac{1}{g}\frac{\partial}{\partial \theta}\left(p\frac{\partial s}{\partial \lambda}\right) - \frac{p}{g}\frac{\partial}{\partial \lambda}\left(\frac{\partial s}{\partial \theta}\right) \\
&= \frac{1}{g}\frac{\partial}{\partial \theta}\left[p\frac{\partial}{\partial \lambda}(\Pi\theta + \phi)\right] - \frac{p}{g}\frac{\partial \Pi}{\partial \lambda} \\
&= \frac{\partial}{\partial \theta}\left(\theta\frac{p}{g}\frac{\partial \Pi}{\partial \lambda} + p\frac{\partial z}{\partial \lambda}\right) - \frac{p}{g}\frac{\partial \Pi}{\partial \lambda} \\
&= \theta\frac{\partial}{\partial \theta}\left[\frac{p}{g}\left(\frac{\partial \Pi}{\partial \lambda}\right)\right] + \frac{\partial}{\partial \theta}\left[p\frac{\partial z}{\partial \lambda}\right].
\end{aligned} \tag{39}$$

When we take the zonal mean of (39), the term $\theta(\partial/\partial\theta)[p/g(\partial\Pi/\partial\lambda)]$ on the bottom line drops out (as you will show in problem 2 at the end of this chapter), and we are left with

$$\boxed{-\overline{\rho_\theta \frac{\partial s}{\partial \lambda}}^{\lambda} = \frac{\partial}{\partial \theta}\left(\overline{p\frac{\partial z}{\partial \lambda}}^{\lambda}\right)}. \tag{40}$$

This relationship was worked out by Klemp and Lilly (1978). Equation (40) says that in the zonal mean, the effect of the *zonal* pressure-gradient force on the mass-weighted angular momentum can be written as the divergence of a *vertical* flux of angular momentum given by $\partial/\partial\theta(\overline{p(\partial z/\partial\lambda)}^{\lambda})$. An interpretation is given below. Substitution of (40) into (37) gives the zonally averaged angular momentum equation in the form

$$\frac{1}{a\cos\varphi}\frac{\partial}{\partial \varphi}(\overline{\rho_\theta v M}^{\lambda,t}\cos\varphi) + \frac{\partial}{\partial \theta}\left(\overline{\rho_\theta \dot{\theta} M}^{\lambda,t} + \overline{F_u}^{\lambda,t} a\cos\varphi - \overline{p\frac{\partial z}{\partial \lambda}}^{\lambda,t}\right) = 0. \tag{41}$$

Vertical integration through the entire atmospheric column gives

$$\frac{\partial}{\partial \varphi}\left(\int_0^\infty \overline{\rho_\theta v M}^{\lambda,t} d\theta \cos\varphi\right) = a\cos\varphi\left[(\overline{F_u}^{\lambda,t})_S a\cos\varphi - \overline{p_S\frac{\partial z_S}{\partial \lambda}}^{\lambda,t}\right]. \tag{42}$$

Here we used $(\rho_\theta)_{\theta=0} = (\rho_\theta)_{\theta\to\infty} = 0$, and $(F_u)_{\theta\to\infty} = 0$.

The term $\overline{p_S(\partial z_S/\partial\lambda)}^{\lambda,t}$ on the right-hand side of (42) represents the effects of *mountain torque*. It vanishes if either p_S or z_S is independent of λ. Figure 5.21 illustrates the mechanism that gives rise to mountain torque. Near a mountain range, we expect higher pressure on the upstream side and lower pressure on the downstream side, as shown in the figure. The spatial correlation shown in the figure need not occur at a particular moment with respect to a particular mountain range, but it does

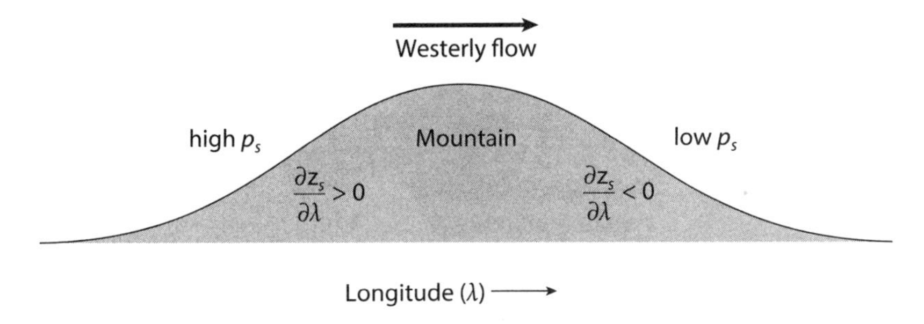

Figure 5.21. Sketch illustrating the concept of "mountain torque." High p_S occurs on the upstream side of the mountain, and low p_S on the downstream side.

hold in the time-averaged zonal mean. The pressure pattern shown in the figure gives rise to a "form drag" that is analogous to the aerodynamic drag on a car or other moving object due to relatively high pressure on the front and lower pressure on the back. The atmosphere and the mountain are exchanging momentum: the atmosphere tries to push the mountain downstream, and (as decreed by Newton) the atmosphere feels an equal and opposite force pushing it back in the upstream direction. As discussed later in this chapter, the force that the atmosphere exerts on the mountain can actually cause the Earth's rate of rotation to speed up or slow down slightly.

In problem 1 at the end of this chapter, you are asked to prove that the mountain torque would vanish if p_S were a function of z_S only, that is, if the data in figure 3.1 fell exactly onto a single curve. This would be expected in the absence of all dynamical processes, that is, if the atmosphere were "just sitting there," in a balanced, resting state.

Surface friction can be described using

$$(\mathbf{F_V})_S = -\rho_S c_D |\mathbf{V}_S| \mathbf{V}_S, \tag{43}$$

which is another bulk aerodynamic formula. The minus sign in (43) means that the near-surface momentum flux has a direction opposite the surface wind. For example, the zonal component of (43) is

$$(F_u)_S = -\rho_S c_D |\mathbf{V}_S| u_S, \tag{44}$$

which says that the flux of zonal momentum is negative, that is, downward, when the zonal component of the surface wind is positive, that is, westerly. Surface friction transfers positive angular momentum from the atmosphere to the ocean–solid Earth system in places where the surface winds are westerly, mostly in the middle latitudes of the winter hemisphere. Conversely, positive angular momentum flows back into the atmosphere in places where the surface winds are easterly, mostly in the tropics. The mountain torques generally have the same sign as the frictional torques.

Figure 5.22 shows maps of the observed magnitude of the surface wind stress over the oceans. These data were obtained using centimeter-wavelength radars on satellites, which can sense stress-induced centimeter-scale *capillary waves* on the sea surface. The retrieval algorithm works over water surfaces, but not over land. Unfortunately, there are no comparable observations of the wind stress over land. The plots in figure 5.22 show *the average of the magnitude* of the vector rather than the magnitude of the average of the vector. The surface stress is particularly strong in the storm-track regions. In chapter 3, we discussed the belt of low pressure and strong low-level

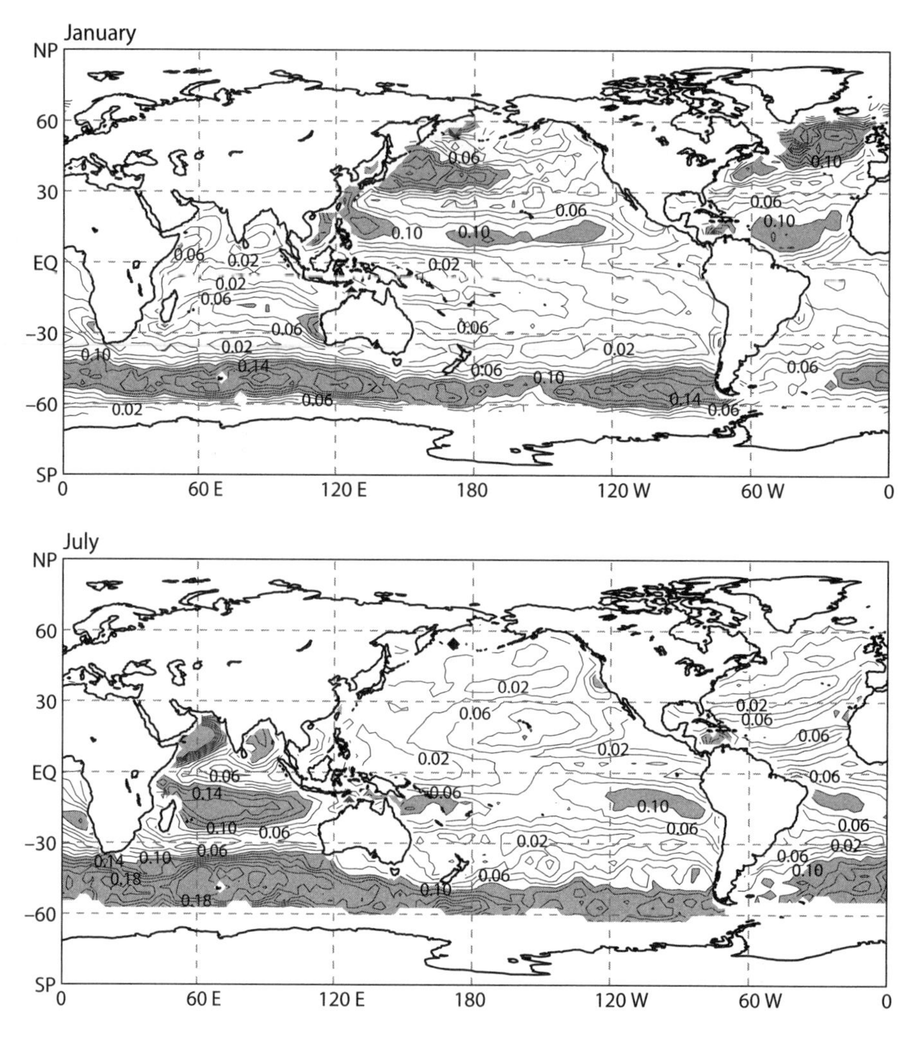

Figure 5.22. Maps of the magnitude of the surface wind stress ***over the oceans only***, based on scatterometer data from ERS-1 (e.g., Liu, 2002). The contour interval is 0.02 Pa. Values higher than 0.1 Pa are shaded.

westerlies north of Antarctica. Figure 5.22 shows that the surface wind stress is particularly strong along that entire latitude belt, especially in July, and of course the time average of the zonal component of the wind stress is particularly strong there. The time-averaged tropical wind stresses are relatively weak, with the exception of a maximum over the Arabian Sea in July, associated with the Somali jet.

Figure 5.23 shows the mountain torque and the frictional torque due to the zonally averaged zonal surface stress. The frictional torque dominates overall, but mountain torque is appreciable in the Northern Hemisphere. In the tropics, the trade winds cause a strong flow of angular momentum into the atmosphere. In the middle latitudes of the Northern Hemisphere, both frictional drag and mountain torque remove angular momentum from the atmosphere. In the middle latitudes of the Southern Hemisphere, strong westerly winds deposit angular momentum in the oceans. The mountains of Greenland and Antarctica add angular momentum to the atmosphere.

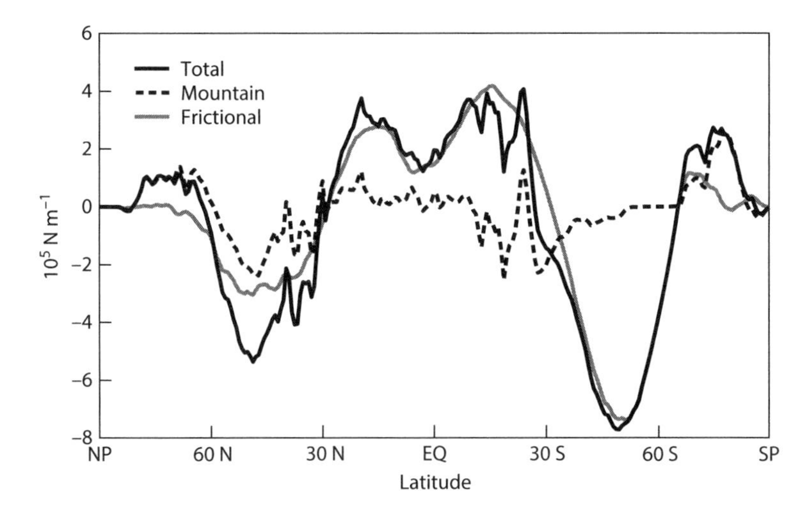

Figure 5.23. The annual mean of the zonally integrated mountain torque (solid line), frictional torque (dashed line), and the sum of the two. Positive values mean that the atmosphere is receiving angular momentum from the surface.

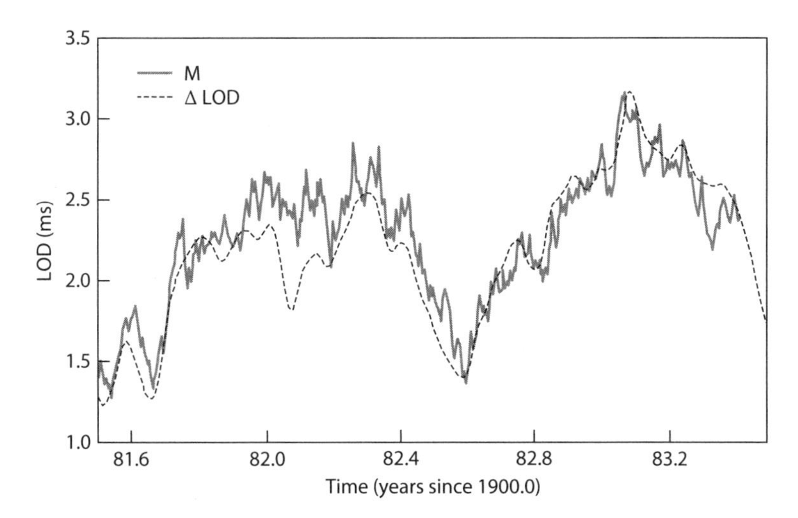

Figure 5.24. Changes in the length of day as inferred from the angular momentum of the atmosphere (solid line) and as determined from astronomical data (dotted line) since July 1981. The mean difference between the two quantities has been subtracted from the angular momentum curve. From Rosen et al. (1984).

The Earth-atmosphere system can exchange angular momentum with other bodies, such as the Moon, through tidal torques. Such torques are causing the Moon to gain angular momentum from the Earth, so that the length of the day and the radius of the Moon's orbit are both slowly increasing, but these are very slow processes. To the extent that we can neglect them, the angular momentum of the Earth-atmosphere system must remain constant with time. Therefore, *if the angular momentum of the atmosphere increases, the Earth's rotation must slow down, and vice versa*. Evidence that this is in fact true is shown in figure 5.24. The data plotted in the figure indicate that changes in the rate of rotation of the solid Earth, that is, the length of day, are highly correlated with changes in the angular momentum of the atmosphere. The length of day was measured by timing the Earth's rotation with respect to distant stars, while the atmospheric angular momentum was measured using radiosondes. The agreement shown is expected but nevertheless amazing.

Averaged over many years, the exchanges of angular momentum between the atmosphere and the solid Earth and oceans are expected to be very small. In equilibrium, there has to be a mixture of easterly and westerly winds at the Earth's surface, distributed in such a way that the net exchange of angular momentum between the atmosphere and the surface is near zero.

We multiply both sides of (41) by $2\pi a$ and then integrate vertically through the entire atmospheric column and also meridionally from the South Pole to an arbitrary latitude φ to obtain

$$\boxed{2\pi a\cos\varphi\int_0^\infty \overline{\rho_\theta vM}^{\lambda,t}d\theta = 2\pi a^2\int_{\frac{\pi}{2}}^{\varphi}\left[(\overline{F}_u^{\lambda,t})_S a\cos\varphi - \overline{p_S\frac{\partial z_S}{\partial\lambda}}^{\lambda,t}\right]\cos\varphi' d\varphi'}. \quad (45)$$

The quantity on the left-hand side of (45) is the total northward transport of angular momentum across latitude φ. After correcting $[(\overline{F}_u^{\,t})_S a\cos\varphi - p_S\overline{(\partial z_S/\partial\lambda)}^{t}]$ so that it has a global mean of zero, we obtain the results shown in figure 5.25 for January, July, and the annual mean. Angular momentum is carried from the tropics, where the surface easterlies extract it from the oceans and solid Earth, to middle latitudes, where the surface westerlies deposit it back into the oceans and the solid Earth. The angular momentum transport passes through zero near the equator. It diverges in the tropics and converges poleward of ±30°, where the subtropical jet streams are found. The transport is very weak poleward of ±60°, because there are two factors of $\cos\varphi$ on the left-hand side of (45): one that is plainly visible outside the integral, and a second built into the definition of M.

To balance the poleward angular momentum transport by the atmosphere, there has to be a return flow of angular momentum from middle latitudes toward the equator, carried by the ocean currents or the solid Earth, or some combination of the two. Oort

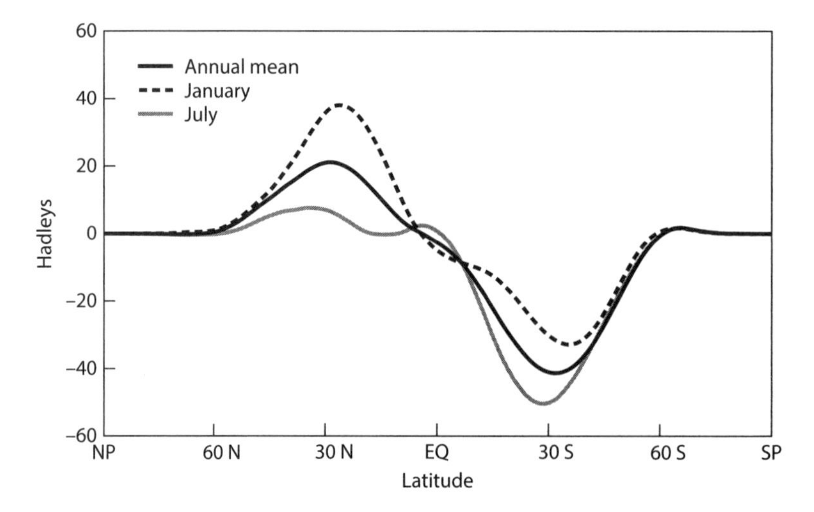

Figure 5.25. The vertically integrated meridional transport of angular momentum, for January, July, and the annual mean. A hadley is 10^{18} N m.

(1989) argues that the return flow occurs largely through stresses in the solid Earth rather than through the circulation of the oceans. These stresses do not necessarily involve movements of the solid earth, as in continental drift, because a rigid material (like a steel ball) can transmit forces spatially. The modeling study of Bryan (1997) supports Oort's claim that meridional angular momentum transport by the oceans is small.

We have attributed the pressure-gradient term of (42) to the form drag on the topography of the Earth's surface. The pressure-gradient term of (41) has a very similar interpretation, as a *vertical exchange of angular momentum within the atmosphere by the form drag on "bumpy" isentropic surfaces.* The isentropic surfaces bulge upward where there are blobs of cold air at a given pressure level and downward where there are blobs of warm air. From the perspective of isentropic coordinates, the upward flux of zonal momentum is associated with the pressure force rather than with a covariance between the "vertical velocity" (which vanishes in isentropic coordinates unless there is heating) and the zonal velocity. A layer of air confined between two isentropic surfaces will feel two momentum fluxes associated with the pressure force: one on its underside and a second on its upper side. It is the difference between these two forces that acts to produce a net acceleration of the layer. That is why we see $\partial/\partial\theta\,[\overline{p(\partial z/\partial\lambda)}^{\lambda}]$ in (41).

Figure 5.26 shows the vertical flux of angular momentum due to isentropic form drag, for January and July. Comparison with figure 5.25 shows that the midlatitude surface sinks of angular momentum due to mountain torque and friction are supplied by the downward momentum transport via isentropic form drag. The flux is particularly strong in the winter hemisphere. This topic is discussed further in chapter 9.

Figure 5.27 shows the total, mean, bolus, and eddy meridional fluxes of angular momentum. The Hadley cells produce large poleward fluxes in the upper troposphere, but they also produce large equatorward fluxes in the lower troposphere. These nearly cancel in a vertical average because, as mentioned earlier, the small value of the vertically integrated meridional mass flux guarantees that the vertically integrated flux of Earth's angular momentum is also small. We return to this point below. In both January and July, the total flux of angular momentum converges in middle latitudes in the winter hemisphere. The mean flux converges from the equatorward side, and the bolus flux converges from the poleward side. The eddy fluxes of angular momentum appear very weak in this figure, but they actually dominate after vertical integration. Figure 5.28 is like figure 5.27 but modified to show the fluxes of relative angular momentum only. The contour interval has been reduced by a factor of 4. The importance of the eddy fluxes is apparent.

Equation (37) can be used to define a streamfunction that depicts the zonally averaged flow of angular momentum through the atmosphere. We call it ψ_M and define it by

$$\left[\rho_\theta v(M - M_{GM})\right]^{\lambda,t} 2\pi a \cos\varphi \equiv -\frac{\partial \psi_M}{\partial \theta}, \tag{46}$$

and

$$\left[\rho_\theta \dot{\theta}(M - M_{GM})^{\lambda,s} - a\cos\varphi F_u\right] 2\pi a^2 \cos\varphi \equiv \frac{\partial \psi_M}{\partial \varphi}, \tag{47}$$

where M_{GM} is the mass-weighted average of M over the entire atmosphere. We can use (46) to determine ψ_M by vertical integration, given the meridional flux on the left-hand side and an upper or lower boundary condition. Since there is no angular momentum flux across the top of the atmosphere, we can choose $\psi_M = 0$ there. Figure

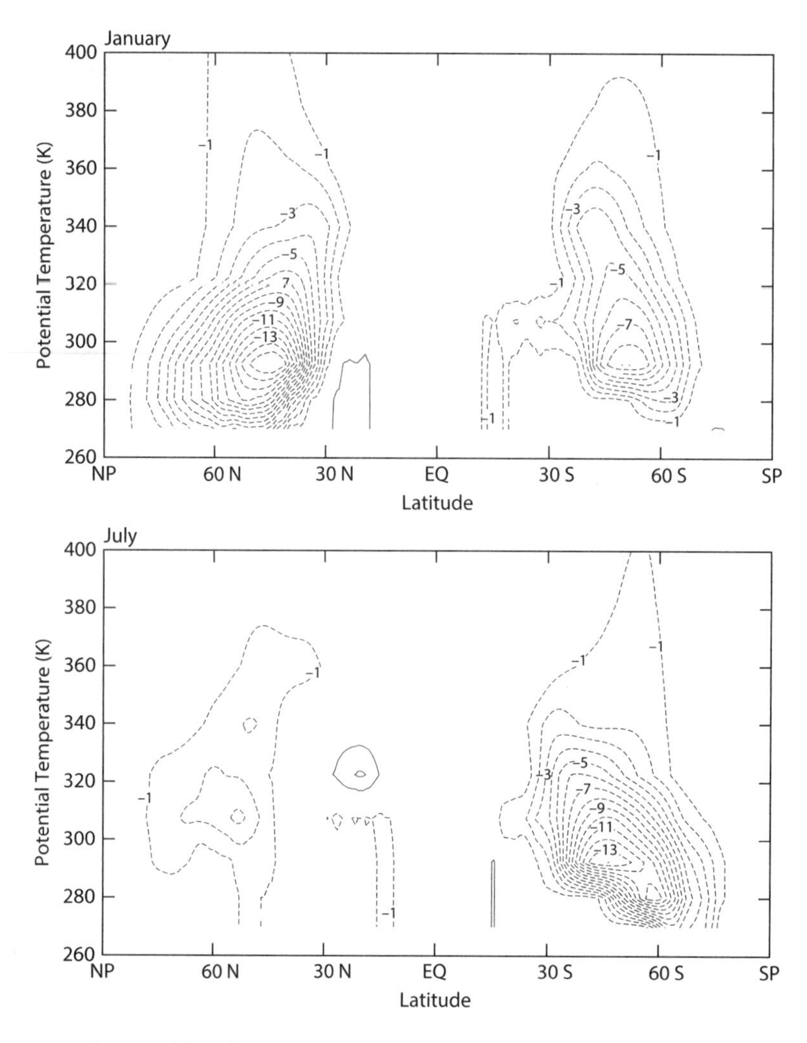

Figure 5.26. The vertical flux of angular momentum due to the isentropic form drag, for January and July. The contour interval is 10^6 N m^{-1}.

5.29 shows plots of ψ_M for January and July. The figures show that angular momentum enters the atmosphere in the tropics and is carried poleward, mainly toward the winter pole, by the upper branches of the Hadley cells. There is very strong downward angular momentum transport between 30° and 60° away from the equator in the winter hemisphere. The streamfunction contours intersect the Earth's surface, indicating that angular momentum leaves the atmosphere there; compare figure 5.29 with figure 5.23. Angular momentum converges meridionally near the 300 K isentropic surface in the same latitude band.

Segue

In this chapter we discussed the meridional transports of mass, total moisture, moist static energy, and angular momentum. Using the implied streamfunction plots, we also inferred the vertical transports of the same quantities.

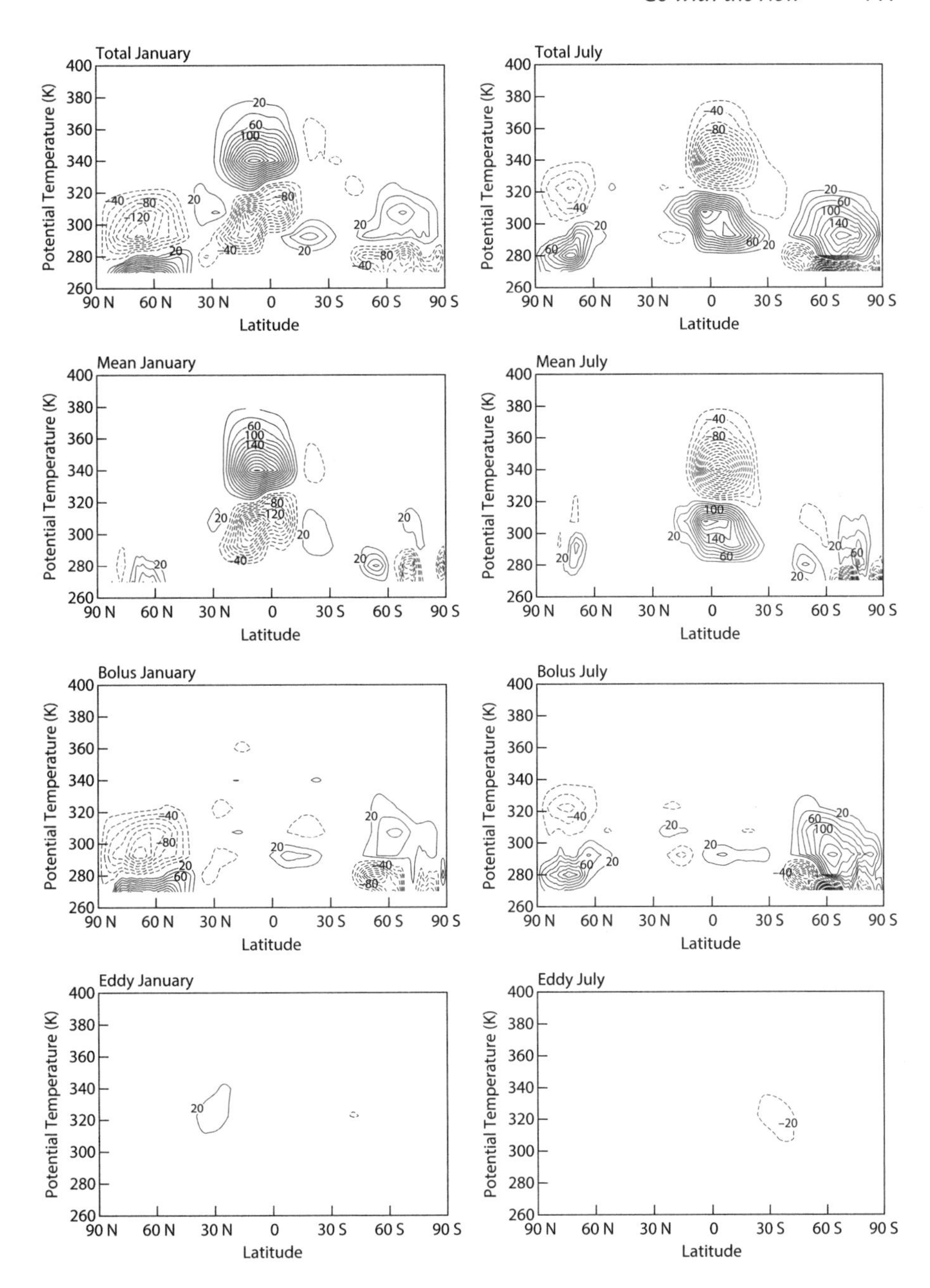

Figure 5.27. Isentropic meridional angular momentum fluxes for January (left) and July (right). The top row shows the total flux, the second row shows the flux associated with the zonally averaged meridional wind, the third row shows the bolus flux, and the bottom row shows the eddy flux. The contour interval is 20×10^9 N m^{-1} K^{-1}.

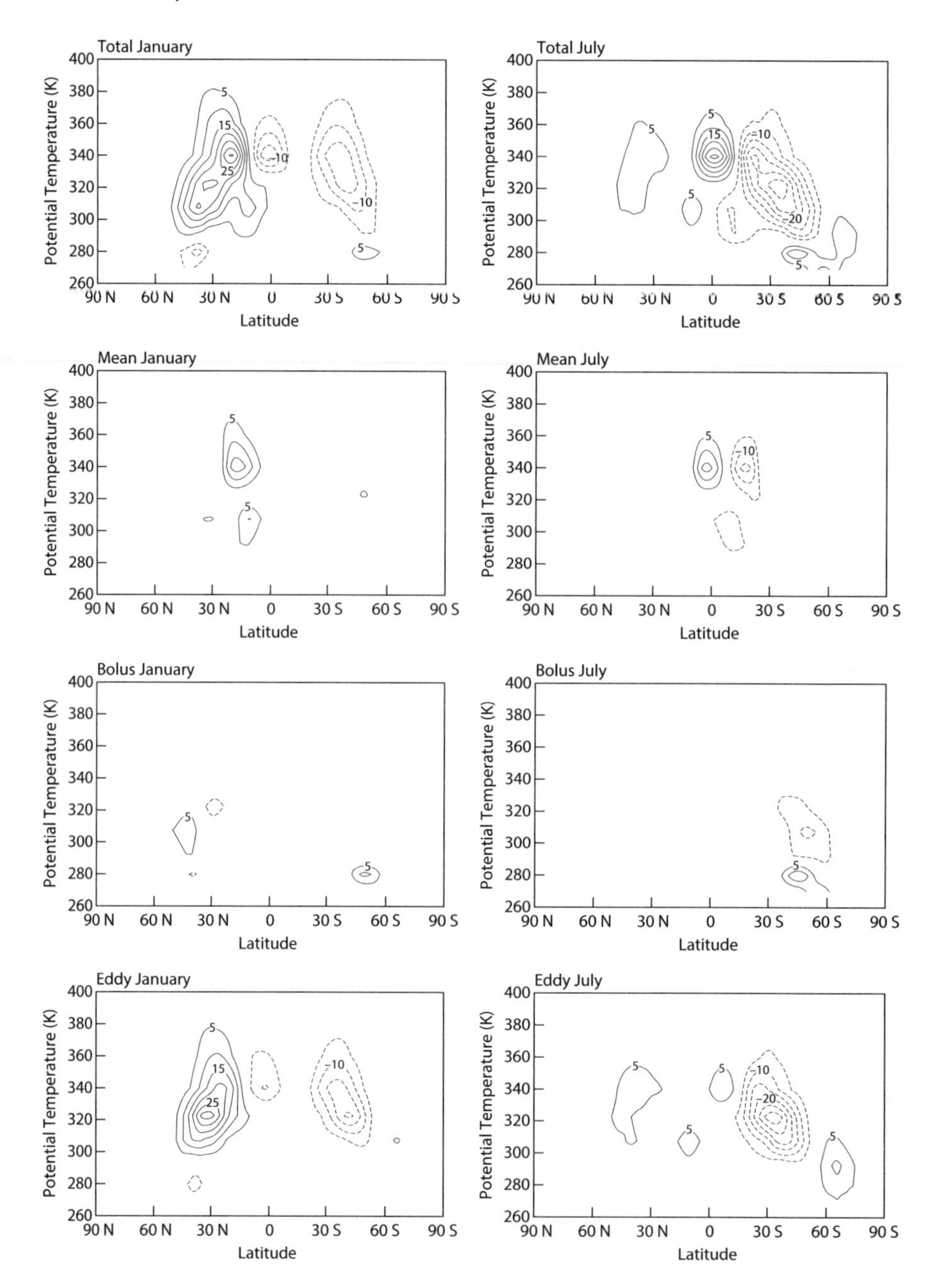

Figure 5.28. Like figure 5.27, but showing only the fluxes of relative angular momentum. The contour interval is 5×10^9 N m^{-1} K^{-1}, four times smaller than in figure 5.27.

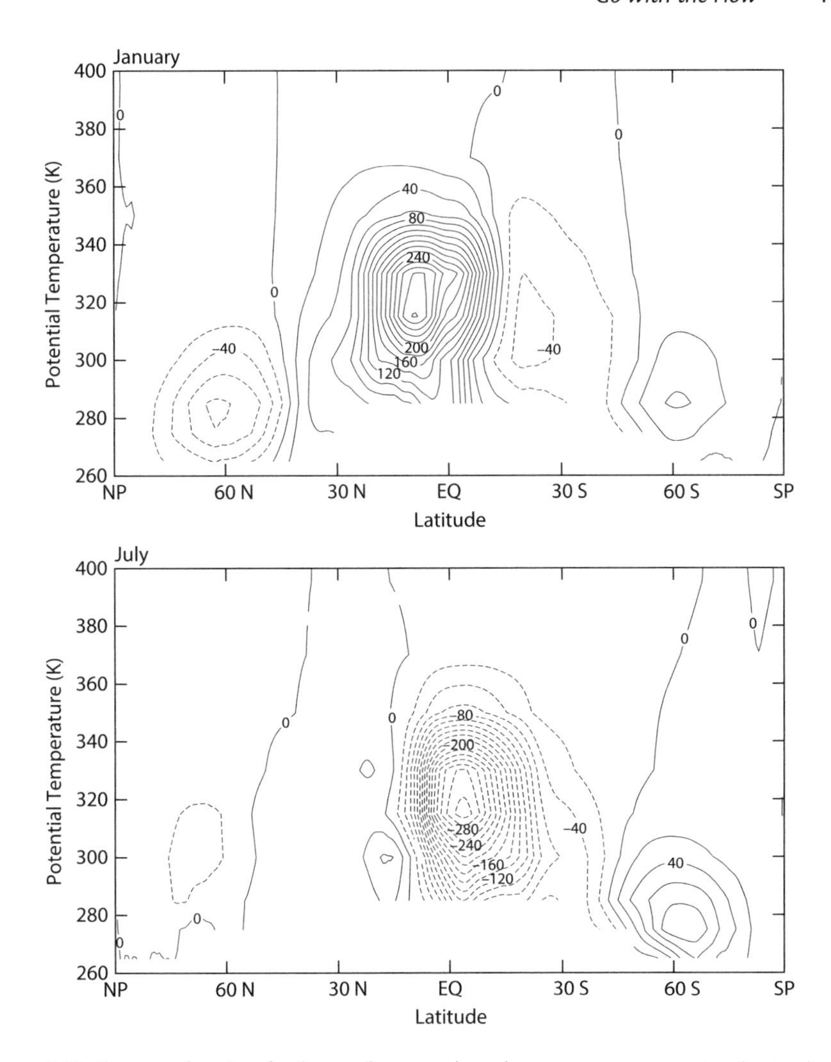

Figure 5.29. The streamfunctions for the zonally averaged angular momentum transport on isentropic surfaces, for January and July. The units are hadleys.

What we have not discussed in detail yet is how the various meridional and vertical transports are actually produced. Most of the rest of the book is devoted to answering this question, starting with chapter 6, which discusses the key role of convection.

Problems

1. Prove that the mountain torque vanishes if p_S depends only on ϕ_S.
2. Prove that $\overline{p(\partial\Pi/\partial\lambda)}^{\lambda}=0$.
3. Suppose that at 40° N the northward energy transport is entirely due to the atmosphere and is produced by the combination of a northward wind around half of the latitude circle and a compensating southward wind in the other half of the latitude circle, with uniform speeds of 5 m s^{-1} each. For simplicity,

suppose that these currents fill the entire depth of the atmosphere and that the surface pressure is a uniform 1000 hPa. Further assume that the atmospheric energy transported consists of internal energy and potential energy only. Assume that the temperature is zonally uniform within each current and that the temperature *difference* between the two currents is uniform with height. Compute the required temperature difference between the northward and southward flows.

CHAPTER 6

Up Moist, Down Dry

Upward Energy Transport by Deep Convection

Riehl and Malkus (1958; see fig. 6.1) argued from the observed energy balance and vertical structure of the tropical atmosphere that deep, penetrative cumulus convection is the primary mechanism for upward energy transport in the tropics. They began by estimating the mass circulation across a latitude 10° on the winter side of the intertropical convergence zone (ITCZ). Recall that the "body" of the main solstitial Hadley cell lies in the winter hemisphere. They neglected the mass transport across the boundary of the ITCZ on the summer side. Riehl and Malkus then attempted to evaluate the lateral energy transports across, into, and out of the ITCZ as functions of height. They considered transports of internal energy, potential energy, and latent energy. Because the low-level inflow is warm and wet, while the upper level outflow is cold and dry, both internal and latent energy flow into the ITCZ; nevertheless, there is a net loss of total energy owing to the export of potential energy in the elevated outflow layer. The net export of energy by the meridional flow implies that there is a compensating net input of energy at the top and bottom of the column. Their estimates of the various quantities are summarized in table 6.1.

Because energy flows into the ITCZ at low levels and out at high levels, Riehl and Malkus inferred that there must be a net upward transport of energy inside the ITCZ. They argued, however, that this upward energy flux cannot be due to the mean vertical motion, because the observed profile of moist static energy has a minimum at midlevels, as discussed in chapter 5. If the mean vertical motion were acting alone, then since h is conserved following parcels, h would become uniform with height throughout the ascending column, which is close to the observed sounding in a thunderstorm updraft but far from the observed large-scale vertical profile of h. Similar reasoning shows that diffusive energy transport cannot explain the observed upward flux of moist static energy. Riehl and Malkus concluded that *the upward energy transport must occur in deep convective clouds that penetrate the troposphere.* More than 50 years later, their conclusion holds up very well.

Neelin and Held (1987) considered the moist static energy budget of the ITCZ from a perspective very similar to that of Riehl and Malkus. They reasoned as follows: In a time average, the vertically integrated moist static energy budget is expressed by

Table 6.1. Lateral Energy Transports on the Poleward Side of the ITCZ

δp 100s mb	v (m s^{-1})	M_0 (10^{13} g s^{-1})	s (J g^{-1})	sM_0 (10^{16} J s^{-1})	Lq (J g^{-1})	LqM_0 (10^{16} J s^{-1})	hM_0 (10^{16} J s^{-1})	Eddy moisture transport (10^{16} J s^{-1})	Total energy transport out of the ITCZ (10^{16} J s^{-1})
10–9	−1.3	−5.2	301.5	−1.56	37.6	−0.20			
9–8	−1.1	−4.4	305.7	−1.34	27.6	−0.12			
8–7	−0.4	−1.6	311.6	−0.49	18.4	−0.03			
7–6	0	0	317.8	0	11.3	0			
6–5	0	0	323.3	0	7.1	0			
5–4	0.3	1.2	329.1	0.39	4.2	0.00			
4–3	0.6	2.4	335.4	0.80	2.1	0.00			
3–2	1.3	5.2	340.8	1.77	0.8	0.00			
2–1.25	0.8	2.4	348.4	0.83	0	0			
10–5				−3.39		−0.35	−3.74	0.07	−3.67
5–1.25				3.79		0.01	3.80	0	3.80
10–1.25				0.40		−0.34	0.06	0.07	0.13

Source: Adapted from Riehl and Malkus (1958).

Figure 6.1. The late Joanne Malkus (later Joanne Simpson) and the late Herbert Riehl. Photo courtesy of Joanne Simpson.

$$g^{-1}\nabla\cdot\left(\int_0^{p_S}\mathbf{V}h\,dp\right)=-(N_S-N_T),\tag{1}$$

where N is the net downward flux of energy due to turbulence, convection, and radiation, and subscripts T and S denote the top of the atmosphere and the surface, respectively. Similarly, mass continuity gives

$$\nabla\cdot\left(\int_0^{p_S}\mathbf{V}\,dp\right)=0.\tag{2}$$

We divide the column into upper and lower portions, and write

$$p_S^{-1}\nabla\cdot\left(\int_0^{p_S}\mathbf{V}h\,dp\right)=\nabla\cdot(\mathbf{V}h)_u+\nabla\cdot(\mathbf{V}h)_l=-g\left(\frac{N_S-N_T}{p_S}\right),\tag{3}$$

$$p_S^{-1}\nabla\cdot\left(\int_0^{p_S}\mathbf{V}\,dp\right)=\nabla\cdot\mathbf{V}_u+\nabla\cdot\mathbf{V}_l=0.\tag{4}$$

The upper and lower portions of the column correspond to the first two "summary" rows of table 6.1. In the tropics, horizontal variations of h are weak, so that it is useful to define h_u and h_l by

$$\nabla\cdot(\mathbf{V}h)_l\equiv h_l(\nabla\cdot\mathbf{V}_l),$$

and

$$\nabla\cdot(\mathbf{V}h)_u\equiv h_u(\nabla\cdot\mathbf{V}_u)=-h_u(\nabla\cdot\mathbf{V}_l),\tag{5}$$

from which it follows that

$$\nabla\cdot(\mathbf{V}h)_u+\nabla\cdot(\mathbf{V}h)_l=-(\nabla\cdot\mathbf{V}_l)(h_u-h_l).\tag{6}$$

Neelin and Held called the quantity h_u-h_l the *gross moist stability*. Equation (6) shows that low-level convergence of mass (i.e., $\nabla\cdot\mathbf{V}_l<0$, as in the ITCZ) leads to a net vertically integrated divergence of moist static energy (i.e., $\nabla\cdot(\mathbf{V}h)_u+\nabla\cdot(\mathbf{V}h)_l>0$) when the gross moist stability is positive. This must be the average state of affairs in the tropics, because the tropical atmosphere exports moist static energy toward higher latitudes. In a region where the gross moist stability is negative, low-level convergence leads to a net vertically integrated convergence of moist static energy. If the troposphere is moistened, without changing the temperature profile, the gross moist stability tends to decrease.

We can also show from (3)–(6) that

$$\nabla\cdot\mathbf{V}_l=\frac{g}{p_S}\left(\frac{N_S-N_T}{h_u-h_l}\right),\tag{7}$$

which implies that low-level convergence ($\nabla\cdot\mathbf{V}_l<0$) must occur where the column is gaining energy ($N_S-N_T<0$), provided that

$$h_u-h_l>0.\tag{8}$$

According to (7), the pattern of the gross moist stability is closely linked to the pattern of low-level convergence, for a given distribution of N_T-N_S. As discussed later [see (48)], we expect the gross moist stability to be small where cumulus convection is active.

The term *gross moist stability* has been redefined in a variety of ways in work that followed that of Neelin and Held. It is important to determine which definition is being used in published research.

The ideas just outlined fit well with the work of Charney (1963), which was discussed in chapter 3. Recall Charney's conclusion that horizontal temperature gradients tend to be weak in the tropics. As discussed in chapter 4, the thermodynamic energy equation can be written as

$$c_p\left(\frac{\partial T}{\partial t} + \mathbf{V}\cdot\nabla T + \omega\frac{\partial T}{\partial p}\right) = \omega\alpha + Q; \tag{9}$$

here the vertical coordinate is pressure, and Q is the heating rate per unit mass. An averaging interval of one month is sufficient to make the time-rate-of-change term of (9) fairly small. If we also invoke Charney's ideas to neglect the horizontal advection term, we find that

$$\boxed{\overline{\omega\frac{\partial s}{\partial p}}^t \cong \overline{Q}^t \text{ in the tropics.}} \tag{10}$$

Here s is the dry static energy, and the hydrostatic equation was used to combine the two terms of (9) that involve ω. Recall that s normally increases with altitude. The vertical motion term of (10) therefore represents cooling $[-\omega(\partial s/\partial p)]$ when the air is rising $(\omega<0)$, and warming $[-\omega(\partial s/\partial p)>0]$ when the air is sinking $(\omega>0)$. Equation (10) simply says that in a time average, *heating has to be balanced by vertical advection.* The only way to balance tropical heating or cooling is through vertical motion, and so clearly there should be a very strong correspondence between the pattern of vertical motion and the pattern of heating. Tropical rising motion occurs almost exclusively where latent and radiative heating are active, that is, where $Q>0$, and tropical sinking motion occurs almost exclusively where radiative cooling is dominant, that is, where $Q<0$.

These conclusions apply not only in the tropics but even in the middle latitudes in summer, simply because horizontal temperature gradients are weak there as well. They definitely do not apply in the middle latitudes in winter, where the horizontal advection term of (9) can be dominant.

Basics of Moist Convection

In atmospheric science, the term *convection* refers to an important type of buoyancy-driven circulation that gives rise to thermals in the boundary layer and cumulus clouds in the free troposphere above. In the older literature, the term *natural convection* is sometimes used instead.

Before discussing the moist convection that gives rise to cumulus clouds, we briefly discuss dry convection. Consider the equations of vertical motion and dry static energy conservation, linearized with respect to a resting, horizontally uniform basic state:

$$\overline{\rho}\frac{\partial w'}{\partial t} = -\frac{\partial p'}{\partial z} - \rho' g, \tag{11}$$

$$\frac{\partial s'}{\partial t} = -w'\frac{\partial \overline{s}}{\partial z}. \tag{12}$$

Here the overbars denote horizontal averages, and primes denote departures from those averages. The gravity term of (11) represents the effects of buoyancy. Equation (12) describes dry adiabatic motion.

We assume for simplicity that $\partial\bar{s}/\partial z$ is independent of height. Also for simplicity, we neglect the perturbation pressure term in (11). Its main effect is to partially cancel the effects of buoyancy. Finally, we use the approximation

$$-\left(\frac{\rho'}{\bar{\rho}}\right) \cong \frac{T'}{\overline{T}} = \frac{s'}{c_p\overline{T}}. \tag{13}$$

With these adjustments, (11) reduces to

$$\frac{\partial w'}{\partial t} = \frac{gs'}{c_p\overline{T}}. \tag{14}$$

Equations (12) and (14) form a closed system. We look for solutions of the form

$$\begin{aligned} w'(t) &= w'(0)\,\mathrm{Re}\{e^{\sigma t}\}, \\ s'(t) &= s'(0)\,\mathrm{Re}\{e^{\sigma t}\}, \end{aligned} \tag{15}$$

where σ may be either real or imaginary. Performing the appropriate substitutions, we find that for nontrivial solutions

$$\sigma^2 = -\frac{g}{c_p\overline{T}}\frac{\partial\bar{s}}{\partial z}. \tag{16}$$

For $\partial\bar{s}/\partial z < 0$, σ is real, and the solution grows exponentially with $\sigma > 0$; this is dry convective instability. We say that

$$\boxed{\frac{\partial\bar{s}}{\partial z} < 0 \text{ is the criterion for dry convective instability}}. \tag{17}$$

It can be seen from either (12) or (14) that in the exponentially growing solution, with $\sigma > 0$, $w'(t)$ and $s'(t)$ have the same sign for all time, so that $\overline{w's'}^t > 0$; that is, *convection transports dry static energy upward*. We show in chapter 7 that an upward temperature flux tends to lower the atmosphere's center of gravity; that is, it reduces the total potential energy of the atmospheric column. The reduction in potential energy coincides with a generation of convective kinetic energy through the work done by the buoyancy force, so that total energy is conserved. The generation of convective kinetic energy through an upward flux of dry static energy can be seen directly by multiplying both sides of (14) by w'.

For $\partial\bar{s}/\partial z > 0$, σ is imaginary, and the solutions are oscillatory; these are gravity waves. Their frequency, N, satisfies $N^2 = (g/c_p\overline{T})\,\partial\bar{s}/\partial z$; this is called the *Brunt-Väisälä frequency*. Using the preceding analysis as a starting point, you should be able to show that $\overline{w's'}^t = 0$ for a gravity wave, where the time average is taken over the period of the wave. This means that gravity waves do not transport dry static energy, and they do not transport moisture either. As discussed in chapter 9, they do transport momentum.

The analysis shows that in the absence of phase changes, convection and gravity waves are mutually exclusive; they cannot occur in the same place at the same time. We will see later that this conclusion does not necessarily apply when phase changes are allowed.

To this point, we have considered dry adiabatic motion. To analyze moist convection, we will assume saturated moist adiabatic motion. As discussed in chapter 4, the moist static energy, h, is approximately conserved under moist adiabatic processes. We replace (12) with

$$\frac{\partial h'}{\partial t} = -w'\frac{\partial \overline{h}}{\partial z}. \tag{18}$$

For saturated motion, the moist static energy must be equal to the saturation moist static energy, h_{sat}, so we can rewrite (18) as

$$\frac{\partial h'_{sat}}{\partial t} = -w'\frac{\partial \overline{h_{sat}}}{\partial z}. \tag{19}$$

Next, we have to relate the buoyancy term of (14) to h'_{sat}. Recall that

$$h_{sat} \equiv c_p T + gz + Lq_{sat}(T,p), \tag{20}$$

where the saturation mixing ratio depends, as indicated, on temperature and pressure. Perturbations at fixed height, and at approximately fixed pressure, satisfy

$$h'_{sat} \cong s'(1+\gamma), \tag{21}$$

where, in the linearization,

$$\gamma \equiv \frac{L}{c_p}\left(\frac{\partial q_{sat}}{\partial T}\right)_p \tag{22}$$

is evaluated using the mean-state temperature and pressure. The nondimensional parameter γ is positive and of order 1.

We now write

$$-\left(\frac{\rho'}{\overline{\rho}}\right) \cong \frac{T'}{\overline{T}} = \frac{h'_{sat}}{c_p\overline{T}(1+\gamma)}. \tag{23}$$

This equation is analogous to (13). Substitution of (23) into the equation of vertical motion gives

$$\frac{\partial w'}{\partial t} = \frac{h'_{sat}}{c_p\overline{T}(1+\gamma)}. \tag{24}$$

We look for exponential solutions of the system (19) and (24), and find that

$$\sigma^2 = -\frac{g}{c_p\overline{T}(1+\gamma)}\frac{\partial \overline{h_{sat}}}{\partial z}, \tag{25}$$

which shows that

$$\boxed{\frac{\partial \overline{h_{sat}}}{\partial z} < 0 \text{ is the criterion for moist convective instability of a saturated atmosphere}}. \tag{26}$$

Compare this result with (17).

Before moving on, we need to do one more thing. The dry adiabatic lapse rate of temperature is given by

$$\Gamma_d \equiv -\left(\frac{\partial T}{\partial z}\right)_{\text{dry adiabatic}} = \frac{g}{c_p}. \tag{27}$$

This is the rate at which temperature decreases with height when the dry static energy is independent of height. We can rewrite (12) as

$$\frac{\partial T'}{\partial t} = w'(\overline{\Gamma} - \Gamma_d) \tag{28}$$

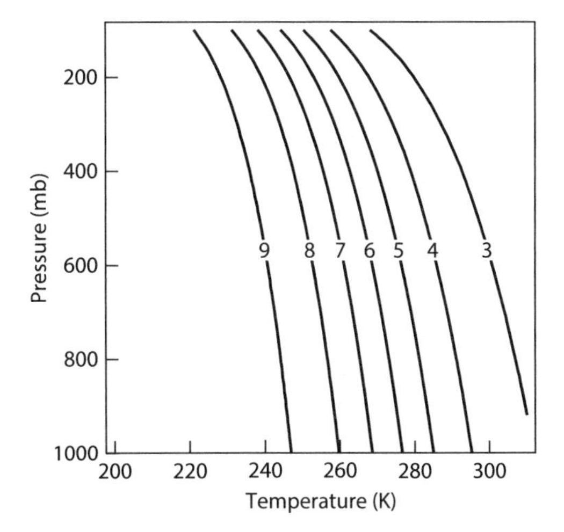

Figure 6.2. A plot of the moist adiabatic lapse rate, Γ_m (in K km^{-1}), as a function of temperature and pressure.

and restate the criterion for dry convective instability as

$$\Gamma > \Gamma_d. \tag{29}$$

Similarly, we can express the criterion for moist convective instability in terms of the moist adiabatic lapse rate, which is given by

$$\begin{aligned} \Gamma_m &\equiv -\left(\frac{\partial \overline{T}}{\partial z}\right)_{\text{moist adiabatic}} \\ &\cong \Gamma_d \left[\frac{1 + \dfrac{L q_{sat}(T,p)}{R_d T}}{1 + \dfrac{L^2 q_{sat}(T,p)}{c_p R_v T^2}} \right]. \end{aligned} \tag{30}$$

Equation (30) is derived in appendix E. The denominator in the square brackets of (30) is larger than the numerator, so $\Gamma_m < \Gamma_d$, although $\Gamma_m \to \Gamma_d$ at cold temperatures. For example, with a pressure of 1000 hPa and a temperature of 288 K, we find that $\Gamma_m = 4.67$ K km^{-1} (see fig. 6.2). As the temperature increases, the moist adiabatic lapse rate decreases. As discussed in chapter 11, this is important for climate change. For saturated motion, we can rewrite (12) as

$$\frac{\partial T'}{\partial t} = w'(\overline{\Gamma} - \Gamma_m). \tag{31}$$

Compare (31) with (28). They will both be used later.

Radiative-Convective Equilibrium

The study of Riehl and Malkus showed that in the tropics (and also in the moist convective regions of the summer hemisphere middle latitudes) upward transport of energy is due to small-scale convection rather than vertical advection by the large-scale vertical motion. Recall that the brightness temperature of the Earth corresponds to a level in the middle troposphere; the brightness temperature of the tropical convective regions corresponds to the actual temperature in the upper troposphere. We can consider that convection transports energy upward, to the middle or upper troposphere, where the energy is then radiated to space.

The simplest model that can represent this process is called a *radiative-convective model.* The basic idea is very simple. First, we assemble physical parameterizations that suffice to determine the time rates of change of the temperature within the atmospheric column and at the Earth's surface, neglecting the effects of large-scale circulations. We then combine the parameterizations into a model and march forward in time using a time step on the order of an hour or so. We keep taking time steps until a steady state is approached to sufficient accuracy. Depending on the initial conditions, convergence can take on the order of 500 simulated days. No significance is ascribed to the time evolution itself; only the steady state is of interest. It is not obvious a priori that a radiative-convective model will actually approach a steady state, but the models discussed below do. Such a state is called *radiative-convective equilibrium*, or RCE.

The physical ingredients of a radiative-convective model include parameterizations of radiation, convection, turbulence, and the processes that determine the change of the surface temperature. We write

$$\rho c_p \frac{\partial T}{\partial t} = \rho LC - \frac{\partial F_s}{\partial z} + Q_R. \tag{32}$$

As a reminder, L is the latent heat of condensation, C is the condensation rate, and F_s is the sensible heat flux. We use Q_R to denote the radiative heating rate. Horizontal and vertical advection are deliberately omitted in (32), because the purpose of a radiative-convective model is to help us to understand what the atmosphere would look like in their absence. To use (32), we have to determine the condensation rate and the convective fluxes. This will entail consideration of the moisture budget. In addition, the vertical distributions of water vapor and clouds are needed to determine Q_R, which satisfies

$$Q_R = \frac{\partial}{\partial z}(S - R). \tag{33}$$

Here S is the net solar radiation (positive downward), and R is the net terrestrial radiation (positive upward).

We also impose an energy budget for the Earth's surface:

$$C_g \frac{\partial T_S}{\partial t} = N_S(T_S). \tag{34}$$

Here C_g is the effective "heat capacity" of the surface, T_S is the surface temperature, and $N_S(T_S)$ denotes the net downward (i.e., a positive value denotes downward) vertical flux of energy due to turbulence, convection, and radiation. Equation (34) can be used to determine T_S, provided that $N_S(T_S)$ is known. The value of C_g determines how rapidly the surface temperature changes in response to a given value of N_S. When C_g is large, T_S changes slowly. When $C_g \to 0$, T_S adjusts instantaneously so as to keep $N_S(T_S) = 0$.

The balance requirements that must be satisfied in RCE can be stated as follows.

As discussed in chapter 2, in a time average there can be no net radiative energy flux at the top of the (globally averaged) atmosphere. We impose that requirement in RCE:

$$N_T = S_T - R_T = 0. \tag{35}$$

Here N_T is also positive downward. The atmospheric column must be in energy balance, so

$$N_S = N_T, \tag{36}$$

where

$$\begin{aligned} N_S &\equiv S_S - R_S - (F_h)_S \\ &= S_S + (R_S)\downarrow - \sigma T_S^4 - (F_h)_S. \end{aligned} \tag{37}$$

From (35) – (36), it follows that in equilibrium, the Earth's surface must also be in energy balance:

$$N_S = 0. \tag{38}$$

We now discuss some results from radiative-convective models. In a series of studies during the 1960s (see the bibliography), Manabe and his colleagues investigated the degree to which pure radiative equilibrium and/or RCE can explain the observed vertical distribution of temperature. These studies led to major advances in our understanding of the vertical structure of the atmosphere. Because little was known about moist physics during the 1960s, Manabe and colleagues did not explicitly represent moist processes in their model; instead, they considered two alternative assumptions for the vertical distribution of moisture (discussed below) and adopted fairly drastic but empirically justified simplifying assumptions to determine the effects of latent heat release and moist convection on the atmospheric temperature profile.

Manabe and colleagues used a time-marching method and some simplifying assumptions to find equilibrium solutions. They first specified the following "initial conditions":

- the temperatures of the atmosphere,
- the temperature of the surface,
- the water vapor content of the atmosphere,
- the distribution of cloudiness as a function of height, and
- the composition of the dry air (including ozone) as a function of height.

They also specified the albedo of the Earth's surface.

The vertical distribution of water vapor has a powerful effect on the radiation budgets of both the atmosphere and the surface. To predict the vertical distribution of water vapor, it is necessary to include in the model representations of turbulence and cumulus convection. Manabe and colleagues did not attempt this. Instead, they prescribed the vertical distribution of water vapor using two alternative methods: (1) fixed specific humidity (which they called "fixed absolute humidity") and (2) fixed relative humidity. The vertical distributions of specific humidity and relative humidity were both prescribed from observations similar to those shown in chapter 3. Although the early model of Manabe and colleagues did not include a water budget, more modern radiative-convective models do. In these models, water vapor is introduced by evaporation from the sea surface; convection and, to a much smaller degree, diffusion carry the moisture upward; precipitation removes it. An example is given later.

Manabe and colleagues also prescribed the vertical distribution of radiatively active cloudiness.

The vertical distribution of ozone is important for the radiation calculation, because the absorption of solar radiation by ozone accounts for the increase of temperature with height in the stratosphere; without ozone, there would be no stratosphere. Manabe and colleagues did not attempt to compute the vertical distribution of ozone; instead, they prescribed it according to observations.

From the vertical profiles of temperature, water vapor, ozone, and cloudiness, it is possible to determine the solar and terrestrial radiative energy fluxes and the net radiative cooling of the atmosphere at each level and then to determine the net atmospheric radiative cooling, which is given by

$$ARC = (R_T - R_S) - (S_T - S_S). \tag{39}$$

Manabe and colleagues used a simple radiative transfer model to compute the *ARC*. So far, so good.

How can we determine $(F_h)_S$? In principle, we could use the bulk aerodynamic formula, as discussed in chapter 5, but the following are problems with that approach:

- Over the ocean, $(F_h)_S$, depends on the low-level wind speed. This could be specified, but we prefer to specify as few things as possible, so this option is not attractive.
- Over land, the surface latent heat flux, which contributes to $(F_h)_S$, depends on the soil moisture content and vegetation cover, among other factors. This is more complexity than we are willing to include in the model.

To avoid such complications, we can assume that on each time step the net radiative cooling of the atmosphere is equal to the net radiative warming of the surface; that is,

$$ARC = S_S - R_S. \tag{40}$$

From (35) and (39) it is clear that (40) must hold in equilibrium, but purely for computational reasons we assume that it also applies during the approach to equilibrium. From (38), (39), and (40), it follows that

$$(F_h)_S = ARC. \tag{41}$$

This means that the surface moist static energy flux is whatever it takes to balance the radiative cooling of the atmospheric column. Again, this is required in equilibrium, but there is no physical reason why it has to be true during the approach to equilibrium. Equation (41) means that the atmospheric column as a whole is in energy balance, even though imbalances occur at particular levels inside the column during the approach to equilibrium.

Equation(41) can be used to determine an updated surface temperature by time-stepping (34) with (37).

An RCE state can be found using the computational procedure summarized in figure 6.3. Initial conditions are given for the vertical profile of atmospheric temperature. After the temperature profile has been modified by radiation, on each time step, the resulting temperature profile is checked for convective stability. Manabe and Wetherald (1967) assumed that moist convective instability exists if the temperature decreases more rapidly than 6.5 K km^{-1} with height. If instability is found, the temperature profile is adjusted so as to restore a lapse rate of 6.5 K km^{-1}. For the case of fixed relative humidity, the vertical distribution of water vapor is then corrected to take into account the temperature change.

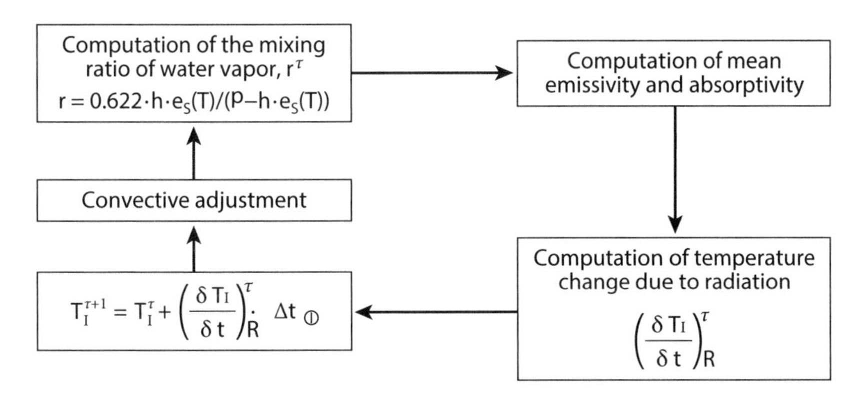

Figure 6.3. Flowchart for the solution of the radiative-convective model of Manabe and Wetherald (1967). © American Meteorological Association. Used with permission.

The assumption that the lapse rate "adjusts" to 6.5 K km^{-1} is based on the physical hypothesis that convection acts to prevent the lapse rate from becoming much steeper than the moist adiabatic lapse rate, which is close to 6.5 K km^{-1} in the tropical lower troposphere (see fig. 6.2). The physical meaning of this important hypothesis is discussed further later in this chapter.

Figure 6.4 shows the results for three cases:

- pure radiative equilibrium of the clear atmosphere with a given distribution of relative humidity,

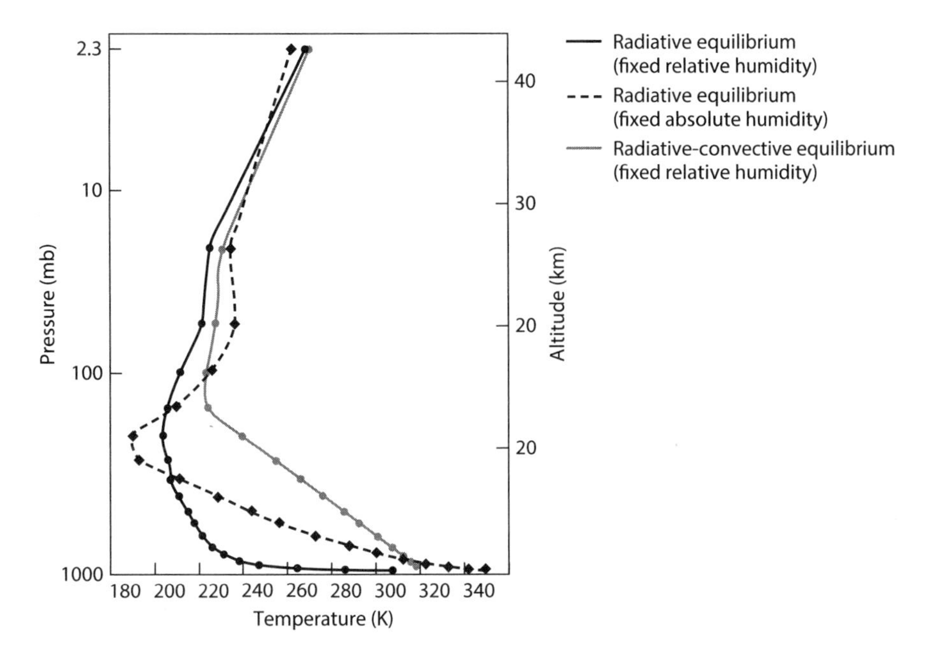

Figure 6.4. Solid black line: radiative equilibrium of the clear atmosphere with a given distribution of relative humidity; dashed line: radiative equilibrium of the clear atmosphere with a given distribution of absolute humidity; solid gray line: radiative-convective equilibrium of the atmosphere with a given distribution of relative humidity. From Manabe and Wetherald (1967). © American Meteorological Association. Used with permission.

- radiative equilibrium of the clear atmosphere with a given distribution of absolute humidity, and
- radiative-convective equilibrium of the atmosphere with a given distribution of relative humidity.

Pure radiative equilibria exhibit a troposphere and a stratosphere, with the tropopause at a fairly realistic height. An unrealistic aspect of both of the radiative equilibria is that the lower troposphere is convectively unstable, even for dry convection. In the RCE calculations, this instability is assumed to be removed by convection. The radiative-convective equilibrium with fixed relative humidity is amazingly realistic, considering that the model ignores all large-scale dynamical processes.

Fels (1985) reported the results of similar RCE calculations on the sphere, but his model included a representation of photochemistry, and he emphasized the structure of the stratosphere and mesosphere, which are often referred to as the *middle atmosphere*. Because the middle atmosphere has a very stable stratification, convection is not active there, and it is not clear that the representation of convection in the troposphere of his model had any significant influence on the results for the middle atmosphere, which are shown in figure 6.5. The observed state of the summer stratosphere (see fig. 6.5B) resembles that predicted by the model, but the winter polar stratosphere simulated by the model is much too cold. Also, the mesosphere of the real world is warm near the winter pole and cold near the summer pole, while the model predicts just the opposite. The differences between the observations and the model results can be attributed to the effects of atmospheric dynamics, which are neglected in the model. Obviously, the motions make quite a difference in the winter middle atmosphere, where they must transport energy poleward to account for the differences between the observations and the results of the radiative-convective model. This factor will be discussed in detail later. It appears that atmospheric motions have little effect on the thermal structure of the stratosphere in summer, however. This is a clue that the summer stratosphere is close to a state of radiative equilibrium.

Some Comments on Radiative-Convective Equilibrium

As originally conceived, RCE means coupling radiation and convection (optionally including the effects of turbulence and radiatively active clouds) but without large-scale circulations. The RCE studies just discussed demonstrate that the vertical structure of the atmosphere is strongly controlled by radiation and convection, at least in the tropics. Early RCE work used parameterized representations of convection and turbulence, but there are now many examples of RCE calculations based on numerical models with grid spacings fine enough to explicitly simulate the growth and decay of individual clouds.

In recent years, RCE studies have increasingly been used to seek basic understanding in an idealized framework, and this work has been quite productive and interesting. There is no question that RCE is a useful concept, but some caveats are needed.

First, large-scale motions do occur in the real world, so RCE can be simulated only as an idealization of reality. *There is no observational basis to say what RCE looks like.* In the real tropical atmosphere, large-scale rising motion can enhance convection, increasing the precipitation rate beyond that expected in RCE, and large-scale sinking motion can suppress convection, cutting off precipitation entirely. We can compare

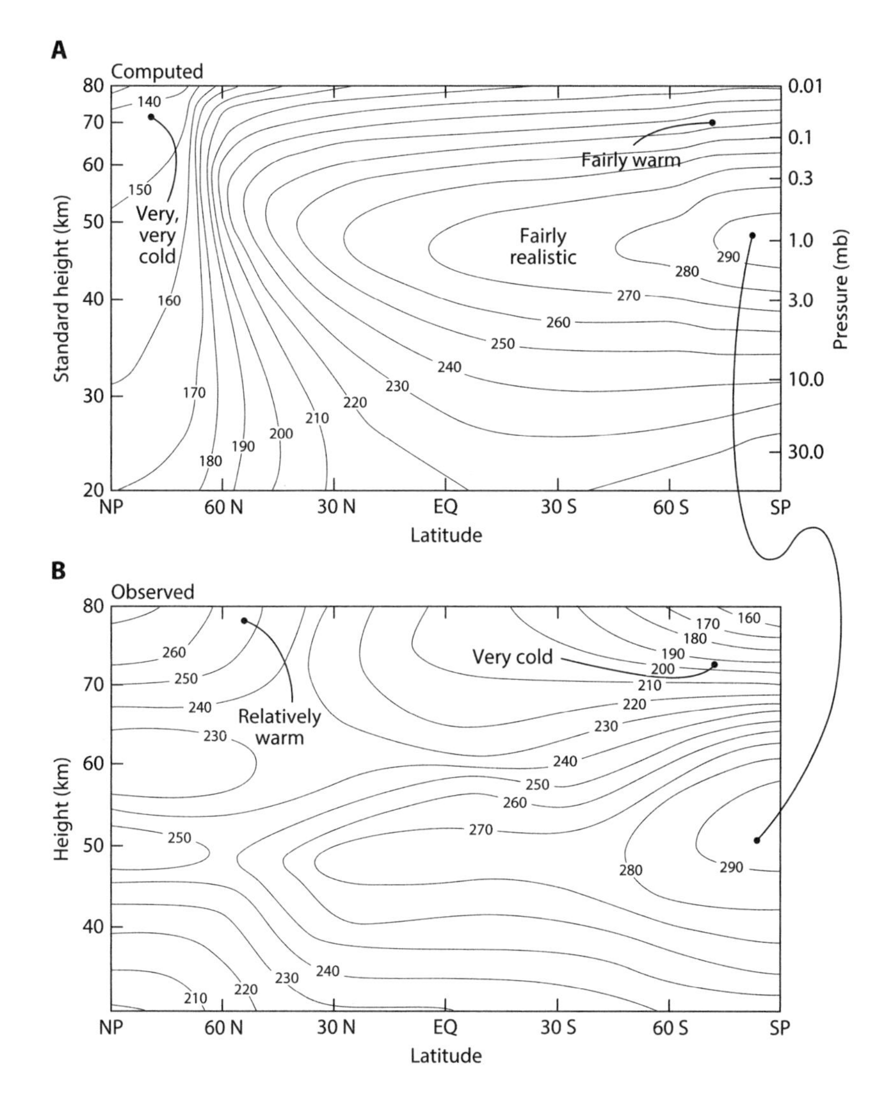

Figure 6.5. (A) Zonal mean temperatures for January 15 calculated by using a time-marched radiative-convective-photochemical model. (B) Zonal mean temperatures for January. From Fels (1985). Reprinted with permission from Elsevier.

RCE simulations with observations from places in the real world, such as the tropical western Pacific Ocean, and the comparison may be informative, but there is no a priori reason to expect agreement.

Second, a simulation of RCE must be referred to a specific "domain" or region. The horizontal width of the domain matters. Studies of RCE typically assume a medium-sized domain, for example, 100 km on a side, but in many cases this assumption is not made explicit. If the domain is very small, for example, 1 m^2, then it is not physically possible for cumulus convection to occur inside. If the domain is very large—for example, the whole global domain—then it can contain some or all of the many and varied weather systems of the global circulation of the atmosphere.

For example, Khairoutdinov and Emanuel (2013) showed that when rotation is included, simulations of RCE over a sufficiently warm prescribed SST, and with a sufficiently large domain, give rise to the formation of tropical cyclones. At what point do simulations of RCE become simulations of weather systems?

Representative Soundings

Figure 6.6 shows the observed vertical structure of the atmosphere in the tropics, the subtropical trade wind regime, and the subtropical marine stratocumulus regime. The quantities plotted are the dry static energy, s, the moist static energy, h, and the saturation moist static energy, which is defined by

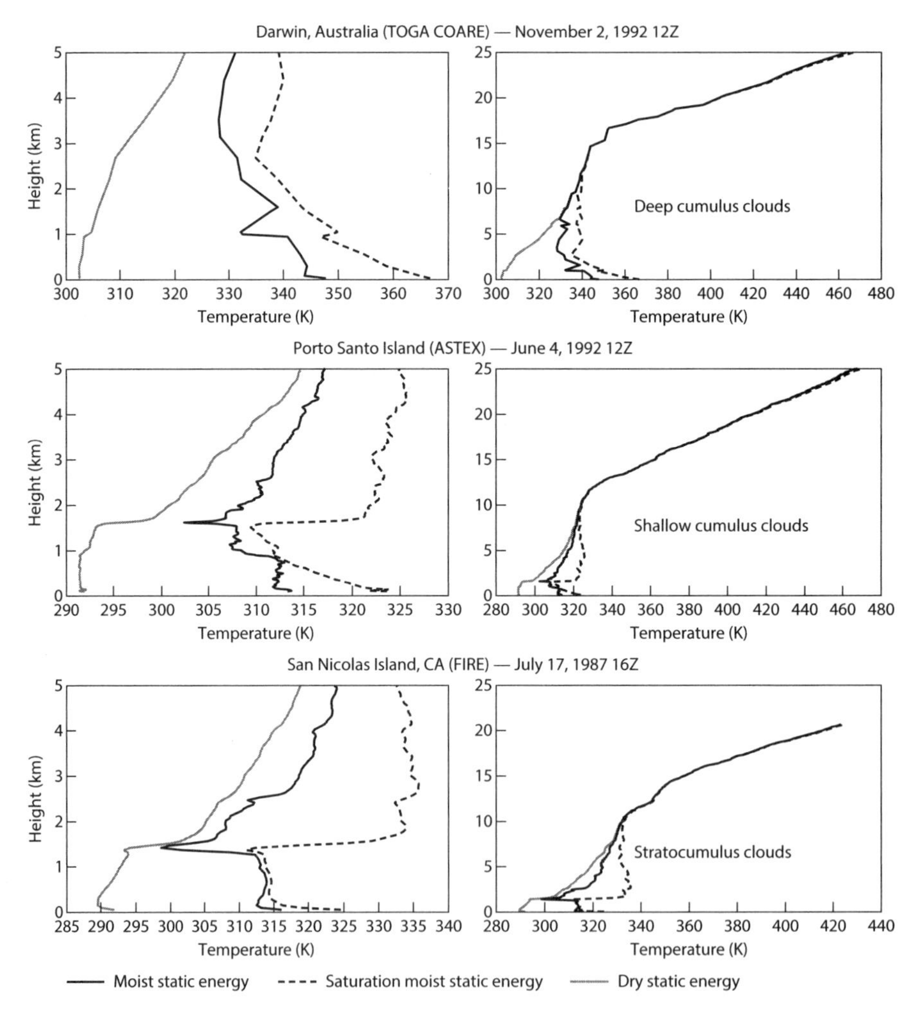

Figure 6.6. Representative observed soundings for Darwin, Australia; Porto Santo Island in the Atlantic trade-wind regime; and San Nicolas Island, in the subtropical marine stratocumulus regime off the coast of Southern California. The curves plotted show the dry static energy, the moist static energy, and the saturation moist static energy. The panels on the right cover both the troposphere and the lower stratosphere, while those on the left zoom in on the lower troposphere to show more detail. Values are divided by c_p to give units in Kelvin.

$$h_{sat} \equiv s + Lq_{sat}, \tag{42}$$

where q_{sat} is the saturation mixing ratio. Quite generally,

$$s \leq h \leq h_{sat}. \tag{43}$$

In cloudy air, $h = h_{sat}$. In very dry air, $h \cong s$. In very cold air, $h_{sat} \cong h \cong s$. It can be shown that the dry static energy increases with height in a statically stable atmosphere. The dry static energy is approximately conserved under dry adiabatic processes, while the moist static energy is approximately conserved under dry adiabatic, moist adiabatic, and pseudoadiabatic processes. The saturation moist static energy is not a conservative variable and, despite its name, does not actually carry any information about the moisture field; the saturation moist static energy sounding is essentially determined by the temperature sounding.

If a parcel of air containing vapor is lifted adiabatically from near the surface, it will eventually become saturated owing to the cooling caused by adiabatic expansion. Before the parcel reaches its lifting condensation level, both its dry static energy and moist static energy will be conserved. Suppose that once the lifting-condensation level has been exceeded, the moist static energy of the now-cloudy parcel remains constant; that is,

$$\frac{\partial h_c}{\partial z} = 0. \tag{44}$$

This will be the case if the effects of turbulent mixing and radiation are negligible. The dry static energy of the cloudy air will increase with height owing to latent heat release. The liquid water mixing ratio will increase, and the water vapor mixing ratio will correspondingly decrease. Under moist adiabatic processes, the total mixing ratio, $q_v + l$, will be conserved.

Conservation of h and $q_v + l$ implies that

$$s_l \equiv h - L(q_v + l) \tag{45}$$

is also conserved. We refer to s_l as the *liquid water static energy*. It is conserved under moist adiabatic processes. Precipitation from a parcel is not a moist adiabatic process, because it involves the removal of mass from the parcel. Precipitation can change the value of s_l, but it does not change the value of h. This means that h is "more conservative" than s_l. Both variables are useful.

The temperature difference between the cloudy air and its environment at the same level is simply proportional to the saturation moist static energy difference; that is,

$$c_p(T_c - \overline{T}) = \frac{(h_{sat})_c - \overline{h_{sat}}}{1+\gamma}, \tag{46}$$

where $\gamma \equiv (L/c_p)(\partial q_{sat}/\partial T)_p$, the subscript c denotes the cloudy air, and an overbar denotes an area average. Because the cloudy air is saturated, however, we can write

$$c_p(T_c - \overline{T}) = \frac{h_c - \overline{h_{sat}}}{1+\gamma}. \tag{47}$$

Equation (47) shows that the buoyancy of the cloudy air, as measured by the difference between its temperature and the temperature of the environment, is proportional to the difference between the moist static energy of the cloudy air and the saturation moist static energy of the environment. Recall, however, that the parcel under consideration is lifted adiabatically from near the surface, conserving its moist static energy. This means that h_c is equal to the low-level moist static energy of the sounding. The cloudy updraft will stop when it encounters a level where $h_c = \overline{h_{sat}}$; if this level is high and cold, then $\overline{h_{sat}} \cong \overline{h}$, and so we expect to find

$$\overline{h}_{\text{tropopause}} \cong \overline{h}_{\text{boundary layer}} \text{ in regions where deep convection is active.} \quad (48)$$

In the parlance of Neelin and Held (1987), equation (48) means that the gross moist stability is small.

Suppose that the air in the cloud is neutrally buoyant with respect to its environment at each level. Equations (44) and (47) imply that

$$\frac{\partial \overline{h_{sat}}}{\partial z} = 0, \quad (49)$$

which is the condition for neutral stability in saturated air, derived earlier. Note, however, that we derived this condition without assuming that the air is saturated; it applies even when the relative humidity of the environment of the cloud is less than 100%. Equation (49) is the condition for neutral stability with respect to a nonentraining cumulus cloud. This topic will be discussed in more detail later.

These ideas can be applied to the tropical (i.e., Darwin) sounding shown in figure 6.6. Near the surface, $h_{sat} > h$, indicating that the air is unsaturated. If a parcel is lifted adiabatically from near the surface, its moist static energy will follow a straight, vertical line in the diagram, starting from a near-surface value. At the same time, the environmental saturation moist static energy decreases with height. After the parcel rises a kilometer or so, the vertical line representing the moist static energy traced out by adiabatic parcel ascent from the surface will be to the right of the observed sounding of saturation moist static energy, so that $h_c > \overline{h_{sat}}$. According to (47), the parcel will then be positively buoyant if it is saturated. Figure 6.6 thus confirms that the Darwin sounding is conditionally unstable. The positive buoyancy of the lifted parcel will continue upward until the vertical line representing constant moist static energy again crosses over to the left side of the curve representing the environmental saturation moist static energy. For the Darwin sounding, this transition occurs at about the 15 km level, near the tropopause. Thus, deep cumulus convection is expected to occur in this sounding, although the mere existence of a conditionally unstable sounding is not enough, in itself, to show that cumulus convection will be significantly active.

Another interesting aspect of the Darwin sounding is that the saturation moist static energy is nearly uniform with height throughout most of the troposphere. Recall that in an atmosphere that is neutrally stable with respect to saturated moist convection, $\partial \overline{h_{sat}} / \partial z = 0$, which may suggest that the Darwin sounding is close to neutrally stable for moist convection. In fact, this is true in a sense that will be explained later.

The subtropical "trade-wind" sounding is also conditionally unstable, but only through a shallow layer. The trade-wind convective layer is capped by a very strong temperature inversion, and the water vapor mixing ratio decreases strongly with height through this "trade inversion." The middle troposphere is much drier in the trade-wind sounding than in the tropical sounding.

The subtropical marine stratocumulus sounding is not conditionally unstable at all. The sounding shows evidence of cloudiness in the lowest kilometer. The cloud layer is capped by a very strong inversion, essentially similar to the trade inversion

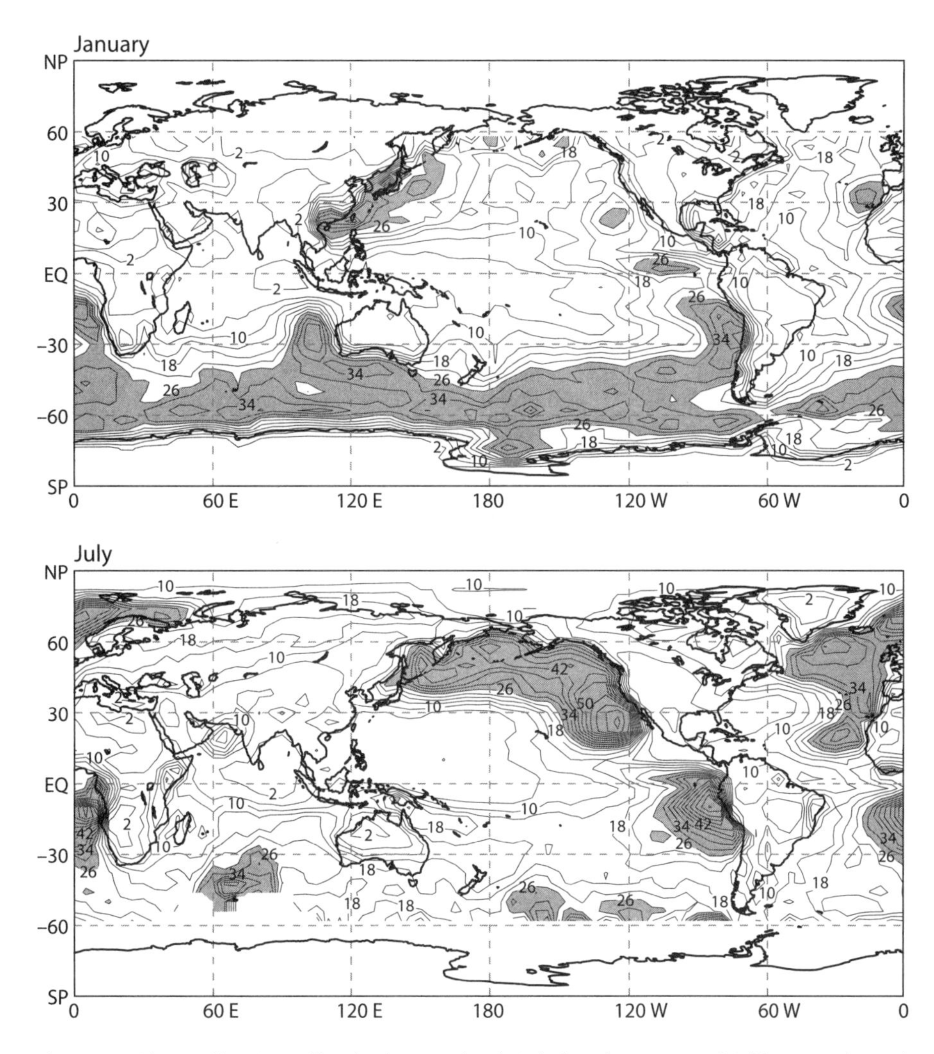

Figure 6.7. Observed locations of low-level stratus clouds, including the one west of California, as observed by ISCCP, the International Satellite Cloud Climatology Project. Cloud amounts greater than 26% are shaded.

but residing at a lower level. Marine stratocumulus regimes occur in several places around the world, typically in association with subtropical highs (see fig. 6.7).

The relationships among the three soundings shown in figure 6.6 are summarized in figure 6.8. The subtropical marine stratocumulus regime is shown on the right side of the figure, in a region of large-scale subsidence. The sea-surface temperatures are relatively cool in such regimes. Toward the left, is the trade-wind cumulus regime, which has weaker subsidence and warmer sea-surface temperatures. Finally, on the left side of the figure is the region of deep convection, characterized by warm sea-surface temperatures and large-scale rising motion.

As the air descends in the subtropical branches of the Hadley cells it is gradually cooled by radiation. As a result, the potential temperature of the air in the subtropical free atmosphere decreases downward, or in other words, it increases with altitude.

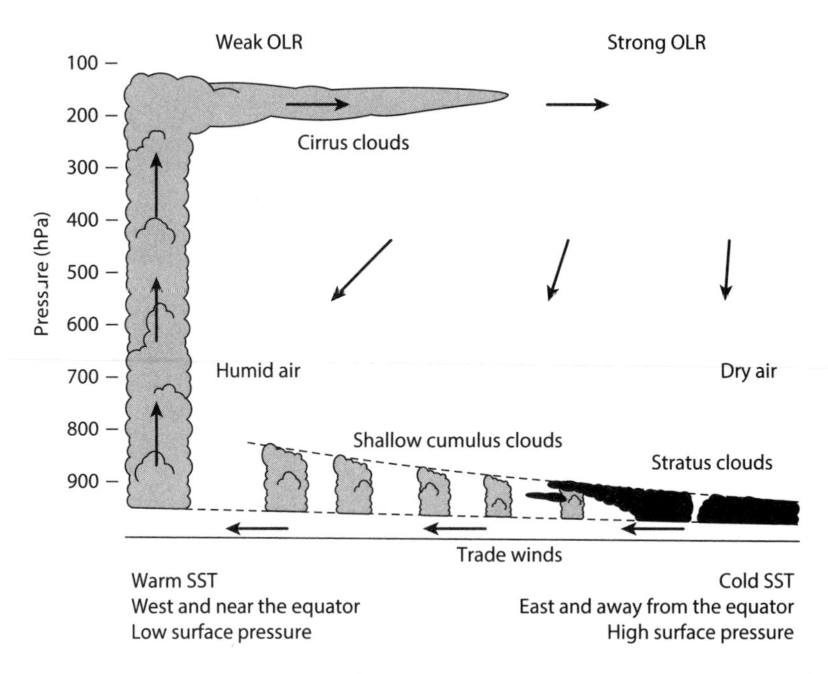

Figure 6.8. Schematic diagram summarizing the relationships among the cloud regimes depicted in figure 6.6 and how they fit into the mean meridional circulation. Adapted from Schubert et al. (1995).

This subject is discussed further in chapter 8.

The lapse rate of the deep convective zones is essentially determined by convection. However, as discussed in chapter 3, the horizontal temperature gradients are weak throughout the tropics and subtropics, for reasons explained by Charney (1963). This means that the lapse rate in the subtropics, above the trade inversion, must be nearly the same as the lapse rate in the tropics. For the reasons discussed earlier in this chapter, we can write an approximate thermodynamic balance for the descending branch of the Hadley cell as

$$\omega \frac{\partial s}{\partial p} = Q_R. \tag{50}$$

Here $Q_R < 0$ is the radiative cooling rate. We see from (50) that the speed of the large-scale sinking motion in the subtropics is essentially determined by the requirement of thermodynamic balance. Typical clear-sky tropical tropospheric radiative cooling rates are on the order of 2 K per day (see fig. 6.9), which corresponds to a few hundred meters of sinking per day.

The sinking air passes through the trade inversion. How does this happen? It is remarkable, for example, that the average mixing ratio of the air suddenly increases from perhaps 1 g kg^{-1} above the trade inversion to 6 or 7 g kg^{-1} below it. After all, it is the same air. How does it suddenly become so moist? The answer is that the convective vertical motions associated with the shallow stratocumulus and/or trade cumulus clouds transport moisture upward and deposit it at the base of the inversion, where it moistens the sinking air.

The air also cools as it descends through the inversion. This cooling is produced by a combination of concentrated radiative cooling near cloud tops, evaporative cooling due

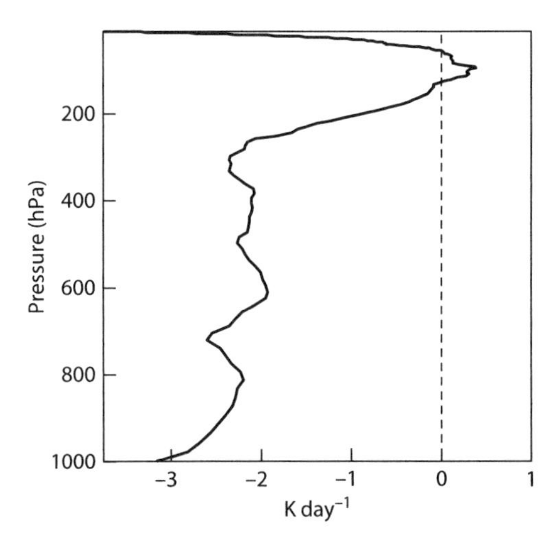

Figure 6.9. Typical tropical clear-sky infrared radiative cooling rate based on a standard tropical sounding.

to the evaporation of liquid deposited at the trade inversion level by the shallow clouds, and a downward flux of sensible heat that cools the air as it crosses the inversion.

A macroscopic view of this entrainment process is as follows. Let A be an arbitrary scalar satisfying a conservation equation that can be written in "flux form" as

$$\frac{\partial(\rho A)}{\partial t} + \nabla \cdot (\rho \mathbf{V} A) + \frac{\partial(\rho w A)}{\partial z} = -\frac{\partial F_A}{\partial z} + S_A, \tag{51}$$

where $F_A \equiv \rho \overline{w'A'}$ is the upward turbulent flux of A; bars are omitted on the mean quantities; and S_A is a source or sink of A, per unit volume. Integrating (51) from just below to just above the inversion, and using Leibniz's rule, we get

$$\begin{aligned}\frac{\partial}{\partial t}\left(\int_{z_B-\varepsilon}^{z_B+\varepsilon} \rho A \, dz\right) - \Delta(\rho A)\frac{\partial z_B}{\partial t} + \nabla \cdot \left(\int_{z_B-\varepsilon}^{z_B+\varepsilon} \rho \mathbf{V} A \, dz\right) - \Delta(\rho \mathbf{V} A) \cdot \nabla z_B + \Delta(\rho w A) \\ = -(F_A)_{B+} + (F_A)_B + \int_{z_B-\varepsilon}^{z_B+\varepsilon} S_A \, dz,\end{aligned} \tag{52}$$

where the indicated terms drop out as the domain of integration shrinks to zero and/or because all the turbulence variables go to zero above the inversion. Here we used the notation $\Delta(\;) \equiv (\;)_{z=z_B+\varepsilon} - (\;)_{z=z_B-\varepsilon} \equiv (\;)_{B+} - (\;)_B$. Henceforth, subscripts $B+$ and B denote levels just above and just below the inversion, respectively. For $A \equiv 1$, (52) reduces to mass conservation in the form

$$\rho_{B+}\left(\frac{\partial z_B}{\partial t} + \mathbf{V}_{B+} \cdot \nabla z_B - w_{B+}\right) = \rho_B\left(\frac{\partial z_B}{\partial t} + \mathbf{V}_B \cdot \nabla z_B - w_B\right) \equiv E, \tag{53}$$

where E is the downward mass flux across the inversion. In essence, (53) says simply that the mass flux is continuous across the PBL top; that is, no mass is created or destroyed between levels B and $B+$. We interpret E as the mass flux due to the turbulent entrainment of free atmospheric air across the inversion. With the definition of E as given by (53), we can simplify (52) to

Figure 6.10. Douglas Lilly, who did important work on a wide range of topics, including cumulus convection, numerical methods, gravity waves, and stratocumulus clouds. Used with permission of the University Corporation for Atmospheric Research (UCAR).

$$-\Delta A E = (F_A)_B + \int_{z_B-\varepsilon}^{z_B+\varepsilon} S_A \, dz. \tag{54}$$

For $S_A = 0$, (54) says simply that the total flux of A must be continuous across the inversion. Notice that for $\Delta A \neq 0$, a mass flux across the inversion is generally associated with the convergence of a turbulent flux of A at level B. This flux convergence changes the A of entering particles from A_{B+} to A_B. Lilly (1968; see fig. 6.10) was the first to derive (54) using this approach.

As a simple example, consider the moistening of the air as it moves down across the inversion. The dry entrained air is moistened by an upward moisture flux that converges "discontinuously" at level B. This process is described by

$$-\Delta q_T E = (F_{q_T})_B, \tag{55}$$

which is a special case of (54). Here $q_T \equiv q_v + l$ is the total water mixing ratio.

At middle and high latitudes, representative summer and winter soundings for Denver, Colorado, and Barrow, Alaska, are shown in figure 6.11. The Denver sounding is conditionally unstable in summer, but the dry near-surface air has to be lifted quite a long way before it can become positively buoyant; high cloud bases are expected. The winter sounding for Denver is strongly stable near the surface. The Barrow sounding is quite stable all year, but especially so in winter. Note that the tropopause is much lower at Barrow than at Denver. In summer, there are low clouds at Barrow.

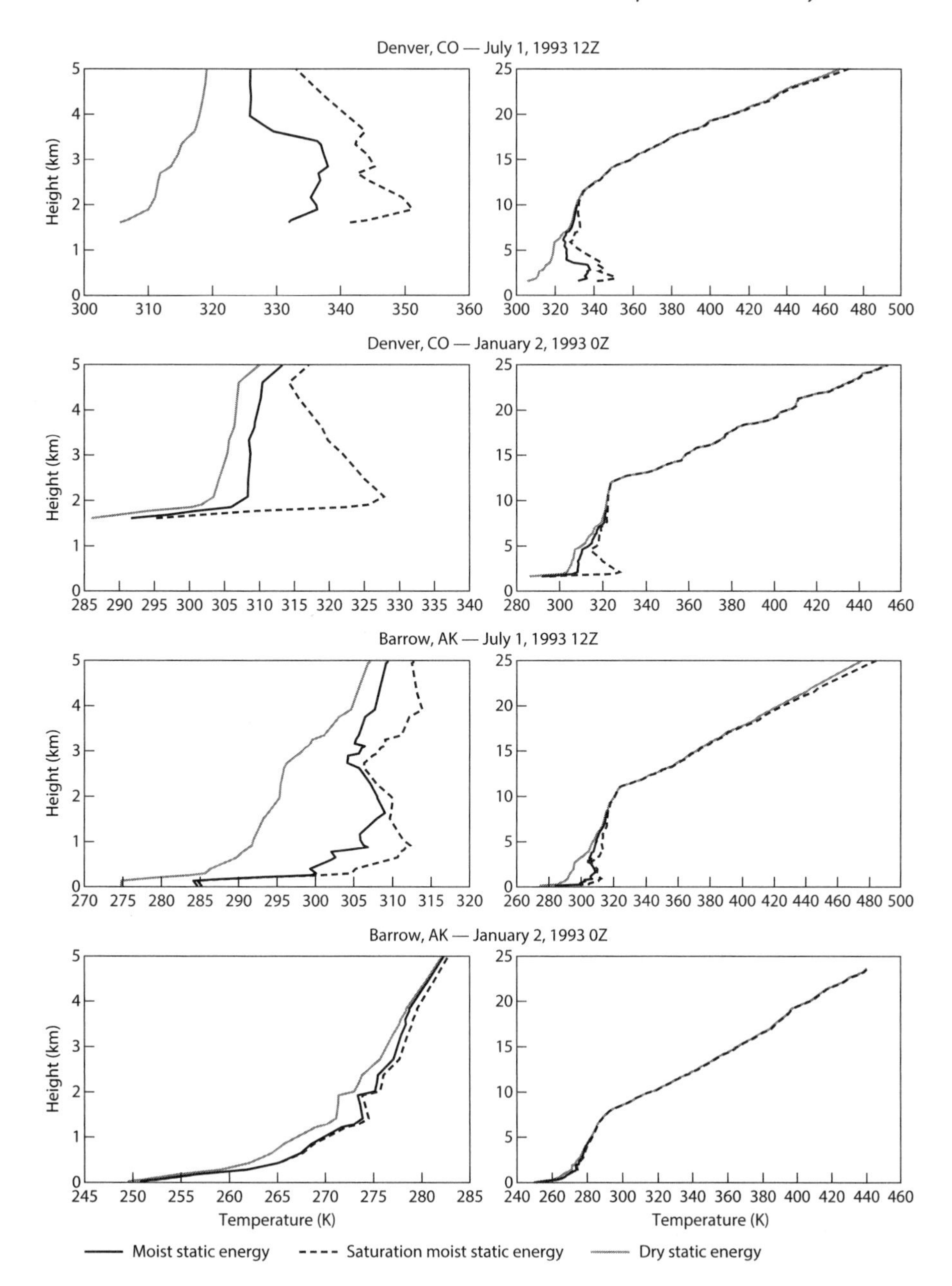

Figure 6.11. Representative observed soundings for Denver, Colorado, and Barrow, Alaska, for July and January. The curves plotted show the dry static energy, the moist static energy, and the saturation moist static energy. The panels on the left cover both the troposphere and lower stratosphere, while those on the right zoom in on the lower troposphere to show more detail. Values are divided by c_p to give units in Kelvin.

The Convective Mass Flux

Moist convection is important in several ways:

- As discussed earlier, Riehl and Malkus (1958) inferred from large-scale observations that moist convection is the primary mechanism for transporting energy upward in the deep tropics. Convection also transports moisture and momentum, as well as various chemical constituents.
- As hypothesized by Manabe and Wetherald (1967), convection acts to prevent the lapse rate in convectively active regions from exceeding a value close to the moist adiabatic lapse rate. Even if the validity of this hypothesis is accepted, the problem becomes one of determining when and where convection is active.

In addition, of course, moist convection is important because

- convection produces a large fraction of the Earth's precipitation; and
- convection generates radiatively important stratiform clouds, especially in the upper troposphere in regions of deep convection.

For the four reasons just summarized, moist convection plays a key role in the global circulation of the Earth's atmosphere. It is no exaggeration to say that the global circulation cannot be understood without considering the interactions between the global circulation and moist convection. There is extensive literature on this subject. Many issues remain unresolved, and some have not even been confronted yet (Randall et al., 2003). Current ideas are that cumulus convection exerts its effects on the large-scale stratification by transporting mass vertically (the "cumulus mass flux") and that the intensity of convection is regulated by the processes that act to produce convective instability; these include radiative cooling of the air relative to the temperature of the lower boundary, surface fluxes of sensible and latent heat, and the effects of both horizontal and vertical advection of temperature and moisture.

The effects of convection on the global circulation can be analyzed following the theory of Arakawa and Schubert (1974; hereafter AS; see fig. 6.12). The theory was developed partly to elucidate mechanisms and partly for use in determining heating and drying rates in numerical models.

Let an overbar denote an average over an area large enough to contain a good sample of cumulus clouds but small enough to contain only a small portion of a large-scale weather system. As pointed out by AS, the existence of such an area is a fundamental assumption of their theory. The area-averaged budget equations for mass, dry static energy, water vapor mixing ratio, and liquid water mixing ratio are as follows:

$$0 = -\nabla \cdot (\rho \overline{\mathbf{V}}) - \frac{\partial(\rho \overline{w})}{\partial z}, \tag{56}$$

$$\rho \frac{\partial \overline{s}}{\partial t} = -\rho \overline{\mathbf{V}} \cdot \nabla \overline{s} - \rho \overline{w} \frac{\partial \overline{s}}{\partial z} + \overline{Q_R} + \rho L \overline{C} - \frac{\partial F_s}{\partial z}, \tag{57}$$

$$\rho \frac{\partial \overline{q_v}}{\partial t} = -\rho \overline{\mathbf{V}} \cdot \nabla \overline{q_v} - \rho \overline{w} \frac{\partial \overline{q_v}}{\partial z} - \rho \overline{C} - \frac{\partial F_{q_v}}{\partial z}, \tag{58}$$

$$\rho \frac{\partial \overline{l}}{\partial t} = -\rho \overline{\mathbf{V}} \cdot \nabla \overline{l} - \rho \overline{w} \frac{\partial \overline{l}}{\partial z} + \rho \overline{C} - \frac{\partial F_l}{\partial z} - \overline{\chi}. \tag{59}$$

Figure 6.12. Professor Akio Arakawa, cruising along in midlecture. This photo was taken at a symposium at UCLA in 1998.

Here ρ is the density of the air, which is presumed to be quasi-constant at each height; $s \equiv c_p T + gz$ is the dry static energy; q is the water vapor mixing ratio; w is the vertical velocity; $\overline{\mathbf{V}}$ is the horizontal velocity; and χ is the rate at which liquid water is being converted into precipitation, which then falls out and so acts as a sink of l. The vertical "eddy fluxes," $F_s \equiv \rho\overline{ws} - \rho\overline{w}\,\overline{s}$ and $F_{q_v} \equiv \rho\overline{wq_v} - \rho\overline{w}\,\overline{q_v}$, can in principle represent quite a variety of physical processes, but here we assume for simplicity that above the boundary layer these fluxes are due only to the vertical currents associated with cumulus convection.

AS used a very simple cumulus cloud model to formulate the eddy fluxes that appear in (57)–(59) in terms of a convective mass flux and the differences between the in-cloud and environmental soundings. The cloud model was also used to formulate the net condensation rate, C, per unit mass flux. AS allowed the possibility that clouds of many different "types" coexist; here a *cloud type* can roughly be interpreted as a cloud size category. We now briefly explain the AS parameterization, using a single cloud type for simplicity.

As a first step, we divide the domain into an arbitrary number N of subdomains, each having a characteristic fractional area σ_i, a characteristic vertical velocity w_i, and corresponding characteristic values of the moist static energy, dry static energy, water vapor mixing ratio, and all the other variables of interest. Some of the subdomains represent cloudy updrafts or downdrafts, while others might represent mesoscale subdomains or the broad "environment" of the clouds. The fractional areas must sum to unity:

$$\sum_{i=1}^{N} \sigma_i = 1. \tag{60}$$

The area-averaged vertical velocity and moist static energy satisfy

$$\sum_{i=1}^{N} \sigma_i w_i = \overline{w}, \tag{61}$$

$$\sum_{i=1}^{N} \sigma_i h_i = \overline{h}, \tag{62}$$

respectively. Other area averages are constructed in a similar way. It then follows that

$$F_h \equiv \rho\overline{wh} - \rho\overline{w}\,\overline{h} = \sum_{i=1}^{N} M_i(h_i - \overline{h}), \tag{63}$$

where

$$M_i \equiv \rho\sigma_i(w_i - \overline{w}) \tag{64}$$

is the *convective mass flux* associated with cloud type i. The convective mass flux is a key concept. It represents the rate at which mass is pumped through the convective circulations—through the updraft, through the compensating sinking motion outside the updraft, and through the horizontal branches of the convective circulation that connect the updraft and the sinking motion.

We assume for simplicity that the effects of mesoscale organization and convective-scale downdrafts can be neglected, so that the cloudy layer consists of concentrated convective updrafts of various sizes and intensities, embedded in a broad uniform environment. We use a tilde to denote an environmental value, and a subscript c to denote the collective properties of the cloudy updrafts. Then (60) simplifies to

$$\sigma_c + \tilde{\sigma} = 1, \tag{65}$$

where

$$\sigma_c \equiv \sum_{\text{all clouds}} \sigma_i \tag{66}$$

is the total fractional area covered by all the convective updrafts, and $\tilde{\sigma}$ is the fractional area of the environment. Similarly, (61) and (62) become

$$\sigma_c w_c + \tilde{\sigma}\tilde{w} = \overline{w}, \tag{67}$$

$$\sigma_c h_c + \tilde{\sigma}\tilde{h} = \overline{h}, \tag{68}$$

where we define

$$w_c \equiv \frac{1}{\sigma_c} \sum_{\text{all clouds}} \sigma_i w_i, \tag{69}$$

$$h_c \equiv \frac{1}{\sigma_c} \sum_{\text{all clouds}} \sigma_i h_i. \tag{70}$$

Why σ_c Is Small and How This Simplifies Things

It is observed that

$$\sigma_c \ll 1 \text{ and } \tilde{\sigma} \cong 1. \tag{71}$$

This means that the cumulus updrafts occupy only a very small fraction of the area. A simple explanation for this important fact was given by Bjerknes (1938).

Suppose that at a certain time the temperature is horizontally uniform, with lapse rate

$$\Gamma \equiv -\frac{\partial T}{\partial z}. \tag{72}$$

Consider temperature changes due to adiabatic vertical motion only. As discussed earlier, in a cloudy region, the temperature satisfies

$$\frac{\partial T_c}{\partial t} = w_c(\Gamma - \Gamma_m), \tag{73}$$

while in a neighboring clear region,

$$\frac{\partial \tilde{T}}{\partial t} = \tilde{w}(\Gamma - \Gamma_d). \tag{74}$$

If the sounding is conditionally unstable, then

$$\Gamma - \Gamma_m > 0 \text{ and } \Gamma - \Gamma_d < 0. \tag{75}$$

Suppose that the cloudy air is rising and the environmental air is sinking, so that $w_c > 0$, $\tilde{w} < 0$, and $w_c - \tilde{w} > 0$. Comparing (75) with (73) and (74), we see that when the sounding is conditionally unstable, both T_c and $\tilde{T}$ will tend to increase with time. The buoyancy of the updraft is proportional to $T_c - \tilde{T}$. An increase in T_c favors convection because it tends to increase the buoyancy of the cloudy air, but an increase in $\tilde{T}$ tends to decrease the buoyancy. Which effect wins out? The answer depends on the value of σ_c.

The mean vertical motion satisfies (67), from which it follows that

$$w_c = \overline{w} + (1 + \sigma_c)(w_c - \tilde{w}), \tag{76}$$

$$\tilde{w} = \overline{w} - \sigma_c(w_c - \tilde{w}). \tag{77}$$

Subtracting (74) from (73), and substituting (76) and (77), we find that

$$\begin{aligned} \frac{\partial}{\partial t}(T_c - \tilde{T}) &= w_c(\Gamma - \Gamma_m) - \tilde{w}(\Gamma - \Gamma_d) \\ &= \overline{w}(\Gamma_d - \Gamma_m) + (w_c - \tilde{w})[(1 - \sigma_c)(\Gamma - \Gamma_m) + \sigma_c(\Gamma - \Gamma_d)]. \end{aligned} \tag{78}$$

Consider the quantity $[(1-\sigma_c)(\Gamma-\Gamma_m)+\sigma_c(\Gamma-\Gamma_d)]$, which multiplies $w_c - \tilde{w}$ on the second line of (78). Inspection shows that $[(1-\sigma_c)(\Gamma-\Gamma_m)+\sigma_c(\Gamma-\Gamma_d)]$ is maximized for $\sigma_c \to 0$. The physical interpretation is simple. With a conditionally unstable sounding, saturated rising motion is aided by positive buoyancy created through condensation, while unsaturated sinking motion must fight against the dry-stable stratification. The rate of temperature increase in the updraft is proportional to the updraft speed, while the rate of temperature increase in the downdraft is proportional to the downdraft speed. Therefore, convection is favored by rapid rising motion in the cloudy region and slow sinking motion in the clear region, both of which can be achieved, for a given value of $w_c - \tilde{w}$, by making the updraft narrow and the downdraft broad.

With this simple idea, Bjerknes (1938) explained the observed smallness of σ_c.

The smallness of σ_c is important for the global circulation because it means that even in the regions where deep convection is most active there is a lot of clear sky where no phase changes occur. The convective clouds "tunnel through" deep layers of (mostly) unsaturated air. Cloud processes would be relatively simple if they involved only uniform cloudiness over large regions. Nature's preference for narrow saturated updrafts in clear environments makes the interaction of moist convection with the global circulation a much more subtle (and interesting) problem than it would otherwise be.

In view of (71), we can write (68) as

$$\tilde{h} \cong \overline{h}, \tag{79}$$

It is *not* true, however, that $\tilde{w} \cong \overline{w}$, because the cumulus updrafts are typically several orders of magnitude stronger than the large-scale vertical motions; that is,

$$w_i \gg \overline{w}. \tag{80}$$

Similarly, it is *not* true in general that $\overline{l} \cong \tilde{l}$, because there may be no liquid water at all in the environment of the convective clouds. Using (80), we can approximate (64) by

$$M_i \cong \rho\sigma_i w_i. \tag{81}$$

From this point, we simplify the discussion by considering only one type of convective cloud, whose properties are denoted by subscript c. We write

$$F_s \cong M_c(s_c - \overline{s}), \tag{82}$$

$$F_{q_v} \cong M_c[(q_v)_c - \overline{q_v}], \tag{83}$$

$$F_l \cong (l_c - \tilde{l}). \tag{84}$$

Here s_c, $(q_v)_c$, and l_c are the in-cloud dry static energy, water vapor mixing ratio, and liquid water mixing ratio, respectively, and

$$M_c \equiv \rho\sigma_c w_c. \tag{85}$$

In (84), we allow the possibility of liquid water in the environment of the cumulus clouds, but we do *not* assume that $\overline{l} \cong \tilde{l}$, because it is quite possible that the only liquid present is in the convective updrafts. It is also possible, however, that the environment is filled or partially filled with stratiform clouds, perhaps created by earlier detrainment from the convective clouds. Finally, we need

$$\overline{s} = (1 - \sigma_c)\tilde{s} + \sigma_c s_c, \tag{86}$$

$$\overline{q_v} = (1 - \sigma_c)\tilde{q}_v + \sigma_c (q_v)_c, \tag{87}$$

$$\overline{l} = (1 - \sigma_c)\tilde{l} + \sigma_c l_c, \tag{88}$$

$$\overline{C} = (1 - \sigma_c)\tilde{C} + \sigma_c C_c, \tag{89}$$

$$\overline{\chi} = (1 - \sigma_c)\tilde{\chi} + \sigma_c \chi_c. \tag{90}$$

With these results, we can now rewrite (67)–(69) as

$$\rho\frac{\partial\overline{s}}{\partial t} = -\rho\overline{\mathbf{V}}\cdot\nabla\overline{s} - \rho\overline{w}\frac{\partial\overline{s}}{\partial z} + Q_R + \rho L(\tilde{C} + \sigma_c C_c) - \frac{\partial}{\partial z}[M_c(s_c - \overline{s})], \tag{91}$$

$$\rho\frac{\partial\overline{q}_v}{\partial t} = -\rho\overline{\mathbf{V}}\cdot\nabla\overline{q}_v - \rho\overline{w}\frac{\partial\overline{q}_v}{\partial z} - \rho(\tilde{C} + \sigma_c C_c) - \frac{\partial}{\partial z}\{M_c[(q_v)_c - \overline{q}_v]\}, \tag{92}$$

$$\rho\frac{\partial\overline{l}}{\partial t} = -\rho\overline{\mathbf{V}}\cdot\nabla\overline{l} - \rho\overline{w}\frac{\partial\overline{l}}{\partial z} + \rho(\tilde{C} + \sigma_c C_c) - \frac{\partial}{\partial z}[M_c(l_c - \overline{l})] - [(1 - \sigma_c)\tilde{\chi} + \sigma_c\chi_c]. \tag{93}$$

The convective condensation rate, C_c, appears in all three of these equations, as would be expected.

A Simple Cumulus Cloud Model

To go further, we need to know the soundings inside the updrafts, so a simple cumulus cloud model is required. We assume that all cumulus clouds originate from the top of the PBL, carrying the mixed-layer properties upward. The mass flux changes with height according to

$$\frac{\partial M_c(z)}{\partial z} = E(z) - D(z). \tag{94}$$

Here E is the entrainment rate, and D is the detrainment rate. The in-cloud profile of moist static energy, $h_c(z)$, is governed by

$$\begin{aligned}\frac{\partial}{\partial z}[M_c(z)h_c(z)] &= E(z)\overline{h}(z) - D(z)h_c(z) \\ &\cong E(z)\overline{h}(z) - D(z)h_c(z).\end{aligned} \tag{95}$$

There are no source or sink terms in (95) because the moist static energy is unaffected by phase changes and/or precipitation processes, and we neglect radiative effects. By combining (94) and (95), we can show that

$$\frac{\partial h_c(z)}{\partial z} = \frac{E(z)}{M_c}[\overline{h}(z) - h_c(z)]. \tag{96}$$

This means that $h_c(z)$ is affected by entrainment, which dilutes the cloud with environmental air, but not by detrainment, which has been assumed to expel from the cloud air that carries the cloud's own moist static energy at each level.

Similarly, we can write

$$\frac{\partial}{\partial z}(M_c s_c) = E\overline{s} - Ds_c + \rho\sigma_c LC_c, \tag{97}$$

$$\frac{\partial}{\partial z}[M_c(q_v)_c] = E\overline{q_v} - D(q_v)_c - \rho\sigma C_c, \tag{98}$$

$$\frac{\partial}{\partial z}(M_c l_c) = E\tilde{l} - Dl_c + \rho\sigma C_c - \chi_c. \tag{99}$$

A simple microphysical model is needed to determine χ_c, that is, to determine how much of the condensed water is converted to precipitation, and the fate of the precipitation. The role of convectively generated precipitation, which drives convective downdrafts and moistens the lower troposphere by evaporating as it falls, is an important issue, but it will not be discussed here.

Compensating Subsidence

By using (97)–(99), we can rewrite the large-scale budget equations in a very interesting way, as follows. First, we consider the dry static energy. We write

$$\frac{\partial}{\partial z}[M_c(s_c - \overline{s})] = \frac{\partial}{\partial z}(M_c s_c) - M_c\frac{\partial \overline{s}}{\partial z} - \overline{s}\frac{\partial M_c}{\partial z}. \tag{100}$$

We substitute from (94) and (97) into (100), to obtain

$$\begin{aligned}\frac{\partial}{\partial z}[M_c(s_c - \overline{s})] &= (E\overline{s} - Ds_c + \rho L\sigma_c C_c) - M_c\frac{\partial \overline{s}}{\partial z} - \overline{s}(E - D) \\ &= -M_c\frac{\partial \overline{s}}{\partial z} - D(s_c - \overline{s}) + \rho L\sigma_c C_c.\end{aligned} \tag{101}$$

This allows us to rewrite (91) as

$$\rho \frac{\partial \bar{s}}{\partial t} = -\rho \bar{\mathbf{V}} \cdot \nabla \bar{s} - \rho \bar{w} \frac{\partial \bar{s}}{\partial z} + \overline{Q_R} + \rho L \tilde{C} + M_c \frac{\partial \bar{s}}{\partial z} + D(s_c - \bar{s}). \tag{102}$$

The last two terms on the right-hand side of (102) represent the cumulus effects, and the first of these in particular is quite interesting. It "looks like" an advection term. It represents the warming of the environment due to the downward advection of air from above, with higher dry static energies, by the sinking motion that compensates for the rising motion in the cloudy updraft. The sinking motion is often called *compensating subsidence*, because it compensates for the concentrated rising motion in the saturated updrafts: up moist, down dry.

The role of compensating subsidence can be seen more explicitly by combining the two "vertical advection" terms of (92) and using (67) to obtain

$$\boxed{\rho \frac{\partial \bar{s}}{\partial t} = -\rho \bar{\mathbf{V}} \cdot \nabla \bar{s} - \tilde{M} \frac{\partial \bar{s}}{\partial z} + \overline{Q_R} + \rho L \tilde{C} + D(s_c - \bar{s})}, \tag{103}$$

where

$$\tilde{M} \equiv \rho \bar{w} - M_c \tag{104}$$

is the *environmental mass flux*. Why does $\tilde{M}$ appear in (103)? The reason is that the environmental subsidence is modifying $\tilde{s}$, but $\bar{s} = \tilde{s}$. The last term on the right-hand side of (103) represents the effects of detrainment. You may be surprised to see that the cumulus condensation rate does not appear in (102) or (103). The reason is that condensation inside the updraft cannot directly warm the environment. Since almost the entire area is the environment, condensation in the updrafts does not, to any significant degree, directly affect the area-averaged dry static energy. Instead, the effects of condensation are felt indirectly, through the compensating subsidence term, as already explained. The physical role of condensation, then, is to make possible the convective updraft that drives the compensating subsidence, which in turn warms the environment. This is how condensation warms indirectly. Note that the vertical profile of the indirect condensation heating rate due to compensating subsidence is, in general, different from the vertical profile of the convective condensation rate itself.

In a similar way, we find that the water vapor budget equation can be rewritten as

$$\boxed{\rho \frac{\partial \overline{q_v}}{\partial t} = -\rho \bar{\mathbf{V}} \cdot \nabla \overline{q_v} - \tilde{M} \frac{\partial \overline{q_v}}{\partial z} - \rho \tilde{C} + D[(q_v)_c - \overline{q_v}]}. \tag{105}$$

Equation (105) describes the convective drying in terms of convectively induced subsidence in the environment, which brings down drier air from aloft. Detrainment of water vapor, liquid, and ice from the convective clouds can tend to moisten the environment, although this is not an entirely straightforward process, for reasons discussed in chapter 8.

Finally, similar methods can be used to write the liquid water budget equation as

$$\rho \frac{\partial \bar{l}}{\partial t} = -\rho \bar{\mathbf{V}} \cdot \nabla \bar{l} - \rho \bar{w} \frac{\partial \bar{l}}{\partial z} + \rho \overline{C} + M_c \frac{\partial \bar{l}}{\partial z} + D(l_c - \tilde{l}) - [(1 - \sigma_c) \tilde{\chi} + \sigma_c \chi_c]. \tag{106}$$

In (106), we cannot combine the two "vertical advection" terms, because one of them involves $\partial \bar{l}/\partial z$, while the other involves $\partial \tilde{l}/\partial z$. Detrained liquid (or ice) can persist in the form of stratiform "anvil" and cirrus clouds.

Within the limits of applicability of the assumptions used above, (103) and (105)–(106) are equivalent to (57)–(59). We will use (103) and (105) in chapter 8.

The Mass Flux Profile

As discussed earlier, the buoyancy of the cloudy air at height z is approximately given by

$$B(z) \cong T_c - \tilde{T} \sim \frac{1}{c_p}\left[h_c(z) - \overline{h_{sat}}(z)\right], \tag{107}$$

where $\overline{h_{sat}}$ is the saturation moist static energy. Because more rapidly entraining clouds lose their buoyancy at lower levels, in effect the cloud types differ according to their cloud-top height, for a given sounding. The cloud top occurs at level $\hat{p}$, where

$$B(\hat{p}) \approx 0. \tag{108}$$

This equation can be used to find $\hat{p}$, after the in-cloud sounding has been determined using (96)–(99).

To determine the entrainment rate, AS assumed that

$$E = \lambda M_c, \tag{109}$$

where λ, which is called the *fractional entrainment rate* and has units of inverse length, is assumed to be a constant (with height) for each cloud type. Larger values of λ mean stronger entrainment; $\lambda = 0$ means no entrainment, and $\lambda < 0$ has no physical meaning and so is not allowed. For a given sounding, clouds with smaller values of λ (weaker fractional entrainment) will have higher tops.

For simplicity, AS assumed that detrainment occurs only at cloud top. This means that the mass flux jumps discontinuously to zero at cloud top. Below the cloud top, there is entrainment but no detrainment, so that (94) reduces to

$$\frac{\partial M_c(z)}{\partial z} = E(z). \tag{110}$$

Combining (109) and (110), and using the assumption that λ is constant with height for each cloud type, we find that

$$M_c(z,\lambda) = M_B(\lambda)\exp(\lambda z), \tag{111}$$

where $M_B(\lambda)$ is the *cloud-base mass flux distribution function*. We define a normalized mass flux, denoted by $\eta(\lambda, z)$; the normalization is in terms of the cloud-base mass flux:

$$M_c(z,\lambda) \equiv M_B(\lambda)\eta(\lambda, z). \tag{112}$$

Note that by virtue of its definition, $\eta(\lambda, z_B) = 1$; here z_B is the cloud-base height.

At this point, we have the equations needed to determine the rates at which cumulus convection warms and dries the large-scale state, *if we can determine $M_B(\lambda)$, which is a measure of convective intensity*. A physical idea is needed to determine $M_B(\lambda)$.

Determining the Intensity of the Convection

To determine the intensity of convective activity, AS proposed a "quasi-equilibrium" hypothesis, according to which the convective clouds quickly convert whatever moist convective available potential energy is present in convectively active atmospheric columns into convective kinetic energy. The starting point for the quasi-equilibrium closure is the recognition that cumulus convection occurs as a result of moist convective instability, in which the potential energy of the mean state is converted into the kinetic energy of cumulus convection.

AS defined the "cloud work function," A, for a cumulus subensemble, as a vertical integral of the buoyancy of the cloud air with respect to the large-scale environment:

$$A(\lambda) = \int_{z_B}^{z_{D(\lambda)}} \frac{g}{c_p \overline{T}(z)} \eta(z,\lambda)\left[s_{vc}(z,\lambda) - \overline{s_v}(z)\right]dz. \quad (113)$$

Here $z_D(\lambda)$ is the height of the detrainment level for cloud type λ, and s_v denotes the virtual static energy. From (113) we see that the function $A(\lambda)$ is a property of the large-scale environment. A positive value of $A(\lambda)$ means that a cloud with fractional entrainment rate can convert the potential energy of the mean state into convective kinetic energy. For $\lambda = 0$, $A(\lambda)$ is equivalent to the *convective available potential energy* (CAPE), as conventionally defined.

Numerical models use the conservation equations for thermodynamic energy and moisture to predict $\overline{T}(z)$ and $\overline{q}(z)$, from which $A(\lambda)$ can be determined; therefore, these models *indirectly* predict $A(\lambda)$. By taking the time derivative of (113) and using the conservation equations for thermodynamic energy and moisture, AS derived an equation that can be written in simplified form as

$$\frac{dA(\lambda)}{dt} = J(\lambda) M_B(\lambda) + F(\lambda). \quad (114)$$

The JM_B term of (114) represents all the terms involving convective processes, each of which turns out to be proportional to M_B. The JM_B term is actually an integral over cloud types and is written here as a product merely to simplify the discussion. The quantity $J(\lambda)$ symbolically represents the kernel of the integral, which is a property of the large-scale sounding (see AS for details). The JM_B term of (114) tends to reduce $A(\lambda)$, because cumulus convection stabilizes the environment, so that $J(\lambda)$ is usually negative. Keep in mind that an equation like (114) holds for each cumulus subensemble.

The $F(\lambda)$ term of (114) represents what AS called the "large-scale forcing," that is, the rate at which the cloud work function tends to increase with time owing to a variety of processes including the following:

- horizontal and vertical advection by the mean flow;
- the surface turbulent fluxes of sensible and latent heat, and the rate of change of the planetary boundary-layer depth;
- radiative heating and cooling; and
- precipitation and turbulence in stratiform clouds.

Note that some of these "forcing" processes, such as those involving boundary-layer turbulence and stratiform clouds, are themselves parameterized processes that may involve fluctuations on small spatial scales; for this reason it seems inappropriate to describe the collection of processes that contribute to F as "large-scale;" a better term would be "nonconvective."

AS assumed quasi-equilibrium (QE) of the cloud work function; that is,

$$\boxed{\frac{d}{dt}A(\lambda) = JM_B(\lambda) + F(\lambda) \cong 0 \text{ when } F(\lambda) > 0}. \quad (115)$$

Equation (115) means that the moist convective instability generated by the forcing, $F(\lambda)$, is very rapidly consumed by cumulus convection; that is, the two terms on the right-hand side of (115) approximately balance each other. In a steady-state situation, this balance is of course trivially satisfied, by definition. The physical content of (115) is therefore the assertion that near-balance is maintained even when $F(\lambda)$ is varying with time, provided that the variations of $F(\lambda)$ are sufficiently slow. The cumulus ensemble thus closely follows the lead of the forcing, like a defensive basketball player (the convection) playing one-on-one against an offensive player (the forcing). However, the forcing depends on the large-scale circulation, which is strongly affected by the convection, just as the play of an offensive basketball player is strongly affected by the moves of his or her defensive opponent. It would be wrong to imagine that the forcing is "given" and that the convection just meekly responds to it. The convection and the forcing evolve together according to the rules defined by the combination of large-scale dynamics and cloud dynamics.

The QE approximation is expected to hold if τ_{LS}, the timescale for changes in $F(\lambda)$, is much longer than the "adjustment time," τ_{adj}, required for the convection to consume the available CAPE; this allows the convection to keep up with the changes in $F(\lambda)$. AS introduced the concept of τ_{adj} by describing what would happen if a conditionally unstable initial sounding was modified by cumulus convection, without any forcing to maintain the CAPE over time. They asserted that the CAPE would be consumed by the convection (i.e., converted into convective kinetic energy) on a timescale they defined as and estimated to be on the order of a couple of hours. Just such an unforced convective situation has been numerically simulated by Soong and Tao (1980) and others, using high-resolution cloud models; their results are consistent with the scenario of AS. If the adjustment time is on the order of 10^3 to 10^4 s, then use of (115) is justified, as an approximation, for the simulation of "weather" whose timescale is on the order of one day or longer, that is, at least one order of magnitude longer than τ_{adj}.

By using (115) together with

$$|JM_B| \sim \frac{A}{\tau_{adj}}, \tag{116}$$

AS found that

$$A \sim \tau_{adj} F \ll \tau_{LS} F, \tag{117}$$

where τ_{LS} is the timescale on which the forcing itself is varying. Equation (117) means that the cloud work function is "small" compared with $\tau_{LS}F$, which is the value that the cloud work function would take if the forcing acted without opposition over a timescale τ_{LS}. Although day-to-day variations of A should be expected, values as large as $\tau_{LS}F$ never occur. This means that A is trapped in the range of values between zero (since by definition A cannot be negative) and $\tau_{adj}F$. In this sense, A is "close to zero" (see also Xu and Emanuel, 1989).

Based on the preceding analysis, it can be asserted that the cloud work function (or the CAPE) "is quasi-invariant with time," that is, that

$$\frac{dA}{dt} \cong 0, \tag{118}$$

which is a shorthand form of (115), and that "the CAPE is small" in convectively active regimes; that is,

$$A \cong 0, \tag{119}$$

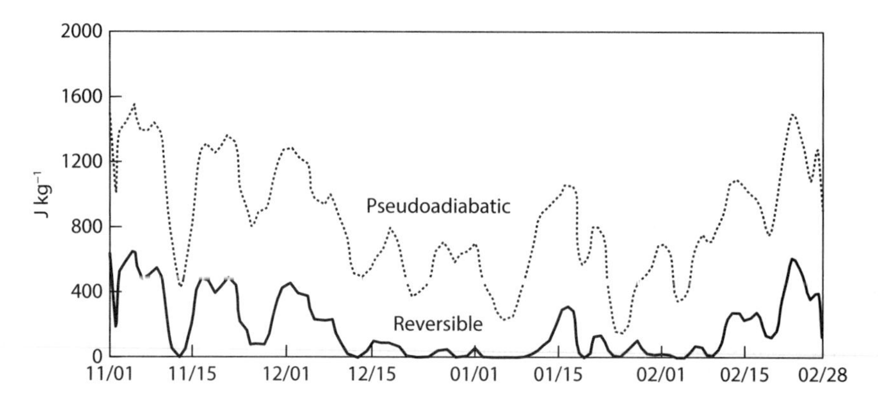

Figure 6.13. Observed time sequence of the CAPE during TOGA COARE, over the Intensive Flux Array. K. Emanuel's code was used to construct these curves. Used with kind permission from Springer Science and Business Media.

which is a short-hand form of (117). A sounding for which $A \cong 0$ tends to follow a saturated moist adiabat throughout the depth of the convective layer. This provides a rationalization for Manabe and Wetherald's (1967) assumption that the lapse rate cannot exceed the moist adiabatic lapse rate, which they approximated as 6.5 K km^{-1}.

Unfortunately, because (118) and (119) are shorthand forms, they are subject to misinterpretation. For example, data like those shown in figure 6.13 are sometimes viewed as being inconsistent with QE. It is natural to wonder how the CAPE can be described as "quasi-invariant" when it is observed to undergo such "large" changes. This point of view appears to be based on the tacit assumption that saying that the changes in the CAPE are "small" means that they are small compared with the time average of the CAPE. In fact, however, it should be clear from the preceding discussion that this is not what is meant at all. Instead, what is meant is that the changes in the CAPE are small compared with those that would occur if the convection were somehow suppressed while the nonconvective processes continued to increase the CAPE with time. Equation (115) does not imply that A is invariant from day to day, and the observed day-to-day changes in the CAPE, such as those shown in figure 6.13, are not in conflict with QE. What QE does imply is that the changes in the CAPE actually seen, from day to day, are much smaller than they would be if the negative convective term of (115) could somehow be suppressed, so that the positive (under disturbed conditions) forcing term had control of the sounding.

A practical application of (115) is to solve it for the convective mass flux as a function of cloud type, λ. After discretization this leads to a system of linear equations (Lord et al. 1982). Although the system is linear, the mass flux distribution function, $M_B(\lambda)$, is required to be nonnegative for all λ. This cannot be guaranteed without making additional assumptions (e.g., Hack et al., 1984). An alternative approach that avoids these difficulties was proposed by Randall and Pan (1993) and Pan and Randall (1998).

The QE hypothesis has been observationally tested by Arakawa and Schubert (1974), Lord and Arakawa (1980), Lord (1982), Kao and Ogura (1987), Arakawa and Chen (1987), Grell et al. (1991), Wang and Randall (1994), and Cripe and Randall (2001), among others, using both tropical and midlatitude data. In addition, idealized tests of QE using a high-resolution cloud ensemble model have been performed by Xu and Arakawa (1992) and Jones and Randall (2011).

For further discussions of cumulus parameterization see the collections of essays edited by Emanuel and Raymond (1993) and Smith (1998), as well as the articles by Arakawa (2004) and Randall et al. (2003, 2013).

In chapter 8 we discuss the role of convection in tropical eddies.

Conditional Symmetric Instability

Deep convection occurs in both the tropics and middle latitudes. Especially in middle latitudes, the convective instability can interact with inertial instability; in fact, there are strong analogies between the two types of instability. Emanuel (1979, 1982) pointed out that when the air is saturated with water vapor the criterion for inertial instability can be expressed in terms of angular momentum variations along surfaces of constant equivalent potential temperature (or moist static energy), rather than surfaces of constant dry potential temperature. For saturated motion, instability can occur when the angular momentum decreases poleward along surfaces of constant-saturation moist static energy. Another way of saying the same thing is that for saturated motion, instability can occur when the saturation moist static energy decreases with height along surfaces of constant angular momentum.

To show how this process works, we now discuss inertial instability and its connections to convection. To avoid unnecessary complications, we also assume that there are no variations with longitude. This assumption can be justified, at least in this pedagogical example, because the physics of inertial instability relate to spatial variations of the angular momentum, which occur mainly with latitude. Inertial instability is sometimes called "symmetric instability" in the atmospheric science literature. This is not meant to suggest that inertial instability has to be or is intrinsically independent of longitude. It can occur, for example, in the vicinity of a local maximum of the zonal wind, that is, a "jet max."

We use θ-coordinates with the quasi-static approximation and assume that there are no variations in the longitudinal direction, and no heating. We can write the Lagrangian time derivative as

$$\frac{D}{Dt} \equiv \frac{\partial}{\partial t} + \frac{v}{a}\frac{\partial}{\partial \varphi}, \tag{120}$$

where the partial derivatives are taken along θ-surfaces. Particles move along isentropic surfaces and so do not experience any buoyancy forces. The equations of horizontal motion are given by

$$\frac{Du}{Dt} - \left(2\Omega + \frac{u}{a\cos\varphi}\right)v\sin\varphi = 0, \tag{121}$$

$$\frac{Dv}{Dt} + \left(2\Omega + \frac{u}{a\cos\varphi}\right)u\sin\varphi + \Omega^2 a\cos\varphi\sin\varphi = -\frac{1}{a}\frac{\partial s}{\partial \varphi}. \tag{122}$$

The term $\Omega^2 a\cos\varphi\sin\varphi$ in (122) represents the centrifugal acceleration. As discussed in chapter 4, equation (121) implies that angular momentum is conserved; that is,

$$\frac{DM}{Dt} = 0, \tag{123}$$

where

$$M \equiv a\cos\varphi\,(u + \Omega a\cos\varphi). \tag{124}$$

Using (124), we can rewrite (122) in the more compact form

$$\frac{Dv}{Dt} + \frac{M^2\sin\varphi}{a^3\cos^3\varphi} = -\frac{1}{a}\frac{\partial s}{\partial\varphi}. \tag{125}$$

We now linearize the system about a basic state in which the flow is purely zonal and in gradient wind balance, so that $\overline{v}^{\lambda} = 0$, and $[(\overline{M}^{\lambda})^2 \sin\varphi / a^3\cos^3\varphi] = -(1/a)(\partial\overline{s}^{\lambda}/\partial\varphi)$. We can write the linearized versions of (125) and (123) as

$$\frac{\partial v'}{\partial t} + \left(\frac{2\overline{M}^{\lambda}\sin\varphi}{a^3\cos^3\varphi}\right)M' = -\frac{1}{a}\frac{\partial s'}{\partial\varphi}, \tag{126}$$

$$\frac{\partial M'}{\partial t} + \frac{v'}{a}\frac{\partial\overline{M}^{\lambda}}{\partial\varphi} = 0. \tag{127}$$

To investigate inertial stability and instability, we neglect the pressure-gradient term on the right-hand side of (126), which plays only a secondary role; the dry static energy fluctuations on an isentropic surface are quite small. If we assume solutions of the form $(v', M') = (\hat{v}, \hat{M})e^{\sigma t}$, our system reduces to

$$\sigma\hat{v} + \left(\frac{2\overline{M}^{\lambda}\sin\varphi}{a^3\cos^3\varphi}\right)\hat{M} = 0, \tag{128}$$

$$\frac{\partial\overline{M}^{\lambda}}{\partial\varphi}\frac{\hat{v}}{a} + \sigma\hat{M} = 0. \tag{129}$$

For nontrivial solutions, we need

$$\begin{aligned}\sigma^2 &= \left(\frac{2\overline{M}^{\lambda}\sin\varphi}{a^3\cos^3\varphi}\right)\frac{1}{a}\frac{\partial\overline{M}^{\lambda}}{\partial\varphi} \\ &= \frac{\sin\varphi}{a^4\cos^3\varphi}\frac{\partial(\overline{M}^{\lambda})^2}{\partial\varphi}.\end{aligned} \tag{130}$$

According to (130), the system is inertially *stable if the angular momentum decreases toward the pole* along isentropic surfaces, and inertially *unstable if the angular momentum increases toward the pole* along isentropic surfaces. In a neutrally stable state, the angular momentum is constant along isentropic surfaces. These conclusions hold in either hemisphere.

The criterion for inertial instability is often expressed in terms of vorticity rather than angular momentum, as follows. In spherical coordinates, the vertical component of the absolute vorticity is given by

$$\zeta + f = \frac{1}{a\cos\varphi}\frac{\partial v}{\partial\lambda} - \frac{1}{a\cos\varphi}\frac{\partial}{\partial\varphi}(u\cos\varphi) + 2\Omega\sin\varphi. \tag{131}$$

However, the meridional derivative of the angular momentum is

$$\frac{1}{a}\frac{\partial M}{\partial\varphi} = \frac{\partial}{\partial\varphi}(u\cos\varphi) - 2\Omega a\cos\varphi\sin\varphi. \tag{132}$$

Comparing (131) and (132), we see that for a purely zonal flow

$$\zeta + f = \frac{-1}{a\cos\varphi}\frac{\partial M}{\partial\varphi}. \tag{133}$$

This allows us to rewrite (130) as

$$\sigma^2 \cong \frac{-2\sin\varphi}{a^3\cos^2\varphi}\overline{M}^{\lambda}(\overline{\zeta}^{\lambda}+f). \tag{134}$$

For the zonally averaged flow we can safely assume that $\overline{M}^{\lambda}>0$, except possibly close to the poles. It follows that the sign of σ^2 is determined by the sign of $-\sin\varphi(\overline{\zeta}^{\lambda}+f)$. Inertial instability occurs when $\sigma^2>0$. In the Northern Hemisphere, where $\sin\varphi>0$, the criterion for instability is satisfied for $\overline{\zeta}^{\lambda}+f<0$, and in the Southern Hemisphere it is satisfied for $\overline{\zeta}^{\lambda}+f>0$. In either hemisphere, inertial instability occurs when the absolute vorticity has the "wrong" sign. In middle latitudes, this situation can occur on small scales (≤ 100 km) on the anticyclonic (i.e., equatorward) side of a jet maximum. Since $\overline{\zeta}^{\lambda}+f$ passes through zero near the equator, inertial instability is relatively likely to occur in the tropics. The criterion for inertial instability can be satisfied when absolute vorticity is advected across the equator (e.g., Thomas and Webster, 1997).

When the zonal wind does not vary with height, surfaces of constant angular momentum are vertical. When the zonal wind does vary with height, however—for example, below a jet stream or in the vicinity of a front—angular momentum surfaces are tilted in the latitude-height plane. It is possible for the saturation moist static energy to decrease along such tilted surfaces of constant angular momentum *even when it does not decrease with height.* Emanuel (1983b) called this *conditional symmetric instability*, a term that is now commonly used in the forecasting community. Emanuel argued that conditional symmetric instability is relevant to extratropical squall lines that form in regions of strong vertical wind shear.

Segue

The observed vertical structure of the atmosphere is controlled, to a remarkable degree, by diabatic processes. For example, the observed height of the tropopause is approximately that predicted by radiative-convective models, which completely ignore the effects of large-scale circulation systems. The message is that although large-scale dynamics is important, it is strongly constrained by radiation and convection. At the same time, the heating due to radiation, convection, and boundary-layer turbulence is strongly controlled by the global circulation. It is impossible to understand the circulation without understanding the heating, and vice versa. These topics are discussed further in later chapters.

In chapter 5, we saw evidence that much of the upward transport of energy and moisture is carried out by small-scale convection. This chapter outlined the mechanisms by which convection produces these transports. The energetics of the convection itself played a key role in the discussion. The next chapter is devoted to the energetics of the global circulation.

Problems

1. Derive

$$F_h \equiv \rho\overline{wh} - \rho\overline{w}\,\overline{h} = \sum_{i=1}^{N} M_i(h_i - \overline{h}).$$

2. Refer to table 6.1.
   a) Consider the value in the bottom right corner of the table, namely, 1.30×10^{15} J s^{-1}. Assuming that there is no net radiative source or sink of moist static energy at any level in the equatorial trough zone, how can you account for this value? Make a simple sketch to explain your answer.
   b) Using the data given in the table, estimate the *total* upward transport of moist static energy across the 500 hPa surface in the equatorial trough zone, in joules per second.
   c) Using the data given in the table, estimate the upward transport of moist static energy across the 500 hPa surface *due to the large-scale rising motion* in the equatorial trough zone, in joules per second.
   d) Using your answers from parts a–c, calculate the numerical value of $(F_h)_{500\,mb}$, the upward flux of moist static energy due to convection at 500 hPa, in watts per square meter. Assume that the area covered by the equatorial trough zone is 4×10^{13} m^2.
   e) Estimate a rough numerical value (in kg m^{-2} s^{-1}) of the convective mass flux, M_c, at the 500 hPa level in the ITCZ. You will need to use

$$F_h \cong M_c(h_c - \overline{h}), \tag{135}$$

   where h_c is the in-cloud moist static energy, and $\overline{h}$ is the large-scale mean moist static energy. State your assumptions.
3. Suppose that moist static energy is simply conserved; that is,

$$\frac{\partial h}{\partial t} = -\nabla \cdot (\mathbf{V}h) - \frac{\partial (wh)}{\partial z}. \tag{136}$$

   The density was omitted here and throughout the rest of this problem for simplicity. The corresponding continuity equation is

$$0 = -\nabla \cdot \mathbf{V} - \frac{\partial w}{\partial z}. \tag{137}$$

   a) By using Reynolds averaging, show that

$$\frac{\partial \overline{h}}{\partial t} = -\nabla \cdot (\overline{\mathbf{V}}\overline{h} + \overline{\mathbf{V}'h'}) - \frac{\partial}{\partial z}(\overline{w}\overline{h} + \overline{w'h'}). \tag{138}$$

   For large-scale averages (138) can be approximated by

$$\frac{\partial \overline{h}}{\partial t} \cong -\nabla \cdot (\overline{\mathbf{V}}\overline{h}) - \frac{\partial}{\partial z}(\overline{w}\overline{h} + \overline{w'h'}). \tag{139}$$

   b) Show that the moist static energy variance, $\overline{h'^2}$, satisfies

$$\begin{aligned} \frac{\partial \overline{h'^2}}{\partial t} = &-\nabla \cdot (\overline{\mathbf{V}}\overline{h'^2} + \overline{\mathbf{V}'h'h'}) - \frac{\partial}{\partial z}(\overline{w}\overline{h'^2} + \overline{w'h'h'}) \\ &- 2\overline{\mathbf{V}'h'} \cdot \nabla \overline{h} - 2\overline{w'h'}\frac{\partial \overline{h}}{\partial z}. \end{aligned} \tag{140}$$

   For large-scale averages, the time-rate-of-change term of (140) is negligible, as are the terms representing horizontal and vertical advection of $\overline{h'^2}$ by the mean flow and the other terms involving horizontal derivatives, so that (140) can be drastically simplified to

$$0 \cong -\frac{\partial(\overline{w'h'h'})}{\partial z} - 2\overline{w'h'}\frac{\partial \overline{h}}{\partial z}. \tag{141}$$

c) Now, suppose that the vertical velocity fluctuations represented by w' are associated with a single family of cumulus updrafts covering fractional area. Show that

$$\overline{w'h'} = \sigma(1-\sigma)(w_u - w_d)(h_u - h_d), \tag{142}$$

$$\overline{w'h'h'} = \sigma(1-\sigma)(1-2\sigma)(w_u - w_d)(h_u - h_d)^2, \tag{143}$$

where w_u and w_d are the updraft and downdraft velocities, respectively, and h_u and h_d are the corresponding values of the moist static energy.

d) Define a "convective mass flux" by

$$M_c \equiv \sigma w_u. \tag{144}$$

You may assume

$$\sigma \ll 1, \tag{145}$$

and correspondingly that

$$w_u \gg w_d \text{ and } h_d \cong \overline{h}. \tag{146}$$

Show that if the convective mass flux is independent of height, then (141) can be approximated by

$$-\frac{\partial(\overline{w'h'})}{\partial z} \cong M_c \frac{\partial \overline{h}}{\partial z}. \tag{147}$$

CHAPTER 7

Heat Where It's Hot, and Cool Where It's Cold

Available Potential Energy

We showed in chapter 4 that under dry adiabatic and frictionless processes, the total energy of the atmosphere is invariant; that is,

$$\frac{d}{dt}\int_V (c_v T + \phi + K)\rho dV = 0. \tag{1}$$

Here the integral is taken over the mass of the whole atmosphere. Also recall from chapter 4 that for each vertical column, hydrostatic balance implies that the vertically integrated sum of the mass-weighted internal and potential energies is equal to the vertical integral of the mass-weighted enthalpy; that is,

$$\int_0^{p_s} (c_v T + \phi)\, dp = \int_0^{p_s} c_p T dp. \tag{2}$$

Let H and K be the total (mass-integrated) enthalpy and kinetic energy of the entire atmosphere. It follows from (1) and (2) that

$$\boxed{\frac{d}{dt}(H+K) = 0 \text{ under dry adiabatic frictionless processes.}} \tag{3}$$

Imagine that we have the power to spatially rearrange the mass of the atmosphere at will, adiabatically and reversibly. As the parcels move, their entropy and potential temperature do not change, but their enthalpy and temperature do. Suppose that we are given a state of the atmosphere—a set of maps, if you like. Starting from this given state, we move parcels around adiabatically and without friction until we find the unique state of the system that minimizes H. This means that we have reduced H as much as possible from its value in the given state. Because $H + K$ does not change, K is maximized in this special state, which Lorenz (1955) called the *reference state*, and which we will call the *A-state*.

You should prove for yourself that the mass-integrated potential energy of the entire atmosphere is lower in the A-state than in the given state. This means that the "center of gravity" of the atmosphere descends as the atmosphere passes from the given state to the A-state.

An adiabatic process that reduces H will also reduce the total potential energy of the atmospheric column. The nonkinetic energy that "disappears" in this process is converted into kinetic energy. Therefore, we can say that *an adiabatic process that reduces H tends to increase the kinetic energy of the atmosphere.* Two very important processes of this type are convective instability and baroclinic instability.

In passing from the given state to the A-state we have

$$\begin{aligned} &H_{\text{given state}} \rightarrow H_{\min}, \\ &K_{\text{given state}} \rightarrow K_{\text{given state}} + \left(H_{\text{given state}} - H_{\min}\right) = K_{\max}, \end{aligned} \quad (4)$$

where $H_{\min}$ is the value of H in the A-state. The nonnegative quantity

$$\boxed{A = H_{\text{given state}} - H_{\min} \geq 0} \quad (5)$$

is called the *available potential energy*, or APE. The APE was first defined by Lorenz (1955; see fig. 7.1). Equation (5) gives the fundamental definition of the APE.

Notice that the APE is a property of the entire atmosphere; it cannot be rigorously defined for a portion of the atmosphere, although the literature does contain studies in which the APE is computed, without rigorous justification, for a portion of the atmosphere, for example, the Northern Hemisphere.

The A-state is invariant under adiabatic processes, because it depends only on the probability distribution of θ over the mass, rather than on any particular spatial arrangement of the air. Therefore,

Figure 7.1. Professor Edward N. Lorenz, who proposed the concept of available potential energy. He also published a great deal of fundamental research on various topics in both the atmospheric sciences and the relatively new science of nonlinear dynamical systems. Figure used with permission of the MIT Museum.

$$\frac{dA}{dt} = \frac{d}{dt}(H_{\text{given state}} - H_{\min}) \\ = \frac{dH_{\text{given state}}}{dt}. \tag{6}$$

Then, (3) implies that

$$\boxed{\frac{d}{dt}(A+K) = 0 \text{ under dry adiabatic frictionless processes.}} \tag{7}$$

The sum of the available potential and kinetic energies is invariant under adiabatic frictionless processes. This means that such processes only convert between A and K.

The APE of the A-state itself is obviously zero.

To compute A, we have to find the A-state and its (minimum) enthalpy. We can deduce the properties of the A-state as follows. There can't be any horizontal pressure gradients in the A-state, because if there were, K could increase. It follows that the potential temperature must be constant on isobaric surfaces, which is of course equivalent to the statement that p is constant on θ-surfaces. This means that both the variance of θ on isobaric surfaces and the variance of p on isentropic surfaces are measures of the APE. There can be no static instability ($\partial\theta/\partial z < 0$) in the A-state for a similar reason. From these considerations we conclude that in the A-state, θ and T are uniform on each pressure surface or, equivalently, that p is uniform on each θ-surface and also that θ does not decrease with height.

It would appear—and to a large extent it is true—that in passing from the given state to the A-state, no mass can cross an isentropic surface, because we allow only dry adiabatic processes for which $\dot{\theta} = 0$. This implies that $\overline{p}^{\theta}$, the average pressure on an isentropic surface, cannot change as the system passes to the A-state. There is an important exception to this rule, however. If $\partial\theta/\partial z < 0$ in the given state, then the average pressure on an isentropic surface will be different in the A-state. This case of static instability is discussed below.

A complication arises: What about θ-surfaces that intersect the ground in the given state? These are treated as though they continue along the ground, as shown in figure 7.2. Because the "layers" between the isentropic surfaces that are following the ground contain no mass, they have no effect on the physics. They are called *massless layers*. Wherever a θ-surface intersects the Earth's surface the pressure is $p = p_S$.

As shown by the arrows in figure 7.3, warm air must rise (move to lower pressure), and cold air must sink (move to higher pressure) to pass from the given state to the A-state. This is what happens as APE is released; to reach the A-state, the isobaric surfaces flatten out to coincide with isentropic surfaces.

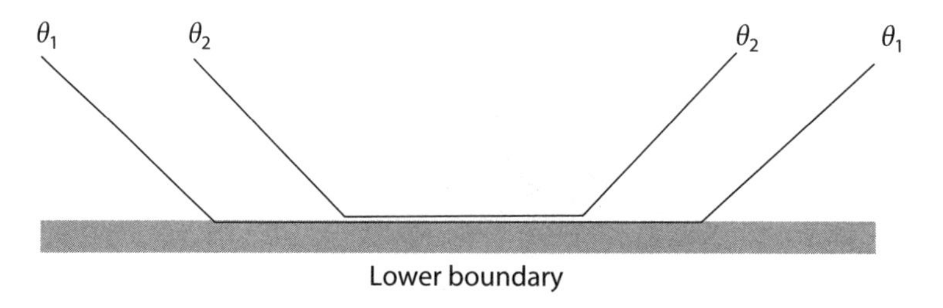

Figure 7.2. Sketch illustrating Lorenz's concept of "massless layers."

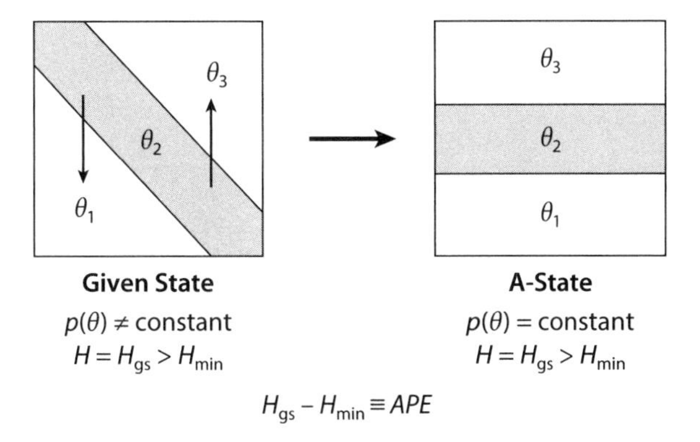

Figure 7.3. Sketch illustrating the transition from a given state to the A-state. Here p increases downward, and we assume that $\theta_1 > \theta_2 > \theta_3$, which means that the atmosphere is statically stable.

The concept of massless layers allows us to write

$$\int_0^{p_s} (\ )dp = \int_0^{\infty} (\ )\frac{\partial p}{\partial \theta} d\theta, \tag{8}$$

where ( ) can be anything. Note that the lower limit of integration on the right-hand side of (8) is zero. Equation (8) will be used further, and it is important for you to understand why it is true.

A useful expression for the APE can be derived as follows. The total enthalpy is given by

$$\begin{aligned} H &= c_p p_0^{-\kappa} \int_M p^\kappa \theta dM \\ &= \frac{c_p a^2}{g p_0^\kappa} \int_{-\pi/2}^{\pi/2} \int_0^{2\pi} \int_0^{P} p^\kappa \theta \cos\varphi dp\, d\lambda d\varphi. \end{aligned} \tag{9}$$

Integration by parts gives

$$H = \frac{c_p a^2}{(1+\kappa) g p_0^\kappa} \int_{-\pi/2}^{\pi/2} \int_0^{2\pi} \int_0^{\infty} p^{1+\kappa} \cos\varphi d\theta\, d\lambda d\varphi. \tag{10}$$

Note that vertical integration is now with respect to θ rather than p and that the lower limit of integration is $\theta = 0$. Let $\overline{p}^\theta$ be the average pressure on an isentropic surface (taking into account intersections with the ground). Recall that $\overline{p}^\theta$ is the same in the A-state as in the given state, so long as there are no regions of dry static instability. Then, use of (10) in (5) gives

$$\boxed{A = \frac{c_p a^2}{(1+\kappa) g p_0^\kappa} \int_{-\pi/2}^{\pi/2} \int_0^{2\pi} \int_0^{\infty} \left[p^{1+\kappa}(\theta) - (\overline{p}^\theta)^{1+\kappa}\right] \cos\varphi d\theta d\lambda d\varphi}. \tag{11}$$

Note that (11) is valid only if $\partial\theta/\partial z \geq 0$ everywhere in the given state, because we assumed that $\overline{p}^\theta$ is the same in the A-state as in the given state. So long as this assumption is satisfied, (11) is exact. The most general expression for the available potential energy is the definition $A \equiv H_{\text{given state}} - H_{\text{min}}$.

Let p' be the departure of p from its average on an isentropic surface, so that $p = \overline{p}^{\theta} + p'$, where $\overline{(p')}^{\theta} = 0$. According to the binomial theorem,

$$\begin{aligned} p^{1+\kappa}(\theta) &= (\overline{p}^{\theta})^{1+\kappa}\left[1 + \frac{p'(\theta)}{\overline{p}^{\theta}}\right]^{1+\kappa} \\ &= (\overline{p}^{\theta})^{1+\kappa}\left\{1 + (1+\kappa)\frac{p'(\theta)}{\overline{p}^{\theta}} + \frac{\kappa(1+\kappa)}{2!}\left[\frac{p'(\theta)}{\overline{p}^{\theta}}\right]^2 + \cdots\right\}. \end{aligned} \tag{12}$$

Lorenz used (12) to write

$$\overline{p^{1+\kappa}(\theta)} \cong (\overline{p}^{\theta})^{1+\kappa}\left\{1 + \frac{\kappa(1+\kappa)}{2!}\overline{\left[\frac{p'(\theta)}{\overline{p}^{\theta}}\right]^2}\right\}, \tag{13}$$

and he showed that this is actually a fairly good approximation. Substitution of (13) into (11) gives

$$A \cong \frac{Ra^2}{2gp_0^{\kappa}} \int_{-\pi/2}^{\pi/2}\int_0^{2\pi}\int_0^{\infty} (\overline{p}^{\theta})^{1+\kappa}\left(\frac{p'}{\overline{p}^{\theta}}\right)^2 \cos\varphi \, d\theta \, d\lambda \, d\varphi. \tag{14}$$

Because he wanted to express his results in terms of perturbations on isobaric surfaces, rather than pressure perturbations on isentropic surfaces, Lorenz also used

$$\begin{aligned} p'(\theta) &\cong \theta'(p)\frac{\partial p}{\partial \theta} \\ &\cong \theta'(p)\frac{\partial \overline{p}^{\theta}}{\partial \theta} \\ &= \theta'(p)\left(\frac{\partial \overline{\theta}}{\partial p}\right)^{-1}, \end{aligned} \tag{15}$$

where, as before, p' represents the departure of p from its global average on an isentropic surface, and θ' represents the departure of θ from $\overline{\theta}$, its global average on a p-surface. Substitution of (15) into (14) gives

$$A \cong \frac{Ra^2}{2gp_0^{\kappa}} \int_{-\pi/2}^{\pi/2}\int_0^{2\pi}\int_0^{\infty} \frac{\overline{\theta}^2}{(p^{1+\kappa})\left(-\frac{\partial \overline{\theta}}{\partial p}\right)}\left[\frac{\theta'(p)}{\overline{\theta}}\right]^2 \cos\varphi \, dp \, d\lambda \, d\varphi. \tag{16}$$

Here the independent variable used for vertical integration was changed from θ to p. Correspondingly, an overbar now represents an average on an isobaric surface, and a prime denotes the departure from an average on an isobaric surface. Equation (16) says that the available potential energy is a weighted average of the square of the departure of θ from its mean on the pressure surface. The average of the square of the departure from the mean is called the *variance about the mean*, or just the variance. The variance is a measure of the variability of a quantity; if the quantity is constant, and so everywhere equal to its mean, then its variance must be zero. If the quantity is not constant, its variance is positive. Because we are interested in variability, variances are quite important in the study of the general circulation.

Finally, Lorenz used the hydrostatic equation in the form

$$\frac{\partial\theta}{\partial p} = -\frac{\kappa\theta}{p}\left(\frac{\Gamma_d - \Gamma}{\Gamma_d}\right), \tag{17}$$

as well as

$$\frac{\theta'}{\overline{\theta}} = \frac{T'}{\overline{T}}, \tag{18}$$

to rewrite (16) as

$$A = \frac{a^2}{2}\int_{-\pi/2}^{\pi/2}\int_{0}^{2\pi}\int_{0}^{p}\frac{\overline{T}}{(\Gamma_d - \Gamma)}\left[\frac{T'(p)}{\overline{T}}\right]^2 \cos\varphi dp d\lambda d\varphi. \tag{19}$$

This result shows that the available potential energy is closely related to the variance of temperature on isobaric surfaces. The available potential energy also increases as the lapse rate increases, that is, as the atmosphere becomes less statically stable in the dry sense.

Observations show that the APE is only about half a percent of $P + I$ and that it is comparable in magnitude to the total kinetic energy. Both the APE and the total kinetic energy are on the order of 10^6–10^7 J m^{-2}.

The Gross Static Stability

The A-state defines a correspondence or mapping between θ and p. For each p there is one possible value of θ (the converse is not necessarily true). We can say that in the A-state, p and θ are perfectly correlated (or, more correctly, perfectly anticorrelated).

Consider the opposite limit, in which θ and p are completely uncorrelated. In this "S- state," all possible values of θ occur, with equal probability, for any given p. This will be the case if the θ-surfaces are vertical, so that $\partial\theta/\partial p = 0$, and the surface pressure is globally uniform (see fig.7.4). A globally uniform surface pressure seems reasonable enough in the absence of topography, but it is a very strange state when topography is present. Lorenz (personal communication, 2003) suggested an alternative definition of the S-state, in which the surface pressure is allowed to vary in a simple and natural

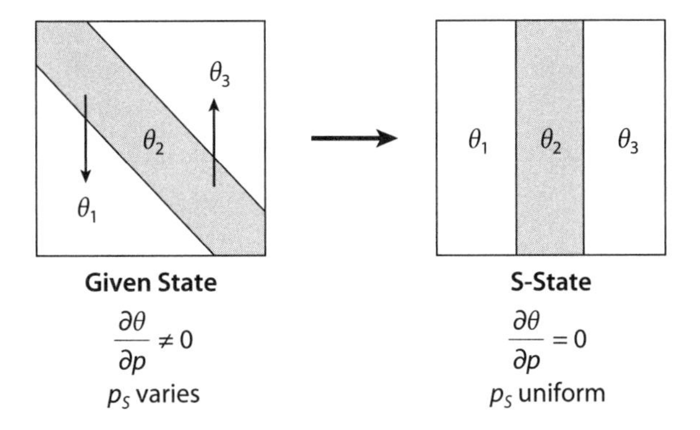

Figure 7.4. Sketch illustrating the transition from a given state to the S-state used to define the gross static stability.

way with the surface height (roughly as shown in fig. 3.1), but the (vertically uniform) potential temperature is geographically distributed so that it is uncorrelated with the surface height (and therefore uncorrelated with the surface pressure).

The system passes from the given state to the S-state by way of an adiabatic process; it follows that the S-state itself is invariant under adiabatic processes. The *gross static stability* is defined to be the enthalpy of the S-state minus the enthalpy of the given state; that is,

$$S \equiv H_{\text{S-state}} - H_{\text{given state}}. \tag{20}$$

The total enthalpy of the S-state is given by

$$\begin{aligned} H_{\text{S-state}} &= c_p p_0^{-\kappa} \int_M p^{\kappa} \theta dM \\ &= \frac{c_p a^2}{g p_0^{\kappa}} \int_{-\pi/2}^{\pi/2} \int_0^{2\pi} \int_0^{p_s} p^{\kappa} \theta \cos\varphi dp d\lambda d\varphi \\ &= \frac{c_p a^2}{g p_0^{\kappa}} \int_{-\pi/2}^{\pi/2} \int_0^{2\pi} \theta \left(\int_0^{p_s} p^{\kappa} dp \right) d\lambda \cos\varphi d\varphi \\ &= \frac{c_p a^2}{g p_0^{\kappa}} \int_{-\pi/2}^{\pi/2} \int_0^{2\pi} \theta \left(\frac{p_S^{1+\kappa}}{1+\kappa} \right) d\lambda \cos\varphi d\varphi. \end{aligned} \tag{21}$$

At this point, we can consider each of the two definitions of the S-state that were introduced. If the surface pressure is globally uniform in the S-state, then we can simply replace p_S with $\overline{p_S}$ in (21) and write

$$\begin{aligned} H_{\text{S-state}} &= \frac{c_p a^2}{g p_0^{\kappa}} \int_{-\pi/2}^{\pi/2} \int_0^{2\pi} \theta \left[\frac{(\overline{p_S})^{1+\kappa}}{1+\kappa} \right] d\lambda \cos\varphi d\varphi \\ &= \frac{c_p a^2}{g p_0^{\kappa}} \left[\frac{(\overline{p_S})^{1+\kappa}}{1+\kappa} \right] \int_{-\pi/2}^{\pi/2} \int_0^{2\pi} \theta \, d\lambda \cos\varphi d\varphi \\ &= \frac{c_p 4\pi a^4}{g p_0^{\kappa}} \left[\frac{(\overline{p_S})^{1+\kappa} \overline{\theta}}{1+\kappa} \right], \end{aligned} \tag{22}$$

where $\overline{p_S}$ is the globally averaged surface pressure, which is the same in the given state and the S-state, and $\overline{\theta}$ is the globally averaged potential temperature in the S-state, which is the same as the mass-averaged potential temperature in the given state.

Alternatively, if θ is uncorrelated with $p_S^{1+\kappa}$ in the S-state, then it follows immediately from (21) that

$$H_{\text{S-state}} = \frac{c_p 4\pi a^4}{g p_0^{\kappa}} \left[\frac{(\overline{p_S})^{1+\kappa} \overline{\theta}}{1+\kappa} \right]. \tag{23}$$

We conclude that the two definitions of the S-state actually give exactly the same result for $H_{\text{S-state}}$.

Why is S is called the "gross static stability?" To see the explanation, we rewrite (21) as

$$H_{\text{S-state}} = \frac{c_p a^2}{g p_0^{\kappa}} \left(\frac{\tilde{p}^{\kappa}}{p_0^{\kappa}} \right) \int_{-\pi/2}^{\pi/2} \int_0^{2\pi} \left(\int_0^{p_s} \theta dp \right) \cos\varphi d\lambda d\varphi, \tag{24}$$

where

$$\tilde{p}^{\kappa} \equiv \frac{(\overline{p_S})^{\kappa}}{1+\kappa}. \tag{25}$$

In (24), the integral over pressure amounts simply to multiplication by $\overline{p_S}$, because θ is independent of height in the S-state. The integral of the potential temperature over the entire mass of the atmosphere must be exactly the same for the given state and the S-state; that is,

$$\begin{aligned} &\frac{a^2}{g}\left[\int_{-\pi/2}^{\pi/2}\int_0^{2\pi}\left(\int_0^{p_S}\theta dp\right)\cos\varphi d\lambda d\varphi\right]_{\text{given state}} \\ &= \frac{a^2}{g}\left[\int_{-\pi/2}^{\pi/2}\int_0^{2\pi}\left(\int_0^{\overline{p_S}}\theta dp\right)\cos\varphi d\lambda d\varphi\right]_{\text{S-state}}. \end{aligned} \tag{26}$$

This allows us to rewrite (24) as

$$H_{\text{S-state}} = \frac{c_p a^2}{g}\left(\frac{\tilde{p}^{\kappa}}{p_0^{\kappa}}\right)\left[\int_{-\pi/2}^{\pi/2}\int_0^{2\pi}\left(\int_0^{p_S}\theta dp\right)\cos\varphi d\lambda d\varphi\right]_{\text{given state}}. \tag{27}$$

Now, we substitute (27) into (20), to obtain

$$\begin{aligned} S \equiv &\frac{c_p a^2}{g}\left(\frac{\tilde{p}^{\kappa}}{p_0^{\kappa}}\right)\left[\int_{-\pi/2}^{\pi/2}\int_0^{2\pi}\left(\int_0^{p_S}\theta dp\right)\cos\varphi d\lambda d\varphi\right]_{\text{given state}} \\ &- \frac{c_p a^2}{g p_0^{\kappa}}\int_{-\pi/2}^{\pi/2}\int_0^{2\pi}\int_0^{p_S} p^{\kappa}\theta\cos\varphi d\lambda d\varphi, \end{aligned} \tag{28}$$

which we can rearrange to

$$S \equiv \frac{c_p a^2}{g p_0^{\kappa}}\left\{\int_{-\pi/2}^{\pi/2}\int_0^{2\pi}\left[\int_0^{p_S}(\tilde{p}^{\kappa} - p^{\kappa})\theta dp\right]\cos\varphi d\lambda d\varphi\right\}_{\text{given state}}. \tag{29}$$

Integration by parts gives

$$\boxed{S \equiv \frac{c_p a^2}{g p_0^{\kappa}(1+\kappa)}\left\{\int_{-\pi/2}^{\pi/2}\int_0^{2\pi}\left[\int_0^{p_S}[(\overline{p_S})^{\kappa}p - p^{1+\kappa}]\left(-\frac{\partial\theta}{\partial p}\right)dp\right]\cos\varphi d\lambda d\varphi\right\}_{\text{given state}}}, \tag{30}$$

which shows that S is a weighted average of $(-\partial\theta/\partial p)$; it is therefore a measure of the static stability, which accounts for its name. Like the available potential energy, the gross static stability is defined only for the atmosphere as a whole.

The globally averaged surface pressure and the probability distribution of θ are invariant under adiabatic processes. This means that the S-state is invariant as well. Because S is defined as the difference between the total enthalpy of the S-state and the total enthalpy of the given state, we can write

$$\frac{dS}{dt} = -\frac{dH}{dt}. \tag{31}$$

Since $(d/dt)(H+K) = 0$, it follows from (31) that

$$\frac{dS}{dt} = \frac{dK}{dt}. \tag{32}$$

Equation (32) shows that when K is produced by conversion from APE, the gross static stability increases. As the system passes from the given state to the A-state, the available potential energy decreases, and the kinetic energy and gross static stability both increase, as would be expected.

The observed state of the atmosphere is "in between" the A-state and the S-state.

Conversion of Potential Energy into Kinetic Energy

As discussed in chapter 6, convection produces an upward flux of dry static energy, which means also an upward flux of θ and T. Suppose that the θ-profile in a particular atmospheric column is altered by a vertical flux of θ; that is,

$$\frac{\partial \theta}{\partial t} = g\frac{\partial F_\theta}{\partial p}. \tag{33}$$

It follows that the enthalpy changes according to

$$c_p\frac{\partial T}{\partial t} = c_p g\frac{\partial}{\partial p}\left[\left(\frac{p}{p_0}\right)^{\kappa} F_\theta\right] - \frac{Rg}{p_0^{\kappa}}\left(\frac{F_\theta}{p^{1-\kappa}}\right). \tag{34}$$

Integrating through the depth of the column, we find that the change of the total enthalpy of the column satisfies

$$\frac{\partial}{\partial t}\left(\int_0^{p_S} c_p T dp\right) = c_p\left[T_S\frac{\partial p_S}{\partial t} + g\left(\frac{p_S}{p_0}\right)^{\kappa}(F_\theta)_S\right] - \frac{Rg}{p_0^{\kappa}}\int_0^{p_S}\left(\frac{F_\theta}{p^{1-\kappa}}\right)dp. \tag{35}$$

The first term inside the square brackets on the right-hand side of (35) is zero if no mass is exchanged with neighboring columns. The second term inside the square brackets is diabatic, because it represents an exchange of energy between the atmosphere and the lower boundary. The remaining term, $-(g\kappa/p_0^{\kappa})/\int_0^{p_S}(F_\theta/p^{1-\kappa})dp$, arises purely from θ redistribution within the column. The form of this redistribution term makes it clear that $F_\theta > 0$, that is, an upward flux of θ, tends to reduce the total enthalpy of the column. It follows that the total potential energy of the column also decreases or, in other words, that the "center of mass" of the column moves downward. The conclusion is that *an adiabatic process that produces an upward flux of* θ *reduces the total enthalpy of the column and so generates kinetic energy*. This is relevant to both small-scale convection and large-scale "baroclinic eddies." Both produce an upward flux of θ and so lower the atmosphere's center of mass. Both convert available potential energy into kinetic energy.

Examples: The Available Potential Energies of Three Simple Systems

Available potential energy can arise in several different ways. We now consider three idealized examples, each of which illustrates a pure, unadulterated form of available potential energy.

Example 1: The APE Associated with Static Instability

Consider a simple system containing two parcels of equal mass. In the given state, parcels with potential temperature θ_1 and θ_2 reside at pressures p_1 and p_2, respectively. We assume that $\theta_1 < \theta_2$ and $p_1 < p_2$, so that the given state is statically unstable. The enthalpy per unit mass of parcel i is $c_p \theta_i (p_i/p_0)^\kappa$. If the parcels are interchanged (or "swapped") so that parcel 2 goes to pressure p_1, and vice versa, the change in the total enthalpy per unit mass is

$$\Delta H = c_p(\theta_1 - \theta_2)\left[\left(\frac{p_2}{p_0}\right)^\kappa - \left(\frac{p_1}{p_0}\right)^\kappa\right], \tag{36}$$

which is negative. This result implies that the total enthalpy has been reduced and so is minimized by the swap; the final state is the A-state, and the change in enthalpy given by (36) is minus the available potential energy of the system, per unit mass. The process just described is an idealization of dry atmospheric convection, which transports potential temperature upward, increasing the static stability and lowering the atmosphere's center of mass.

Lorenz (1978) generalized the concept of APE for a moist atmosphere, in which moist adiabatic processes are acknowledged to be, well, adiabatic. Randall and Wang (1992) showed that this moist APE can be used to define a generalized CAPE that represents the potential energy available for conversion into the kinetic energy of cumulus convection.

Example 2: The APE Associated with Meridional Temperature Gradients

Consider an idealized planet, with no orography and a uniform surface pressure in the given state. Suppose that the potential temperature of the given state is a function of latitude only:

$$\theta_{\text{given state}}(\mu) = \theta_0(1 - \Delta_H \mu^2). \tag{37}$$

Here Δ_H is a constant, and $\mu \equiv \sin\varphi$. For realistic states, $0 < \Delta_H < 1$. The available potential energy of this idealized given state arises solely from the meridional temperature gradient. To obtain an expression for the available potential energy of the given state, we need to compute the total enthalpies of the given state and the A-state, and subtract them. The first step is to find the A-state. Note that in this idealized example, the given state is identical to the S-state.

The mass in a latitude belt of width $d\varphi$ is

$$\begin{aligned} dm &= 2(2\pi a \cos\varphi)\left(\frac{p_S}{g}\right)(a d\varphi) \\ &= \frac{4\pi a^2}{g} p_S d\mu, \end{aligned} \tag{38}$$

where $\mu \equiv \sin\varphi$, and $d\mu = \cos\varphi d\varphi$. In (38) the leading factor of 2 is included because there is symmetry across the equator, so when we increment latitude in one hemisphere we actually pick up mass from two "rings" of air, one in each hemisphere. The rate of change of θ as we add mass is

$$\frac{d\theta}{dm} = \left(\frac{d\theta}{d\mu}\right)_{gs} \left(\frac{dm}{d\mu}\right)^{-1}. \tag{39}$$

Combining (38) and (39), we get

$$\frac{d\theta}{dm} = \frac{g}{4\pi a^2 p_S}\left(\frac{d\theta}{d\mu}\right)_{\text{given state}}. \tag{40}$$

In the A-state, the θ-surfaces are flat, so the meridional temperature gradient has been reduced to zero. For this to happen, the warm air in the given state has to rise and spread out toward the poles, while the cold air descends and spreads out toward the equator. Potential temperature is thus transported both upward and poleward. In the A-state, the warm air fills the upper portion of the domain, and the cold air fills the lower portion.

To find the structure of the A-state, we start by writing the increment of mass between two θ-surfaces, which is

$$dm = \frac{4\pi a^2}{g}dp, \tag{41}$$

or

$$dm = \frac{4\pi a^2}{g}\left(\frac{dp}{d\theta}\right)_{rs}d\theta, \tag{42}$$

from which it follows that

$$\left(\frac{dp}{d\theta}\right)_{rs} = \frac{d\theta}{dm}\frac{4\pi a^2}{g}. \tag{43}$$

The subscript *rs* denotes the A-state. Next, we substitute $d\theta/dm$ from (40) into (43). We can do this because the distribution of θ over the mass must be the same in the A-state as in the given state. The result is

$$\left(\frac{dp_*}{d\theta}\right)_{rs} = \left(\frac{d\theta}{d\mu}\right)_{\text{given state}}, \tag{44}$$

where

$$p_* \equiv \frac{p}{p_S}. \tag{45}$$

We have to be careful when looking at this equation. The left-hand side refers to the distribution of θ with pressure in the A-state. The right-hand side refers to the distribution of θ with μ in the given state. Keep in mind that in this idealized example, θ does not vary with pressure in the given state, and it does not vary with latitude in the A-state.

From (45), we see that p_* plays the same role in the A-state as μ plays in the given state. We also recall that the potential temperature must increase with height in the A-state. Referring to (37), we conclude that

$$\theta_{rs} = \theta_0(1 - \Delta_H p_*^2), \tag{46}$$

which is the desired formula for the distribution of potential temperature in the A-state. It can be verified that (46) gives the correct values of θ at $p = 0$ and $p = p_S$.

Now all that is necessary to find the APE is to calculate the enthalpy in the given state and the A-state, and subtract. In the given state, the potential temperature is independent of height, so (37) gives

$$
\begin{aligned}
H_{\text{given state}} &= \int_{-1}^{1}\int_{0}^{p_S} 2\pi a^2 c_p \theta_0 (1 - \Delta_H \mu^2)\left(\frac{p}{p_0}\right)^{\kappa} \frac{dp}{g} d\mu \\
&= 4\pi a^2 c_p \theta_0 \left(\frac{p_S}{g}\right)\left(\frac{p_S}{p_0}\right)^{\kappa}\left(\frac{1 - \frac{1}{3}\Delta_H}{1+\kappa}\right).
\end{aligned}
\tag{47}
$$

Similarly, the total enthalpy of the A-state is

$$
\begin{aligned}
H_{\min} &= \int_{-1}^{1}\int_{0}^{p_S} 2\pi a^2 c_p \theta_0 \left[1 - \Delta_H \left(\frac{p}{p_S}\right)^2\right]\left(\frac{p}{p_0}\right)^{\kappa} \frac{dp}{g} d\mu \\
&= 4\pi a^2 c_p \theta_0 \left(\frac{p_S}{g}\right)\left(\frac{p_S}{p_0}\right)^{\kappa}\left[\left(\frac{1}{1+\kappa}\right) - \left(\frac{\Delta_H}{3+\kappa}\right)\right].
\end{aligned}
\tag{48}
$$

Finally, we obtain

$$
A = H_{\text{given state}} - H_{\min} = 4\pi a^2 c_p \theta_0 \Delta_H \left(\frac{p_S}{g}\right)\left(\frac{p_S}{p_0}\right)^{\kappa}\left[\frac{2\kappa}{3(3+\kappa)(1+\kappa)}\right]. \tag{49}
$$

Note that A is proportional to Δ_H, as might be expected.

Now, consider a meridional transport process:

$$
\begin{aligned}
\frac{\partial \theta_{\text{given state}}}{\partial t} &= -\frac{1}{a\cos\varphi}\frac{\partial}{\partial \varphi}(F_\theta \cos\varphi) \\
&= -\frac{1}{a}\frac{\partial}{\partial \mu}\left(F_\theta \sqrt{1-\mu^2}\right).
\end{aligned}
\tag{50}
$$

Here F_θ is a meridional flux of potential temperature, which we regard as given and assume to be independent of height and longitude and symmetrical about the equator. Note that

$$
F_\theta = 0 \text{ at both poles.} \tag{51}
$$

We assume that the surface pressure does not change with time at any latitude. What is the time rate of change of the APE associated with this meridional redistribution of potential temperature?

To answer this question, we first note that the time rate of change of the total enthalpy of the given state is

$$
\begin{aligned}
\frac{\partial H_{\text{given state}}}{\partial t} &= \int_{-1}^{1}\int_{0}^{p_S} 2\pi a^2 c_p \left(\frac{\partial \theta}{\partial t}\right)\left(\frac{p}{p_0}\right)^{\kappa}\frac{dp}{g} \\
&= \frac{2\pi a^2 c_p}{(1+\kappa)}\left(\frac{p_S}{g}\right)\left(\frac{p_S}{p_0}\right)^{\kappa}\int_{-1}^{1}\frac{\partial \theta}{\partial t} d\mu \\
&= \frac{2\pi a^2 c_p}{(1+\kappa)}\left(\frac{p_S}{g}\right)\left(\frac{p_S}{p_0}\right)^{\kappa}\int_{-\pi/2}^{\pi/2}\frac{1}{a\cos\varphi}\frac{\partial}{\partial \varphi}(F_\theta\cos\varphi)\cos\varphi d\varphi \\
&= 0.
\end{aligned}
\tag{52}
$$

Here we used the facts that $\partial\theta/\partial t$ is independent of height and that p_S is independent of latitude, and we substituted from (50). According to (52), the specified transport process has no effect on the total enthalpy of the given state. The reason is that the average θ on each pressure surface is unchanged, and it follows that the average temperature on each pressure surface is unchanged.

The meridional transport process can, however, alter the total enthalpy of the A-state. Here again there is a possibility of confusion. As already mentioned, the specified meridional transport process does not alter the average value of θ on a pressure surface. This statement seems to imply that the process is isentropic, and we already know that isentropic processes do not alter the A-state. The entropy is proportional to $\ln(\theta)$, however, and the average value of $\ln(\theta)$ is altered by the transport process. Another point of view is that, generally speaking, the specified transport process is not reversible. For example, if F_θ is a downgradient flux due to diffusive mixing, then it could, in principle, homogenize θ throughout the atmosphere. This process would clearly be irreversible. Following such homogenization, the A-state would be the same as the (homogenized) given state and would therefore be different from the A-state found previously.

From (52), it follows that

$$\frac{dA}{dt} = -\frac{d}{dt}H_{\min}. \tag{53}$$

Now, we recall that, based on comparison of (37) and (46), p_* plays the same role in the A-state as μ plays in the given state. In particular, $p_* = 1$ (the surface) corresponds to $\mu = 1$ (the pole), and $p_* = 0$ (the "top of the atmosphere") corresponds to $\mu = 0$ (the equator). Our goal is to determine the time rate of change of θ in the A-state at a particular instant, namely, the time when the distribution of θ satisfies (37) [and (46)], and the time rate of change of θ in the given state satisfies (50). We can find the time rate of change of θ in the A-state at this instant by using our expression for the time rate of change of θ in the given state and simply replacing μ with p_*, everywhere. The time rate of change of θ in the A-state thus satisfies

$$\frac{\partial \theta_{rs}}{\partial t} = -\frac{1}{a}\frac{\partial}{\partial p_*}\left(F_\theta \sqrt{1-p_*^2}\right). \tag{54}$$

Again, there is a possibility of confusion. We have specified that F_θ is not a function of height, although of course it does depend on latitude. It would thus appear that we can pull F_θ out of the derivative in (54), but this is not correct. The reason is that when we replaced μ with p_* we also replaced the μ-dependence of F_θ with a corresponding p_*-dependence. Thus, in (54), F_θ should be regarded as a function of p_* but not as a function of latitude! This is understandable, because F_θ is acting to change $\theta_{rs}(p_*)$. As an example, suppose that F_θ is symmetric about the equator, and poleward in both hemispheres, and that $\Delta_H > 0$, so that the poles are in fact colder than the tropics. Then F_θ tends to warm the poles and cool the tropics, reducing Δ_H and, we expect, reducing A. As the tropics cool and the poles warm in the given state, the A-state evolves in a corresponding way, so that θ_{rs} cools aloft and warms at the lower levels.

We now write the time rate of change of the total enthalpy in the A-state as

$$\begin{aligned}\frac{\partial H_{\min}}{\partial t} &= \int_{-1}^{1}\int_{0}^{p_S} 2\pi a^2 c_p \left(\frac{\partial \theta}{\partial t}\right)\left(\frac{p}{p_0}\right)^{\kappa}\frac{dp}{g}\,d\mu \\ &= \frac{4\pi a^2 c_p p_S}{g}\left(\frac{p_S}{p_0}\right)^{\kappa}\int_0^1 \left(\frac{\partial \theta}{\partial t}\right)(p_*)^{\kappa}\,dp_* \\ &= \frac{4\pi a^2 c_p p_S}{g}\left(\frac{p_S}{p_0}\right)^{\kappa}\int_0^1 \left[\frac{\partial}{\partial p_*}\left(F_\theta\sqrt{1-p_*^2}\right)\right]p_*^{\kappa}\,dp_*.\end{aligned} \tag{55}$$

We cannot evaluate the integral on the last line of (55), because the form of F_θ has not been specified. The last step would be to substitute (55) into (53).

This example is relevant to the process by which midlatitude baroclinic eddies transport potential temperature upward and poleward, reducing the meridional temperature gradient and increasing the static stability.

Example 3: The APE Associated with Surface-Pressure Variations

This example is designed to illustrate that APE can occur in the presence of surface-pressure gradients, even when there are no potential temperature gradients; this is analogous to the APE of shallow water with a nonuniform free-surface height. To explore this possibility in a simple framework, we consider a planet with an atmosphere of uniform potential temperature, θ_0. The surface pressure, $p_S(\lambda,\varphi)$, is given as a function of longitude λ and latitude φ. For simplicity we assume that the Earth's surface is flat, although a similar but more complicated analysis can be developed for the case of arbitrary surface topography.

We begin with (5), the basic definition of the APE. The total enthalpy satisfies

$$H = \int_{-\pi/2}^{\pi/2}\int_{0}^{2\pi}\int_{0}^{p_S} c_p\theta\left(\frac{p}{p_0}\right)^{\kappa}\frac{dp}{g}\,a\cos\varphi d\lambda a d\varphi. \tag{56}$$

Because $\theta = \theta_0$ everywhere in this example, we can simplify (56) considerably:

$$H = \frac{c_p\theta_0 a^2}{g(1+\kappa)p_0^{\kappa}}\int_{-\pi/2}^{\pi/2}\int_{0}^{2\pi} p_S^{1+\kappa}\cos\varphi d\lambda d\varphi. \tag{57}$$

Equation (57) can be applied to both the given state and the A-state, so that the APE is given by

$$A = \frac{c_p\theta_0 a^2}{g(1+\kappa)p_0^{\kappa}}\int_{-\pi/2}^{\pi/2}\int_{0}^{2\pi}\left[(p_S)_{\text{given state}}^{1+\kappa} - (p_S)_{rs}^{1+\kappa}\right]\cos\varphi d\lambda d\varphi. \tag{58}$$

Before we can evaluate the double integrals in (58), we must substitute for $[p_S(\lambda,\varphi)]_{\text{given state}}$ and $[p_S(\lambda,\varphi)]_{rs}$. The former is assumed to be known. Our problem thus reduces to finding $[p_S(\lambda,\varphi)]_{rs}$. This is very simple, because (since by assumption there are no mountains) the surface pressure is globally uniform in the A-state and equal to the globally averaged surface pressure in the given state. We denote this globally averaged surface pressure by $\overline{p_S}$ and rewrite (55) as

$$A = \frac{c_p\theta_0 a^2}{g(1+\kappa)p_0^{\kappa}}\int_{-\pi/2}^{\pi/2}\int_{0}^{2\pi}\left[(p_S)_{\text{given state}}^{1+\kappa} - (\overline{p_S})_{rs}^{1+\kappa}\right]\cos\varphi d\lambda d\varphi. \tag{59}$$

As an exercise, prove that (59) gives $A \geq 0$.

This example can be generalized to include topography.

Variance Budgets

Equation (19) shows that the available potential energy is closely related to the spatial variance of temperature or potential temperature on pressure surfaces. We now examine a conversion process that couples the variance associated with the meridional gradient of the zonally averaged potential temperature with the eddy variance of potential temperature. This same process is closely related to the conversion between the zonal available potential energy, A_Z, and eddy available potential energy, A_E. We will show that the eddy potential temperature variance interacts with the meridional gradient of the zonally averaged potential temperature through

$$\frac{\partial}{\partial t}\left(\frac{1}{2}\overline{\theta^{*2}}^{\lambda}\right) \sim \frac{-\overline{\theta^* v^*}^{\lambda}}{a}\frac{\partial \overline{\theta}^{\lambda}}{\partial \varphi}. \tag{60}$$

The term shown on the right-hand side of (60) can be called the *meridional gradient-production term*. There are several additional terms, as will be discussed further.

To gain an intuitive understanding of the gradient-production terms, consider the simple example illustrated in figure 7.5. State A consists of two latitude belts of equal mass, each with a different (but uniform) θ. State B is obtained by homogenizing state A, without changing the average. Consider the average of the square of θ, for each state. For state A,

$$\overline{\theta^2} = \frac{1}{2}(\theta_1^2 + \theta_2^2). \tag{61}$$

Here the overbar denotes an average over both latitude belts. For state B,

$$\begin{aligned}\overline{\theta^2} = \left[\frac{1}{2}(\theta_1 + \theta_2)\right]^2 &= \frac{1}{4}(\theta_1^2 + 2\theta_1\theta_2 + \theta_2^2)\\ &= \frac{1}{2}(\theta_1^2 + \theta_2^2) - \frac{1}{4}(\theta_1^2 + 2\theta_1\theta_2 + \theta_2^2)\\ &= \frac{1}{2}(\theta_1^2 + \theta_2^2) - \frac{1}{4}(\theta_1 - \theta_2)^2\\ &\leq \frac{1}{2}(\theta_1^2 + \theta_2^2).\end{aligned} \tag{62}$$

Comparison of (61) and (62) shows that mixing reduces the average of the square, which appears as a "dissipation" of the mean-state variance.

To show the origin of (60), we start from the conservation equation for potential temperature, which we can write in spherical coordinates as

$$\frac{\partial \theta}{\partial t} + \frac{1}{a\cos\varphi}\frac{\partial}{\partial \lambda}(u\theta) + \frac{1}{a\cos\varphi}\frac{\partial}{\partial \varphi}(v\theta\cos\varphi) + \frac{\partial}{\partial p}(\omega\theta) = \dot{\theta}. \tag{63}$$

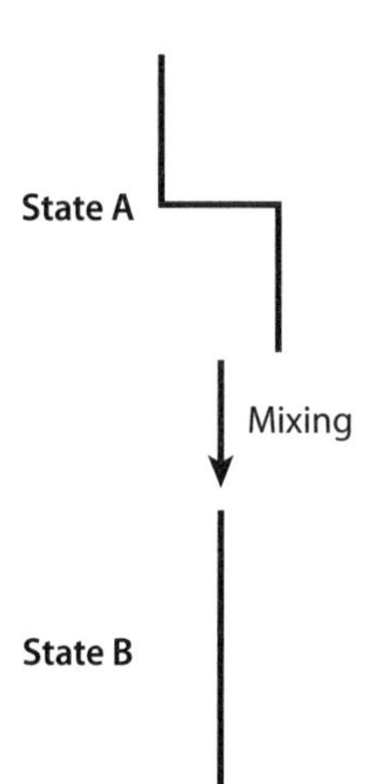

Figure 7.5. A simple sketch to explain the idea of gradient production. State B is obtained by homogenizing state A. The average of the square decreases in passing from state A to state B.

We also need mass continuity, in the form

$$\frac{1}{a\cos\varphi}\frac{\partial u}{\partial\lambda}+\frac{1}{a\cos\varphi}\frac{\partial}{\partial\varphi}(v\cos\varphi)+\frac{\partial\omega}{\partial p}=0. \tag{64}$$

We zonally average (63) and (64) to obtain

$$\frac{\partial\overline{\theta}^{\lambda}}{\partial t}+\frac{1}{a\cos\varphi}\frac{\partial}{\partial\varphi}\left(\overline{v\theta}^{\lambda}\cos\varphi\right)+\frac{\partial}{\partial p}\left(\overline{\omega\theta}^{\lambda}\right)=\overline{\dot{\theta}}^{\lambda}, \tag{65}$$

$$\frac{1}{a\cos\varphi}\frac{\partial}{\partial\varphi}\left(\overline{v}^{\lambda}\cos\varphi\right)+\frac{\partial\overline{\omega}^{\lambda}}{\partial p}=0. \tag{66}$$

Subtracting (65) and (66) from (63) and (64), respectively, we find that

$$\begin{aligned}
&\frac{\partial\theta^{*}}{\partial t}+\frac{1}{a\cos\varphi}\frac{\partial}{\partial\lambda}(u\theta)+\frac{1}{a\cos\varphi}\frac{\partial}{\partial\varphi}\left\{\left(v\theta-\overline{v\theta}^{\lambda}\right)\cos\varphi\right\}+\frac{\partial}{\partial p}\left(\omega\theta-\overline{\omega\theta}^{\lambda}\right)\\
&=\frac{\partial\theta^{*}}{\partial t}+\frac{1}{a\cos\varphi}\frac{\partial}{\partial\lambda}\left(u^{*}\overline{\theta}^{\lambda}+\overline{u}^{\lambda}\theta^{*}+u^{*}\theta^{*}\right)+\frac{1}{a\cos\varphi}\frac{\partial}{\partial\varphi}\left[\left(v^{*}\overline{\theta}^{\lambda}+\overline{v}^{\lambda}\theta^{*}+v^{*}\theta^{*}-\overline{v^{*}\theta^{*}}^{\lambda}\right)\cos\varphi\right]\\
&\quad+\frac{\partial}{\partial p}\left(\omega^{*}\overline{\theta}^{\lambda}+\overline{\omega}^{\lambda}\theta^{*}+\omega^{*}\theta^{*}-\overline{\omega^{*}\theta^{*}}^{\lambda}\right)\\
&=\frac{\partial\theta^{*}}{\partial t}+\frac{1}{a\cos\varphi}\frac{\partial}{\partial\lambda}\left(\overline{u}^{\lambda}\theta^{*}\right)+\frac{1}{a\cos\varphi}\frac{\partial}{\partial\varphi}\left[\left(\overline{v}^{\lambda}\theta^{*}\right)\cos\varphi\right]+\frac{\partial}{\partial p}\left(\overline{\omega}^{\lambda}\theta^{*}\right)\\
&\quad+\frac{1}{a\cos\varphi}\frac{\partial}{\partial\lambda}\left(u^{*}\theta^{*}\right)+\frac{1}{a\cos\varphi}\frac{\partial}{\partial\varphi}\left(v^{*}\theta^{*}\cos\varphi\right)+\frac{\partial}{\partial p}\left(\omega^{*}\theta^{*}\right)\\
&\quad+\frac{1}{a\cos\varphi}\frac{\partial}{\partial\lambda}\left(u^{*}\overline{\theta}^{\lambda}\right)+\frac{1}{a\cos\varphi}\frac{\partial}{\partial\varphi}\left(v^{*}\overline{\theta}^{\lambda}\cos\varphi\right)+\frac{\partial}{\partial p}\left(\omega^{*}\overline{\theta}^{\lambda}\right)\\
&\quad-\frac{1}{a\cos\varphi}\frac{\partial}{\partial\varphi}\left(\overline{v^{*}\theta^{*}}^{\lambda}\cos\varphi\right)-\frac{\partial}{\partial p}\left(\overline{\omega^{*}\theta^{*}}^{\lambda}\right)\\
&=\dot{\theta}^{*}.
\end{aligned} \tag{67}$$

$$\frac{1}{a\cos\varphi}\frac{\partial u^{*}}{\partial\lambda}+\frac{1}{a\cos\varphi}\frac{\partial}{\partial\varphi}\left(v^{*}\cos\varphi\right)+\frac{\partial\omega^{*}}{\partial p}=0. \tag{68}$$

To obtain the first equality of (67) we used

$$\begin{aligned}
v\theta&=\left(\overline{v}^{\lambda}+v^{*}\right)\left(\overline{\theta}^{\lambda}+\theta^{*}\right)\\
&=\overline{v}^{\lambda}\overline{\theta}^{\lambda}+v^{*}\overline{\theta}^{\lambda}+\overline{v}^{\lambda}\theta^{*}+v^{*}\theta^{*},
\end{aligned} \tag{69}$$

$$\overline{v\theta}^{\lambda}=\overline{v}^{\lambda}\overline{\theta}^{\lambda}+\overline{v^{*}\theta^{*}}^{\lambda}, \tag{70}$$

$$v\theta-\overline{v\theta}^{\lambda}=v^{*}\overline{\theta}^{\lambda}+\overline{v}^{\lambda}\theta^{*}+v^{*}\theta^{*}-\overline{v^{*}\theta^{*}}^{\lambda}, \tag{71}$$

and so on. We can use (66) and (68) to rewrite (67) as follows:

$$\begin{aligned}
&\left(\frac{\partial}{\partial t}+\frac{\overline{u}^{\lambda}}{a\cos\varphi}\frac{\partial}{\partial\lambda}+\frac{\overline{v}^{\lambda}}{a}\frac{\partial}{\partial\varphi}+\overline{\omega}^{\lambda}\frac{\partial}{\partial p}\right)\theta^{*}+\left(\frac{u^{*}}{a\cos\varphi}\frac{\partial}{\partial\lambda}+\frac{v^{*}}{a}\frac{\partial}{\partial\varphi}+\omega^{*}\frac{\partial}{\partial p}\right)\theta^{*}\\
&\qquad+\frac{v^{*}}{a}\frac{\partial\overline{\theta}^{\lambda}}{\partial\varphi}+\omega^{*}\frac{\partial\overline{\theta}^{\lambda}}{\partial p}\\
&=\frac{1}{a\cos\varphi}\frac{\partial}{\partial\varphi}\left(\overline{v^{*}\theta^{*}}^{\lambda}\cos\varphi\right)+\frac{\partial}{\partial p}\left(\overline{\omega^{*}\theta^{*}}^{\lambda}\right)+\dot{\theta}^{*}.
\end{aligned} \tag{72}$$

Multiplying (69) by θ^{*}, and using (68) again, we obtain

$$\begin{aligned}&\left(\frac{\partial}{\partial t}+\frac{\overline{u}^{\lambda}}{a\cos\varphi}\frac{\partial}{\partial\lambda}+\frac{\overline{v}^{\lambda}}{a}\frac{\partial}{\partial\varphi}+\overline{\omega}^{\lambda}\frac{\partial}{\partial p}\right)\left(\frac{1}{2}\theta^{*2}\right)\\&+\frac{1}{a\cos\varphi}\frac{\partial}{\partial\lambda}\left[u^{*}\left(\frac{1}{2}\theta^{*2}\right)\right]+\frac{1}{a\cos\varphi}\frac{\partial}{\partial\varphi}\left[v^{*}\left(\frac{1}{2}\theta^{*2}\right)\cos\varphi\right]+\frac{\partial}{\partial p}\left[\omega^{*}\left(\frac{1}{2}\theta^{*2}\right)\right]\\&=\theta^{*}\left[\frac{1}{a\cos\varphi}\frac{\partial}{\partial\varphi}\left(\overline{v^{*}\theta^{*}}^{\lambda}\cos\varphi\right)+\frac{\partial}{\partial p}\left(\overline{\omega^{*}\theta^{*}}^{\lambda}\right)\right]-\frac{v^{*}\theta^{*}}{a}\frac{\partial\overline{\theta}^{\lambda}}{\partial\varphi}-\omega^{*}\theta^{*}\frac{\partial\overline{\theta}^{\lambda}}{\partial p}+\dot{\theta}^{*}\theta^{*}.\end{aligned}\tag{73}$$

Zonally averaging (73) gives us

$$\begin{aligned}&\left(\frac{\partial}{\partial t}+\frac{\overline{v}^{\lambda}}{a}\frac{\partial}{\partial\varphi}+\overline{\omega}^{\lambda}\frac{\partial}{\partial p}\right)\left(\frac{1}{2}\overline{\theta^{*2}}^{\lambda}\right)+\frac{1}{a\cos\varphi}\frac{\partial}{\partial\varphi}\left[\overline{v^{*}\left(\frac{1}{2}\theta^{*2}\right)}^{\lambda}\cos\varphi\right]+\frac{\partial}{\partial p}\left[\overline{\omega^{*}\left(\frac{1}{2}\theta^{*2}\right)}^{\lambda}\right]\\&\qquad=-\frac{\overline{v^{*}\theta^{*}}^{\lambda}}{a}\frac{\partial\overline{\theta}^{\lambda}}{\partial\varphi}-\overline{\omega^{*}\theta^{*}}^{\lambda}\frac{\partial\overline{\theta}^{\lambda}}{\partial p}+\overline{\dot{\theta}^{*}\theta^{*}}^{\lambda}.\end{aligned}\tag{74}$$

Finally, we can use (66) to rewrite (74) in flux form:

$$\boxed{\begin{aligned}&\frac{\partial}{\partial t}\left(\frac{1}{2}\overline{\theta^{*2}}^{\lambda}\right)+\frac{1}{a\cos\varphi}\frac{\partial}{\partial\varphi}\left(\overline{v}^{\lambda}\frac{1}{2}\overline{\theta^{*2}}^{\lambda}\cos\varphi\right)+\frac{\partial}{\partial p}\left(\overline{\omega}^{\lambda}\frac{1}{2}\overline{\theta^{*2}}^{\lambda}\right)\\&\qquad\underbrace{+\frac{1}{a\cos\varphi}\frac{\partial}{\partial\varphi}\left(\overline{v^{*}\frac{1}{2}\theta^{*2}}^{\lambda}\cos\varphi\right)+\frac{\partial}{\partial p}\left(\overline{\omega^{*}\frac{1}{2}\theta^{*2}}^{\lambda}\right)}_{\text{eddy transport}}\\&\qquad=\underbrace{-\frac{\overline{v^{*}\theta^{*}}^{\lambda}}{a}\frac{\partial\overline{\theta}^{\lambda}}{\partial\varphi}-\overline{\omega^{*}\theta^{*}}^{\lambda}\frac{\partial\overline{\theta}^{\lambda}}{\partial p}}_{\text{gradient production}}+\overline{\dot{\theta}^{*}\theta^{*}}^{\lambda}.\end{aligned}}\tag{75}$$

According to (75), $(1/2)\overline{\theta^{*2}}^{\lambda}$ can change owing to advection by the mean meridional circulation, transport by the eddies themselves, or "gradient production."

Equation (75) governs the eddy variance at a particular latitude. The meridional and vertical gradients of $\overline{\theta}^{\lambda}$ also contribute to the *global* variance of θ. To derive an equation for this part of the global variance of θ, we start by using (66) to rewrite (65) as

$$\left(\frac{\partial}{\partial t}+\frac{\overline{v}^{\lambda}}{a}\frac{\partial}{\partial\varphi}+\overline{\omega}^{\lambda}\frac{\partial}{\partial p}\right)\overline{\theta}^{\lambda}=-\frac{1}{a\cos\varphi}\frac{\partial}{\partial\varphi}\left(\overline{v^{*}\theta^{*}}^{\lambda}\cos\varphi\right)-\frac{\partial}{\partial p}\left(\overline{\omega^{*}\theta^{*}}^{\lambda}\right)+\overline{\dot{\theta}}^{\lambda}.\tag{76}$$

Multiplication by $\overline{\theta}^{\lambda}$ gives

$$\left(\frac{\partial}{\partial t}+\frac{\overline{v}^{\lambda}}{a}\frac{\partial}{\partial\varphi}+\overline{\omega}^{\lambda}\frac{\partial}{\partial p}\right)\frac{1}{2}\left(\overline{\theta}^{\lambda}\right)^{2}=-\frac{\overline{\theta}^{\lambda}}{a\cos\varphi}\frac{\partial}{\partial\varphi}\left(\overline{v^{*}\theta^{*}}^{\lambda}\cos\varphi\right)-\overline{\theta}^{\lambda}\frac{\partial}{\partial p}\left(\overline{\omega^{*}\theta^{*}}^{\lambda}\right)+\overline{\theta}^{\lambda}\overline{\dot{\theta}}^{\lambda}.\tag{77}$$

We can rearrange (77) to

$$\begin{aligned}&\left(\frac{\partial}{\partial t}+\frac{\overline{v}^{\lambda}}{a}\frac{\partial}{\partial\varphi}+\overline{\omega}^{\lambda}\frac{\partial}{\partial p}\right)\frac{1}{2}\left(\overline{\theta}^{\lambda}\right)^{2}\\&=-\frac{1}{a\cos\varphi}\frac{\partial}{\partial\varphi}\left(\overline{\theta}^{\lambda}\overline{v^{*}\theta^{*}}^{\lambda}\cos\varphi\right)-\frac{\partial}{\partial p}\left(\overline{\theta}^{\lambda}\overline{\omega^{*}\theta^{*}}^{\lambda}\right)+\frac{\overline{v^{*}\theta^{*}}^{\lambda}}{a}\frac{\partial\overline{\theta}^{\lambda}}{\partial\varphi}+\overline{\omega^{*}\theta^{*}}^{\lambda}\frac{\partial\overline{\theta}^{\lambda}}{\partial p}\overline{\theta}^{\lambda}\overline{\dot{\theta}}^{\lambda}.\end{aligned}\tag{78}$$

Converting (78) to flux form, we find that

$$\boxed{\begin{aligned}&\frac{\partial}{\partial t}\left[\frac{1}{2}\left(\overline{\theta}^{\lambda}\right)^{2}\right]+\frac{1}{a\cos\varphi}\frac{\partial}{\partial\varphi}\left[\overline{v}^{\lambda}\frac{1}{2}\left(\overline{\theta}^{\lambda}\right)^{2}\cos\varphi\right]+\frac{\partial}{\partial p}\left[\overline{\omega}^{\lambda}\frac{1}{2}\left(\overline{\theta}^{\lambda}\right)^{2}\right]\\&=-\frac{1}{a\cos\varphi}\frac{\partial}{\partial\varphi}\left(\overline{\theta}^{\lambda}\overline{v^{*}\theta^{*}}^{\lambda}\cos\varphi\right)-\frac{\partial}{\partial p}\left(\overline{\theta}^{\lambda}\overline{\omega^{*}\theta^{*}}^{\lambda}\right)+\frac{\overline{v^{*}\theta^{*}}^{\lambda}}{a}\frac{\partial\overline{\theta}^{\lambda}}{\partial\varphi}+\overline{\omega^{*}\theta^{*}}^{\lambda}\frac{\partial\overline{\theta}^{\lambda}}{\partial p}+\overline{\theta}^{\lambda}\overline{\dot{\theta}}^{\lambda}.\end{aligned}}\tag{79}$$

When we add (79) and (75), the gradient production terms cancel, which shows that those terms represent a "conversion" between $(1/2)\left(\overline{\theta}^{\lambda}\right)^{2}$ and $\overline{\theta^{2}}^{\lambda}=\left(\overline{\theta}^{\lambda}\right)^{2}+\overline{\theta^{*2}}^{\lambda}$.

Note that $(1/2)\overline{\theta^{*2}}^{\lambda}$, that is, the zonal average of the square is the sum of the square of the zonal average and the square of the departure from the zonal average. Similarly, $\overline{\theta\dot{\theta}}^{\lambda} = \overline{\theta}^{\lambda}\overline{\dot{\theta}}^{\lambda} + \overline{\dot{\theta}_* \theta_*}^{\lambda}$. We obtain

$$\begin{aligned}\frac{\partial}{\partial t}\left(\frac{1}{2}\overline{\theta}^{2\lambda}\right) + \frac{1}{a\cos\varphi}\frac{\partial}{\partial\varphi}\left[\left(\overline{v}^{\lambda}\frac{1}{2}\overline{\theta}^{2\lambda} + \overline{v^{*}\frac{1}{2}\theta^{*2}}^{\lambda} + \overline{\theta}^{\lambda}[v^{*}\theta^{*}]\right)\cos\varphi\right] \\ + \frac{\partial}{\partial p}\left(\overline{\omega}^{\lambda}\frac{1}{2}\overline{\theta}^{2\lambda} + \overline{\omega^{*}\frac{1}{2}\theta^{*2}}^{\lambda} + \overline{\theta}^{\lambda}\overline{\omega^{*}\theta^{*}}^{\lambda}\right) = \overline{\theta\dot{\theta}}^{\lambda}.\end{aligned} \tag{80}$$

Finally, integration of (80) over the entire atmosphere gives

$$\boxed{\frac{d}{dt}\int_{M}\left(\frac{1}{2}\overline{\theta}^{2\lambda}\right)dM = \int_{M}\overline{\theta\dot{\theta}}^{\lambda}\,dM}. \tag{81}$$

This result shows that the total variance changes owing only to the covariance between temperature and the heating rate. The gradient-production terms have no effect on the total variance.

Generation of Available Potential Energy and Its Conversion into Kinetic Energy

Earlier we derived (19), Lorenz's approximate expression for the available potential energy of a statically stable atmosphere, which is repeated here for convenience:

$$A = \frac{a^2}{2}\int_{-\pi/2}^{\pi/2}\int_{0}^{2\pi}\int_{0}^{P}\frac{\overline{T}}{(\Gamma_d - \Gamma)}\left(\frac{T'}{\overline{T}}\right)^2\cos\varphi\, dp\, d\lambda\, d\varphi. \tag{82}$$

Recall that in this equation an overbar represents a global mean on a pressure surface, and a prime denotes a departure from the global mean. The APE is an integral of the variance of the temperature about its global mean on pressure surfaces. In the previous section we derived an equation for the time rate of change of the potential temperature variance. We now work out an approximate equation for the time rate of change of A due to generation and conversion to or from kinetic energy.

Let the subscript GM denote a global mean on an isobaric surface; that is,

$$(\;)_{GM} \equiv \frac{1}{4\pi a^2}\int_{-\pi/2}^{\pi/2}\int_{0}^{2\pi}(\;)a^2\cos\varphi\, d\lambda\, d\varphi. \tag{83}$$

We can show that for any two quantities α and β,

$$(\alpha\beta)_{GM} = \left(\overline{\alpha}^{\lambda}\overline{\beta}^{\lambda} + \overline{\alpha^{*}\beta^{*}}^{\lambda}\right)_{GM}, \tag{84}$$

and

$$\begin{aligned}\{(\alpha - \alpha_{GM})(\beta - \beta_{GM})\}_{GM} &= (\alpha\beta)_{GM} - \alpha_{GM}\beta_{GM} \\ &= \left(\overline{\alpha}^{\lambda}\overline{\beta}^{\lambda} + \overline{\alpha^{*}\beta^{*}}^{\lambda}\right)_{GM} - \alpha_{GM}\beta_{GM}.\end{aligned} \tag{85}$$

As a special case of (85), the variance of an arbitrary quantity α about its global mean is given by

$$\begin{aligned}\alpha_{Var} &\equiv (\alpha - \alpha_{GM})^2_{GM} \\ &= \left[(\overline{\alpha}^{\lambda})^2 + \overline{\alpha^{*2}}^{\lambda}\right]_{GM} - \alpha^2_{GM}.\end{aligned} \tag{86}$$

Using the notation introduced above, we can approximate the expression for the available potential energy given by (82) as

$$\begin{aligned} A &\cong 2\pi a^2 \int_0^P \frac{T_{GM}}{(\Gamma_d - \Gamma_{GM})}\left(\frac{T_{Var}}{T_{GM}^2}\right)dp \\ &= 2\pi a^2 \int_0^P \frac{T_{GM}}{(\Gamma_d - \Gamma_{GM})}\left(\frac{\theta_{Var}}{\theta_{GM}^2}\right)dp. \end{aligned} \tag{87}$$

We now work out an equation for dA/dt, based on (87). The global means of (63) and (64) are

$$\frac{\partial \theta_{GM}}{\partial t} + \frac{\partial(\omega\theta)_{GM}}{\partial p} = \dot{\theta}_{GM}, \tag{88}$$

and

$$\frac{\partial \omega_{GM}}{\partial p} = 0, \tag{89}$$

respectively. Since $\omega = 0$ at $p = 0$, it follows from (89) that

$$\omega_{GM} = 0 \text{ for all } p. \tag{90}$$

This result allows us to write

$$\begin{aligned} (\omega\theta)_{GM} &= \omega_{GM}\theta_{GM} + \{(\omega - \omega_{GM})(\theta - \theta_{GM})\}_{GM} \\ &= \{\omega(\theta - \theta_{GM})\}_{GM}. \end{aligned} \tag{91}$$

Area-weighted integration of (80) over all latitudes, at a given pressure level, leads to

$$\begin{aligned} &\frac{\partial}{\partial t}\frac{1}{2}\left[(\overline{\theta}^\lambda)^2 + \overline{\theta^{*2}}^\lambda\right]_{GM} + \frac{\partial}{\partial p}\left\{\overline{\omega}^\lambda \frac{1}{2}\left[(\overline{\theta}^\lambda)^2 + \overline{\theta^{*2}}^\lambda\right] + \overline{\omega^* \frac{1}{2}\theta^{*2}}^\lambda + \overline{\theta}^\lambda \overline{\omega^*\theta^*}^\lambda\right\}_{GM} \\ &\quad = \left(\overline{\theta}^\lambda \overline{\dot{\theta}}^\lambda + \overline{\dot{\theta}^*\theta^*}^\lambda\right)_{GM}. \end{aligned} \tag{92}$$

Here, as usual, we ignore the complications arising from the fact that some pressure surfaces intersect the Earth's surface. From (86), we see that

$$\frac{\partial \theta_{Var}}{\partial t} = \frac{\partial}{\partial t}\left[(\overline{\theta}^\lambda)^2 + \overline{\theta^{*2}}^\lambda\right]_{GM} - 2\theta_{GM}\frac{\partial \theta_{GM}}{\partial t}. \tag{93}$$

Substituting into (93) from (88) and (92), and using (85) and (91), we find that

$$\begin{aligned} \frac{\partial}{\partial t}\left(\frac{\theta_{Var}}{2}\right) &= -\frac{\partial}{\partial p}\left\{\overline{\omega}^\lambda \frac{1}{2}\left[(\overline{\theta}^\lambda)^2 + \overline{\theta^{*2}}^\lambda\right] + \overline{\omega^* \frac{1}{2}\theta^{*2}}^\lambda + \overline{\theta}^\lambda \overline{\omega^*\theta^*}^\lambda\right\}_{GM} + \theta_{GM}\frac{\partial(\omega\theta)_{GM}}{\partial p} \\ &\quad + \left[\overline{\theta}^\lambda \overline{\dot{\theta}}^\lambda + \overline{\dot{\theta}\theta^*}^\lambda\right] \\ &= -\frac{\partial}{\partial p}\left\{\overline{\omega}^\lambda \frac{1}{2}\left[(\overline{\theta}^\lambda)^2 + \overline{\theta^{*2}}^\lambda\right] + \overline{\omega^* \frac{1}{2}\theta^{*2}}^\lambda + \left[(\overline{\theta}^\lambda \overline{\omega^*\theta^*}^\lambda)_{GM} - \theta_{GM}(\omega\theta)_{GM}\right]\right\}_{GM} \\ &\quad - (\omega\theta)_{GM}\frac{\partial \theta_{GM}}{\partial p} + \left[(\theta - \theta_{GM})(\dot{\theta} - \dot{\theta}_{GM})\right]_{GM} \\ &= -\frac{\partial}{\partial p}\left\{\overline{\omega}^\lambda \frac{1}{2}\left[(\overline{\theta}^\lambda)^2 + \overline{\theta^{*2}}^\lambda\right] + \overline{\omega^* \frac{1}{2}\theta^{*2}}^\lambda + \left[(\overline{\theta}^\lambda \overline{\omega^*\theta^*}^\lambda)_{GM} - \theta_{GM}(\omega\theta)_{GM}\right]\right\}_{GM} \\ &\quad - \left[\omega(\theta - \theta_{GM})\right]_{GM}\frac{\partial \theta_{GM}}{\partial p} + \left[(\theta - \theta_{GM})(\dot{\theta} - \dot{\theta}_{GM})\right]_{GM}. \end{aligned} \tag{94}$$

We can recognize the various processes in this equation. Vertical transport is clearly visible, as are gradient production and "Heat where it's hot, cool where it's cold." Use of (17) in the approximate form

$$\frac{\partial \theta_{GM}}{\partial p} \cong -\frac{\kappa \theta_{GM}}{p}\left(\frac{\Gamma_d - \Gamma_{GM}}{\Gamma_d}\right) \tag{95}$$

allows us to rewrite (94) as

$$\begin{aligned}\frac{\partial}{\partial t}\left(\frac{\theta_{Var}}{2}\right) = &-\frac{\partial}{\partial p}\left\{\overline{\omega}^{\lambda}\frac{1}{2}\left[(\overline{\theta}^{\lambda})^2 + \overline{\theta^{*2}}^{\lambda}\right] + \overline{\omega^{*}\frac{1}{2}\theta^{*2}}^{\lambda} + \left[(\overline{\theta}^{\lambda}\overline{\omega^{*}\theta^{*}}^{\lambda})_{GM} - \theta_{GM}(\omega\theta)_{GM}\right]\right\}_{GM} \\ &+ \left[\omega(\theta - \theta_{GM})\right]_{GM}\frac{\kappa\theta_{GM}}{p}\left(\frac{\Gamma_d - \Gamma_{GM}}{\Gamma_d}\right) + \left[(\theta - \theta_{GM})(\dot{\theta} - \dot{\theta}_{GM})\right]_{GM}.\end{aligned} \tag{96}$$

Vertical integration of (96) gives

$$\begin{aligned}\frac{d}{dt}\left(\int_0^{p_s}\frac{\theta_{Var}}{2}dp\right) = &\int_0^{p_s}\{\omega(\theta - \theta_{GM})\}_{GM}\frac{\kappa\theta_{GM}}{p}\left(\frac{\Gamma_d - \Gamma_{GM}}{\Gamma_d}\right)dp \\ &+ \int_0^{p_s}\{(\theta - \theta_{GM})(\dot{\theta} - \dot{\theta}_{GM})\}_{GM}dp.\end{aligned} \tag{97}$$

Here we write a total time derivative, d/dt, because we have now integrated over all three space variables. With the use of (87), we can approximate (97) by

$$\frac{dA}{dt} = 4\pi a^2\int_0^{p_s}\left[\omega\left(\frac{\theta - \theta_{GM}}{\theta_{GM}}\right)\right]_{GM}\frac{\kappa T_{GM}}{p\Gamma_d}dp + 4\pi a^2\int_0^{p_s}\frac{\{(\theta - \theta_{GM})(\dot{\theta} - \dot{\theta}_{GM})\}_{GM}T_{GM}}{\theta_{GM}^2(\Gamma_d - \Gamma_{GM})}dp. \tag{98}$$

Finally, we note that

$$\frac{\theta - \theta_{GM}}{\theta_{GM}} = \frac{\alpha - \alpha_{GM}}{\alpha_{GM}} = \frac{T - T_{GM}}{T_{GM}}, \tag{99}$$

so that

$$\begin{aligned}\frac{dA}{dt} &= 4\pi a^2\int_0^{p_s}\left\{\omega\left(\frac{\alpha - \alpha_{GM}}{\alpha_{GM}}\right)\right\}_{GM}\frac{\kappa T_{GM}}{p\Gamma_d}dp + 4\pi a^2\int_0^{p_s}\frac{\{(\theta - \theta_{GM})(\dot{\theta} - \dot{\theta}_{GM})\}_{GM}T_{GM}}{\theta_{GM}^2(\Gamma_d - \Gamma_{GM})}dp \\ &= \frac{4\pi a^2}{g}\int_0^{p_s}\{\omega(\alpha - \alpha_{GM})\}_{GM}dp + 4\pi a^2\int_0^{p_s}\frac{\left\{\left(\frac{\theta - \theta_{GM}}{\theta_{GM}}\right)\left(\frac{T_{GM}}{\theta_{GM}}\right)(\dot{\theta} - \dot{\theta}_{GM})\right\}_{GM}}{(\Gamma_d - \Gamma_{GM})}dp.\end{aligned} \tag{100}$$

To obtain the second line of (100), we used $\Gamma_d = g/c_p$.

Recall that C represents conversion between kinetic and nonkinetic energy, so by inspection of (100) we can identify the rate of conversion of kinetic energy (KE) into APE as

$$\boxed{C \cong \frac{4\pi a^2}{g}\int_0^{p_s}\{\omega(\alpha - \alpha_{GM})\}_{GM}dp}. \tag{101}$$

The rate of generation or destruction of APE by heating is

$$\boxed{G \cong 4\pi a^2\int_0^{p_s}\frac{\left\{\left(\frac{\theta - \theta_{GM}}{\theta_{GM}}\right)\left(\frac{T_{GM}}{\theta_{GM}}\right)(\dot{\theta} - \dot{\theta}_{GM})\right\}}{(\Gamma_d - \Gamma_{GM})}dp}. \tag{102}$$

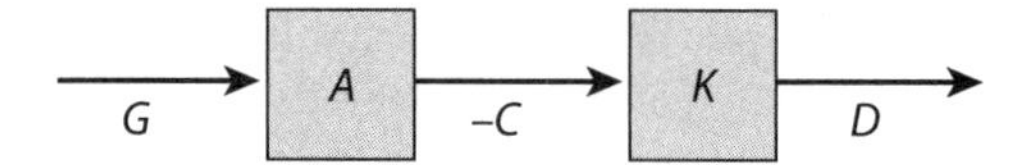

Figure 7.6. Sketch illustrating the flow of energy through the atmospheric general circulation. Generation produces APE, which is converted to KE, which in turn is dissipated.

Note that $G > 0$ if we "heat where it's hot and cool where it's cold." This should sound familiar. A heating field that generates APE destroys entropy.

We can summarize (100) as

$$\boxed{\frac{dA}{dt} = C + G}. \tag{103}$$

Similarly, we can show that

$$\boxed{\frac{dK}{dt} = -C - D}, \tag{104}$$

where K is the globally integrated kinetic energy (e.g., in joules), and D is the globally integrated dissipation rate. We can also conclude that C is negative from the fact that $D > 0$. There must be a net conversion of APE into KE, to supply the KE that is destroyed by dissipation. Because this conversion depletes the APE, the generation of APE must be positive. In other words, energy must flow as shown in figure 7.6.

The Governing Equations for the Eddy Kinetic Energy, Zonal Kinetic Energy, and Total Kinetic Energy

We now present a discussion of the eddy kinetic energy, zonal kinetic energy, and total kinetic energy equations. The derivations of these equations follow methods similar to those used to derive the conservation equation for the potential energy variance. The details are given in appendix F.

We define the eddy kinetic energy per unit mass by

$$KE \equiv \frac{1}{2}\left[\overline{(u^*)^2}^\lambda + \overline{(v^*)^2}^\lambda\right]. \tag{105}$$

We can show that it satisfies the following equation:

$$\boxed{\begin{aligned}
&\frac{\partial}{\partial t}KE \\
&+\frac{1}{a\cos\varphi}\frac{\partial}{\partial\varphi}\left\{\overline{v}^\lambda KE\cos\varphi + \frac{1}{2}\left[\overline{v^*(u^{*2}+v^{*2})}^\lambda\right] + \overline{v^*\phi^*}^\lambda\cos\right\} \\
&+\frac{\partial}{\partial p}\left\{\overline{\omega}^\lambda KE + \frac{1}{2}\left[\overline{\omega^*(u^{*2}+v^{*2})}^\lambda\right] + \overline{\omega^*\phi^*}^\lambda\right\} \\
&+\frac{\overline{u^*v^*}^\lambda}{a}\frac{\partial\overline{u}^\lambda}{\partial\varphi} + \frac{\overline{v^*v^*}^\lambda}{a}\frac{\partial\overline{v}^\lambda}{\partial\varphi} + \overline{u^*\omega^*}^\lambda\frac{\partial\overline{u}^\lambda}{\partial p} + \overline{\omega^*v^*}^\lambda\frac{\partial\overline{v}^\lambda}{\partial p} \\
&= \left(-\overline{u}^\lambda\overline{u^*v^*}^\lambda + \overline{u^*u^*}^\lambda\overline{v}^\lambda\right)\frac{\tan\varphi}{a} \\
&-\overline{\omega^*\alpha^*} \\
&+\overline{u^*g\frac{\partial F_u^*}{\partial p}}^\lambda + \overline{v^*g\frac{\partial F_v^*}{\partial p}}^\lambda.
\end{aligned}} \tag{106}$$

Gradient-production terms appear on the fourth line of (106). They represent the conversion between the kinetic energy of the mean flow and that of the eddies, in the sense that the eddy kinetic energy increases when the eddy momentum flux is "down the gradient," that is, when it is from higher mean momentum to lower mean momentum. The $\omega^* \alpha^*$ term represents eddy kinetic energy generation from eddy available potential energy, while the terms involving ϕ^* represent the effects of "pressure work."

The appearance of the metric terms in (106) may be somewhat surprising. They arise because we defined *eddies* in terms of departures from the zonal mean, so that a particular latitude-longitude coordinate system is implicit in the very definition of *KE*. Obviously, there cannot be any metric terms in the equation for the total kinetic energy per unit mass, which we denote by *K*.

We define the zonal kinetic energy, *KZ*, by

$$KZ \equiv \frac{1}{2}\left[(\overline{u}^{\lambda})^2 + (\overline{v}^{\lambda})^2\right] \tag{107}$$

and note that

$$\overline{K}^{\lambda} = KZ + KE. \tag{108}$$

All three quantities in (108) are independent of longitude. The zonal kinetic energy satisfies

$$\boxed{\begin{aligned}
&\frac{\partial}{\partial t} KZ \\
&\quad + \frac{1}{a\cos\varphi}\frac{\partial}{\partial\varphi}\left[\left(\overline{v}^{\lambda} KZ + \overline{u}^{\lambda}\,\overline{u^* v^*}^{\lambda} + \overline{v}^{\lambda}\,\overline{v^* v^*}^{\lambda} + \overline{v}^{\lambda}\overline{\phi}^{\lambda}\right)\cos\varphi\right] \\
&\quad + \frac{\partial}{\partial p}\left(\overline{\omega}^{\lambda} KZ + \overline{u}^{\lambda}\,\overline{\omega^* u^*}^{\lambda} + \overline{v}^{\lambda}\,\overline{\omega^* v^*}^{\lambda} + \overline{\omega}^{\lambda}\overline{\phi}^{\lambda}\right) \\
&= \frac{\overline{u^* v^*}^{\lambda}}{a}\frac{\partial \overline{u}^{\lambda}}{\partial\varphi} + \frac{\overline{v^* v^*}^{\lambda}}{a}\frac{\partial \overline{v}^{\lambda}}{\partial\varphi} + \overline{\omega^* u^*}^{\lambda}\frac{\partial \overline{u}^{\lambda}}{\partial p} + \overline{\omega^* v^*}^{\lambda}\frac{\partial \overline{v}^{\lambda}}{\partial p} \\
&\quad + \left(\overline{u}^{\lambda}\,\overline{u^* v^*}^{\lambda} - \overline{v}^{\lambda}\,\overline{u^* u^*}^{\lambda}\right)\frac{\tan\varphi}{a} \\
&\quad - \overline{\omega}^{\lambda}\overline{\alpha}^{\lambda} \\
&\quad + \overline{u}^{\lambda} g\frac{\partial \overline{F_u}^{\lambda}}{\partial p} + \overline{v}^{\lambda}\frac{\partial \overline{F_v}^{\lambda}}{\partial p}.
\end{aligned}} \tag{109}$$

Notice that the "gradient-production" terms of (106) appear with the opposite sign in (109). They represent conversions between *KE* and *KZ*. Adding the equations for *KZ* and *KE*, we obtain the equation for the zonally averaged total kinetic energy:

$$\begin{aligned}
&\frac{\partial \overline{K}^{\lambda}}{\partial t} \\
&+ \frac{1}{a\cos\varphi}\frac{\partial}{\partial\varphi}\left\{\left(\overline{v}^{\lambda}\overline{K}^{\lambda} + \frac{1}{2}\left[\overline{v^*\left(u^{*2} + v^{*2}\right)}^{\lambda}\right] + \overline{u}^{\lambda}\,\overline{u^* v^*}^{\lambda} + \overline{v}^{\lambda}\,\overline{v^* v^*}^{\lambda} + \overline{v}^{\lambda}\overline{\phi}^{\lambda} + \overline{v^*\phi^*}^{\lambda}\right)\cos\varphi\right\} \\
&+ \frac{\partial}{\partial p}\left(\overline{\omega}^{\lambda}\overline{K}^{\lambda} + \frac{1}{2}\left[\overline{\omega^*\left(u^{*2} + v^{*2}\right)}^{\lambda}\right] + \overline{u}^{\lambda}\,\overline{\omega^* u^*}^{\lambda} + \overline{v}^{\lambda}\,\overline{\omega^* v^*}^{\lambda} + \overline{\phi}^{\lambda}\overline{\omega}^{\lambda} + \overline{\omega^*\phi^*}^{\lambda}\right) \\
&= -\overline{\omega}^{\lambda}\overline{\alpha}^{\lambda} - \overline{\omega^*\alpha^*} \\
&+ \overline{u^* g\frac{\partial F_u^*}{\partial p}}^{\lambda} + \overline{v^* g\frac{\partial F_v^*}{\partial p}}^{\lambda} + \overline{u}^{\lambda} g\frac{\partial \overline{F_u}^{\lambda}}{\partial p} + \overline{v}^{\lambda}\frac{\partial \overline{F_v}^{\lambda}}{\partial p}.
\end{aligned} \tag{110}$$

As expected, the metric terms canceled; they cannot affect the zonally averaged *total* kinetic energy.

Observations of the Energy Cycle

Arpé et al. (1986) discussed the observed energy cycle of the atmosphere, based on ECMWF analyses. They wrote the following equations for the energy cycle:

$$\frac{d}{dt}KZ = -\sum_m CK(m) + CZ - DZ, \tag{111}$$

$$\frac{d}{dt}AZ = -\sum_m CA(m) - CZ + GZ, \tag{112}$$

$$\frac{d}{dt}KE(m) = CK(m) + LK(m) + CE(m) - DE(m), \tag{113}$$

$$\frac{d}{dt}AE(m) = CA(m) + LA(m) - CE(m) - GE(m). \tag{114}$$

Here m is the zonal wavenumber. The eddy kinetic energy and eddy available potential energy are defined as functions of the zonal wavenumber, and the contributions for the individual waves have been determined. The terms $LK(m)$ and $LA(m)$ represent wave-wave interactions due to nonlinear processes. For example, if there is a "kinetic energy cascade" from lower wavenumbers to higher wavenumbers, then $LK(m)$ will represent a flow of energy from larger scales to smaller scales. If (111)–(114) are added together, all terms on the right-hand side of the result cancel, except for *DZ*, *GZ*, *DE*, and *GE*.

A direct mean meridional circulation, such as the Hadley circulation, converts *AZ* into *KZ* and so is associated with positive values of *CZ*.

Baroclinic instability of the zonal flow is represented by the combination of positive *CA* and positive *CE*; the first of these represents the conversion of *AZ* to *AE*, and the second represents the conversion of *AE* to *KE*. The net effect is thus conversion of *AZ* to *KE*. When *AZ* is converted to *AE*, an eddy temperature variance is created by conversion from the global temperature variance associated with the meridional gradient of the zonally averaged temperature. This is the gradient-production process analyzed earlier in this chapter. When *AE* is converted to *KE*, warm air rises and cool air sinks, potential temperature is transported upward, and the center of gravity of the atmosphere is lowered.

A tendency for eddies to pump momentum into the jet, increasing its strength, would be represented by negative values of *CK*. Mathematically, this process is represented by the gradient-production term in the eddy kinetic energy equation.

It is important to understand the meaning of each term of (111) through (114), as outlined above. It is also important, however, to notice that certain terms are *not* present, which means that certain processes do not exist. For example, there is no process that directly converts *AZ* into $KE(m)$. Such conversion can occur only indirectly, in two steps, for example, first $AZ \rightarrow AE$ and then $AE \rightarrow KE$.

Figure 7.7 shows the observed energy cycle for (northern) winter and summer and for the two hemispheres separately. The figure is arranged so that the "summer hemispheres" are on the right (for both seasons) and the "winter hemispheres" are on the left. The numbers in the boxes represent amounts of energy, and the numbers on the arrows between boxes represent energy conversions or processes that generate or destroy energy. Note that Arpé et al. defined the APE for the Northern and Southern Hemispheres separately; as discussed earlier, this is not strictly correct.

The arrows leading into *AZ* and *AE* from the left represent generation (an arrow leading out simply represents negative generation), and the arrows leading out of *KE* and *KZ* to the right represent dissipation. These arrows can also represent interactions

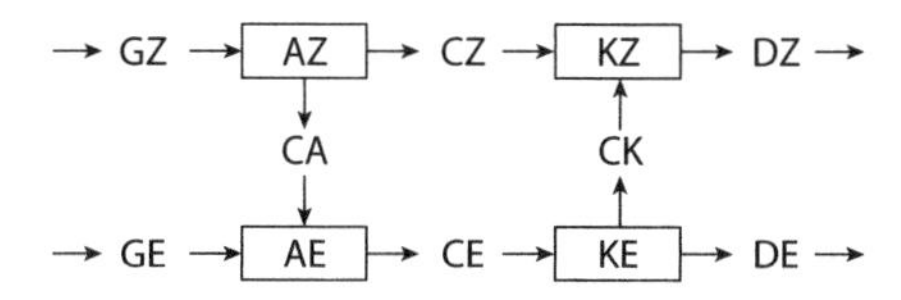

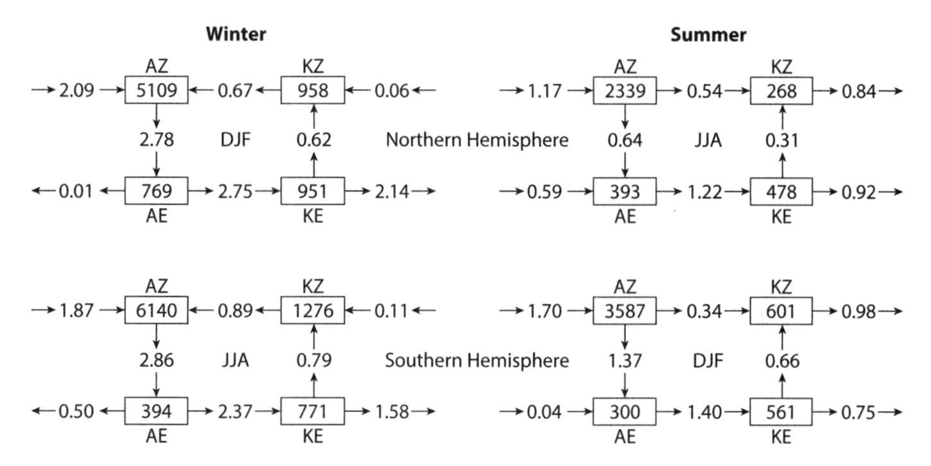

Figure 7.7. Energy cycle during winter (left) and summer (right) for both hemispheres. Integrals between 1000 and 50 hPa are presented. Data are calculated from 00 GMT initialized analyses. Energy amounts are given in kJ m^{-2}, and conversion rates in W m^{-2}. From Arpé et al. (1986). Used with permission of Schweizerbart Science Publishers (www.schweizerbart.de).

between the hemispheres. For example, the arrow leading into *KZ* from the right for the Northern Hemisphere in winter apparently indicates a physically impossible negative dissipation rate but actually represents a gain of *KZ* in the Northern Hemisphere via energy exchanges with the Southern Hemisphere.

Figure 7.7 shows that, especially in the Northern Hemisphere, the energy flows of the atmosphere are much more vigorous in winter than in summer. Note that *AZ* is several times larger than *AE* or *KZ* or *KE*. For the Northern Hemisphere in winter, *AZ* is strongly generated; this energy is converted to *AE*, leading to a production of *KE* via baroclinic instability. The eddies act to increase *KZ* by transporting angular momentum into the jet stream. Simmons and Hoskins (1978) found that this happens as occluding baroclinic eddies near the end of their life cycles. To maintain thermal wind balance with the stronger jet, *KZ* is converted back to *AZ* by strengthening the meridional temperature gradient. Meanwhile, *KE* and *KZ* are both dissipated. This suggests that the mean meridional circulation is overall "indirect"; that is, a net conversion of *AZ* into *KZ* is not seen, as it would be if the direct Hadley circulation were dominating the energetics.

Note, however, that in the summer hemisphere (in both DJF and JJA) the mean meridional circulation is direct, and the eddies are much less active. From the point of view of energetics, then, the winter hemisphere is dominated by eddy processes, while the summer hemisphere is dominated by the mean meridional circulation. Of the four conversion processes—*CA*, *CZ*, *CK*, and *CE*—only the hemispheric values of *CZ* change sign seasonally; the others fluctuate in magnitude but not in sign.

In all cases, *KE* is supplied from *AE*; baroclinic instability is the dominant mechanism for generating eddies. Figure 7.8 shows the annual cycles of energy conversions and amounts. The various panels show the zonal means and the Northern and Southern Hemispheres separately. The global means are relatively constant throughout the

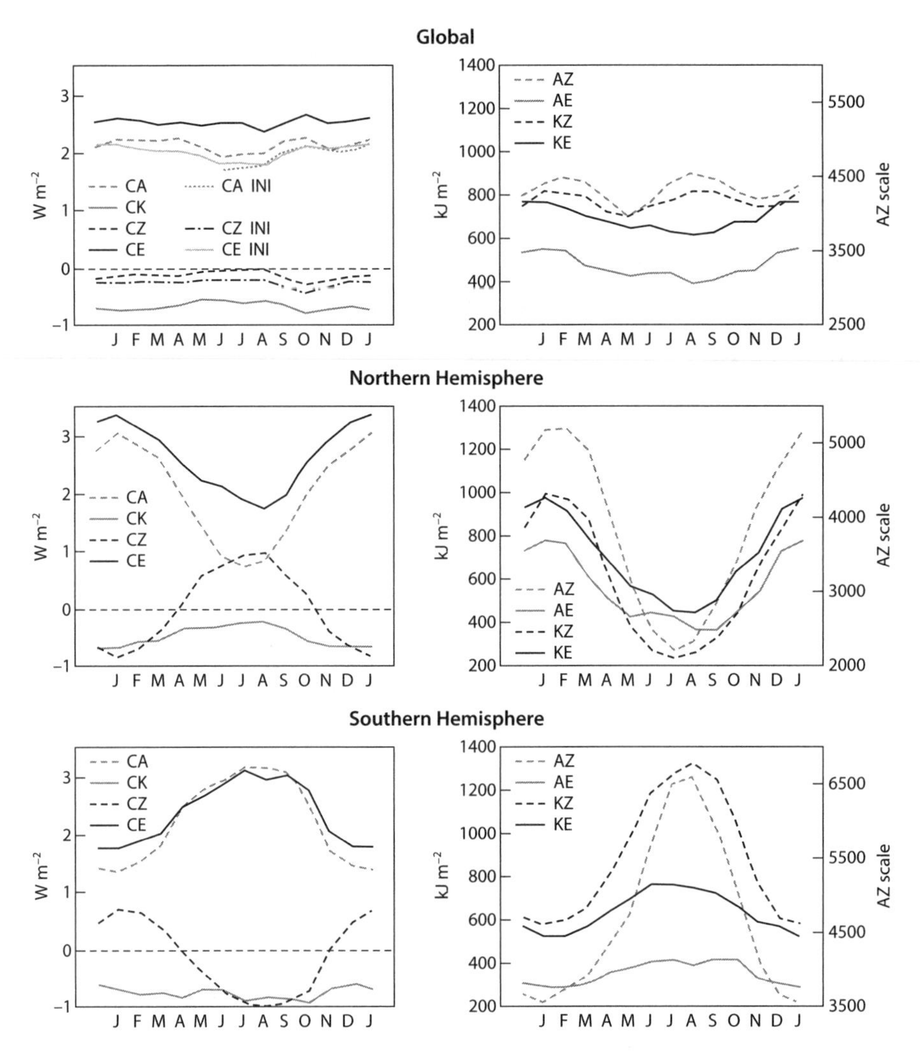

Figure 7.8. Mean annual cycles of energy conversions, and amounts for global and hemispheric averages. Data are calculated from 12-hour forecasts and for global averages. Data from initialized analyses are given as well (curves labeled "INI"). In the left three panels, the zero line is drawn for reference only. Note that the right three panels have two vertical scales. From Arpé et al. (1986). Used with permission of Schweizerbart Science Publishers (www.schweizerbart.de).

year, while the individual hemispheres show large seasonal cycles. Within each hemisphere the winter is much more active, in all respects, than the summer. *CZ* changes sign seasonally. Recall that when *CZ* is positive the MMC is direct overall, and that when *CK* is positive the eddies are deriving kinetic energy from the jet, thus tending to weaken it (rather than acting to increase the kinetic energy of the jet). The figure makes it clear that the energetics of the summer and winter hemispheres are drastically different. The global *AZ* is a maximum shortly after the two solstices and a minimum shortly after the two equinoxes.

Figure 7.9 shows the vertical and meridional distributions of the zonally averaged eddy kinetic energy for January and August. The Northern Hemisphere shows a strong seasonal cycle, while the Southern Hemisphere does not. Wavenumbers 10 and above play little role.

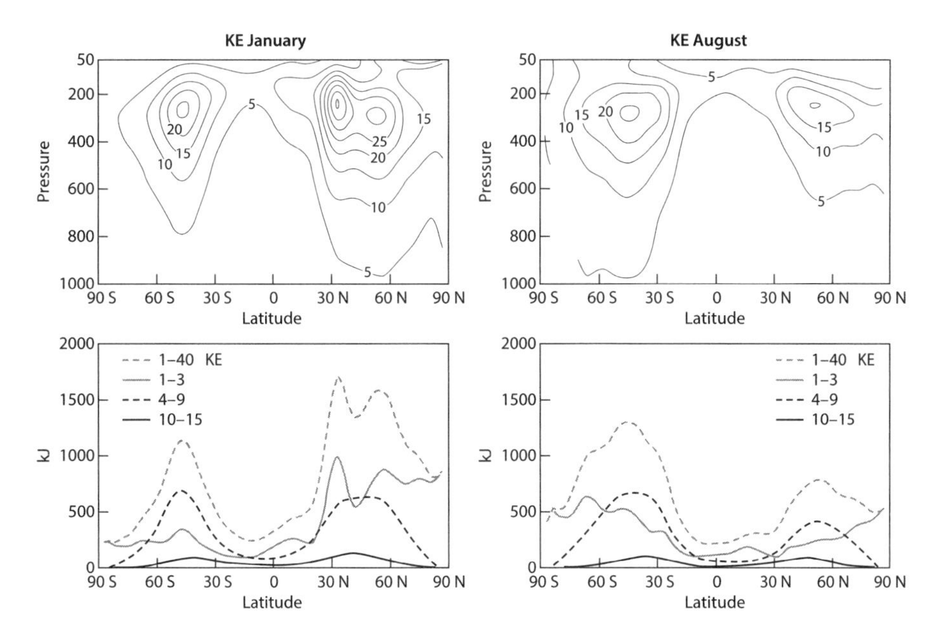

Figure 7.9. Vertical and meridional distribution of zonal mean eddy kinetic energy together with contributions by wavenumber groups to vertical integrals in January and August. Units in the cross sections are J $(m^2\,Pa)^{-1}$ = 100 kJ $(m^2\,bar)^{-1}$. From Arpé et al. (1986). Used with permission of Schweizerbart Science Publishers (www.schweizerbart.de).

Figure 7.10 again shows the mean annual cycles of the energy amounts and conversions for the two hemispheres, but this time the information is given for several distinct wavenumber groups. Wavenumbers 10–15 are quite unimportant, while wavenumbers 4–9 tend to be the most active energetically, as would be expected from the theory of baroclinic instability.

The Role of Heating

We have seen that heating generates APE only when it occurs where the temperature is warm. Heating at cold temperatures actually destroys APE, because it reduces temperature contrasts in the atmosphere. In many cases, heating is in fact a response to a lowering of the temperature. For example, large-surface sensible heat fluxes heat the cold air rushing from the continents out over warm ocean currents in the winter. Similarly, deep moist convection heats the upper troposphere when cooling occurs aloft, for example, because of large-scale lifting or the horizontal advection of cold air aloft. Such examples illustrate that heating does not necessarily promote a more vigorous circulation.

Moist Available Energy

Lorenz (1978, 1979) extended the concept of available potential energy by allowing moist adiabatic processes to occur during the transition to the A-state. To do this, he had to replace the dry enthalpy, c_pT, with a moist enthalpy, which is approximately given by

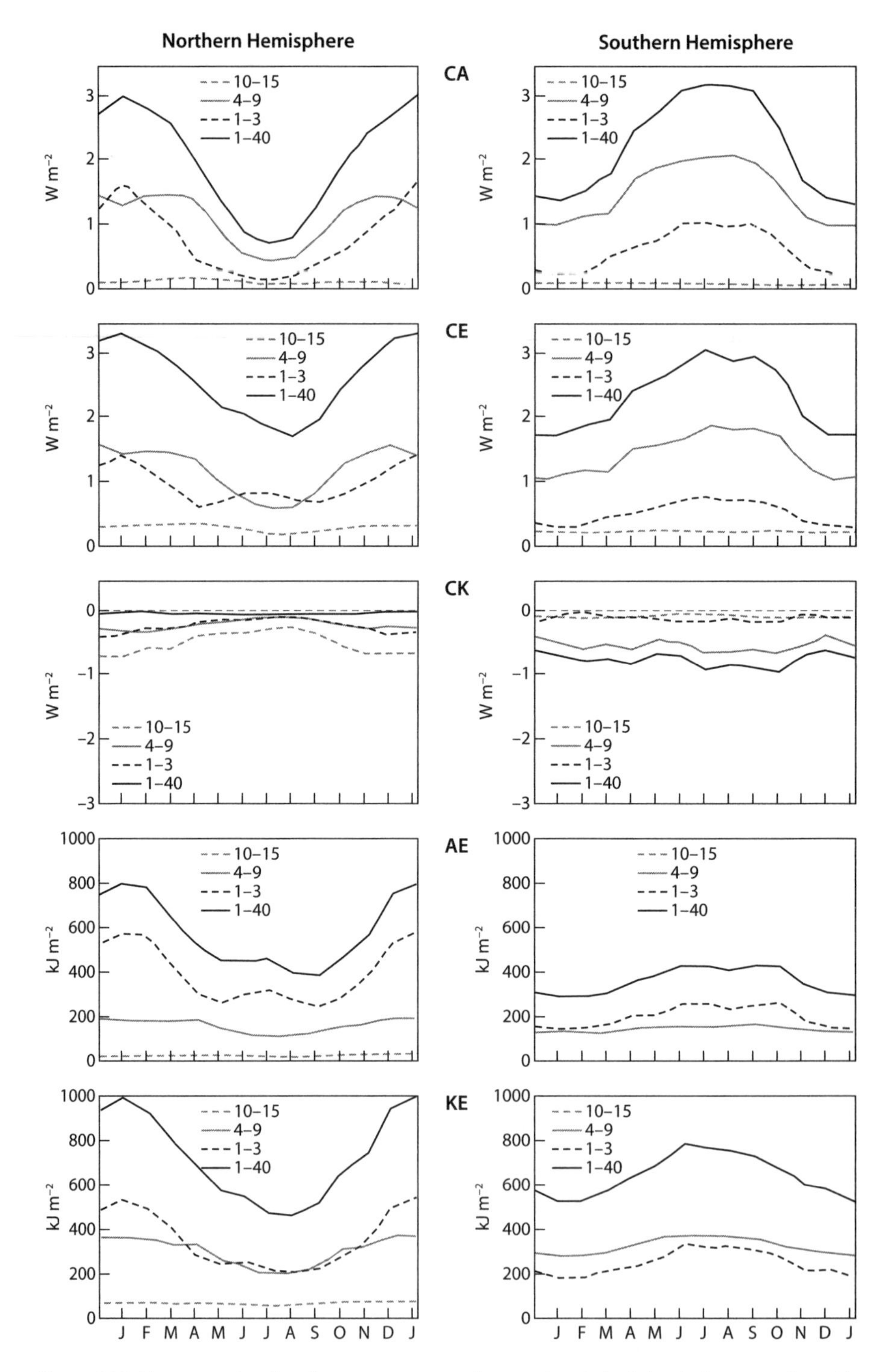

Figure 7.10. Mean annual cycles of energy amounts and conversions in both hemispheres together with contributions from wavenumber groups calculated from 12-hour forecasts. From Arpé et al. (1986). Used with permission of Schweizerbart Science Publishers (www.schweizerbart.de).

$$h \cong c_p T - Ll. \tag{115}$$

As before, it can be shown that the integral of the moist enthalpy over the whole atmosphere, added to the integral of the kinetic energy over the whole atmosphere, is invariant under moist adiabatic, frictionless processes. In other words, (3) still applies, with a suitable redefinition of H, based on (115). This idea can be used to generalize the concept of convective available potential energy, which was discussed in chapter 6 (Randall and Wang, 1992).

With the conventional *dry available energy*, which we have thus far been calling the available potential energy, processes that involve phase changes can either create or destroy A. In the case of moist available energy, phase changes have no direct effect, but the surface fluxes of moisture due to evaporation and precipitation can be quite important (Lorenz, 1979). The concept of moist available energy has not been developed much till now. It is ripe for further investigation.

Segue

We defined the available potential energy and the gross static stability, and studied the generation and conversion of available potential energy. Finally, we presented observations of the atmospheric energy cycle.

The observations of Arpé et al. reminded us again of the wide range of eddy scales that are simultaneously at work in the global circulation of the atmosphere. All these eddies undergo their life cycles in the presence of the same mean meridional circulation. The complicated nonlinear interactions among the eddies are the subject of chapter 10, in which we view the global circulation as a kind of large-scale turbulence.

Problems

1. Prove that the mass-integrated potential energy of the entire atmosphere is lower in the A-state than in the given state.
2. Prove that

$$\frac{\partial \theta}{\partial p} = -\frac{\kappa\theta}{p}\left(\frac{\Gamma_d - \Gamma}{\Gamma_d}\right).$$

3. Prove that

$$\int_0^{p_S} p^{\kappa}\theta dp = \frac{1}{1+\kappa}\int_0^{\infty} p^{1+\kappa} d\theta.$$

   Note that on the right-hand side the lower limit of integration is zero. This result was used to derive (10) from (9).
4. Prove that the gross static stability cannot be negative. State any assumptions.
5. As discussed in chapter 4, the shallow water equations are

$$\frac{\partial h}{\partial t} + \nabla \cdot (h\mathbf{V}) = 0,$$

$$\frac{\partial \mathbf{V}}{\partial t} + (\zeta + f)\mathbf{k} \times \mathbf{V} + \nabla[K + g(h + h_T)] = 0,$$

where $\zeta \equiv \mathbf{k} \cdot (\nabla \times \mathbf{V})$, $K \equiv 1/2(\mathbf{V} \cdot \mathbf{V})$, h is the depth of the water, and h_T is the height of the lower boundary.

a) Show that the available potential energy of the system, per unit area, is $A = (1/2)g[\overline{(h_{fs})^2} - (\overline{h_{fs}})^2]$, where $h_{fs} \equiv h + h_T$ is the height of the free surface, and the bar represents an average over the whole domain.

b) Prove that

$$\frac{d}{dt}(A + \overline{hK}) = 0.$$

6. a) For the example given in the discussion beginning with (37), calculate the variance of θ for both the given state and the A-state, and demonstrate that the two variances are equal.

   b) Continuing the example, assume that

$$F_\theta = -\frac{D}{a}\frac{\partial \theta}{\partial \varphi},$$

   where D is a positive diffusion coefficient. Derive expressions for the time rates of change of the APE and the potential temperature variance, valid at the instant when $\theta(\varphi)$ satisfies (37).

   c) Suppose that a "heating" term is added to the potential temperature equation, so that

$$\frac{\partial \theta}{\partial t} = -\frac{1}{a\cos\varphi}\frac{\partial}{\partial \varphi}(F_\theta \cos\varphi) + Q(\varphi).$$

   Continue to use the form of F_θ given in part b. Find the form of $Q(\varphi)$ needed to maintain a steady state. Plot $Q(\varphi)$. Find the global mean of $Q(\varphi)$. Find the covariance of $Q(\varphi)$ and θ. Discuss.

7. Suppose that

$$\frac{d}{dt}AZ = GZ - \frac{AZ}{\tau_Z} \cong 0,$$

$$\frac{d}{dt}KE = CE - \frac{KE}{\tau_E} \cong 0.$$

   Estimate the values of the timescales τ_Z and τ_E by using the numerical values given in figure 7.7. Compare GZ and CE with the actual rates of change of AZ and KE as shown in figure 7.8.

8. In passing from a given state to the A-state, does the average of θ on an isobaric surface remain the same? Explain why or why not, as clearly as you can.

CHAPTER 8

A Taxonomy of Eddies

Not All Eddies Are Waves

This chapter describes the observed climatological distribution of eddy activity and offers some theories of the mechanisms that produce both stationary and transient eddies.

As discussed in chapter 5, the term *eddies* is used for all circulation features that vary with longitude. Virtually all waves are eddies, but only a subset of eddies are waves: *a wave can be defined as a process that transports energy and momentum without transporting mass*. Consistent with this definition, a particle does not undergo any net displacement during the passage of a small-amplitude wave; it moves along a closed path and returns to its starting point. For example, a waiting surfer gently bobs up and down as small-amplitude waves pass by.

We know, however, that a skillful surfer can catch a ride on a large-amplitude water wave. The surfer can be viewed as a particle of mass that is, in fact, transported by the large-amplitude wave, which illustrates that large-amplitude (often called *finite-amplitude*) waves can transport mass and so can also transport any intensive property that is "attached" to the mass. If a wave is defined as a process that transports energy but not mass, then beyond some limit a large-amplitude wave is not really a wave at all. Finite-amplitude waves are doomed to "break," however; they self-destruct into turbulence.

Because waves as defined here do not transport mass, they cannot spatially redistribute any property that is attached to the mass. Examples of such quantities include the mixing ratios of water vapor and other minor atmospheric constituents, the potential temperature (under dry adiabatic processes), and the potential vorticity. Mass can be transported by other types of eddies, such as traveling vortices (e.g., Willoughby, 1998; Provenzale et al., 2008; Zhang et al., 2014), but not by waves. Nevertheless, waves have the power to spatially redistribute both momentum and energy without transporting mass, because momentum and energy can be exchanged between fluid particles through the pressure-gradient force.

Eddies appear in the winds, temperatures, geopotential heights, surface pressure, water vapor mixing ratio, and all the other fields that characterize the circulation. Eddies can be produced by a variety of mechanisms (see fig. 5.1), including flow over

Figure 8.1. A cartoon that appeared in Morel (1973). The man on the diving board is Jule Charney. The other three people are presumably MIT graduate students of the early 1970s. Used with kind permission from Springer Science and Business Media.

mountains, heating localized at particular longitudes, and baroclinic and barotropic instability. In the case of an eddy forced by heating, the wave itself may or may not alter the heating.

Eddies are important aspects of the circulation in their own right; and in addition, they are of interest because they can affect the zonally averaged flow by producing fluxes and flux convergences. Chapter 9 discusses effects of eddies on the zonally averaged flow.

As discussed in this chapter, the atmosphere supports a wide variety of wave motions, including Rossby waves, Kelvin waves, inertia-gravity waves, mixed Rossby-gravity waves and of course sound waves. Wavelengths range from the circumference of the Earth, to the width of North America, to a few millimeters. Periods range from thousandths of a second (for some sound waves) to weeks (for some Rossby waves). The spectral distributions of energy and fluxes over these various scales are complicated and interesting.

Rossby waves, which arise from conservation of potential vorticity on the sphere, are often called *planetary waves*, because they depend for their existence on rotation in the presence of spherical geometry and so are characteristic of planetary atmospheres and oceans, although of course they also occur in stars. Rossby waves propagate westward relative to the mean flow, so it is possible for them to be stationary (with respect to the surface) in a westerly regime. They can be excited in many ways, including by interactions of the mean flow with mountains, convective events, and instabilities of various kinds. The most energetic Rossby waves have very large horizontal scales. To the extent that they are excited at low levels, they can propagate energy upward. Such upward-propagating Rossby waves are believed to play an important role in sudden stratospheric warmings. Meridionally propagating Rossby waves carry angular momentum between the tropics and middle latitudes. Rossby waves are discussed further in chapter 9.

Gravity waves were discussed briefly in chapter 6. They depend on the action of the buoyancy force under stable stratification and occur on many scales. Gravity waves with sufficiently long wavelengths are significantly influenced by rotation; these are called *inertia-gravity waves*. They can be produced by many mechanisms, including topographic forcing and convection. As discussed near the beginning of chapter 9, vertically propagating gravity waves are thought to produce important vertical momentum transports that affect the large-scale circulation, especially in the stratosphere and above. Today, the role of gravity waves in the global circulation is an active area of research.

The tropical atmosphere is home to two special classes of equatorially trapped waves, namely, mixed Rossby-gravity waves, also called *Yanai* waves, which propagate westward; and Kelvin waves, which propagate eastward. As described later, these two types of waves have been implicated in the physical mechanism that drives the Quasi-Biennial Oscillation, although inertia-gravity waves are now believed to be important, too. Kelvin waves are also believed to play important roles in other major tropical phenomena discussed later in this chapter. Observations and theory also deal with equatorially trapped Rossby waves, which are for the most part similar to mid-latitude Rossby waves.

Free and Forced Small-Amplitude Oscillations of a Thin Spherical Atmosphere

We begin with a general introduction to small-amplitude wave motions on the sphere, based on the work of Pierre-Simon Laplace (see fig. 8.2), a French mathematician and astronomer whose name is connected with many important ideas. His remarkably prescient study of the free and forced oscillations of a thin atmosphere on a spherical

Figure 8.2. A portrait of Pierre-Simon Laplace (1749–1827). This image was kindly provided by the University of Sevilla from Louis Figuier, The Marvels of Industry, or a Description of the Principal Modern Industries, Paris: Furne, Jouvet (ca. 1873–1877).

planet was originally published in French in 1799; an English translation followed in 1832. The 200-year-old paper is still very relevant today. Here we briefly outline his work, omitting the mathematical details. You can find a more detailed discussion of this topic in Lindzen's (1990) book.

Laplace considered a spherical planet without mountains and with a highly idealized basic state:

$$\overline{\mathbf{V}}_h = 0, \overline{\omega} = 0, \frac{\partial \overline{\phi}}{\partial p} = -\overline{\alpha}, p\overline{\alpha} = R\overline{T}(p), p_S = p_0 = \text{constant}. \tag{1}$$

Here $\overline{T}(p)$ is an arbitrary function of p only; in particular, $\overline{T}$ does not depend on latitude. This simple basic state has no meridional temperature gradient and no mean flow. It is, of course, a balanced, self-consistent solution of the equations.

The linearized governing equations are

$$\frac{\partial u'}{\partial t} = (2\Omega \sin\varphi)\, v' - \frac{1}{a\cos\varphi}\frac{\partial \phi'}{\partial \lambda}, \tag{2}$$

$$\frac{\partial v'}{\partial t} = -(2\Omega \sin\varphi)\, u' - \frac{1}{a}\frac{\partial \phi'}{\partial \varphi}, \tag{3}$$

$$\frac{1}{a\cos\varphi}\left[\frac{\partial u'}{\partial \lambda} + \frac{\partial}{\partial \varphi}(v'\cos\varphi)\right] + \frac{\partial \omega'}{\partial p} = 0, \tag{4}$$

$$\frac{\partial}{\partial p}\left(\frac{\partial \phi'}{\partial t}\right) + S_p\,\omega' = -\frac{R}{c_p}\frac{Q}{p}, \tag{5}$$

where

$$S_p \equiv -\frac{\overline{\alpha}}{\overline{\theta}}\frac{\partial \overline{\theta}}{\partial p} \tag{6}$$

is the static stability (first defined in chapter 4), which is assumed to depend only on p, and Q is the heating. Friction has been neglected in the momentum equations. Also,

$$\phi' = gz' + \Phi(\lambda, \varphi, t) \tag{7}$$

where $\Phi(\lambda, \varphi, t)$ is the external gravitational tidal potential due to the Moon and/or Sun. In (7), we recognize that the atmosphere experiences gravitational accelerations owing to the pulls of the Moon and Sun, in addition to that of the Earth. The variation of Φ with p is negligible, because the atmosphere is thin compared with the distances to the Sun and Moon. Note that these equations are valid only for atmospheres that are shallow compared with the planetary radius, a.

The solutions, and also the externally imposed thermal and gravitational forcing, are assumed to have the separable form

$$\begin{bmatrix} u' \\ v' \\ \omega' \\ \phi' \\ Q' \\ \Phi' \end{bmatrix} = \sum_n \left\{ \begin{bmatrix} U_n^{\sigma,m}(p) \\ V_n^{\sigma,m}(p) \\ W_n^{\sigma,m}(p) \\ Z_n^{\sigma,m}(p) \\ J_n^{\sigma,m}(p) \\ G_n^{\sigma,m}(p) \end{bmatrix} \Theta_n^{\sigma,s}(\varphi) \right\} \exp\underbrace{[i(m\lambda + \sigma t)]}_{phase}, \tag{8}$$

where the $\Theta_n^{\sigma,m}(\varphi)$ are as-yet-undetermined functions of latitude only; m is the zonal wavenumber, assumed to be nonnegative; and σ is the frequency. With this convention,

$$\sigma < 0 \rightarrow \text{eastward moving, and } \sigma > 0 \rightarrow \text{westward moving.} \tag{9}$$

Equation (8) expresses each field in terms of a function of pressure, multiplied by a function of latitude (the same function in all cases), multiplied by functions of longitude and time. With a more realistic basic state, such separable solutions would not satisfy the equations. The superscripts σ, m simply denote the particular frequency and zonal wavenumber associated with each mode. The subscript n is introduced to recognize the possibility of multiple solutions, and the summation over n represents a superposition of these solutions. The parameter n is sometimes called the *wave type*. As mentioned earlier, at this point we do not know what meridional structures are represented by the $\Theta_n^{\sigma,m}(\varphi)$. They are called *Hough functions* and will be discussed shortly. It can be shown that the set $\{\Theta_n^{\sigma,s}(\varphi)\}$ for all n is complete for $-\pi/2 \leq \varphi \leq \pi/2$. This means that an expansion in terms of Hough functions can represent any meridional structure.

Several pages of manipulation lead to the following two equations:

$$\boxed{F(\Theta_n^{\sigma,m}) = -\varepsilon_n \Theta_n^{\sigma,m}}, \tag{10}$$

$$\boxed{\frac{d^2 W_n^{\sigma,m}}{dp^2} + \frac{S}{gh_n} W_n^{\sigma,m} = -\frac{R}{gh_n c_p}\left(\frac{J_n^{\sigma,m}}{p}\right)}. \tag{11}$$

Here F is a linear operator, defined by

$$\boxed{F \equiv \frac{d}{d\mu}\left(\frac{1-\mu^2}{v^2-\mu^2}\frac{d}{d\mu}\right) - \frac{1}{(v^2-\mu^2)}\left[\frac{m}{v}\left(\frac{v^2-\mu^2}{v^2-\mu^2}\right) + \frac{m^2}{1-\mu^2}\right]}; \tag{12}$$

$v \equiv \sigma/(2\Omega)$ is the nondimensional frequency; and $\mu \equiv \sin\varphi$, so that $d\mu \equiv \cos\varphi d\varphi$. On the right-hand side of (10), we have introduced the nondimensional quantity

$$\varepsilon_n \equiv \frac{4\Omega^2 a^2}{gh_n}, \tag{13}$$

which is called *Lamb's parameter* (or sometimes "the terrestrial constant"). The quantity h_n, which appears in both (11) and (13), can be called a "separation constant," because it arises during the separation of variables. It has units of length and is called the *equivalent depth*. Equation (10) can be called the meridional structure equation, and (11) is called the vertical structure equation. Equation (10) is also called the *Laplace tidal equation*, or LTE. The LTE was derived by Laplace about 200 years ago. All information about the planetary radius, rotation rate, and gravity is buried in the parameters ε_n and v. The solutions of the LTE are the Hough functions.

In the preceding derivation, we assumed, among other things, that the basic state is at rest and that the temperature depends on pressure (i.e., height) only. Separation of variables is not possible if the basic state is made more realistic, for example, if the observed zonally averaged temperature and winds are used.

In many studies, and in the current discussion, the heating amplitude $J_n^{\sigma,m}$ is treated as a known quantity, so that (11) contains the single unknown $W_n^{\sigma,m}$. The assumption that $J_n^{\sigma,m}$ is known makes sense only if $J_n^{\sigma,m}$ is at least approximately independent of the motion. Such an assumption would be reasonable (but approximate) for heating due to absorption of solar radiation by ozone. It would be completely inappropriate for cumulus heating.

The LTE is a second-order ordinary differential equation, and so two boundary conditions are needed; these are simply that the $\Theta_n^{\sigma,m}$ are bounded at the poles, that is, at $\mu = -1$ and 1. The vertical structure equation, (11), is a second-order ordinary differential equation for $W_n^{\sigma,m}(p)$, so it also needs two boundary conditions. At the top of the atmosphere we use

$$\boxed{W_n^{\sigma,m} = 0 \text{ at } p = 0}. \tag{14}$$

This is exact. The exact lower boundary condition (in the absence of mountains) is $w \equiv (Dz/Dt)$ at $p = p_S(\lambda, \varphi, t)$. We use the linearized lower boundary condition

$$\frac{Dz'}{Dt} \cong \left(\frac{\partial z'}{\partial t}\right)_p + \omega' \frac{\partial \bar{z}}{\partial p} = 0 \text{ at } p = p_0. \tag{15}$$

Here p_0 is the spatially and temporally constant value of p_S in the basic state. Because

$$gz' = \phi' - \Phi, \tag{16}$$

where $\Phi(\lambda, \varphi, t)$ is known, and using the hydrostaticity of the basic state, as expressed by

$$g\left(\frac{\partial z}{\partial p}\right)_{p=p_0} = -\bar{\alpha}_0 = -\frac{R\bar{T}_0}{p_0} \equiv -g\frac{H_0}{p_0}, \tag{17}$$

we can rewrite the linearized lower boundary condition (15) as

$$\frac{\partial \phi'}{\partial t} - \omega' g \frac{H_0}{p_0} = \frac{\partial \Phi}{\partial t} \text{ at } p = p_0. \tag{18}$$

Equation (18) involves both ϕ' and ω'. After some additional algebra to eliminate ϕ', we can finally express the lower boundary condition entirely in terms of $W_n^{\sigma,m}$ as

$$\boxed{\frac{dW_n^{\sigma,m}}{dp} - \frac{H_0}{h_n}\frac{W_n^{\sigma,m}}{p_0} = \frac{i\sigma}{gh_n} G_n^{\sigma,m} \text{ at } p = p_0}. \tag{19}$$

Note that the gravitational forcing, $G_n^{\sigma,m}$, enters the problem through the lower boundary condition on the vertical structure equation. The thermal forcing enters through the vertical structure equation itself. Neither type of forcing appears in the LTE.

The LTE and its boundary conditions are satisfied by the trivial solution $\Theta_n^{\sigma,m} \equiv 0$. Nontrivial solutions do exist but only for particular choices of the parameters v and ε_n. The method of solution differs depending on whether nonzero forcing is included. A *free oscillation* is one for which there is no thermal or gravitational forcing; in this context, "free" means "not forced." By definition, a free oscillation is therefore able to persist in the absence of forcing, although it may be damped by friction. A free oscillation is analogous to the tone emitted by a bell. For the case of free oscillations, the solution procedure is as follows:

1. Solve the vertical structure equation, with the equivalent depths as eigenvalues.
2. Solve the LTE, with the nondimensional frequencies as eigenvalues.

This is a double eigenvalue problem.

Atmospheric tides are examples of *forced oscillations*, in which the forcing is either thermal or gravitational. A key point is that *the frequency and zonal wavenumber of the tidal forcing are known.* Within the limits of applicability of the linear equations, the frequency and zonal wavenumber of the atmospheric response must be the same as those of the forcing; that is, *the frequency and zonal wavenumber of the forced solution are the same as those of the forcing*. The solution procedure for forced oscillations is:

1. Solve the LTE (the meridional structure equation) as an eigenvalue problem for the equivalent depth.
2. Expand the gravitational and thermal forcing in terms of Hough functions to obtain their vertical structures $G_n^{\sigma,m}(p)$ and $J_n^{\sigma,m}(p)$, respectively.
3. For each wave type, solve the vertical structure equation for the response to the forcing.

This is a single eigenvalue problem.

It is of course possible for free oscillations and forced oscillations to coexist. We will not discuss forced oscillations in this book. For more information, see chapter 9 of Lindzen's (1990) book.

For the case of free oscillations, the vertical structure equation (11) reduces to

$$\frac{d^2W}{dp^2} + \frac{S}{gh}W = 0. \tag{20}$$

Here the superscripts (σ, m) and the subscript n have been dropped for simplicity. When there is no gravitation forcing, the surface boundary condition (19) can be simplified to

$$\frac{dW}{dp} - \frac{H_0}{h}\frac{W}{p_0} = 0 \text{ at } p = p_0. \tag{21}$$

We also have

$$W = 0 \text{ at } p = 0. \tag{22}$$

As discussed earlier, the system (20)–(22) has nontrivial solutions only for special values of h, which we denote by $\hat{h}$. For $h \neq \hat{h}$, the only solution is $W(p) \equiv 0$. To find the $\hat{h}$ and the corresponding nontrivial solutions for $W(p)$, we have to specify the static stability S as a function of height. Different choices for S_p will give different values of $\hat{h}$ and $W(p)$.

As a simple example, suppose that $S = 0$. This means that the potential temperature is uniform with height. Then, we find from (20) that

$$\frac{d^2W}{dp^2} = 0. \tag{23}$$

A solution of (23) that is consistent with the upper boundary condition (22) is

$$W = Ap, \tag{24}$$

where A is an arbitrary constant. Use of (24) in the lower boundary condition (21) gives

$$\hat{h} = H_0. \tag{25}$$

This is the only possible equivalent depth for free oscillations of an isentropic atmosphere. With more general stratifications there can be many (infinitely many) equivalent depths; as mentioned earlier, this is why the subscript n is required. The procedure used to find the equivalent depth in this simple example can also be used with other stratifications. In one of the problems at the end of this chapter, you are asked to find the equivalent depth of an isothermal atmosphere.

With $\hat{h}$ given by (25), nontrivial solutions of the LTE, (10), exist only when there is a special relation, that is, a dispersion relation, among v, m, and n. The Hough functions were tabulated by Longuet-Higgins (1968); from the perspective of the twenty-first century, it is strange to see a paper that is filled with tables of numbers. In this book we consider only some limiting cases. First suppose that there is *no rotation*, so that $v = \sigma/(2\Omega) \rightarrow \infty$ and $\varepsilon \rightarrow 0$. We continue to assume that $S = 0$, so that (25) applies. For this case, we find that

$$v^2 F \rightarrow \frac{d}{d\mu}\left[(1-\mu^2)\frac{d}{d\mu}\right] - \frac{m^2}{1-\mu^2}, \tag{26}$$

and

$$v^2 \varepsilon \rightarrow \frac{\sigma^2 a^2}{gh} = \frac{\sigma^2 a^2}{gH_0}. \tag{27}$$

Then, we can write the LTE

$$\frac{d}{d\mu}\left[(1-\mu^2)\frac{d\Theta}{d\mu}\right] + \left(\frac{\sigma^2 a^2}{gH_0} - \frac{m^2}{1-\mu^2}\right)\Theta = 0. \tag{28}$$

It can be shown that (28) has solutions that are bounded as $\mu \rightarrow \pm 1$ if and only if

$$\frac{\sigma^2 a^2}{gH_0} = n(n+1), \quad n = 1,2,3\ldots \tag{29}$$

This is a dispersion equation that relates frequency and wavenumber. The quantity $n(n+1)/a^2$ is essentially the square of the total horizontal wavenumber. To have $\sigma = 0$, that is, a stationary solution, would require that $n = 0$, which means no spatial structure—that is, a trivial solution.

For this case of no rotation, the eigenfunctions—that is, the nontrivial solutions of (28)—turn out to be the so-called associated Legendre functions of order n and rank m, denoted by

$$\Theta_n = P_n^m(\mu), \text{ for } n \geq m. \tag{30}$$

In other words, for the special case of no rotation, the Hough functions reduce to the associated Legendre functions. Note that n and m are both integers such that $n \geq m$. The P_n^m are discussed in appendix G. By combining (30) with the longitudinal structure shown in (7), we obtain the two-dimensional horizontal structure of the waves:

$$Y_n^m(\mu,\lambda) = P_n^m(\mu)\exp(im\lambda). \tag{31}$$

As discussed in the appendix, these are spherical harmonics. Here n is the total number of nodal circles, m is the zonal wavenumber, and $n - m$ is the number of nodes in the meridional direction, also known as the *meridional nodal number*. The solutions found here are external gravity waves. They are called "external" because they have no nodes in the vertical and do not propagate vertically. An isentropic atmosphere cannot support internal gravity waves, because when θ is uniform with height a vertically displaced particle does not experience a gravitational restoring force (see the discussion near the beginning of chapter 6). A stratified (i.e., nonisentropic) atmosphere can support both external and internal gravity waves.

By rearranging (29), we can write the frequencies of the external gravity waves as

$$\sigma = \pm\frac{\sqrt{n(n+1)gH_0}}{a}; \tag{32}$$

they depend on the wave's horizontal scale through the two-dimensional index, n, but they are independent of m. For example, when $n=1$, m can be either 0 or 1 (because all m are allowed in the range $n \geq m \geq 0$), but both modes have the same frequency. We show below that this is not true when rotation is present, because with rotation the zonal direction (in which scale is measured by m) becomes physically "different" from the meridional direction.

Now, we consider $\Omega \neq 0$, still for an isentropic atmosphere, and neglect all details. We will obtain the solution more directly, without using the LTE or the vertical structure equation. We define a streamfunction ψ and a velocity potential χ, so that

$$\begin{aligned} u' &= -\frac{1}{a}\left(\frac{\partial \psi}{\partial \varphi}\right)_p + \frac{1}{a\cos\varphi}\left(\frac{\partial \chi}{\partial \lambda}\right)_p, \\ v' &= -\frac{1}{a\cos\varphi}\left(\frac{\partial \psi}{\partial \lambda}\right)_p + \frac{1}{a}\left(\frac{\partial \chi}{\partial \varphi}\right)_p. \end{aligned} \tag{33}$$

The vorticity is then $\xi_p = \mathbf{k}\cdot(\nabla_p \times \mathbf{V}_h) = \nabla_p^2 \psi$, and the divergence is $\delta_p = \nabla_p \cdot \mathbf{V}_h = \nabla_p^2 \chi$. We can differentiate the equation of horizontal motion to obtain

$$\frac{\partial}{\partial t}\nabla_p^2 \psi + \beta v' + f\nabla_p^2 \chi = 0 \tag{34}$$

(the vorticity equation) and

$$\frac{\partial}{\partial t}\nabla_p^2 \chi + \beta u' - f\nabla_p^2 \psi = -g\nabla_p^2 z' \tag{35}$$

(the divergence equation), where, as before,

$$\beta \equiv \frac{2\Omega\cos\varphi}{a} = \frac{1}{a}\frac{df}{d\varphi} \tag{36}$$

is the rate of change of the Coriolis parameter with latitude. We can also show (see problem 3 at the end of this chapter) that for the special case of an isentropic atmosphere

$$\frac{\partial z'}{\partial t} + H_0 \nabla_p^2 \chi = 0. \tag{37}$$

Equations (33) through (37) form a closed set that can be solved for ψ, χ, and z', as well as u' and v'.

Without rotation, there were no nontrivial stationary solutions. From (34) and (37), it should be apparent that when the atmosphere is rotating, stationary motion cannot exist unless $v'=0$. For nontrivial stationary motion with $v'=0$, it follows from the equation of zonal motion, (2), that $m=0$; that is, the motion must be both purely zonal (because $v'=0$) and zonally uniform. In other words, *for a rotating planet, the only stationary solutions are zonally uniform zonal currents.* We do see persistent zonal currents in the Earth's atmosphere and also in the atmospheres of the "gas giant" planets in our solar system, although of course they are not completely steady.

Margules (1893) and Hough (1898) showed that the LTE has two classes of solutions, which they named *free oscillations of the first and second classes*. For the case of small ε_n, which means weak rotation or large equivalent depth, approximate solutions of (34) and (35) can be obtained by expanding in spherical harmonics (Longuet-Higgins, 1968). These free oscillations of the first class (FOFC) are essentially gravity waves, satisfying

$$\chi \cong A_n^m P_n^m(\mu) e^{i(m\lambda + \sigma t)}, \psi \cong \text{constant (irrotational)}, \tag{38}$$

and the dispersion equation (32), which was obtained for $\Omega = 0$. Haurwitz (1937) derived a more accurate expression for the frequencies of the FOFC, including the effects of rotation:

$$\sigma \cong \frac{\Omega m}{n(n+1)} \pm \sqrt{\frac{\Omega^2 m^2}{n^2(n+1)^2} + n(n+1)\frac{gH_0}{a^2}}. \tag{39}$$

This dispersion equation governs *inertia-gravity waves,* which are gravity waves modified by rotation. It can be compared with (32). The additional terms in (39) involve Ω and vanish for $\Omega = 0$. For large n, (39) reduces to (32), which means that waves with small horizontal scales are hardly affected by rotation. As an example, for $n \geq 4$, the error that comes from using (32) instead of (39) is less than 1%. From (39) we see that eastward-propagating inertia-gravity waves have frequencies slightly different from those of westward-propagating inertia-gravity waves with the same m and n. The difference is due to rotation; after all, it is planetary rotation that gives meaning to the words *east* and *west.*

The free oscillations of the second class (FOSC) are also called *Rossby-Haurwitz waves,* or planetary waves. They are almost nondivergent, and approximate solutions can be found by assuming $\chi = \text{constant}$ (i.e., strict nondivergence) initially. For nondivergent flow, the vorticity equation (34) reduces to

$$\frac{\partial}{\partial t}\nabla^2\psi + \frac{2\Omega}{a^2}\frac{\partial\psi}{\partial\lambda} = 0. \tag{40}$$

Using the form of the Laplacian in spherical coordinates, we can write (40) as

$$i\sigma\left\{\frac{1}{a^2\cos^2\varphi}\left[\cos\varphi\frac{d}{d\varphi}\left(\cos\varphi\frac{\partial}{\partial\varphi}\right) - m^2\right]\right\}\hat{\psi} + \frac{2\Omega i m}{a^2}\hat{\psi} = 0 \tag{41}$$

or

$$\left\{\frac{d}{d\mu}\left[(1-\mu^2)\frac{d}{d\mu}\right] + \left(\frac{2\Omega m}{\sigma} - \frac{m^2}{1-\mu^2}\right)\right\}\hat{\psi} = 0, \tag{42}$$

which is very similar to (28). It can be shown that nontrivial solutions of (42) exist when $2\Omega m/\sigma = n(n+1)$, which should be compared with (29) and (39). Solving for the frequency, we obtain

$$\boxed{\sigma \cong \frac{2\Omega m}{n(n+1)} > 0}. \tag{43}$$

Whereas inertia-gravity waves can propagate either east or west, Rossby-Haurwitz waves always propagate westward, with zonal phase speed $-2\Omega/n(n+1)$. "Longer" waves, with smaller values of m, propagate faster. The solutions take the form $\psi \cong B_n^{\sigma,m} P_n^m(\mu) e^{i(m\lambda+\sigma t)}$. Equation (43) shows that for Rossby-Haurwitz waves, unlike pure gravity waves, σ does depend explicitly on the zonal wavenumber, m. This is natural, because the very existence of Rossby-Haurwitz waves depends on rotation, as can be seen from the factor of Ω in the numerator on the right-hand side of (43).

The westward propagation of Rossby-Haurwitz waves is due to the Earth's sphericity, which is associated with variations of the Coriolis parameter with latitude. To demonstrate this, we rewrite (40) as

$$\frac{\partial\zeta}{\partial t} = -\beta v, \tag{44}$$

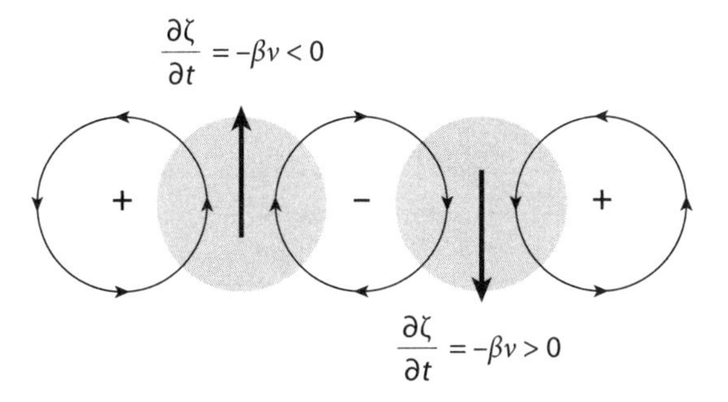

Figure 8.3. Chain of vortices along a latitude circle, illustrating the westward propagation of Rossby waves.

where

$$\beta \equiv \frac{2\Omega \cos\varphi}{a} = \frac{1}{a}\frac{df}{d\varphi} \geq 0. \tag{45}$$

We consider a chain of vortices along a latitude circle, as shown in figure 8.3. Where $v > 0$, to the west of the place where $\zeta < 0$, we have $\beta v > 0$, so we get $\partial\zeta/\partial t < 0$. This occurs to the west of the place where $\zeta < 0$. Similarly, where $v < 0$, to the west of the place where $\zeta > 0$, we have $\beta v < 0$, so we get $\partial\zeta/\partial t > 0$. This is a simple way to see why Rossby waves propagate westward relative to the mean flow. All other factors being equal, the waves propagate westward more quickly near the equator, where β is large, than at higher latitudes. Rossby wave propagation on the sphere is discussed by Grose and Hoskins (1979).

A direct test of the theory of nondivergent Rossby waves was made by Eliassen and Machenhauer (1965) and Deland (1965). They used spherical harmonics to analyze the 500 hPa streamfunction, isolating transient waves by taking the difference in 24 hours. Their results, illustrated in figure 8.4, show westward propagation, as expected. Table 8.1 compares the computed and observed phase speeds, in degrees

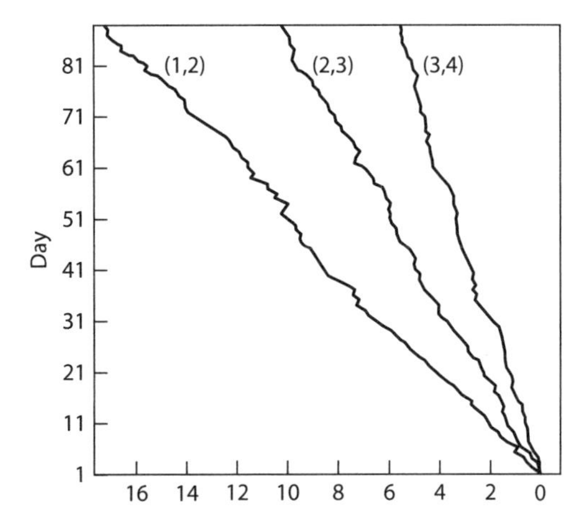

Figure 8.4. Successive daily values of the phase angle for the 24-hour tendency field, at the 500 hPa level for the components $(m,n) = (1,2), (2,3), (3,4)$, during the 90-day period beginning December 1, 1956. The abscissa represents the number of westward circulations around the Earth after the first passage of the Greenwich meridian, and so it is a measure of longitudinal phase propagation. From Eliassen and Machenhauer (1965).

Table 8.1. A Comparison of the Observed and Computed Westward Phase Speeds of the Planetary Waves

m, *n*	Observed phase speed	Computed phase speed
1, 2	70	115
2, 3	40	53
3, 4	20	28
1, 4	20	28
2, 5	12	16
3, 6	8	9

Source: From Eliassen and Machenhauer (1965).
Notes: The units are degrees of longitude per day.
The computed phase speeds are for nondivergent Rossby waves.

of longitude per day. The computed phase speeds are larger than the observed phase speeds, especially for the longer waves, because divergence (which, in reality, is small but nonzero) acts to slow the waves down. *The model overpredicts the westward phase speeds*, owing to the neglect of the effects of divergence.

In the Earth's atmosphere, Rossby waves can be excited in a variety of ways, including flow over topography (discussed later in this chapter), baroclinic instability in middle and high latitudes, and cumulus heating. Once excited, the waves can propagate zonally, meridionally, and vertically.

Propagation of Planetary Waves

The following discussion is based on the famous paper by Charney and Drazin (1961). Closely related work can be found in Dickinson (1968) and Matsuno (1970).

As discussed in chapter 4, the quasi-geostrophic form of the potential vorticity equation is

$$\left(\frac{\partial}{\partial t} + \mathbf{V}_g \cdot \nabla_p\right) Z_{QG} = 0, \tag{46}$$

where

$$Z_{QG} = \frac{\nabla^2 \phi}{f_0} + \beta_0 y + \frac{\partial}{\partial p}\left(\frac{f_0}{S}\frac{\partial \phi}{\partial p}\right) \tag{47}$$

is the quasi-geostrophic pseudopotential vorticity (QGPPV),

$$\psi \equiv \frac{\phi}{f_0} \tag{48}$$

is called the *geostrophic streamfunction*, $S \equiv -(\alpha_{bs}/\theta_{bs})(\alpha\theta_{bs}/\partial p)$ is the static stability, and $f_0$and β_0 are representative values of the Coriolis parameter and its meridional derivative, respectively. We adopt the "log pressure" coordinate

$$z(p) \equiv -\left(\frac{RT_0}{g}\right)\ln\left(\frac{p}{p_0}\right), \tag{49}$$

where T_0 is a constant reference temperature, and we define the Brunt-Väisälä frequency by

$$N^2 \equiv \frac{g}{\theta_{bs}} \frac{\partial \theta_{bs}}{\partial z}. \tag{50}$$

Using (49) and (50), we can rewrite the QGPPV as

$$Z_{QG} = \nabla^2 \psi + \beta_0 y + \frac{1}{\rho_{bs}} \frac{\partial}{\partial z}\left(\rho_{bs} \frac{f_0^2}{N^2} \frac{\partial \psi}{\partial z}\right). \tag{51}$$

Recall that

$$v_g = \frac{\partial \psi}{\partial x} \text{ and } u_g = -\frac{\partial \psi}{\partial y}. \tag{52}$$

Here we use Cartesian coordinates for simplicity. Linearizing (46) about the zonal-mean state, we get

$$\left(\frac{\partial}{\partial t} + \overline{u}^{\lambda} \frac{\partial}{\partial x}\right) Z_{QG}^{*} + v_g^{*} \frac{\partial}{\partial y} \overline{Z_{QG}}^{\lambda} = 0. \tag{53}$$

We look for separable solutions of the form

$$\psi^{*} = \operatorname{Re}\left\{\hat{\psi}(y,z)\, e^{ik(x-ct)}\right\}, \tag{54}$$

$$Z_{QG}^{*} = \operatorname{Re}\left\{\hat{Z}(y,z)\, e^{ik(x-ct)}\right\}. \tag{55}$$

Substituting (51), (54), and (55) into (53), we obtain

$$(\overline{u}^{\lambda} - c)\hat{Z} + \hat{\psi} \frac{\partial \overline{Z}^{\lambda}}{\partial y} = 0, \tag{56}$$

where

$$\hat{Z} = -k^2 \hat{\psi} + \frac{\partial^2 \hat{\psi}}{\partial y^2} + \frac{1}{\rho_{bs}} \frac{\partial}{\partial z}\left(\rho_{bs} \frac{f_0^2}{N^2} \frac{\partial \hat{\psi}}{\partial z}\right). \tag{57}$$

Using (57), we can rewrite (56) as

$$\boxed{\frac{\partial^2 \hat{\psi}}{\partial y^2} + \frac{1}{\rho_{bs}} \frac{\partial}{\partial z}\left(\rho_{bs} \frac{f_0^2}{N^2} \frac{\partial \hat{\psi}}{\partial z}\right) = -\left[\frac{1}{(\overline{u}^{\lambda} - c)} \frac{\partial \overline{Z_{QG}}^{\lambda}}{\partial y} - k^2\right] \hat{\psi}}. \tag{58}$$

This is a fairly general form of the quasi-geostrophic wave equation. We will simplify it considerably before analyzing it.

As wave energy propagates up to higher levels, it encounters decreasing values of ρ_{bs}. The energy density (energy per unit volume) scales as $\rho_{bs}(k\psi)^2$, so if the energy density is constant with height, *$\hat{\psi}$ must increase with height in proportion to* $1/\sqrt{\rho_{bs}}$. Because of this effect, the equations become simpler if we introduce a scaled value of $\hat{\psi}$:

$$\psi \equiv \frac{\sqrt{\rho_S}}{N} \hat{\psi}. \tag{59}$$

Note that here ψ (no hat) is the scaled value; the meaning of ψ now departs from that used in (48). We also note that

$$
\begin{aligned}
\frac{1}{\rho_{bs}}\frac{\partial}{\partial z}\left(\rho_{bs}\frac{f_0^2}{N^2}\frac{\partial\hat{\psi}}{\partial z}\right) &= \frac{f_0^2}{\rho_S}\frac{\partial}{\partial z}\left[\frac{\sqrt{\rho_{bs}}}{N}\frac{\partial}{\partial z}\left(\frac{\sqrt{\rho_{bs}}}{N}\hat{\psi}\right)-\frac{\sqrt{\rho_{bs}}}{N}\hat{\psi}\frac{\partial}{\partial z}\left(\frac{\sqrt{\rho_{bs}}}{N}\right)\right] \\
&= \frac{f_0^2}{\rho_{bs}}\frac{\partial}{\partial z}\left[\frac{\sqrt{\rho_{bs}}}{N}\frac{\partial\psi}{\partial z}-\psi\frac{\partial}{\partial z}\left(\frac{\sqrt{\rho_{bs}}}{N}\right)\right] \\
&= \frac{f_0^2}{\rho_{bs}}\left[\cancel{\frac{\partial}{\partial z}\left(\frac{\sqrt{\rho_{bs}}}{N}\right)\frac{\partial\psi}{\partial z}}+\frac{\sqrt{\rho_{bs}}}{N}\frac{\partial^2\psi}{\partial z^2}-\cancel{\frac{\partial\psi}{\partial z}\frac{\partial}{\partial z}\left(\frac{\sqrt{\rho_{bs}}}{N}\right)}-\psi\frac{\partial^2}{\partial z^2}\left(\frac{\sqrt{\rho_{bs}}}{N}\right)\right] \\
&= \frac{f_0^2}{\rho_{bs}}\left[\frac{\sqrt{\rho_{bs}}}{N}\frac{\partial^2\psi}{\partial z^2}-\psi\frac{\partial^2}{\partial z^2}\left(\frac{\sqrt{\rho_{bs}}}{N}\right)\right].
\end{aligned} \tag{60}
$$

Substituting from (59) and (60), we can rewrite (58) as

$$
\boxed{\frac{\partial^2\psi}{\partial y^2}+\frac{f_0^2}{N^2}\frac{\partial^2\psi}{\partial z^2}=-\left(\frac{f_0^2}{N^2}\frac{n^2}{4H_0^2}\right)\psi}, \tag{61}
$$

where

$$
\boxed{n^2 \equiv \frac{4N^2H_0^2}{f_0^2}\left[\frac{1}{(\overline{u}^\lambda - c)}\frac{\partial\overline{Z_{QG}}^\lambda}{\partial y}-k^2-\frac{f_0^2}{\sqrt{\rho_{bs}}N}\frac{\partial^2}{\partial z^2}\left(\frac{\sqrt{\rho_{bs}}}{N}\right)\right]} \tag{62}
$$

is called the *index of refraction*. Here $H_0 \equiv RT_0/g$, where T_0 is the reference temperature used in (49). Equation (61) is a form of the quasi-geostrophic wave equation. *When* $n^2 > 0$, ψ *is oscillatory (propagating), and when* $n^2 < 0$, ψ *is "evanescent" (exponentially decaying away from the source of excitation).*

Comparing (61)–(62) with (58), we see that the left-hand side of (61) has become simpler, but the expression for the index of refraction has become more complicated. Using some idealizations, we can drastically simplify (62) without altering its basic meaning. First, we consider the special case of an isothermal atmosphere with $T_S(p) \cong T_0 =$ constant. This is not unrealistic for the lower stratosphere. For an isothermal atmosphere $N^2 = g^2/c_pT_0 =$ constant and $\rho_{bs} \sim e^{z/H_0}$, so that (62) reduces to

$$
n^2 \cong \frac{4N^2H_0^2}{f_0^2}\left[\frac{1}{(\overline{u}^\lambda - c)}\frac{\partial\overline{u}^\lambda}{\partial y}-k^2\right]-1. \tag{63}
$$

Inspection of (63) shows that $\overline{u}^\lambda - c > 0$ is necessary for $n^2 > 0$, that is, to have propagating solutions. Now, we simplify further by assuming stationary waves, for which the phase speed, c, is zero. As discussed later in this chapter, stationary waves can be forced by flow over mountains, or by land-sea contrast. For the special case of stationary waves, (63) becomes

$$
n^2 = \frac{4N^2H_0^2}{f_0^2}\left(\frac{1}{\overline{u}^\lambda}\frac{\partial\overline{Z_{QG}}^\lambda}{\partial y}-k^2\right)-1. \tag{64}
$$

To simplify n^2 even further, we note from (51) that

$$
\frac{\partial\overline{Z_{QG}}^\lambda}{\partial y}=\beta_0-\frac{\partial^2\overline{u_g}^\lambda}{\partial y^2}-\frac{1}{\rho_{bs}}\frac{\partial}{\partial z}\left(\rho_{bs}\frac{f_0^2}{N^2}\frac{\partial\overline{u_g}^\lambda}{\partial z}\right). \tag{65}
$$

When the meridional and vertical shears of $\overline{u}^\lambda$ are not too strong,

$$
\frac{\partial\overline{Z_{QG}}^\lambda}{\partial y}\cong\beta_0\geq 0. \tag{66}
$$

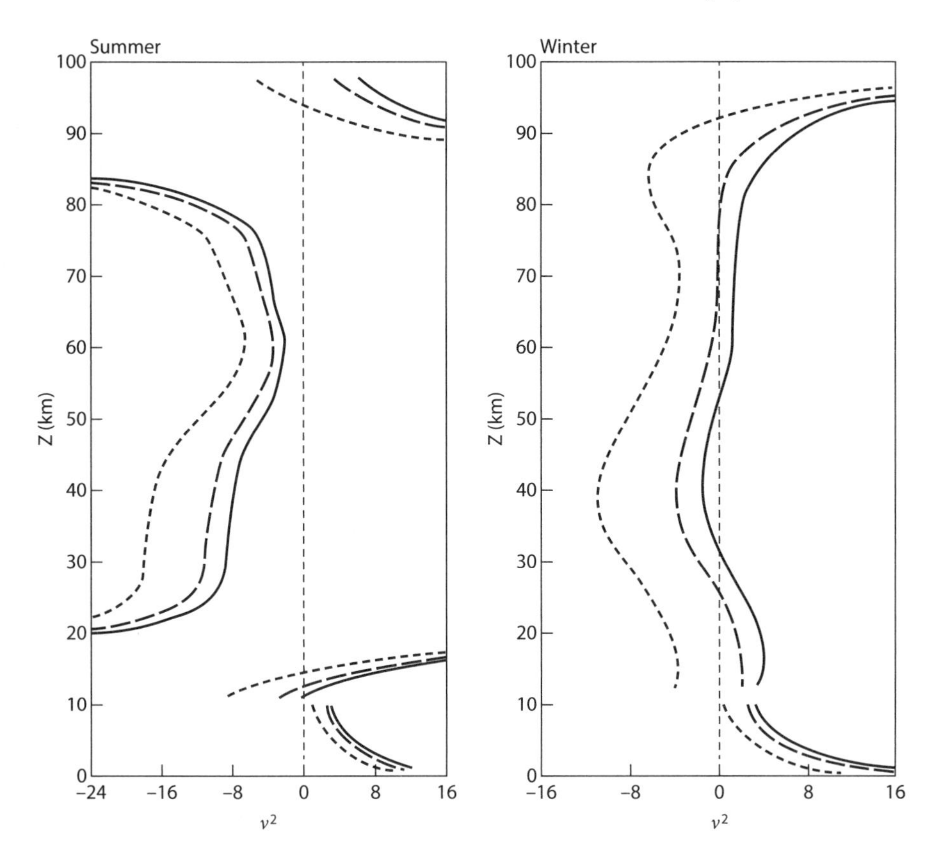

Figure 8.5. The square of the index of refraction for summer and winter, averaged between 30° and 60° N, for waves of different wavelengths, L. The short-dashed lines correspond to L = 6000 km, the long-dashed lines correspond to L = 10,000 km, and the solid lines correspond to L = 14,000 km. From Charney and Drazin (1961). Copyright © 1961 by the American Geophysical Union.

Using (66) in (64), we finally obtain

$$n^2 \cong \frac{4N^2H_0^2}{f_0^2}\left(\frac{\beta_0}{\overline{u}^\lambda} - k^2\right) - 1. \tag{67}$$

From (67), we see the following:

- Propagation ($n^2 > 0$) requires that $\beta_0/\overline{u}^\lambda > 0$. Because $\beta_0 > 0$, $\overline{u}^\lambda$ *must be positive* (*westerly*). Stationary Rossby waves cannot exist in easterlies, simply because they propagate westward relative to the air, so that easterlies cannot hold them in place. Recall that the summer hemisphere stratosphere is dominated by easterlies, while the winter hemisphere stratosphere is dominated by westerlies. Note, however, that large positive $\overline{u}^\lambda$ also makes $n^2 < 0$. Stationary waves cannot propagate through very strong westerlies, which would sweep them downstream. Figure 8.5, from Charney and Drazin (1961), shows the vertical distribution of n^2 for summer and winter, averaged over the Northern Hemisphere middle latitudes, for stationary waves with three different wavelengths.

- Even when $\beta_0/\overline{u}^\lambda > 0$, for a given $\overline{u}^\lambda$, waves with large k (i.e., sufficiently short zonal wavelength) cannot propagate. Short waves are therefore "trapped" near their excitation levels. Since $\overline{u}^\lambda$ has a maximum near the tropopause in middle latitudes, many short waves are trapped in the troposphere, even in winter. Only longer waves can propagate to great heights. This suggests that long waves will dominate in the stratosphere and mesosphere even more than they do in the troposphere.
- A level where $\overline{u}^\lambda = 0$ is called a *critical level* for stationary waves. This is the same terminology used earlier in our discussion of internal gravity waves forced by flow over topography. Equation (38) shows that at a critical level, $n^2 \to \infty$. Suppose that $\overline{u}^\lambda > 0$ below a critical level, and $\overline{u}^\lambda < 0$ above. Then, for waves excited at the lower boundary (e.g., by flow over topography), upward propagation will be completely blocked at the critical level, and no wave activity will be seen above the critical level. Earlier in this chapter we showed a similar result for vertically propagating internal gravity waves. The propagating waves can also be blocked at a critical latitude, where $\overline{u}^\lambda = 0$. As discussed in chapter 3, such critical latitudes are often found in the subtropics. In general, we can speak of *critical lines* in the latitude-height plane, along which $\overline{u}^\lambda = 0$. If we allowed $c \neq 0$ we would find that the critical surfaces are those for which $\overline{u}^\lambda - c = 0$.

Figure 8.6 provides evidence that the theory is correct. It shows the geopotential height fields at 500 hPa, 100 hPa, and 10 hPa, for Northern Hemisphere summer and winter. In winter, planetary waves clearly propagate upward to the 10 hPa level, while in summer they do not. Note that the apparent horizontal scale of the dominant eddies increases with height, in winter. This behavior is consistent with the theory, which predicts that the shorter modes are trapped at lower levels, while longer modes can continue to propagate upward to great heights.

Matsuno (1970) used the observed winds for the Northern Hemisphere winter to compute $\partial \overline{Z_{QG}}^\lambda / \partial \varphi$, the index of refraction, and the energy flow in the latitude-height plane for zonal wavenumber 1. His results are shown in figure 8.7. The upward-propagating waves are directed equatorward by the variations of the index of refraction.

Stationary and Transient Eddies in Middle Latitudes

Blackmon (1976) discussed the observed eddy activity in the Northern Hemisphere as seen in the 500 hPa geopotential height. He used a 10-year record and considered both summer and winter conditions. The data were available twice per day, at 00 Z and 12 Z.

Blackmon filtered the data in both space and time, to isolate particular space-time scales. He expanded the height fields into spherical harmonics Y_n^m, where superscript m denotes the zonal wavenumber, and subscript n denotes the *two-dimensional index*. The number of nodes in the meridional direction, that is, the meridional nodal number is $n - m \geq 0$; note that $m \leq n$ is required. The largest value of n that Blackmon considered was $n = 18$. Note that $m = 18$ corresponds to a zonal wavelength of 20° of longitude, which in middle latitudes corresponds to a wavelength of roughly 1000 km.

As discussed in appendix G, the spherical harmonics form a complete orthonormal basis that can be used to represent an arbitrary function on the sphere:

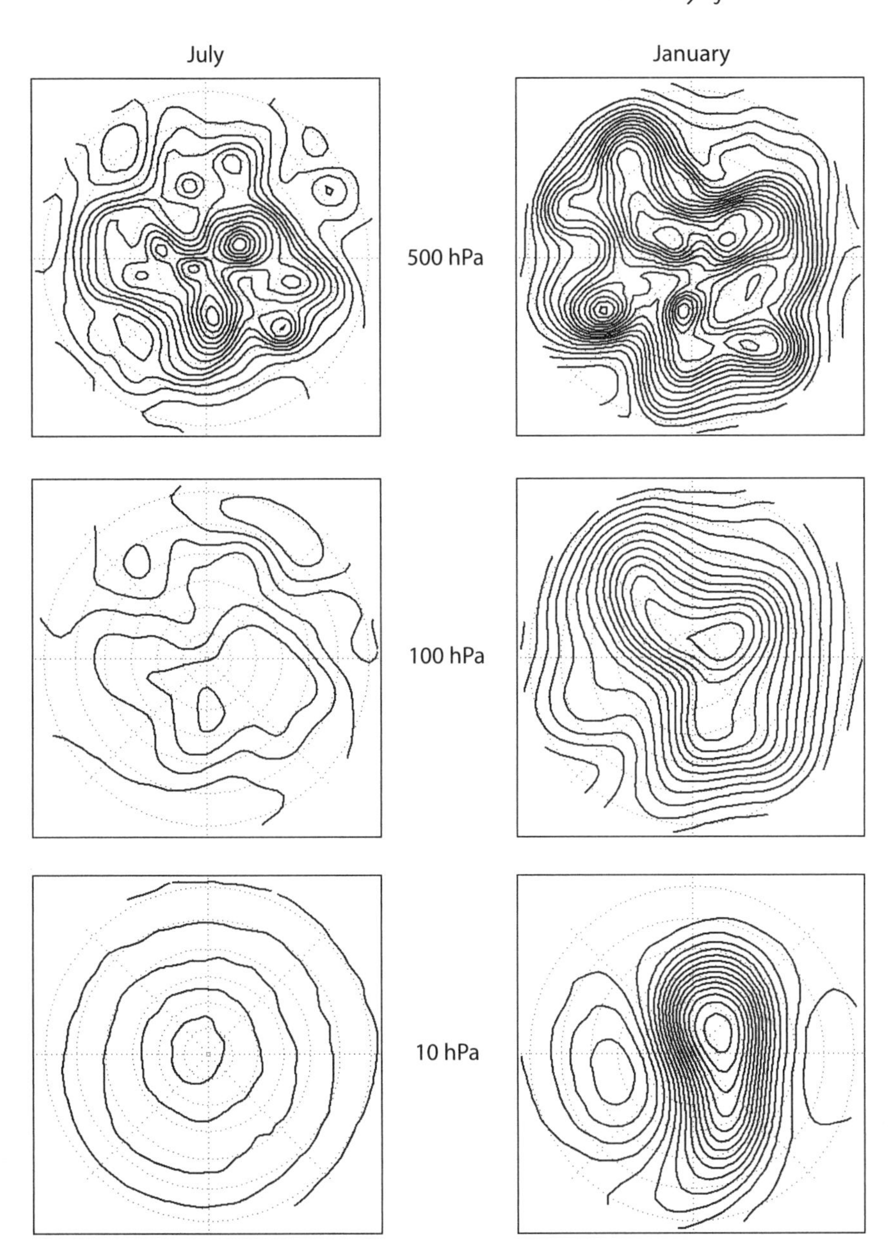

Figure 8.6. These Northern Hemisphere data were collected during the International Geophysical Year. Geopotential heights for July 15, 1958, are shown on the left, and those for January 15, 1959, are shown on the right. The levels plotted are 500 hPa, 100 hPa, and 10 hPa. From Charney (1973) in the book edited by Morel (1973). With kind permission from Springer Science and Business Media.

$$Z(\lambda,\varphi) = \sum_{m=-M}^{M} \left(\sum_{n=|m|}^{M} C_n^m Y_n^m \right). \tag{68}$$

Here the C_n^m are the expansion coefficients, and M is a suitably chosen positive integer. Larger values of M allow a more detailed representation of $Z(\lambda,\varphi)$. The zonal wavenumber is m, and the parameter n is the two-dimensional index, as previously noted.

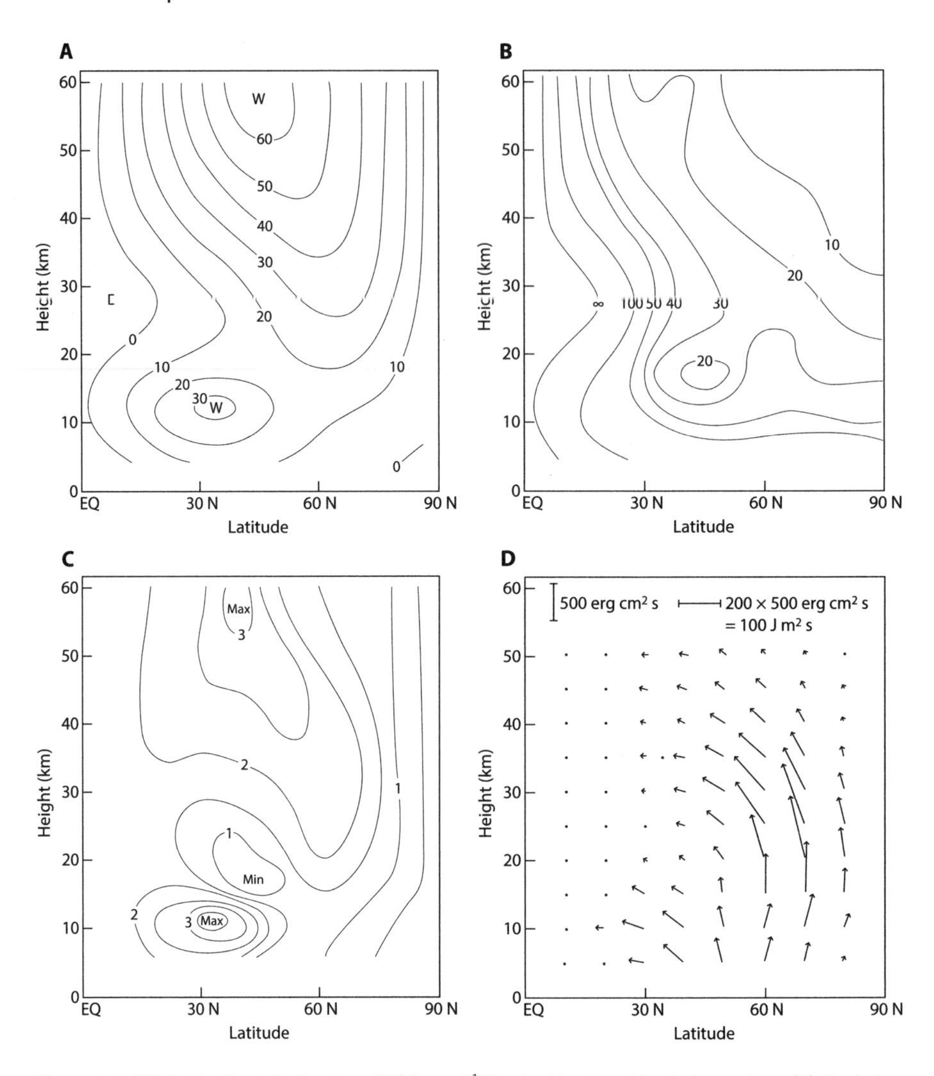

Figure 8.7. (A) An idealized distribution of $\overline{u}^{\lambda}$ (in m s^{-1}) for the Northern Hemisphere winter. (B) The latitudinal gradient of the potential vorticity, $\partial\overline{Z_{QG}}^{\lambda}/\partial\varphi$, expressed as a multiple of the Earth's rotation rate. (C) The refractive index square, n^2, for the $k = 0$ wave. (D) Computed distribution of energy flow in the meridional plane associated with zonal wavenumber 1. From Matsuno (1970). © American Meteorological Association. Used with permission.

Blackmon defined three categories of spatial scales, based on the two-dimensional index:

Regime I: $0 \le n \le 6$, or "long waves";
Regime II: $7 \le n \le 12$, or "medium-scale waves"; and
Regime III: $13 \le n \le 18$, or "short waves."

Because Blackmon's truncation scheme is based on the two-dimensional index, a particular wave can have nodes in the zonal direction or in the meridional direction; most have both. Note that for all three regimes $0 \le m \le n$. All three regimes therefore contain modes with small values of m, that is, long zonal scales.

The expansion coefficients for each set of waves can be determined for each observation time. The sets of waves were filtered in time, using three filters:

Low-pass: admits periods in the range longer than or equal to 10 days;
Medium-pass: admits periods in the range 2.5 days to 6 days;
High-pass: admits periods in the range 1 day to 2 days.

Note that a period of one day represents the most rapidly fluctuating wave that can be captured by the twice-a-day data used in Blackmon's study.

The lower panel of figure 8.8 shows the time-averaged 500 hPa height averaged over nine winters. The features seen here are stationary waves. Note the two prominent troughs, one near the east coast of North America, and the other near Japan. These same features can be seen in the figures of chapter 3, of course. The upper panel of figure 8.8 shows the total root-mean-square (rms) geopotential height for winter without time or space filtering. There are three prominent "centers of action," in the North Pacific, the North Atlantic, and over Siberia.

Figure 8.9 shows low-pass-filtered (long period) rms winter heights for all spatial scales (top left) for Regime I (top right), for Regime II (bottom left), and for Regime III (bottom right). It is clear that Regime III contributes very little, whereas both Regimes I and II contribute significantly. The Regime II contribution shows maxima in regions where blocking commonly occurs. Blocking will be discussed later.

Figure 8.10 is similar to figure 8.9 but for medium-pass eddies, that is, those of "synoptic" periods in the range of 2.5 to 6 days. For this range of periods, most of the action comes from Regime II and Regime III; the long waves do not contribute much. Note, however, that the contour interval is smaller than that used in figure 8.9, so that less total activity is portrayed in figure 8.10 than in figure 8.9. The strong signal seen over the North Pacific and North Atlantic Oceans in figure 8.10 is associated with the "storm tracks." Winter storms often form on the western sides of the ocean basins, near the east coasts of the continents, where the horizontal temperature gradient is strong. The storms move eastward and poleward, creating the tracks seen in the figure. This topic is discussed further later in this chapter.

Even less power resides in high-pass eddies (not shown).

Table 8.2 shows that in winter the greatest power is found along a "ridge" running from lower left (small n, low frequency) to upper right (large n, high frequency). By far the largest power occurs in the low frequencies and for two-dimensional index n on the order of 6. The diurnal tide is also apparent near the bottom right, with $n = 2$.

The corresponding results for the summer season are omitted here for brevity. In summer, the wave amplitudes are greatly reduced, as is the strength of the mean flow, and the centers of action are shifted toward the pole. As in winter, long waves and low frequencies dominate.

One conclusion from Blackmon's study is that most of the transient eddy energy resides at low frequencies and long wavelengths. This has been known for a long time. For example, Wiin-Nielsen et al. (1963) analyzed the total heat and momentum transport across a latitude circle as a function of wavenumber for selected latitudes. As shown in figure 8.11, the strongest contributions come from zonal wavenumbers less than 5. For middle latitudes, this corresponds to wavelengths longer than 4000 km.

Chang et al. (2002) discuss the observed climatology of the Northern Hemisphere storm tracks seen in figure 8.10. Figure 8.12, taken from their paper, shows that the storm tracks are associated with strong conversion from eddy available potential energy to eddy kinetic energy; that is, the storms are baroclinic eddies.

A

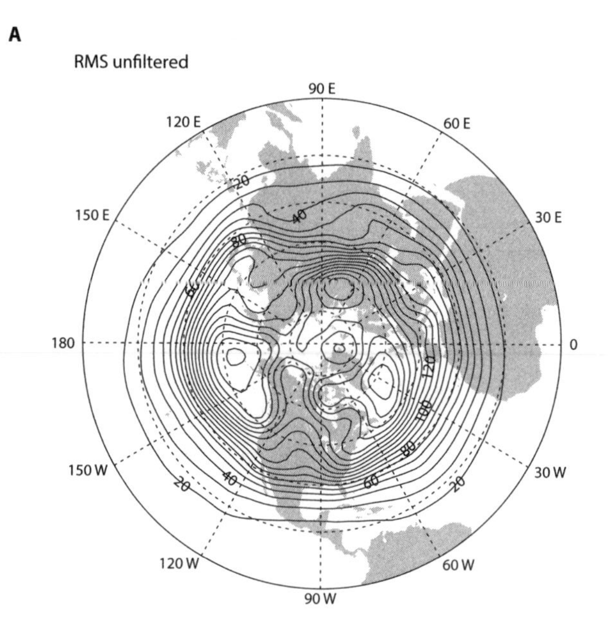

B

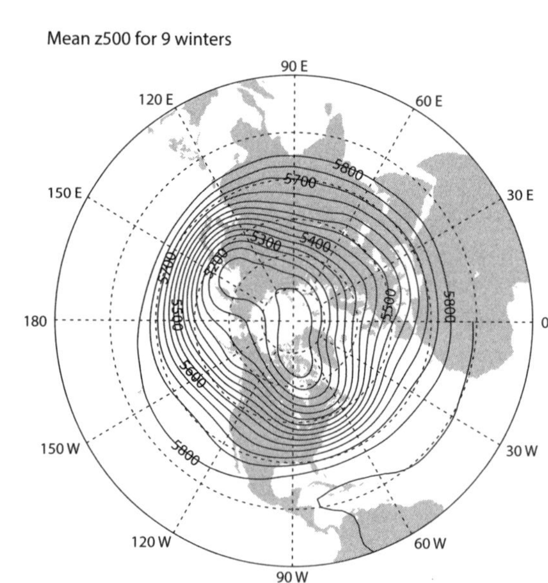

Figure 8.8. Maps of the geopotential height for nine winters: (A) rms unfiltered, contour interval 10 m; (B) average, contour interval 50 m. From Blackmon (1976).

Annular Modes

Thompson and Wallace (1998, 2000) and Limpasuvan and Hartmann (2000) found that variability of the winter sea-level pressure is dominated by zonally symmetric structures, which they called *annular modes*. The Northern Annular Mode, or NAM, had been noticed by Lorenz (1951) and is associated with the North Atlantic

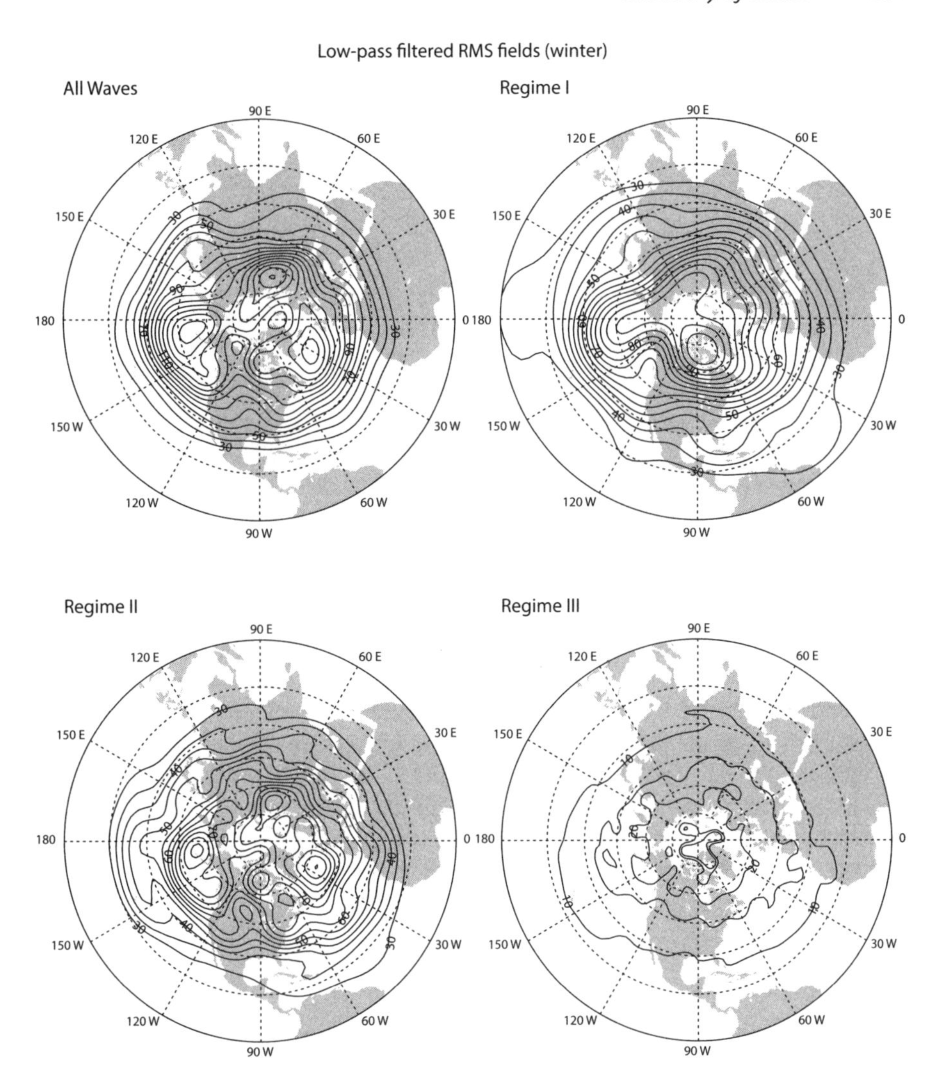

Figure 8.9. Maps of the low-pass-filtered rms fields (winter): (A) all eddies, contour interval 10 m; (B) eddies in Regime I, contour interval 5 m; (C) eddies in Regime II, contour interval 5 m; (D) eddies in Regime III, contour interval 5 m. From Blackmon (1976).

Oscillation (Walker and Bliss, 1932; Wallace and Gutzler, 1981), which strongly influences European weather. The Southern Annular Mode (SAM) (Rogers and van Loon, 1982; Hartmann and Lo, 1998; Thompson et al., 2005) is also called the Antarctic Oscillation. Both the NAM and the SAM have vertically uniform (i.e., barotropic) vertical structures. They contribute about a quarter of the total temporal variance of the winds and pressure (or geopotential height) in the middle latitudes of their respective hemispheres, on a wide range of timescales.

Recently, Thompson and Woodworth (2014) and Thompson and Barnes (2014) identified a Baroclinic Annular Mode (BAM) in the middle latitudes of the Southern Hemisphere. The BAM oscillates with a period of about four weeks. Thompson and Barnes showed that the oscillations involve a feedback loop in which a period of

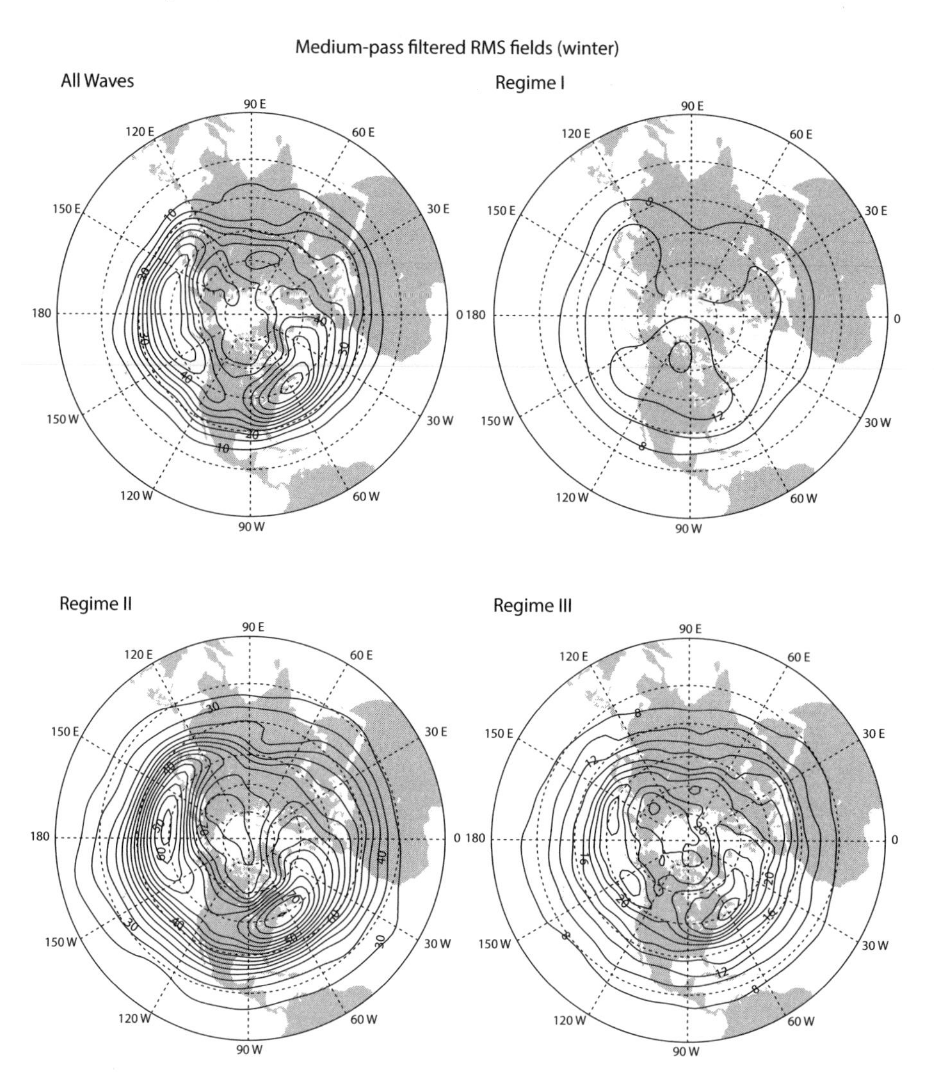

Figure 8.10. Maps of the medium-pass-filtered rms fields (winter): (A) all eddies, contour interval 5 m; (B) eddies in Regime I, contour interval 2 m; (C) eddies in Regime II, contour interval 2 m; (D) eddies in Regime III, contour interval 2 m. From Blackmon (1976).

gradually increasing meridional temperature gradient is followed by an increase in the transient eddy meridional energy flux, which then reduces the meridional temperature gradient, which in turn leads to a decrease in the transient eddy meridional energy flux.

Theory of Orographically Forced Stationary Waves

Stationary waves are forced by mechanical and/or thermal effects that are anchored to the Earth's surface. Mountain ranges can produce waves either by blocking the flow or by acting as elevated heat sources. Geographically fixed thermal forcing is also

Table 8.2. Power as a Function of Wavenumber and Frequency for the Winter Season

	0	1/15	2/15	3/15	4/15	5/15	6/15	7/15	8/15	9/15	10/15	11/15	12/15	13/15	14/15	15/15
18	51.7	49.8	45.8	39.3	37.0	26.7	21.2	17.3	13.7	10.1	7.2	5.3	4.1	3.4	3.2	3.4
17	68.7	66.5	58.6	48.1	40.1	34.0	27.9	22.1	15.8	11.0	7.6	5.4	4.3	3.1	3.2	3.3
16	98.5	97.5	82.4	69.6	56.5	45.0	34.8	26.0	18.4	12.2	7.9	5.5	4.1	3.3	3.3	3.5
15	141.6	131.0	108.1	85.9	70.3	58.6	75.6	32.3	19.7	11.6	7.3	5.1	4.1	3.3	3.3	3.4
14	210.1	195.5	158.9	122.3	99.5	79.2	53.3	32.3	18.2	10.8	6.9	4.9	3.5	3.3	3.3	3.4
13	302.8	272.5	213.2	164.7	129.0	92.6	56.1	31.6	16.8	9.4	5.7	4.4	3.2	2.9	2.9	3.3
12	502.6	435.2	311.1	225.5	159.9	99.8	53.5	26.8	13.6	8.0	5.3	3.9	3.2	3.0	3.0	3.4
11	746.3	639.7	443.3	290.6	176.5	94.4	43.8	21.1	10.6	6.0	4.1	3.6	3.0	2.8	3.0	3.4
10	1190.1	943.3	545.9	322.8	181.0	80.5	33.6	15.6	8.3	5.4	3.9	3.3	3.0	2.9	3.1	3.6
9	1815.2	1417.6	739.3	323.8	135.8	55.0	23.2	11.4	6.1	4.1	3.1	2.8	2.4	2.7	2.9	3.6
8	3305.7	2175.9	732.7	245.2	104.5	42.0	17.3	9.1	4.3	3.6	2.3	2.6	2.2	2.7	2.5	3.4
7	3732.6	2370.4	698.2	191.2	71.2	29.3	12.9	7.7	3.8	3.6	2.3	2.9	1.9	2.6	2.5	4.3
6	3731.5	2374.2	657.5	149.3	51.0	20.6	9.0	5.8	2.4	2.8	1.3	2.3	1.5	2.4	2.3	4.2
5	3563.5	2192.6	522.5	114.0	39.1	16.2	7.1	4.9	3.4	2.7	2.2	2.2	2.2	2.2	2.6	4.8
4	2553.1	1600.8	420.2	95.1	34.5	13.8	5.7	4.2	2.1	2.6	1.5	2.1	1.3	2.0	2.1	3.8
3	1448.2	992.6	329.2	79.7	27.0	9.4	4.6	3.3	2.1	1.8	1.4	1.5	1.2	1.5	5.3	4.6
2	1115.3	642.0	146.3	41.8	18.5	8.3	3.3	2.3	1.6	1.6	1.1	1.3	1.2	1.4	2.2	3.7
1	294.5	221.7	105.9	45.1	15.8	4.1	2.0	1.7	1.2	1.0	0.9	1.0	1.0	1.1	6.0	12.4
0	168.6	92.6	17.1	4.7	2.5	1.4	1.0	0.8	0.6	0.6	0.6	0.6	0.6	0.6	1.8	3.5

Source: Adapted from Blackmon (1976).

Note: Units are m2 rad−1 (15 days)−1.

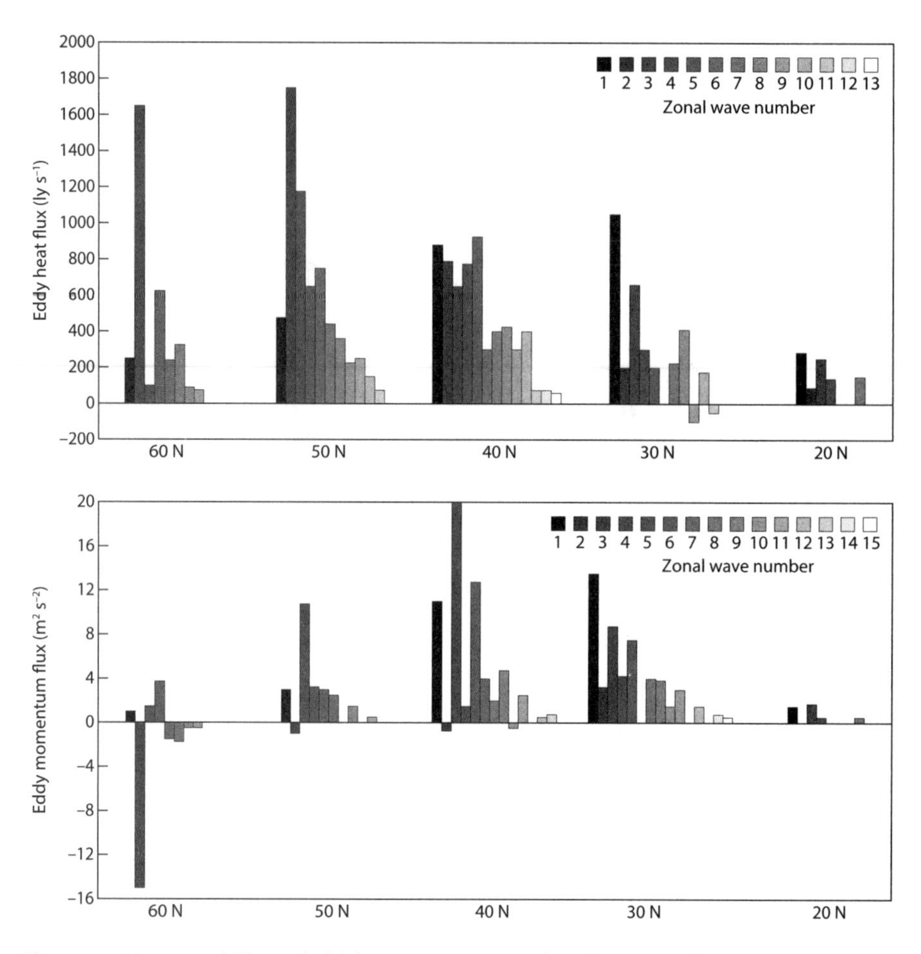

Figure 8.11. Upper panel: The total eddy heat transport across a latitude circle as a function of wavenumber for selected latitudes, for January 1962. Lower panel: The eddy momentum transport averaged with respect to pressure, as a function of zonal wavenumber, for selected latitudes, based on observations for January 1962. In both panels the gray scale denotes the zonal wavenumber. The units are ly s^{-1} in the upper panel and $m^2\, s^{-2}$ in the lower panel. Based on Wiin-Nielsen et al. (1963).

associated with land-sea contrasts, sea-surface temperature gradients, and the like. Held (1983) summarized the work of Charney and Eliassen (1949), who studied the effects of orographic barriers on stationary waves in middle latitudes.

We start by considering the conservation of potential vorticity in shallow water equations introduced in chapter 4 and also using the QG approximation discussed there; that is,

$$\left(\frac{\partial}{\partial t} + \mathbf{V}_g \cdot \nabla\right) Z_{SWQG} = -\frac{r\zeta}{h}, \tag{69}$$

where

$$Z_{SWQG} \equiv \frac{\zeta + f}{h}. \tag{70}$$

Here h is the depth of the water, and h_T is a nonnegative Rayleigh friction coefficient. The geostrophic wind is $\mathbf{V}_g = (g/f_0)\mathbf{k} \times \nabla h_{fs}$, where $h_{fs} \equiv h_T + h$ is the height

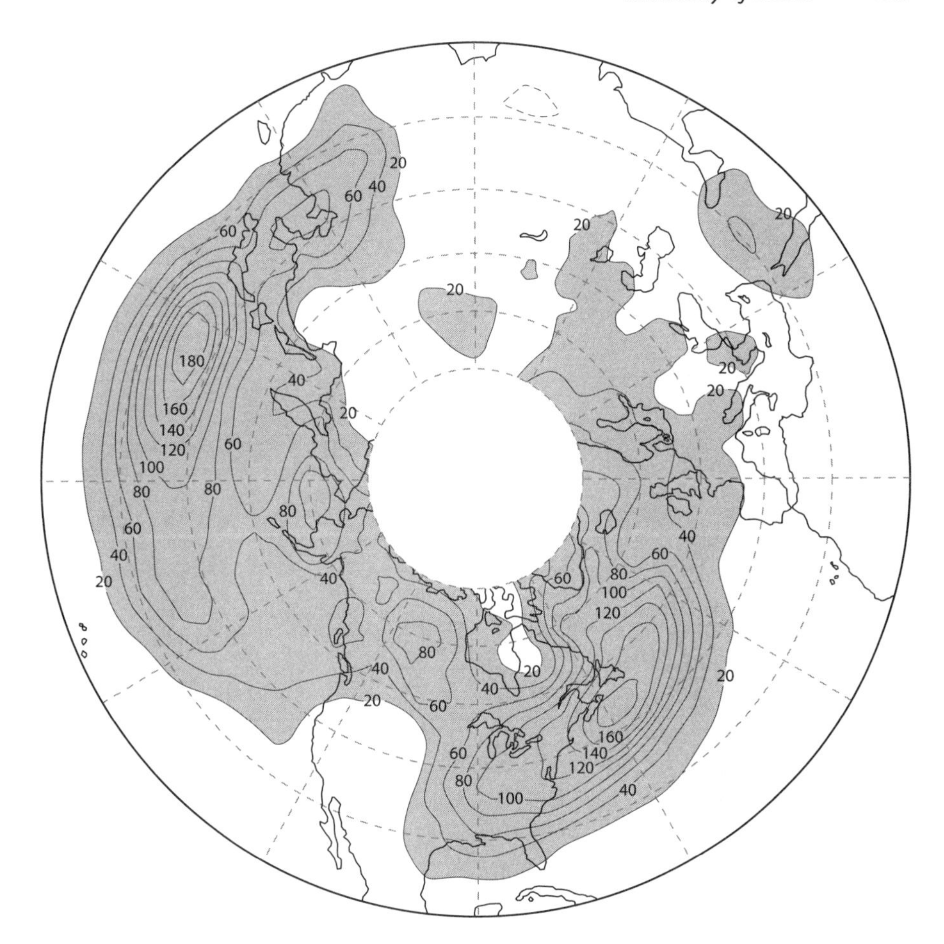

Figure 8.12. The vertically averaged rate of conversion of eddy available potential energy into eddy kinetic energy in Northern Hemisphere winter. The contour interval is 20 $m^2 s^{-2} day^{-1}$. From Chang et al. (2002). © American Meteorological Association. Used with permission.

of the free surface, and h_T is the height of the topography. Quasi-geostrophy allows us to write

$$\zeta = \frac{g}{f_0} \nabla^2 h_{fs}. \tag{71}$$

We linearize about a stationary, geostrophically balanced, zonally averaged state. We assume that $\overline{u}^\lambda$ is independent of position and time, and that $\overline{h_T}^\lambda = 0$. The basic-state zonal wind satisfies

$$\overline{u}^\lambda = \frac{g}{f_0} \frac{\partial \overline{h}^\lambda}{\partial y}. \tag{72}$$

Here we use Cartesian coordinates for simplicity. Then, the PV gradient of the zonally averaged flow is given by

$$\overline{h}^\lambda \frac{\partial}{\partial y} \overline{Z_{SWQG}}^\lambda = \beta + \frac{f_0^2 \overline{u}^\lambda}{g \overline{h}^\lambda}. \tag{73}$$

In linearizing about the basic state given by (72) and (73), we assume not only that the perturbation vorticity and height are small but also that h_T is small compared with $\overline{h}^{\lambda}$, so that we can neglect any products of h_T with a perturbation quantity. The linearized version of (69) is then

$$\left(\frac{\partial}{\partial t} + \overline{u}^{\lambda}\frac{\partial}{\partial x}\right) Z^{*}_{SWQG} + v^{*}\frac{\partial \overline{Z_{SWQG}}^{\lambda}}{\partial y} = -\frac{r\zeta^{*}}{\overline{h}^{\lambda}}. \tag{74}$$

As usual with QG models, all perturbation quantities can be expressed in terms of h^{*}, as follows:

$$v^{*} = \frac{g}{f_0}\frac{\partial h^{*}_{fs}}{\partial x}, \tag{75}$$

$$\zeta^{*} = \frac{g}{f_0}\nabla^2 h^{*}_{fs}, \tag{76}$$

$$\begin{aligned} \overline{h}^{\lambda} Z^{*}_{SWQG} &= \zeta^{*} - f_0\frac{h^{*}}{\overline{h}^{\lambda}} \\ &= \frac{g}{f_0}\nabla^2 h^{*}_{fs} - f_0\frac{h^{*}}{\overline{h}^{\lambda}}. \end{aligned} \tag{77}$$

By substituting from (73) and (75)–(77), we can rewrite (74) as

$$\left(\frac{\partial}{\partial t} + \overline{u}^{\lambda}\frac{\partial}{\partial x}\right)\left(\frac{g}{f_0}\nabla^2 h^{*}_{fs} - f_0\frac{h^{*}}{\overline{h}^{\lambda}}\right) + \frac{g}{f_0}\frac{\partial h^{*}_{fs}}{\partial x}\left(\beta + \frac{f_0^2\overline{u}^{\lambda}}{g\overline{h}^{\lambda}}\right) = -r\frac{g}{f_0}\nabla^2 h^{*}_{fs}, \tag{78}$$

or, after rearranging,

$$\frac{\partial}{\partial t}\left(\nabla^2 h^{*}_{fs} - d^{-2}h^{*}_{fs}\right) + \overline{u}^{\lambda}\frac{\partial}{\partial x}\left(\nabla^2 h^{*}_{fs}\right) + \beta\frac{\partial h^{*}_{fs}}{\partial x} + r\nabla^2 h^{*}_{fs} = -\overline{u}^{\lambda}d^{-2}\frac{\partial h_T}{\partial x}, \tag{79}$$

where d is the *radius of deformation*, which is defined by

$$d^2 \equiv g\overline{h}^{\lambda}/f_0^2, \tag{80}$$

and can be interpreted as the distance that a gravity wave travels before it feels the effects of rotation. The "topographic forcing" term, involving $\partial h_T/\partial x$, was placed on the right-hand side of (79), to set it apart. Equation (79) can describe waves that are "forced" by topography, which enters mathematically through the inhomogeneous term on the right-hand side. In the presence of such forcing, a nonzero h^{*} is demanded by (79). Homogeneous "free wave" solutions also exist, but they will be damped by the friction term, as discussed below.

We assume that the perturbations have the form

$$h^{*}_{fs} = \mathrm{Re}\left\{\hat{h}\exp\left[i(kx + ly - \sigma t)\right]\right\}, \tag{81}$$

where $\hat{h}$ is a constant, and also that the topography satisfies

$$h_T = \mathrm{Re}\left\{\hat{h}_T\exp\left[i(kx + ly)\right]\right\}. \tag{82}$$

In (81), we assumed that the wavenumbers k and l are nonnegative, which means that the sign of σ determines the direction of propagation; a positive value of σ corresponds to eastward propagation. Substitution of (81) and (82) into (79) gives

$$\left[-\sigma(K^2 + d^{-2}) + k(\overline{u}^{\lambda}K^2 - \beta) - irK^2\right]\hat{h}e^{-i\sigma t} = \overline{u}^{\lambda}d^{-2}k\hat{h}_T. \tag{83}$$

Here K is the total wavenumber, which is defined by

$$K^2 \equiv k^2 + l^2. \tag{84}$$

We see directly from (83) that the amplitude of the waves, as measured by $\hat{h}$, is proportional to the amplitude of the forcing, as measured by $\hat{h}_T$. The proportionality factor is rather complicated, however.

First, we consider the special case in which a free wave exists in the absence of topographic forcing. Then, (83) reduces to a dispersion formula, which can be written as

$$-\sigma(K^2 + d^{-2}) + k(\overline{u}^{\lambda} K^2 - \beta) - irK^2 = 0, \tag{85}$$

or

$$\sigma = \frac{k(\overline{u}^{\lambda} K^2 - \beta) - irK^2}{K^2 + d^{-2}}. \tag{86}$$

Equation (86) is the dispersion equation for a damped free Rossby wave in a balanced mean flow. The wave dies out after a finite time, because there is no forcing to sustain it against the frictional damping. To see how the friction leads to damping, we write

$$\sigma = \sigma_0 - i(\tau_f)^{-1}, \tag{87}$$

where

$$\sigma_0 \equiv \frac{k(\overline{u}^{\lambda} K^2 - \beta)}{K^2 + d^{-2}}, \tag{88}$$

and

$$(\tau_f)^{-1} \equiv \frac{rK^2}{K^2 + d^{-2}}. \tag{89}$$

Then, we can rewrite (81) as

$$h^*_{fs}(x, y, t) = e^{-t/\tau_f} \mathrm{Re}\left\{\hat{h} \exp\left[i(kx + ly - \sigma_0 t\right]\right\}. \tag{90}$$

The wavelike solution, with period σ_0, decays with e-folding time τ_f. In other words, the free waves are killed off by friction, as mentioned earlier.

A stationary wave is one for which $\mathrm{Re}\{\sigma\} = 0$. Under what conditions can a free Rossby wave be stationary? Equation (86) shows that a stationary free wave is possible only for $\overline{u}^{\lambda} > 0$. The reason is that the Rossby wave propagates toward the west relative to the mean flow, so to hold the wave stationary relative to the Earth's surface the mean flow must be from the west. In such a case, the total wavenumber of the stationary wave is

$$K^2 = \beta / \overline{u}^{\lambda} \equiv K_S^2. \tag{91}$$

Here the subscript S stands for "stationary."

With this preparation, we return now to the topographically forced case and assume a stationary, neutral wave; that is, $\omega = 0$. For the case of no friction ($r = 0$), we find from (81) that

$$\hat{h} = \frac{\hat{h}_T}{d^2(K^2 - K_S^2)}. \tag{92}$$

According to (92), in the absence of friction the forced wave has "infinite amplitude" for $K^2 = K_S^2$. This is the phenomenon of resonance, familiar from introductory physics. The infinity can be avoided by turning on the friction, that is, allowing r to be positive. With friction, (92) is replaced by

$$\hat{h} = \frac{\hat{h}_T}{d^2\left(K^2 - K_S^2 - i\frac{rK^2}{k\overline{u}^\lambda}\right)}. \tag{93}$$

Equation (93) shows that, mathematically, friction makes the amplitude complex. Clearly, the denominator on the right-hand side of (93) cannot become zero so long as $r > 0$. For $K^2 = K_S^2$, (93) simplifies to

$$\hat{h} = \hat{h}_I\left[\frac{ik(\overline{u}^\lambda)^2}{d^2 r\beta}\right]. \tag{94}$$

Figure 8.13 shows how the squared amplitude of the steady-state wave response varies as the zonal wind speed changes, for different values of r. Resonance occurs for a wind speed close to 20 m s^{-1}. Sufficiently strong damping leads to smooth behavior near resonance.

As seen in figure 8.14, the Charney-Eliassen model, despite its extreme simplicity, can explain reasonably well the observed zonal structure of the 500 hPa height field at 45° N in January. This finding strongly suggests that the observed midlatitude stationary eddies in winter are forced primarily by topography. A similar conclusion was reached by Manabe and Terpstra (1974) through numerical experiments with a global circulation model.

Figure 8.15 shows the observed time-latitude sections of the zonal and meridional contributions to the stationary wave kinetic energy, that is, $\overline{u^{*2}}^\lambda$ and $\overline{v^{*2}}^\lambda$, respectively. In winter, the zonal component is strongest in middle latitudes, which can be considered as corresponding to the orographically forced waves just analyzed, although of course there is also a thermally forced component. In summer, there is a subtropical maximum of stationary eddy kinetic energy associated with the monsoons, which are discussed later in this chapter.

Orographically and thermally forced stationary Rossby waves are discussed in much more detail by Held et al. (2002).

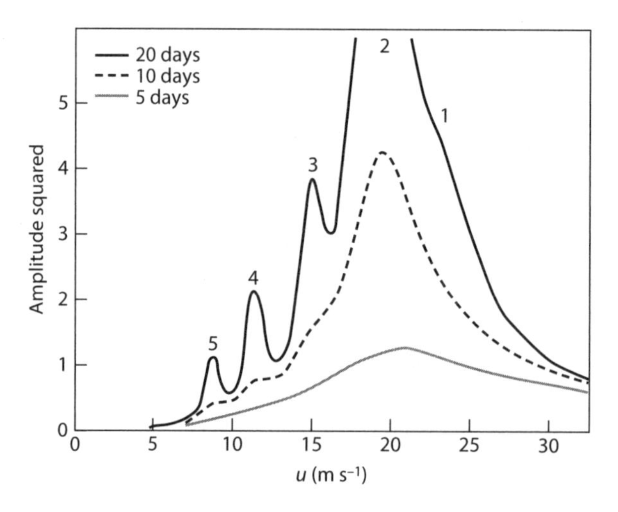

Figure 8.13. The mean-square height response, in the Charney-Eliassen model, as a function of $\overline{u}^\lambda$, for different values of the Rayleigh friction coefficient. The units are 10^4 m^2. The integers written above the curves indicate the values of $\overline{u}^\lambda$ for which particular zonal wavenumbers resonate. From Held (1983). Copyright © 1983 by Academic Press Ltd.

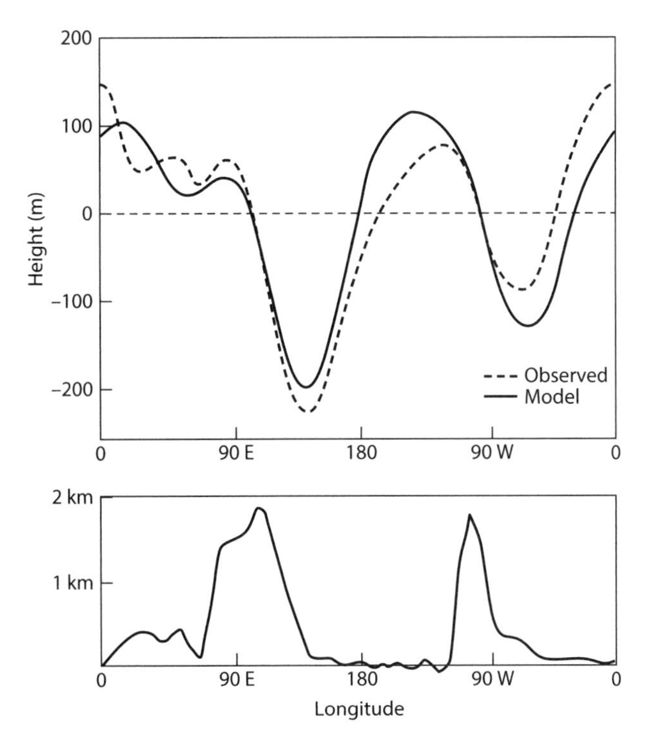

Figure 8.14. Upper panel: The height response as a function of longitude in the Charney-Eliassen model (solid line) and the observed climatological 500 hPa eddy heights at 45° N in January (dashed line). The lower panel shows the topography used. From Held (1983). Copyright © 1983 by Academic Press Ltd.

Tropical Waves

In his doctoral thesis, Taroh Matsuno (1966; see fig. 8.16) studied the linearized shallow water equations applied to the equatorial β-plane, that is, a region centered on the equator that is meridionally narrow enough that β can be treated as a constant. When Matsuno began this work, his motivation was to investigate to what extent near-equatorial motions are geostrophic. In the process, however, he discovered two new classes of tropical waves, which soon after were detected in the observations. These same waves are actually among the solutions found by Laplace, but this was not recognized until later (Lindzen, 1967). The model studied by Matsuno turns out to be relevant to a wide variety of phenomena, including monsoons, the Madden-Julian Oscillation, the Quasi-Biennial Oscillation, and El Niño. This wide applicability is amazing, in view of the model's extreme simplicity.

The shallow water equations on the equatorial β-plane, linearized about a state of rest, are

$$\begin{aligned} \frac{\partial u}{\partial t} - fv + g\frac{\partial h}{\partial x} &= 0, \\ \frac{\partial v}{\partial t} + fu + g\frac{\partial h}{\partial y} &= 0, \\ \frac{\partial h}{\partial t} + H\left(\frac{\partial u}{\partial x} + \frac{\partial v}{\partial y}\right) &= 0. \end{aligned} \tag{95}$$

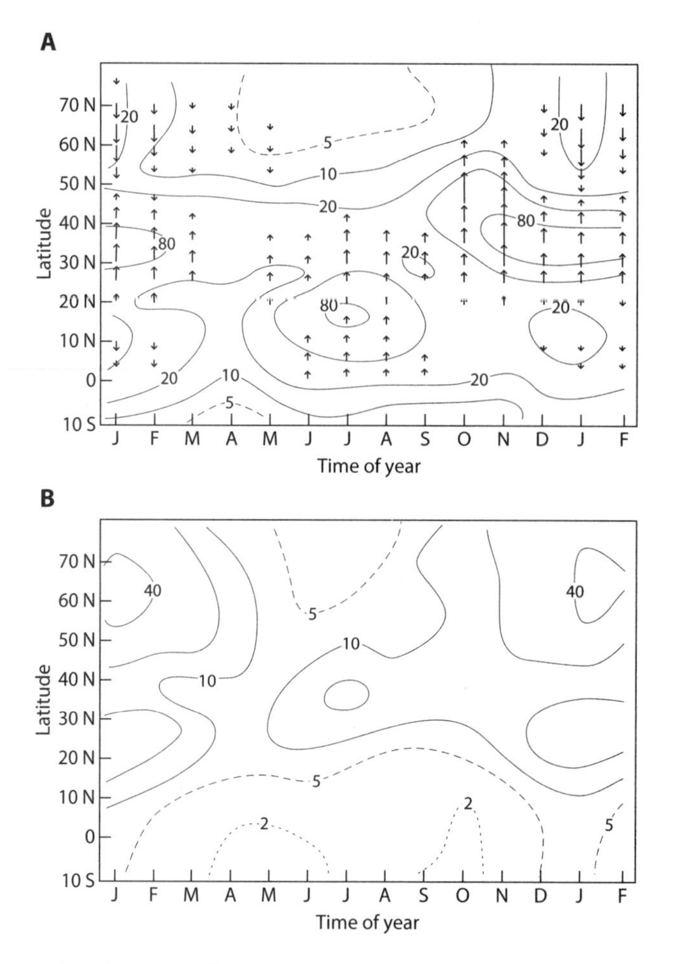

Figure 8.15. Time-latitude sections of Northern Hemisphere climatological mean stationary wave kinetic energy at the 200 hPa level, based on data of Oort and Rasmusson (1971). (A) Zonal wind component; (B) meridional wind component, in units of $m^2\ s^{-2}$. Arrows in panel A denote the direction and relative magnitude of the meridional flux of zonal momentum by the stationary waves. From Wallace (1983). Copyright © 1983 by Academic Press Ltd.

See chapter 4 for an explanation of the shallow water equations. We do not consider topographic effects in this discussion. In (95), $f \equiv \beta y$, where y is distance in the meridional direction, measured from $y = 0$ at the equator (i.e., $y = a\varphi$), and $\beta \equiv df/dy$ is approximated by a constant value. Matsuno defined a timescale, $T \equiv \sqrt{1/(c\beta)}$, and a length scale, $L \equiv \sqrt{c/\beta}$. Here $c \equiv \sqrt{gH}$ is the phase speed of a pure gravity wave. With these length- and timescales, the velocity scale is simply c. The length L can be interpreted as the "equatorial radius of deformation." For a representative phase speed of $c = 10\ \mathrm{m\ s^{-1}}$, we find that $L = 1000$ km and $T \cong 1$ day. Nondimensionalizing the governing equations using T and L, we obtain

$$
\begin{aligned}
&\frac{\partial u}{\partial t} - yv + \frac{\partial \phi}{\partial x} = 0, \\
&\frac{\partial v}{\partial t} + yu + \frac{\partial \phi}{\partial y} = 0, \\
&\frac{\partial \phi}{\partial t} + \frac{\partial u}{\partial x} + \frac{\partial v}{\partial y} = 0.
\end{aligned}
\tag{96}
$$

Figure 8.16. Professor Taroh Matsuno. Photo kindly provided by Prof. Matsuno.

Here ϕ is the nondimensional form of gh.

As a side comment, we note that these equations can in fact apply to a model with vertical structure (e.g., McCreary, 1981; Fulton and Schubert, 1985) and so are more readily applicable to the real atmosphere than one might guess. As a very simple example, consider a two-level model, governed by

$$\frac{\partial \mathbf{V}_1}{\partial t} + f\mathbf{k} \times \mathbf{V}_1 + \nabla\phi_1 = 0,$$
$$\frac{\partial \mathbf{V}_3}{\partial t} + f\mathbf{k} \times \mathbf{V}_3 + \nabla\phi_3 = 0, \tag{97}$$
$$\frac{\partial}{\partial t}(\phi_3 - \phi_1) + S\Delta p\omega_2 = 0.$$

As shown in figure 8.17, subscript 1 denotes the upper level, and subscript 3 denotes the lower level. The vertical velocity is defined in between, at level 2. We use $\Delta p \equiv p_3 - p_1$ to denote the pressure thickness between the two layers, and S is the static stability of the basic state. We let

$$\mathbf{V}_d \equiv \mathbf{V}_3 = \mathbf{V}_1, \tag{98}$$

and

$$\phi_d \equiv \phi_3 - \phi_1 \tag{99}$$

be the vertical shear (actually, difference) of the horizontal wind between the two layers, and the thickness between the two layers, respectively. Then (97) implies that

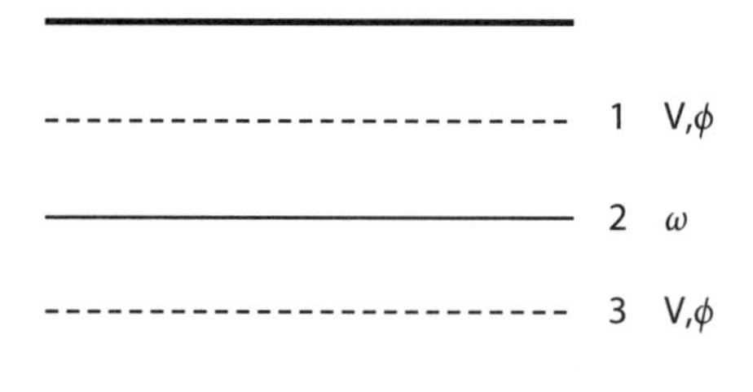

Figure 8.17. Schematic representing the two-level model represented by (97).

$$\frac{\partial \mathbf{V}_d}{\partial t} + f\mathbf{k} \times \mathbf{V}_d + \nabla \phi_d = 0, \tag{100}$$

and

$$\frac{\partial \phi_d}{\partial t} + \frac{S\Delta p^2}{2} \nabla \cdot \mathbf{V}_d = 0, \tag{101}$$

which are identical to the shallow water equations, and we can identify

$$c_i \equiv \Delta p \sqrt{\frac{S}{2}} \tag{102}$$

as the phase speed of gravity waves that derive their restoring force from the static stability.

We return now to our discussion of (96). We assume solutions of the form

$$\begin{aligned} u &= \hat{u}(y)e^{i(kx+\sigma t)}, \\ v &= \hat{v}(y)e^{i(kx+\sigma t)}, \\ \phi &= \hat{\phi}(y)e^{i(kx+\sigma t)}. \end{aligned} \tag{103}$$

If we adopt the convention that k is positive, then $\sigma > 0$ corresponds to westward propagation, and $\sigma < 0$ to eastward propagation. Substitution into (96) gives

$$\begin{aligned} i\sigma\hat{u} - y\hat{v} + ik\hat{\phi} &= 0, \\ i\sigma\hat{v} + y\hat{u} + \frac{d\hat{\phi}}{dy} &= 0, \\ i\sigma\hat{\phi} + ik\hat{u} + \frac{d\hat{v}}{dy} &= 0. \end{aligned} \tag{104}$$

We can solve the first of these equations for $\hat{u}$ in terms of $\hat{v}$ and $\hat{\phi}$, and we can use the result to eliminate $\hat{u}$ in the other two equations. Then, we can rewrite the system (81) as

$$\begin{aligned} \sigma\hat{u} + k\hat{\phi} &= -iy\hat{v}, \\ \left(ky - \sigma\frac{d}{dy}\right)\hat{\phi} &= i(\sigma^2 - y^2)\hat{v}, \\ (\sigma^2 - k^2)\hat{\phi} &= i\left(ky + \sigma\frac{d}{dy}\right)\hat{v}. \end{aligned} \tag{105}$$

Before considering the general case, we discuss a special solution, called the *equatorial Kelvin wave*, for which the meridional wind is identically zero. Setting $\hat{v} = 0$ in (105), we obtain

$$\begin{aligned} k\hat{\phi} + \sigma\hat{u} &= 0, \\ ky\hat{\phi} - \sigma\frac{d\hat{\phi}}{dy} &= 0, \\ (\sigma^2 - k^2)\hat{\phi} &= 0. \end{aligned} \tag{106}$$

The third of these equations implies that for nontrivial solutions,

$$\sigma = \pm k, \tag{107}$$

where σ and k are nonzero. With the use and inclusion of (107), the system (106) simplifies to

$$\begin{aligned} \hat{\phi} \pm \hat{u} &= 0, \\ \left(-y \pm \frac{d}{dy}\right)\hat{\phi} &= 0, \\ \sigma &= \pm k. \end{aligned} \tag{108}$$

The second equation in (108) determines the meridional structure of ϕ. Its solution is

$$\hat{\phi} = e^{\pm y^2/2}. \tag{109}$$

If we choose the plus sign, we get solutions that grow exponentially away from the equator, which is unacceptable, especially since the equatorial β-plane approximation is useful only near the equator. We therefore choose the minus sign, which gives a bell-shaped solution that has a maximum on the equator and decays strongly away from the equator. We can now write the solution for the equatorial Kelvin wave as

$$\begin{aligned} \hat{u} &= e^{-y^2/2}, \\ \hat{\phi} &= e^{-y^2/2}, \\ \sigma &= -k. \end{aligned} \tag{110}$$

Note that $\sigma = -k$ implies eastward propagation. An equatorial Kelvin wave always propagates towards the east. Kelvin waves are symmetrical across the equator.

Returning now to the general case, we can eliminate $\hat{\phi}$ between the second and third equations of (105) to obtain

$$ky\left(ky\hat{v} + \sigma\frac{d\hat{v}}{dy}\right) - \sigma\frac{d}{dy}\left(ky\hat{v} + \sigma\frac{d\hat{v}}{dy}\right) = (\sigma^2 - k^2)(\sigma^2 - y^2)\hat{v}, \tag{111}$$

which we can simplify to

$$\boxed{\frac{d^2\hat{v}}{dy^2} + \left(\sigma^2 - k^2 + \frac{k}{\sigma} - y^2\right)\hat{v} = 0}. \tag{112}$$

The substitution used to eliminate $\hat{\phi}$ between the second and third equations of (105) is valid only for $\sigma^2 - k^2 \neq 0$, so (112) does not apply to the Kelvin wave, for which $\sigma^2 - k^2 = 0$.

We expect the solutions of (112) to have oscillatory behavior for $\sigma^2 - k^2 + (k/\sigma) - y^2 > 0$ and exponential behavior for $\sigma^2 - k^2 + (k/\sigma) - y^2 < 0$. Because of the $-y^2$ term, exponential behavior is guaranteed to emerge sufficiently far from the equator, and, as with the Kelvin wave, we need exponential decay rather than exponential growth. Therefore, as boundary conditions, we use

$$\hat{v} \to 0 \text{ as } y \to \pm\infty. \tag{113}$$

It can be shown that nontrivial solutions satisfying these boundary conditions exist when

$$\boxed{\sigma^2 - k^2 + \frac{k}{\sigma} = 2n + 1 \text{ for } n = 0,1,2\ldots}. \tag{114}$$

This is a dispersion equation. The expression on the right-hand side of (114) generates all positive odd integers, so that (114) is equivalent to the statement that $\sigma^2 - k^2 + (k/\sigma)$ is a positive odd integer. We can confirm by substitution that the solutions of (112) are

$$\hat{v}(y) = Ce^{-\frac{1}{2}y^2} H_n(y), \tag{115}$$

where C is an arbitrary real constant and $H_n(y)$ is the nth Hermite polynomial, which is given by

$$H_n(y) \equiv (-1)^n e^{y^2} \frac{d^n}{dy^n}(e^{-y^2}) \tag{116}$$

(see appendix H). Just as for the equatorial Kelvin wave, the factor $e^{-(1/2)y^2}$ ensures that these modes decay rapidly away from the equator. For realistic values of the dimensional parameters, the dimensional e-folding distance turns out to be about 1000 km.

Because the solutions, (115), involve Hermite polynomials, and because the different Hermite polynomials are distinguished by their values of the parameter n, we can think of the meridional shapes of the equatorial waves as being determined by the value of n. Solutions with the same value of n will have the same meridional shape. Although (114) does not apply to the Kelvin wave and is valid only for nonnegative values of n, it turns out that $\sigma = -k$ (the dispersion relation for the Kelvin wave) is a solution of (114) if we set $n = -1$. For this reason, the Kelvin wave is often called the $n = -1$ solution of Matsuno's model.

The dispersion equation, (114), is cubic in σ, which means that there are three σs for each (k, n) pair. Two of these correspond to inertia-gravity waves. They can be approximated by

$$\sigma_{1,2} \cong \pm\sqrt{k^2 + 2n + 1}. \tag{117}$$

The third root corresponds to a Rossby wave. It can be approximated by

$$\sigma_3 \cong \frac{k}{\sqrt{k^2 + 2n + 1}}. \tag{118}$$

For the special case $n = 0$, the dispersion equation (114) can be *factored*, to give

$$(\sigma - k)(\sigma^2 + k\sigma - 1) = 0. \tag{119}$$

Matsuno showed that for this case the three roots can be interpreted as follows:

$$\text{Eastward gravity wave:} \quad \sigma_1 = -\frac{k}{2} - \sqrt{\left(\frac{k}{2}\right)^2 + 1}, \tag{120}$$

$$\text{Westward gravity wave:} \quad \sigma_2 = \begin{cases} \sqrt{\left(\frac{k}{2}\right)^2 + 1} - \frac{k}{2} & \text{for } k \leq \frac{1}{\sqrt{2}} \\ k & \text{for } k \geq \frac{1}{\sqrt{2}} \end{cases}, \tag{121}$$

$$\text{Rossby wave:} \quad \sigma_3 = \begin{cases} k & \text{for } k \leq \frac{1}{\sqrt{2}} \\ \sqrt{\left(\frac{k}{2}\right)^2 + 1} - \frac{k}{2} & \text{for } k \geq \frac{1}{\sqrt{2}} \end{cases}. \tag{122}$$

Notice that (with $n=0$) the westward gravity wave and the Rossby wave are not truly distinct; they coincide for $k=1/\sqrt{2}$. For reasons that were discussed earlier, the root $\sigma=k$ has to be discarded in (121) and (122). Matsuno concluded, therefore, that for $n=0$ there are actually only two waves: an eastward-moving gravity wave; and a "mixed Rossby-gravity wave," which is also known as the *Yanai wave*. The Yanai wave behaves like a gravity wave for $k<1/\sqrt{2}$ and like a Rossby wave for $k>1/\sqrt{2}$. The dispersion relation for the Yanai wave is

$$\sigma = \sqrt{\left(\frac{k}{2}\right)^2 + 1} - \frac{k}{2}. \tag{123}$$

Because $H_0(y)=1$, (115) reduces to

$$\hat{v}(y) = Ce^{-\frac{1}{2}y^2} \tag{124}$$

for the Yanai wave, which shows that in Yanai waves the meridional velocity has the same sign on both sides of the equator and is a maximum on the equator. The Yanai wave consists of a chain of vortices propagating westward along the equator. It bears some resemblance to the so-called easterly waves that are found off the equator near the latitude of the ITCZ and that sometimes transition into tropical cyclones.

The dispersion relations associated with the various solutions of Matsuno's model are depicted in figure 8.18. Recall that positive values of the frequency correspond

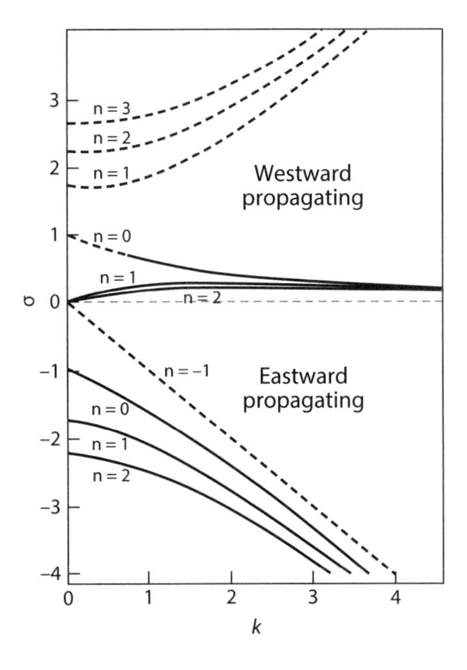

Figure 8.18. Frequencies as functions of zonal wavenumber. Positive frequencies correspond to westward propagation. The eastward-propagating modes labeled $n = 0$, $n = 1$, and $n = 2$ (solid lines) are eastward-propagating inertia-gravity waves. The westward-propagating modes labeled $n = 3$, $n = 2$, and $n = 2$ (dashed lines) are westward-propagating inertia-gravity waves. The westward-propagating modes labeled $n = 1$ and $n = 2$ (solid lines) are Rossby waves. The eastward-propagating mode labeled $n = -1$ (straight dashed line) is the Kelvin wave. The westward-propagating wave labeled $n = 0$ is the mixed Rossby-gravity or Yanai wave. It behaves like a gravity wave and is denoted by a dashed line for $k < 1/\sqrt{2}$, and it behaves like a Rossby wave and is denoted by a solid line for $k > 1/\sqrt{2}$. From Matsuno (1966).

to westward-propagating waves, and negative values to eastward-propagating waves. The dashed line extending downward toward the right from the origin represents the Kelvin wave. The solid curves arcing upward from the origin represent Rossby waves, with positive values of σ. The dashed curves in the upper part of the diagram correspond to westward-propagating inertia-gravity waves, and the solid curves in the lower part of the diagram correspond to eastward-propagating inertia-gravity waves. The westward-propagating wave represented by the curve that is partly solid and partly dashed is the Yanai wave. The dashed portion of this curve, plotted for $k < 1/\sqrt{2}$, represents those wavenumbers for which the Yanai wave behaves like a westward-propagating gravity wave. The solid portion of the curve, for $k > 1/\sqrt{2}$, represents those wavenumbers for which the Yanai wave behaves like a Rossby wave.

For $n = 0$, the eastward-moving inertia-gravity wave and westward-moving Yanai wave have the structures shown in the upper and middle panels of figure 8.19. For a pure gravity wave the winds are expected to be perpendicular to the isobars. When rotation is dominant, the winds are parallel to the isobars. The waves shown look like pure gravity waves near the equator. For $n = 0$ and $k = 1$, the Yanai wave takes on the characteristics of a Rossby wave, as shown in the lower panel.

Solutions for $n = 1$ are shown on the left side of figure 8.20. The corresponding results for $n = 2$ are shown on the right side of the figure. Recall that the subscript n denotes the solution whose meridional structure is described by the nth Hermite polynomial. As can be seen in the figure, higher values of n correspond to more nodes in the meridional direction.

The very simple structure of the Kelvin wave is shown in figure 8.21. The velocity vectors are purely zonal, as expected, and the zonal wind is in phase with the pressure, as in a gravity wave.

The equatorially trapped waves found by Matsuno correspond to solutions of Laplace's equations, if allowances are made for the effects of stratification and sphericity. A vertically continuous (i.e., non–shallow-water) version of Matsuno's analysis was presented by Lindzen (1967).

Matsuno's theoretical discoveries were confirmed by observations in remarkably short order. Maruyama and Yanai (1966) observed the Yanai wave not long after Matsuno had predicted its existence, and Wallace and Kousky (1968) found the Kelvin wave soon thereafter. Yanai and Matsuno shared an office when they were students.

Tropical waves continue to be the subject of many observational, theoretical, and numerical studies (Kiladis et al., 2009). In a particularly famous study, Wheeler and Kiladis (1999) examined the space-time variability of the tropical outgoing longwave radiation. In figure 8.22, the data have been separated into modes that are symmetric across the equator (right panel), such as the Kelvin wave, and modes that are antisymmetric across the equator (left panel), such as the mixed Rossby-gravity wave. By using additional filtering procedures motivated by Matsuno's results, Wheeler and Kiladis were able to show the longitudinal propagation of various types of equatorially trapped disturbances (see figs. 8.23 and 8.24).

Matsuno's model describes free waves that can exist without any forcing, although as discussed in the next section, he also considered stationary forced solutions. Both dry and moist equatorial waves are observed, and the latter are often described as "convectively coupled," which can mean various things. As the waves propagate across the tropics they inevitably influence the ambient convection that is supported by surface evaporation and radiative cooling. For example the waves can promote or suppress convective activity in regions of wave-induced upward or downward

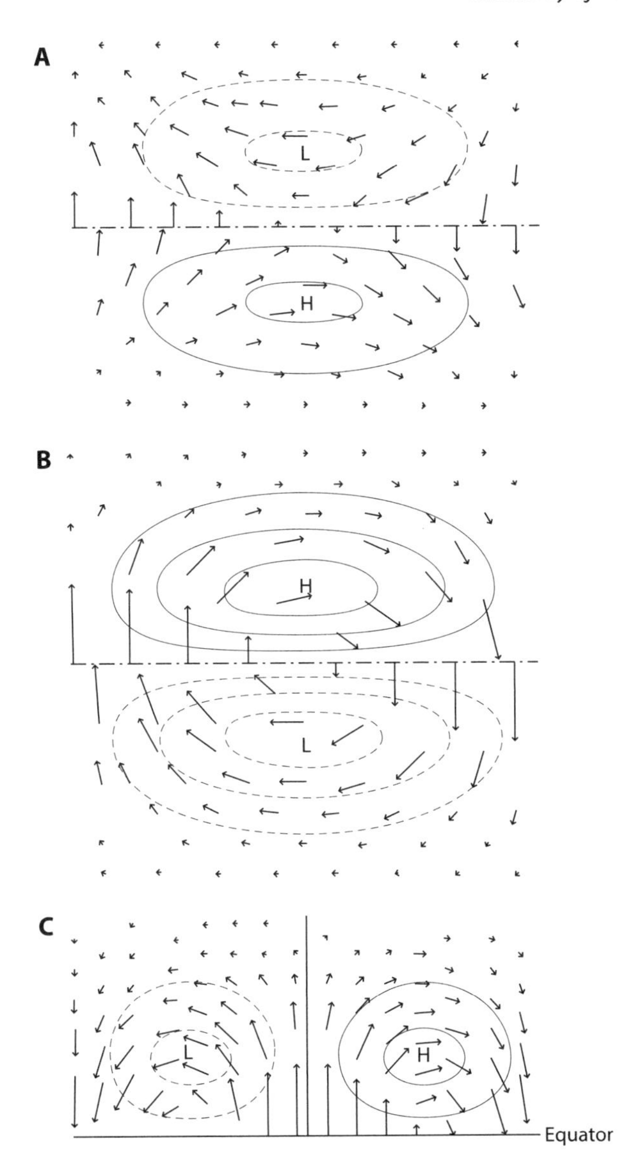

Figure 8.19. Pressure and velocity distributions of solutions for $n = 0$ and $k = 0.5$. (A) Eastward-moving inertia-gravity wave. (B) Westward-moving Yanai wave, which for this value of k behaves like an inertia-gravity wave. (C) The structure of the Yanai wave for $n = 0$ and $k = 1$, in which case the Yanai wave acts like a Rossby wave. For each mode, v is a maximum on the equator and does not pass through zero anywhere. This is characteristic of $n = 0$. From Matsuno (1966).

motion. The convection also influences the waves. For example, convective heating in regions of wave-induced upward motion reduces the cooling that would otherwise occur there, so that the waves feel an "effective static stability" that is somewhat weaker than the actual static stability, and this reduces their phase speed. Interactions of this type do represent coupling of the waves with the convection, but it is *incidental coupling* in the sense that the free-wave solutions of Matsuno's model can exist without the convection (although real-world waves are naturally modified by convection),

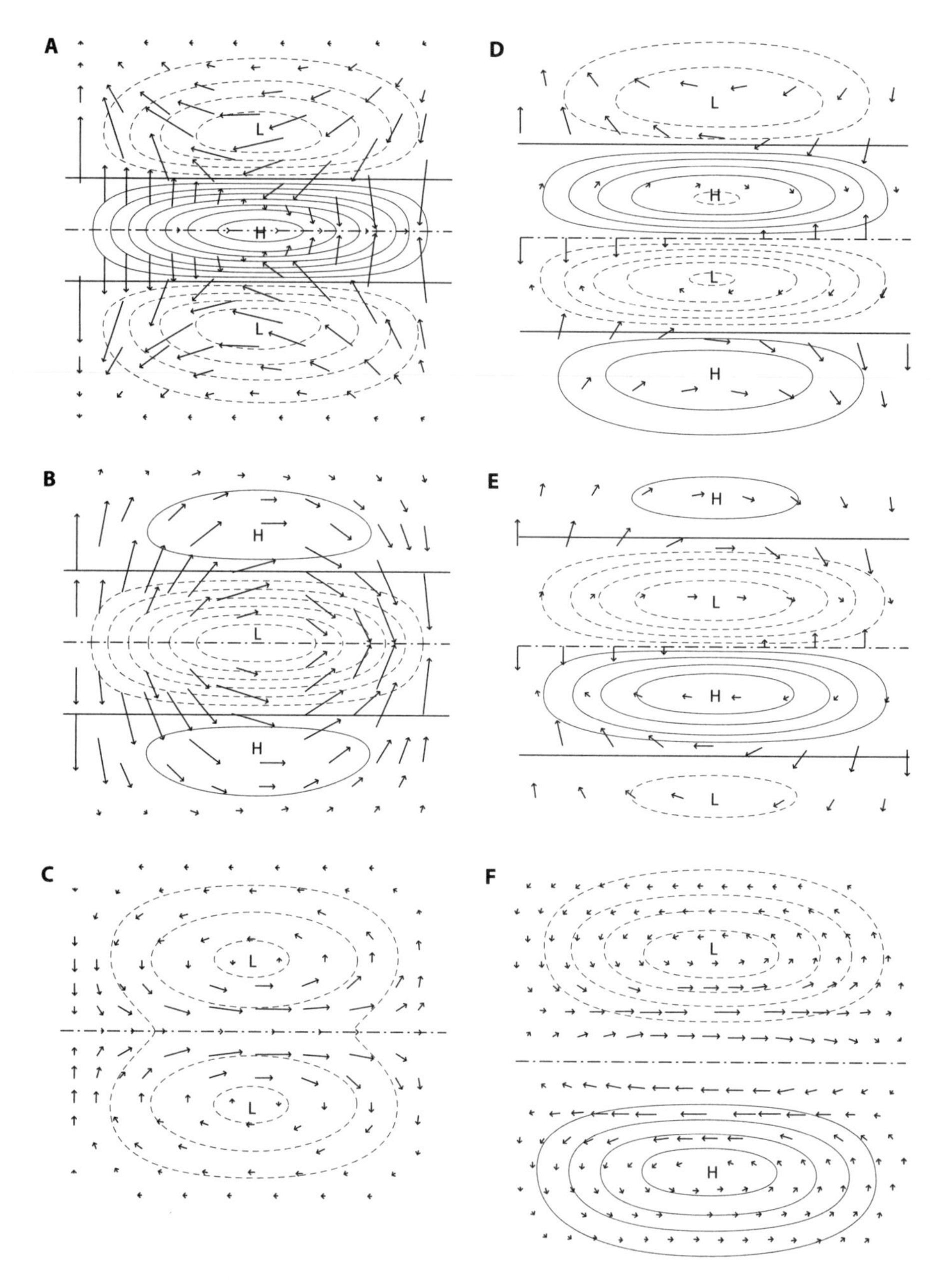

Figure 8.20. Left side: Pressure and velocity distributions of solutions for n = 1. (A) Eastward-propagating inertia-gravity wave. For each mode, u = 0 on the equator, as expected for n = 1. (B) Westward-propagating inertia-gravity wave. (C) Rossby wave. Right side: Corresponding results for n = 2. For each mode, u is symmetric across the equator, as expected for n = 2. From Matsuno (1966).

and the convection can exist without the waves (although it is modified by the waves). Convective excitation of the waves can be quasi-random, much like the notes produced by a cat walking on the keys of a piano.

We can also imagine a more *essential coupling* of convection with tropical eddies, in which the very existence of the eddies depends on the convective heating. Tropical cyclones and the Madden-Julian Oscillation are essentially coupled with convection;

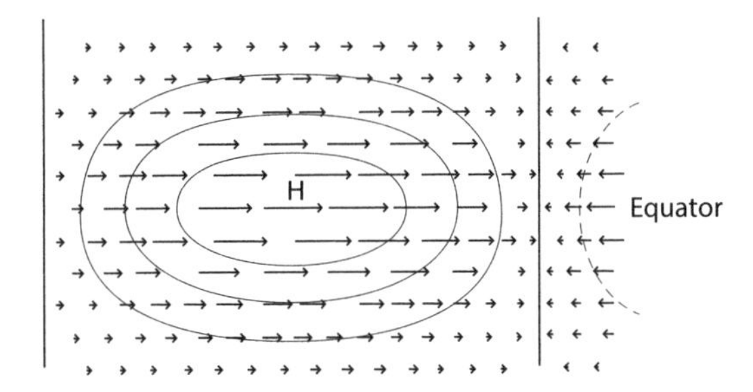

Figure 8.21. Pressure and velocity distributions for $n = -1$ and $k = 0.5$. This is the Kelvin wave. From Matsuno (1966).

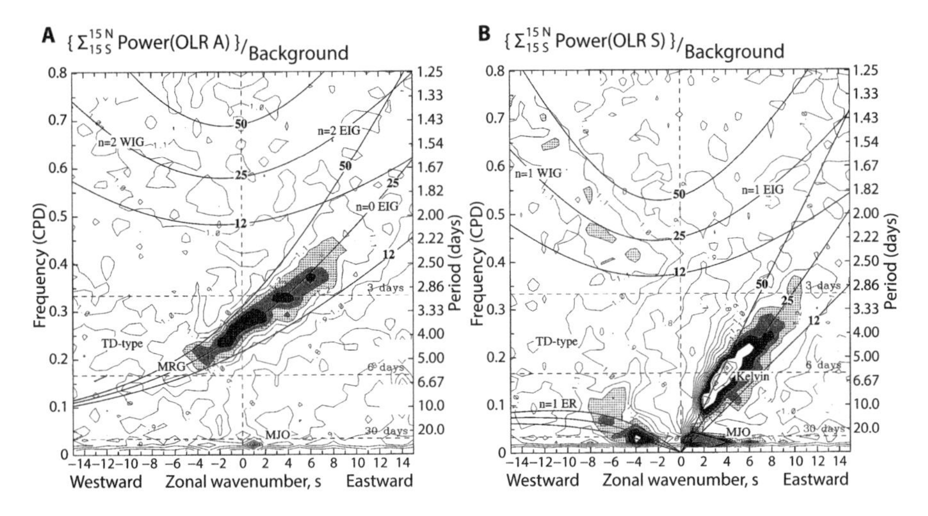

Figure 8.22. The variability of the outgoing longwave radiation (OLR) as a function of frequency and zonal wavenumber for modes that are symmetric across the equator (right panel) and antisymmetric (left panel). Eastward propagation is associated with positive wavenumbers. The boxes indicated with dashed lines select particular wave types. These dispersion diagrams are analogous to those used by Matsuno, as shown in figure 8.18, but the conventions are different. With Matsuno's convention the zonal wavenumber is nonnegative, and the sign of the frequency determines the direction of zonal propagation, while in the figures shown here the frequency is nonnegative, and the sign of the zonal wavenumber determines the direction of zonal propagation. From Wheeler and Kiladis (1999). © American Meteorological Association. Used with permission.

they could not exist without it. The Madden-Julian Oscillation is discussed later in this chapter.

The Response of the Tropical Atmosphere to Stationary Heat Sources and Sinks

The preceding discussion concerned "free waves." Forced solutions to Matsuno's model are also very relevant to observations of monsoons and the Madden-Julian oscillation, both of which are discussed later in this chapter. Figure 8.25, which appears near the end of the paper by Matsuno (1966), shows the stationary circulation driven

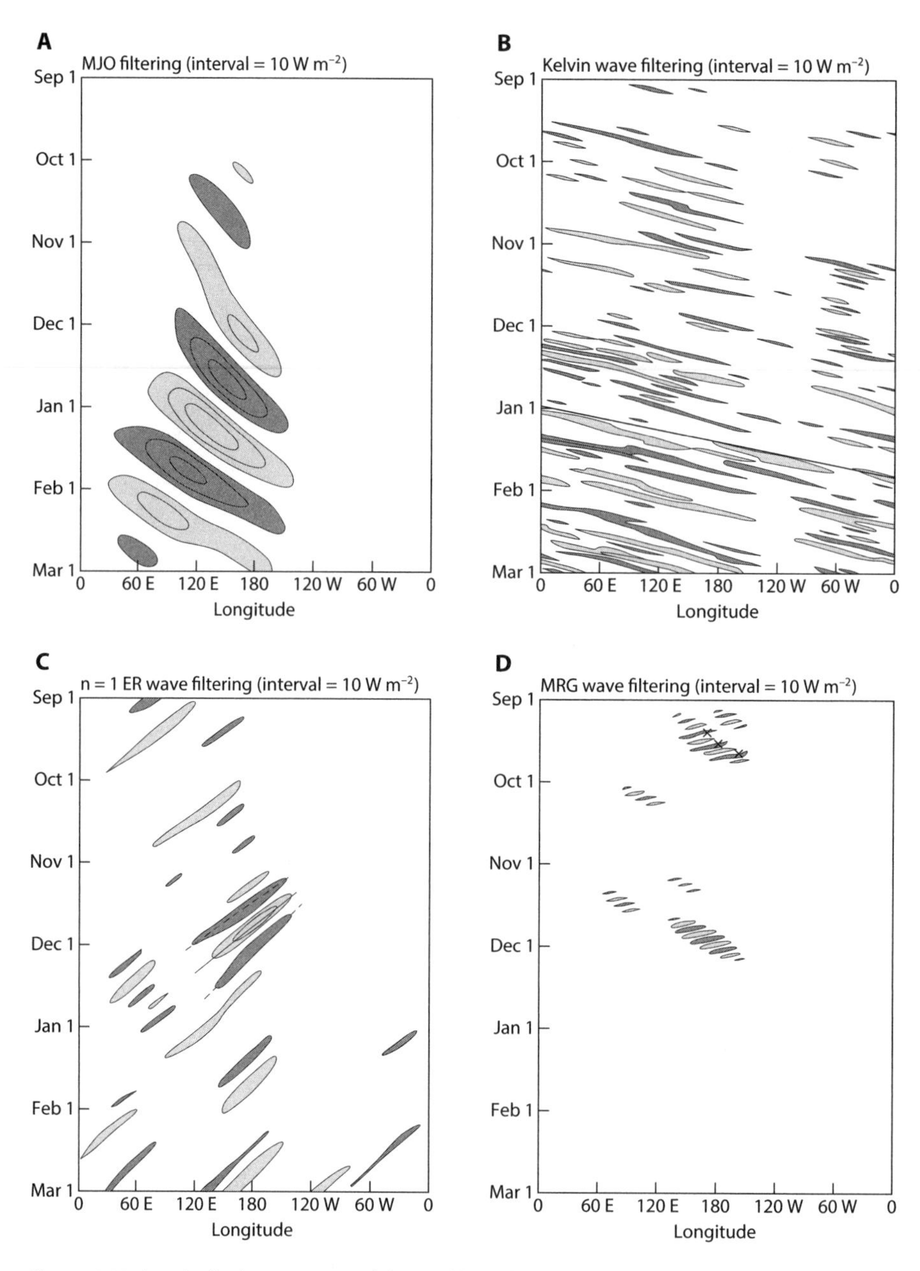

Figure 8.23. Longitudinal propagation of the Madden-Julian oscillation (MJO), Kelvin waves, equatorial Rossby (ER) waves, and mixed-Rossby-gravity (MRG) waves, as seen in the OLR. The zero contour has been omitted. The various modes are selected by including only the contributions from wavenumbers and frequencies that fall within the corresponding boxes in figure 8.22. This is what is meant by *filtering*. From Wheeler and Kiladis (1999). © American Meteorological Association. Used with permission.

by a mass source and sink on the equator. Let us consider this figure in terms of the low-level flow. The mass sink can be interpreted as a region of rising motion, where the air is converging at low levels—for example, in the western equatorial Pacific. The mass source can be interpreted as a region of sinking motion, where the air is diverging at low levels—for example, in the eastern equatorial Pacific. (Unfortunately the

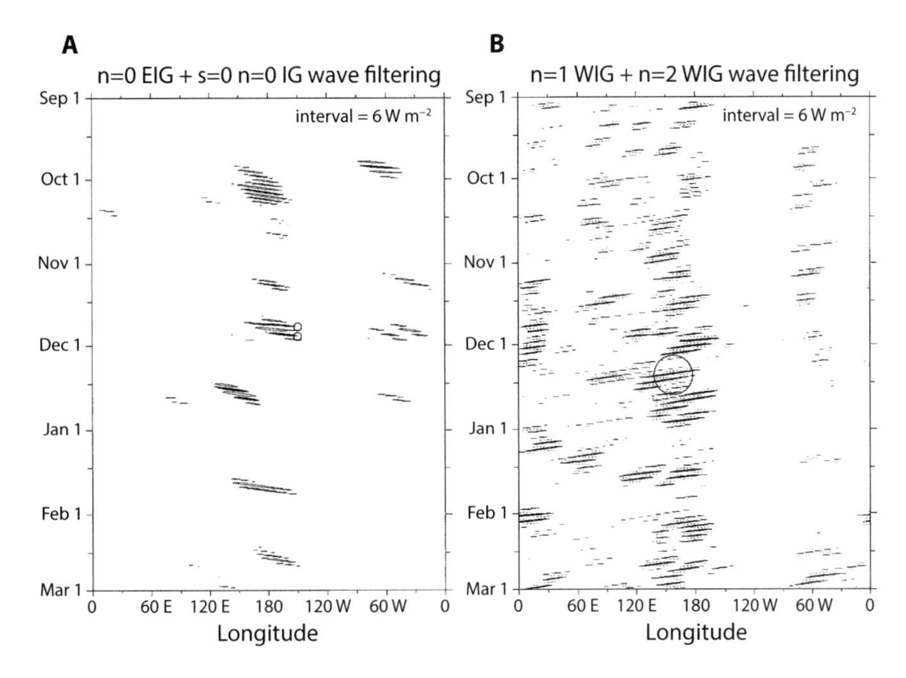

Figure 8.24. Longitudinal propagation of eastward- and westward-propagating inertia-gravity waves, as seen in the OLR. The zero contour has been omitted. As in figure 8.23, the various modes are selected by including only the contributions from wave numbers and frequencies that fall in the corresponding boxes in figure 8.22. From Wheeler and Kiladis (1999). © American Meteorological Association. Used with permission.

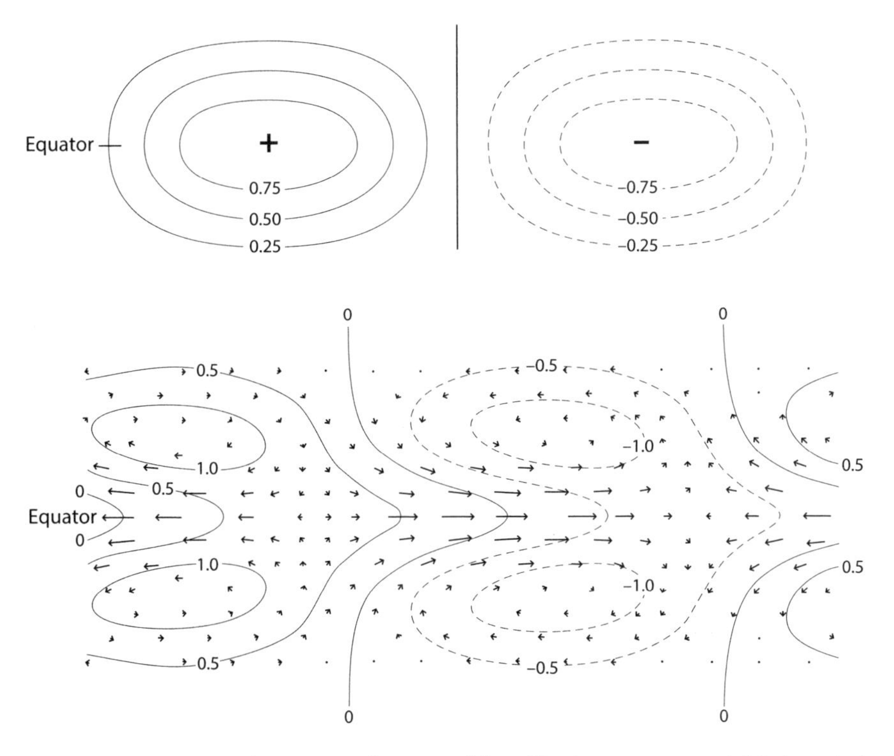

Figure 8.25. Stationary circulation pattern (lower panel) forced by the mass source and sink shown in the upper panel. From Matsuno (1966).

mass sink is plotted on the east side, and the mass source is plotted on the west side, but this does not matter, because the solution is periodic in the zonal direction.) The model predicts strong westerlies converging (from the west, of course) at low levels into the region of rising motion, and low-level easterlies converging on the east side of the region of low-level convergence. The easterlies can be interpreted as the trades and as the lower branch of the Walker circulation, which is discussed later in this chapter. The westerlies can be interpreted as a "monsoon-like" westerly inflow to a region of heating. Monsoons are also discussed later in this chapter.

Webster (1972) and Gill (1980) followed Matsuno's lead by developing simple analytic models of the response of a resting tropical atmosphere to heat sources and sinks. Since much of the convective heating in the tropics is confined over three relatively small land regions (Africa, South America, and the Indonesian region), Gill examined the atmospheric response to a relatively small-scale heating source that is centered on the equator. If the atmosphere is abruptly heated at some initial time, Kelvin waves propagate rapidly eastward and generate easterly trade winds to the east of the heating. Thus, the easterly trade winds in the Pacific could result from Kelvin waves produced by convective heating over Indonesia. Similarly, equatorial Rossby waves propagate westward and generate westerlies to the west of the heating. Because the fastest Rossby wave travels at only one-third the speed of the Kelvin wave, the effects of the Rossby waves would be expected to reach only one-third as far as those of the Kelvin wave. Gill interpreted the westerlies over the Indian Ocean as a response to Rossby waves generated by convective heating over Indonesia.

Gill (1980) studied what amounts to a steady-state version of Matsuno's model and introduced forcing in the form of mass sources and sinks, along with very simple damping. Corresponding to (96), we have

$$\begin{aligned} \varepsilon u - yv + \frac{\partial \phi}{\partial x} &= 0, \\ yu + \frac{\partial \phi}{\partial y} &= 0, \\ \varepsilon \phi + \frac{\partial u}{\partial x} + \frac{\partial v}{\partial y} &= -Q. \end{aligned} \tag{125}$$

And as a purely diagnostic relation,

$$w = \varepsilon \phi + Q. \tag{126}$$

The wind components u and v represent the lower-tropospheric variables. In (125) and (126), ε^{-1} is a dissipation time scale, and Q is a "heating rate" that must be specified. The variables ϕ, w, and Q are defined in the middle troposphere. Gill included dissipation in the form of Rayleigh friction and Newtonian cooling, and for simplicity assumed that the timescales, given by ε^{-1}, are equal. Rayleigh friction is a simple parameterization of friction in which the velocity is divided by a frictional timescale. The friction term is neglected in the meridional momentum equation of (125); see Gill (1980) for an explanation.

Gill focused primarily on cases for which the heating is symmetric or antisymmetric about the equator. The solution for symmetric heating resembles a Walker circulation, with lower-tropospheric inflow into the heating region and upper-tropospheric outflow. The Walker circulation is discussed in detail later in this chapter. The surface easterlies cover a larger area than the surface westerlies because the phase speed of the eastward-propagating Kelvin wave is three times faster than that of the westward-moving Rossby wave.

By using (125) to form a vorticity equation for the case of no damping and then substituting from the continuity equation, Gill found that

$$v = yQ. \tag{127}$$

This equation is closely related to what is sometimes called *Sverdrup balance*, in which the "meridional advection of the Coriolis parameter," that is, the so-called β-term of the vorticity equation, is balanced by the divergence term, which is represented by the heating rate on the right-hand side of (127). According to (127), v changes sign across the equator in a region where Q has a single sign. In a region of cooling ($Q<0$), the flow is toward the equator on both sides, and in a region of heating it is away from the equator on both sides.

For $Q>0$, (127) implies poleward motion in the lower layer and equatorward motion in the upper layer, which suggests that in regions of heating, such as the western Pacific, the Walker circulation produces a north-south circulation that opposes the Hadley circulation. Geisler (1981) found the same result. For $Q<0$, the low-level motion is equatorward; this is what is seen in the subtropical highs, for example, in the eastern Pacific.

The solution for antisymmetric heating consists of a mixed Rossby-gravity wave and a Rossby wave. There is no Kelvin-wave response because the Kelvin wave is intrinsically symmetric across the equator. Long mixed Rossby-gravity waves do not propagate, and so the response of this wave type is largely confined to the region of heating. Owing to the westward propagation of Rossby waves, no response is generated to the east of the forcing region. To the west, the region of westerly flow into the heating region is limited because the Rossby modes travel slowly and so are dissipated before they can propagate far to the west. Gill interpreted the asymmetric case as a simulation of the Hadley circulation, and the symmetric case as a simulation of the Walker circulation.

For heating centered on the equator, as in figure 8.26, Gill found strong westerlies on the west side and strong easterlies on the east side combining to give strong zonal convergence on the heating. The westerlies can be interpreted as the time-averaged response to westward-propagating Rossby waves excited by the heating, and the easterlies can be interpreted as the time-averaged response to eastward-propagating Kelvin waves excited by the heating. This combination implies an east-west overturning circulation, shown in panel C of the figure, and a surface pressure field with a minimum pressure slightly to the west of the heating, depicted by the contours in panel B. Later in this chapter we will discuss the Walker circulation, which resembles panel C in figure 8.26.

When the heating is antisymmetric across the equator, as in figure 8.27, the model produces something like a Hadley circulation, with a low-level cyclonic circulation on the side with positive heating, and a low-level anticyclone on the other side. When the symmetric and antisymmetric heatings are combined, as in figure 8.28, the model produces a circulation that looks remarkably similar to that of the Asian summer monsoon, as discussed in the next subsection.

Although Gill demonstrated that heating of limited extent generates tropical waves that produce broad wind and pressure fields resembling the observations, his results must be viewed with caution owing to several limitations. First, the model was linearized about a specified basic state and so it cannot explain the basic state. Second, results were generated for a specified rather than predicted heating of the tropical troposphere, and so ocean-atmosphere interactions and feedbacks involving moist convection were excluded. Third, the model includes neither a moisture budget nor cloud radiative effects.

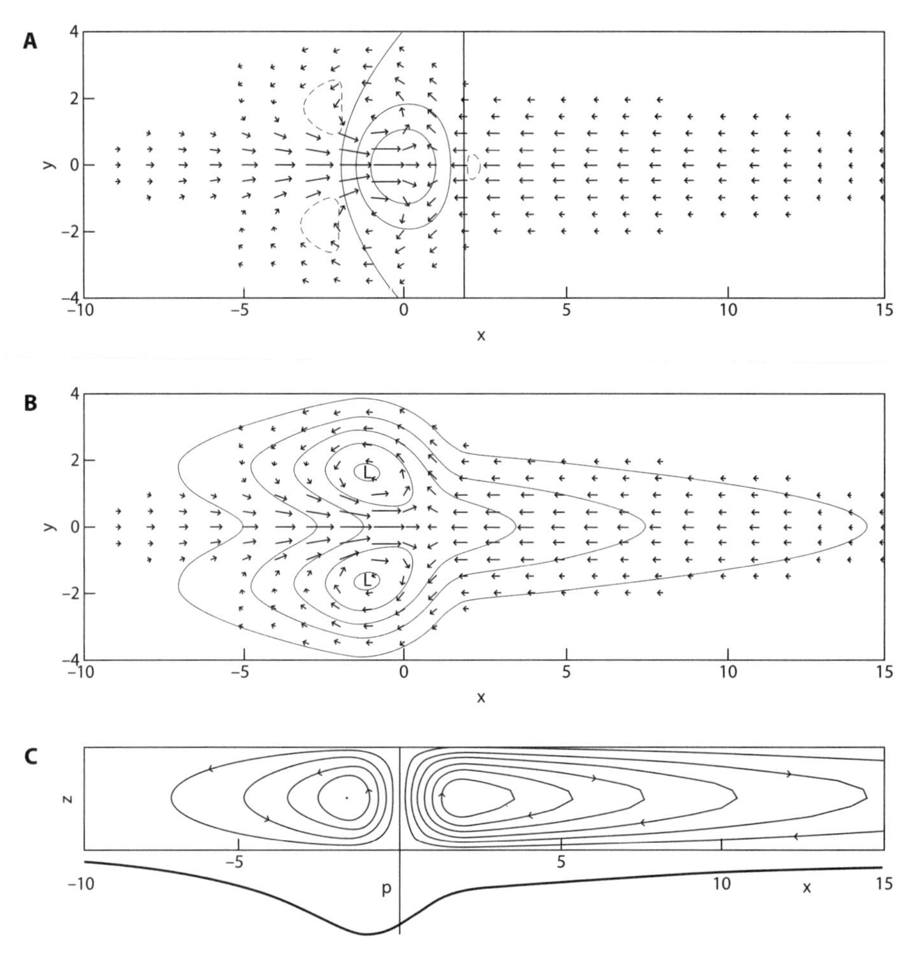

Figure 8.26. Solution of Gill's model for the case of heating symmetric about the equator. Panel A shows the heating field and the low-level wind field. Panel B shows the perturbation pressure field, which features low pressure along the equator generally, with twin cyclones slightly off the equator. Panel C shows the implied vertical motion and the zonal variation of the pressure along the equator. (Panel C is very famous.) From Gill (1980). © Quarterly Journal of Royal Meteorological Society.

Monsoons

Monsoons occur in many parts of the world. In the older literature, a *monsoon* is defined in terms of a dramatic seasonal reversal of the low-level prevailing winds (Lighthill and Pearce, 1981). From a more modern perspective, a monsoon can be viewed as a thermally forced stationary eddy associated with land-sea contrast. Comprehensive overviews are given by Webster et al. (1998) and Chang et al. (2011).

The most spectacular monsoon on Earth is the one associated with Asia, Earth's largest continent. The Asian monsoon can be defined in a number of ways. The winds near the surface reverse from the northeast in winter to the southwest in summer, as seen in figure 8.29. The 15 m s^{-1} low-level southwest wind that crosses the equator and flows from the coast of east Africa to the shores of India is known as the *Somali jet*. It

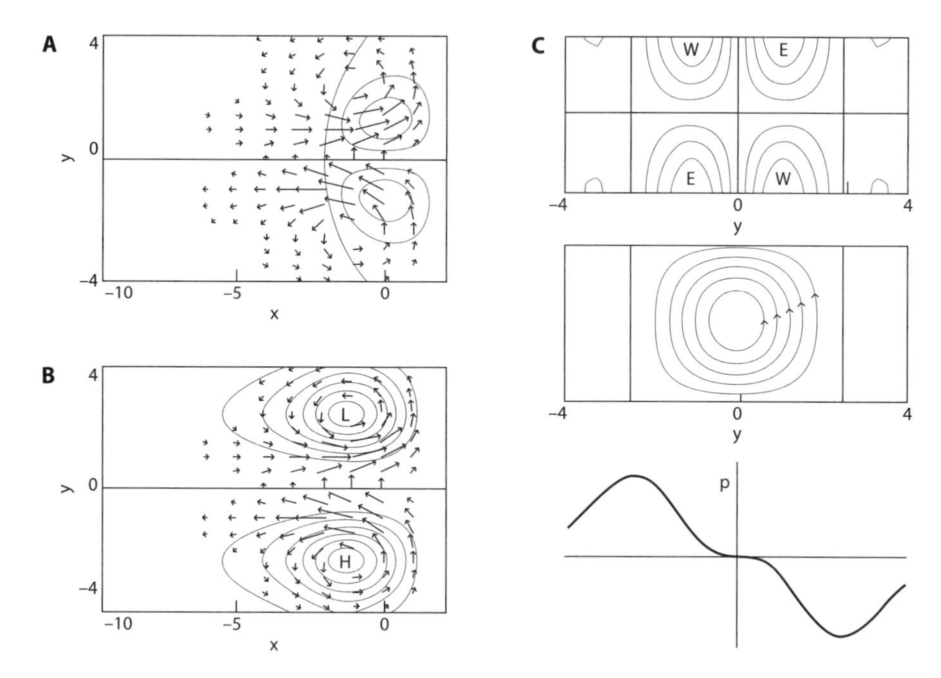

Figure 8.27. The response to antisymmetric heating. On the left side, panel A shows contours of the mid-level vertical velocity superimposed on the horizontal wind vectors for the lower layer. Panel B shows contours of the perturbation surface pressure, again with the lower-layer horizontal wind field superimposed. The right-hand panels show the zonally integrated solution corresponding to the results in the left-hand panels. The upper right-hand panel shows the latitude-height distributions of the zonal velocity and the streamfunction of the mean meridional circulation, as well as the meridional profile of the surface pressure. From Gill (1980). © Quarterly Journal of Royal Meteorological Society.

is one of the strongest low-level jets in the world. There are other seasonal changes of wind in the world, but none has the geographic scope or the socioeconomic impact of the Asian monsoon.

The basic trigger for the monsoon is a contrast between the temperature of the Asian continent and that of the surrounding ocean. The thermal anomaly in the middle troposphere is enhanced by the spectacular topography of the Tibetan Plateau (see fig. 8.30), which extends upward to about the 500 hPa level. Much of the "surface" heating associated with the summer monsoon actually occurs in the middle troposphere, because it is located on the Tibetan Plateau, which towers above the surrounding land surface, with average elevations over the central plateau of over 3000 m. The observed JJA-mean 500 hPa temperature for the monsoon region is shown in figure 8.31. An "island" of warm air is centered over the Tibetan Plateau. The warm air is in contact with the mountainous terrain, which is heated by the sun.

Naturally, the lowest surface air temperatures in the monsoon region occur at the highest elevations. There is a large area of low (less than 280 K) surface air temperature on the Tibetan Plateau, and a large area of high (greater than 305 K) surface air temperatures on the Arabian Peninsula. After the seasonal snow on the plateau has melted in late spring and early summer, the surface and the air above it are heated to a temperature higher than that of the surrounding atmosphere. Rising motion balances this heating, forcing convergence in the lower and middle troposphere and

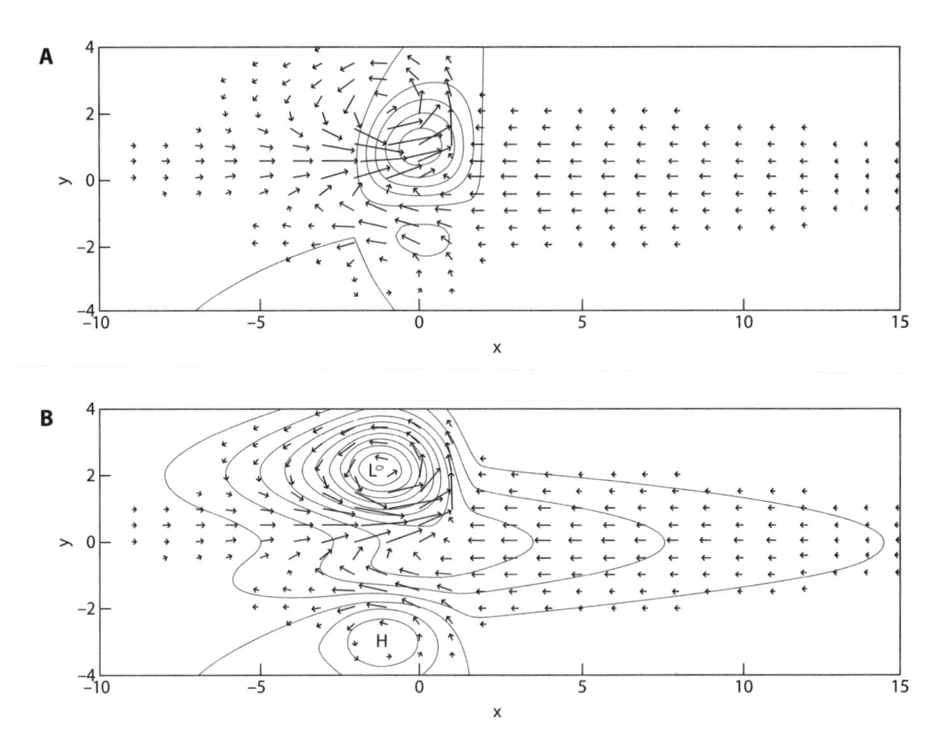

Figure 8.28. The response of Gill's model to a combination of symmetric and antisymmetric heating. (From Gill [1980].) Panel A shows the heating, and panel B shows the surface pressure. The low-level winds are shown in both panels.

compensating divergence aloft. As the seasonal heating builds, a trough forms over southern India in late May and subsequently moves north and west. The monsoon trough is one of the most prominent signatures of the Asian monsoon. It can extend across the Indian Ocean, the Indian subcontinent, and the Arabian Sea, from Bangladesh to the Arabian Peninsula. The sea-level pressure field is dominated by the monsoon trough and an area of high pressure over the Tibetan Plateau.

Because of the agricultural importance of monsoon precipitation, and because of the vast populations living in Asia, the onset of the Asian summer monsoon is one of the most anticipated events in the world. The onset can be defined as the beginning of consistent rainfall of the monsoon season. Figure 8.32 shows typical dates of onset. Cloudiness and precipitation usually begin to increase at the southern tip of India in late May. Nearly the entire subcontinent is receiving rainfall by the end of June. The onset of the summer monsoon brings cooler surface temperatures to India and other areas that receive monsoonal precipitation, owing to the increase in clouds as well as the increase in soil moisture that accompanies the precipitation.

The monsoon trough moves northward episodically, as shown in figure 8.33. In some cases, the progression of the trough is expedited by the passage of a tropical or extratropical cyclone to the north (Mooley and Shukla, 1987). The Ganges valley receives copious amounts of rain from monsoon depressions that form in the Bay of Bengal and propagate northward and westward. An example is shown in figure 8.33A. Many areas within the monsoon region also receive significant amounts of precipitation from tropical cyclones. Figure 8.33B shows the daily precipitation totals

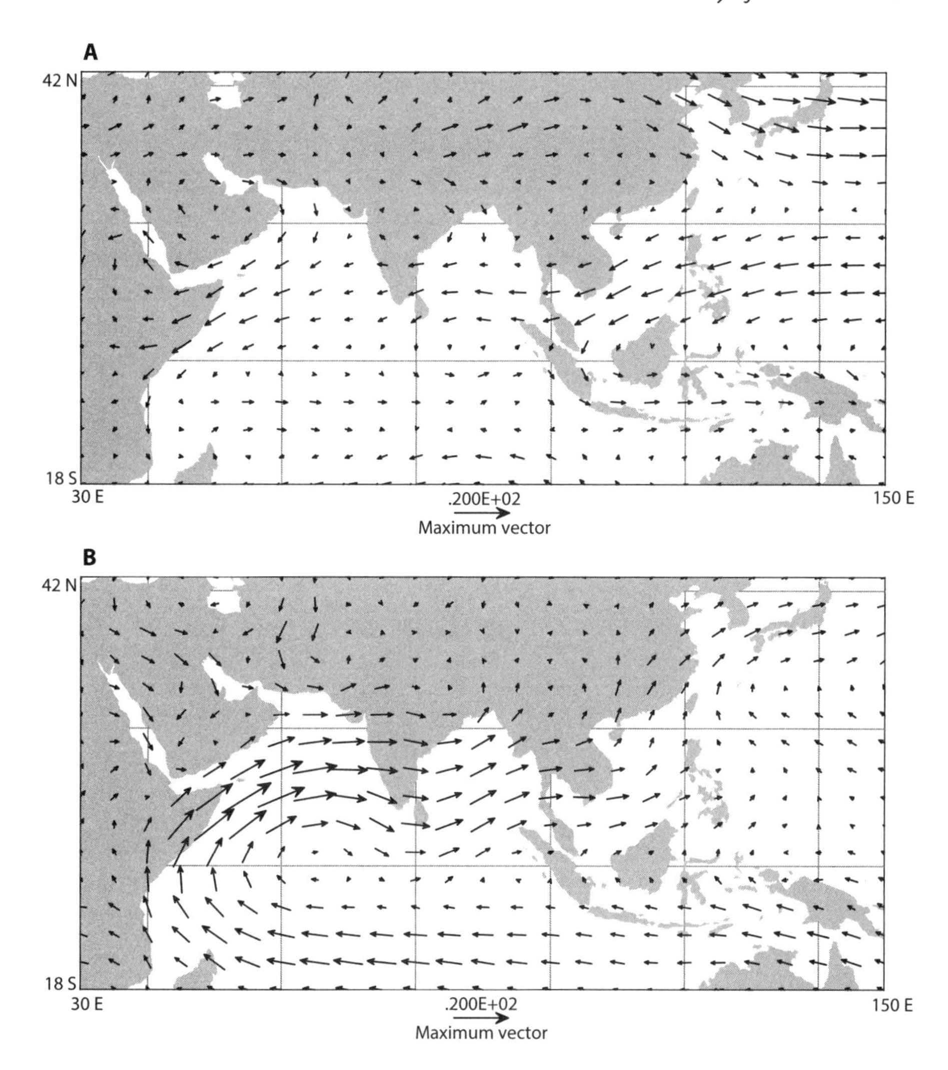

Figure 8.29. Observed 850 hPa wind vectors for (A) January and (B) July.

at an average of several stations on the southwest coast of India and gives a sense of the observed intraseasonal variability of monsoon precipitation. Much of the precipitation of the Asian monsoon is forced by southwest winds flowing over the western shores of India and Southeast Asia, as well as the foothills of the Tibetan Plateau (Johnson and Houze, 1987).

Figure 8.34 shows summer-season means of the precipitation across the monsoon region. There are two major precipitation maxima, one west of the southwest coast of India, and the other west of southern Myanmar. Both these areas receive strong onshore flow, indicating that the ocean is the source of the moisture. Minima occur near Sri Lanka and the east coast of Vietnam. These areas appear to be in orographic rain shadows. The northern and western parts of the monsoon region are quite dry, receiving less than 2 mm day^{-1} of rain. The observed 850 hPa wind analysis shows that these areas do not receive much moisture from the Indian Ocean during JJA. Only a small portion of the monsoon region receives more than 20 mm day^{-1} of precipitation.

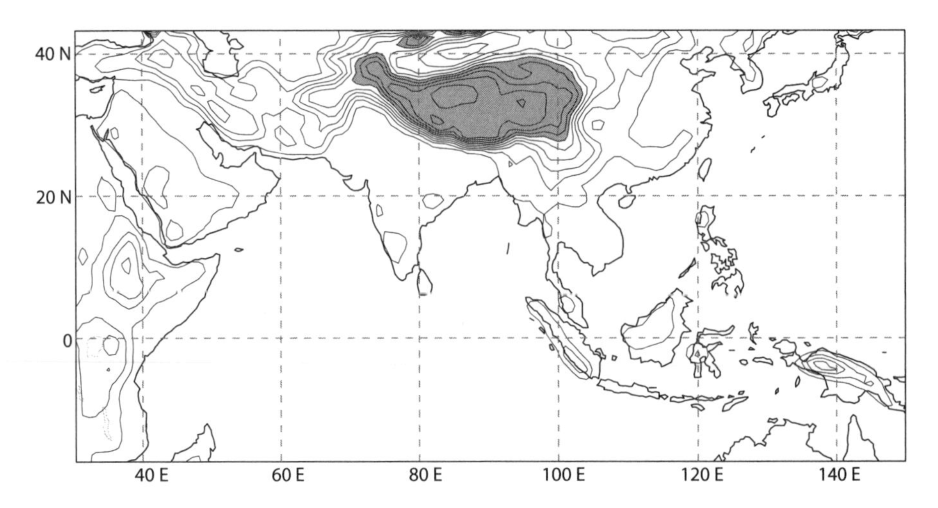

Figure 8.30. Average elevation of the monsoon region. Data were averaged to 1° × 1°, and then 9-point smoothed. Terrain over 3000 m high is shaded.

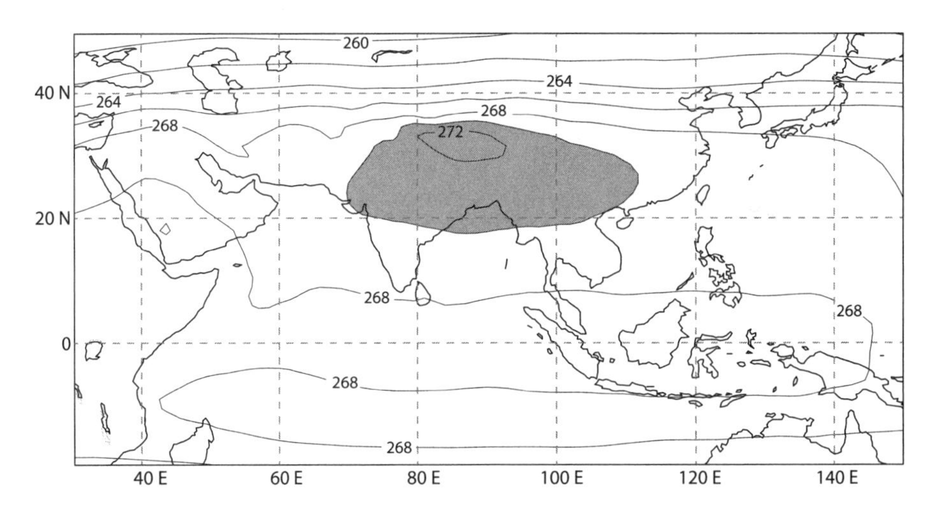

Figure 8.31. Observed JJA climatological 500 hPa temperatures. The contour interval is 2 K; values greater than 270 K are shaded.

The precipitation in much of the monsoon region varies on many timescales, including the diurnal cycle (Johnson, 2011). The individual disturbances that cause the precipitation associated with the Asian monsoon last only a few days at any single location. However, there are also prominent variations known as "breaks," which occur at periods of approximately 10–20 days and 40–50 days (Webster, 1987). The 10–20 day variation is related to the periodic northward propagation of the intertropical convergence zone (ITCZ), which begins near the equator and progresses to the foothills of the Tibetan Plateau in roughly 15 days, as seen in figure 8.35. The ITCZ usually reforms in the south after it has progressed to the foothills of the Tibetan Plateau, but occasionally it stays near the plateau for an "extended break" period of 40–50 days. These

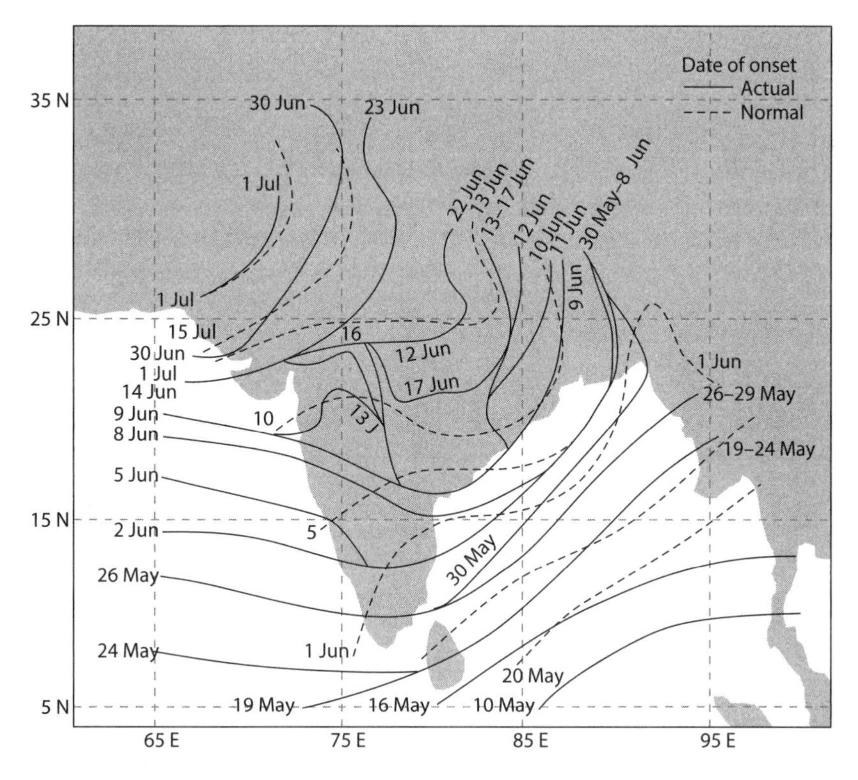

Figure 8.32. Dates of onset of the Asian monsoon near India in 1988 (actual) and mean (normal). From Krishnamurti et al. (1990). Used with kind permission from Springer Science and Business Media.

extended break periods have been linked (Webster, 1987) to the oscillation discovered by Madden and Julian (1972), which is discussed later in this chapter.

The heating of the middle troposphere by the Tibetan Plateau induces convergence in the middle and lower troposphere (Yanai, et al., 1992), largely through the Somali jet. To balance the convergence at the lower levels of the atmosphere, there must be large-scale rising motion, and divergence aloft. The observed JJA mean 500 hPa vertical velocity for the monsoon region is shown in figure 8.36. The strongest areas of rising motion are over southwestern India and the Bay of Bengal. These areas both receive copious amounts of precipitation in JJA. Two of the strongest areas of sinking motion are over the eastern Mediterranean Sea and northern China. Both areas are quite dry in JJA, receiving less than 2 mm day^{-1} of rain. The area-mean vertical velocity is upward, at the rate of about -10 hPa day^{-1}.

The upper-level divergence is associated with a broad, strong anticyclonic circulation over the plateau at 200 hPa, as shown in figure 8.37. The strong upper-level easterlies are consistent with the thermal wind relationship, because the lower and middle tropospheric temperatures actually increase toward the north in the Northern Hemisphere (Yanai et al., 1992; Murakami, 1987a; Yanai and Li, 1993). There is a slight northerly component to the 200 hPa winds at the equator, especially on the eastern side of the region. The winds shift abruptly from easterlies to westerlies at the northern and southern fringes of the monsoon region, with westerlies of up to

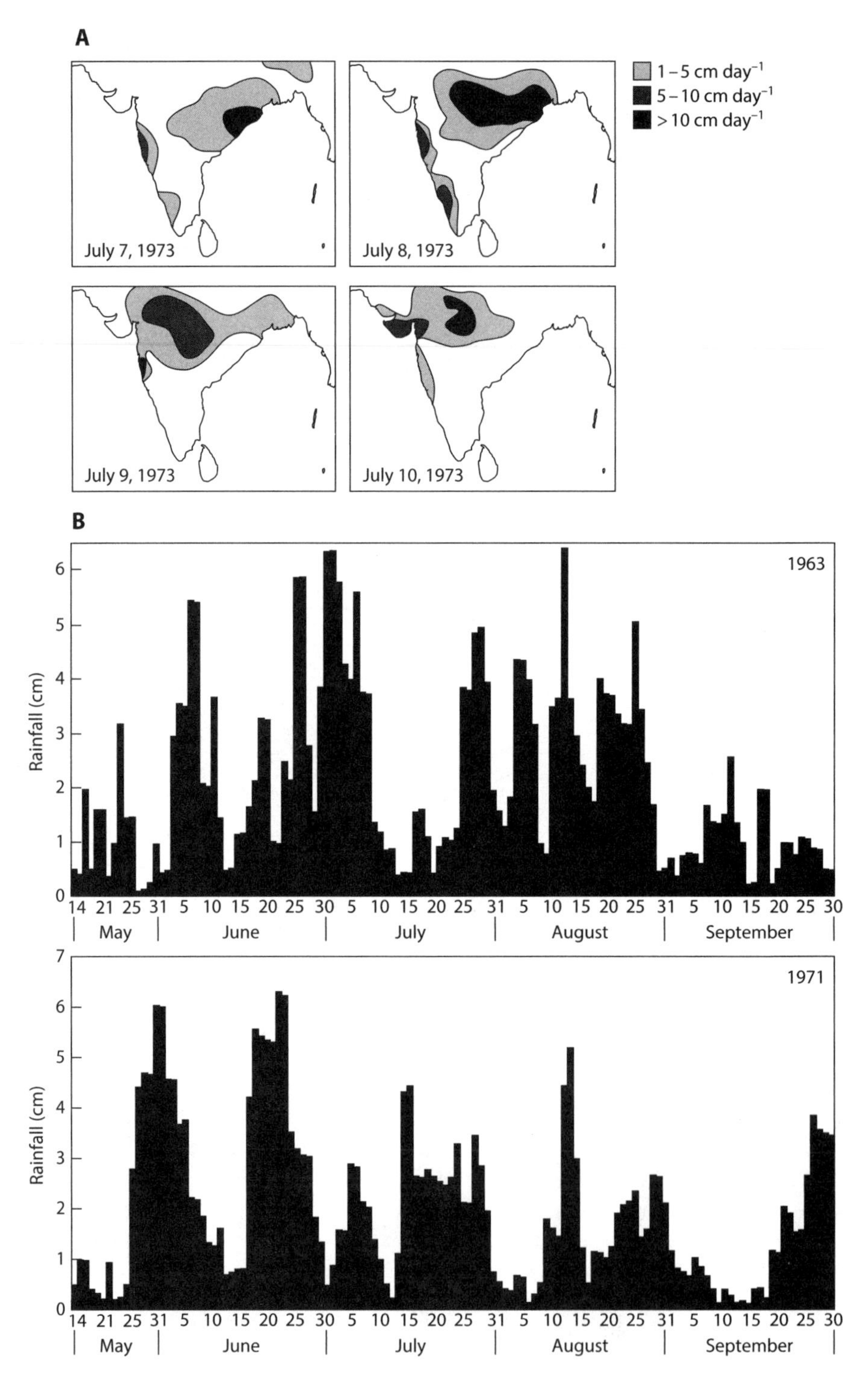

Figure 8.33. (A) An example of the progress of a monsoon depression across India. Many such depressions occur throughout the summer monsoon. (B) Observed daily rainfall along the southwest coast of India for the summer monsoon seasons of 1963 and 1971. From Webster (1987).

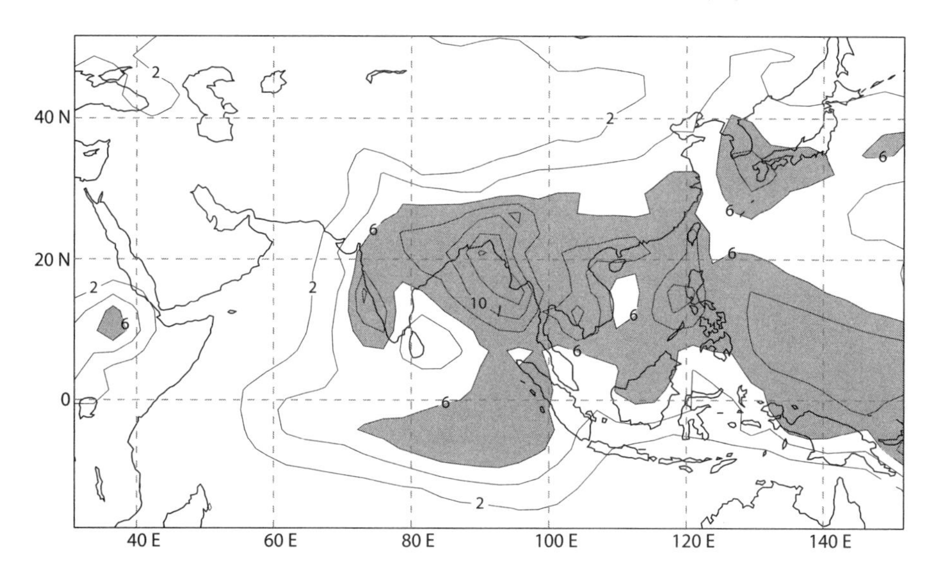

Figure 8.34. The observed JJA climatological precipitation from the Global Precipitation Climatology Project. The contour interval is 2 mm day^{-1}; values greater than 6 mm day^{-1} are shaded.

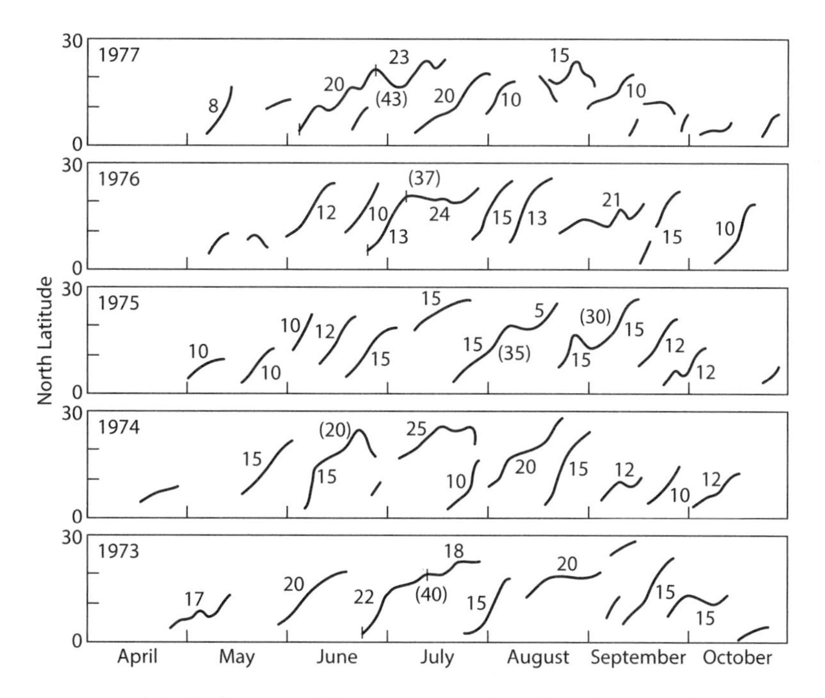

Figure 8.35. Mean latitudinal position of the monsoon trough in the Indian Ocean for the summers of 1973–1977, as obtained from the maximum cloudiness zone and the 700 hPa trough. Numbers refer to longevity of a particular cloudiness zone, with extended break periods indicated by parentheses. From Webster (1987), Webster (1983), and Sikka and Gadgil (1980).

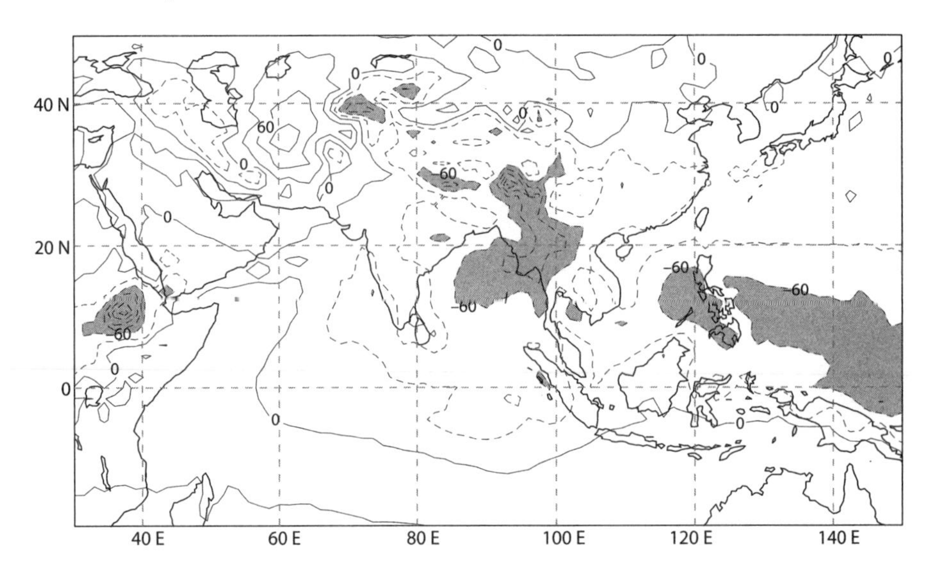

Figure 8.36. Observed JJA climatological 500 hPa vertical velocity over the Asian monsoon region. Contour interval is 20 hPa day^{-1}. Areas with vertical velocities more negative than–60 hPa day^{-1} are shaded.

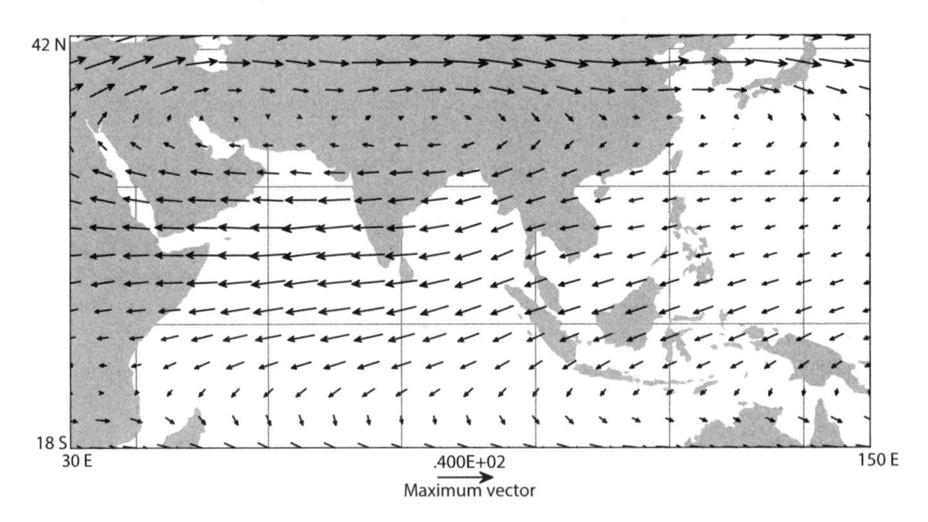

Figure 8.37. JJA climatological 200 hPa winds. The scale vector is 50 m s^{-1}.

30 m s^{-1} at about 35° N. Figure 8.38 shows a latitude-pressure cross section of the observed zonal wind at 77.5° E.

These observations show that the monsoon can be interpreted as a locally enhanced Hadley circulation (Webster, 1987) with a meridional low-level branch that moves toward a warm, convectively active ascending branch, and an upper-level return flow that feeds a relatively cool descending branch. In fact, in the northern summer the Hadley circulation is more or less contained within the longitudes of the monsoon region. The monsoon is thus a direct circulation that converts potential energy to kinetic energy.

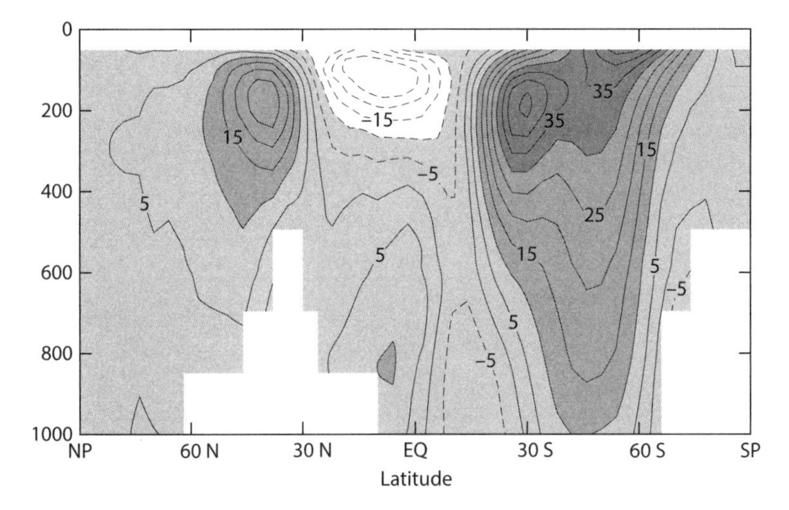

Figure 8.38. Latitude-pressure plot of the northern summer-season climatological zonal winds at 77.5° E. The contour interval is 5 m s^{-1}. Areas of high terrain are blocked out.

Observations show strong interannual variability of the Asian monsoon. Charney and Shukla (1981) argued on the basis of modeling studies that such variations are due to changes in the "boundary conditions," such as the sea-surface temperature (SST). They speculated that if these boundary conditions could be specified without error, then seasonal predictions of quantities such as monthly-mean monsoon precipitation would be possible. Later work has strongly supported this idea. Dry Asian summer monsoons have been statistically correlated to anomalously warm SSTs in the eastern Pacific (El Niño), and wet monsoons have been correlated to anomalously cold SSTs in the eastern Pacific (Philander, 1990). The interannual variability of monsoons has also been linked to variations in Tibetan snow cover (Yanai and Li, 1994; Barnett et al., 1989).

The Walker Circulation

Recall from introductory dynamics that friction tends to make the wind near the surface depart from geostrophic balance and flow down the pressure gradient, that is, toward low pressure. The mechanism is illustrated in figure 8.39. Because the air turns toward low pressure, surface lows tend to be regions of low-level convergence, and surface highs tend to be regions of low-level divergence. Of course, convergence near the surface must be balanced by divergence aloft, and vice versa. In this way, surface friction can influence the upper-tropospheric winds.

The surface winds can create ocean currents by pushing on the water. The currents are typically very slow (centimeters per second) compared with the usual near-surface wind speeds, so for practical purposes the oceans can be considered to be at rest when air-sea momentum exchanges are considered. As discussed in chapter 2, the property of the oceans that most directly affects the atmosphere is the sea-surface temperature, which affects the upward longwave radiation, the sensible heat flux, and the latent heat flux.

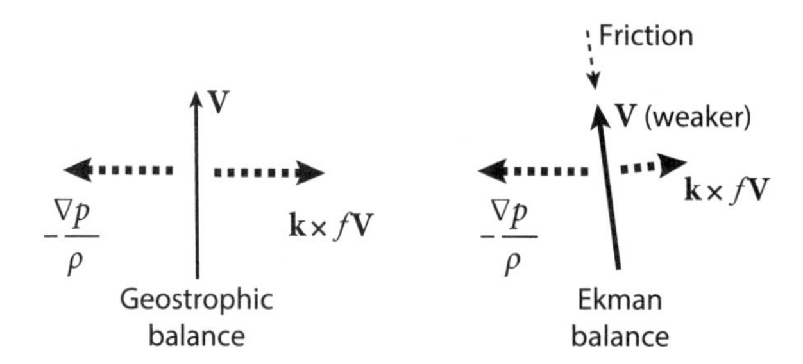

Figure 8.39. Momentum balance without friction (left) and with friction (right). Dashed lines represent forces. Without friction there is geostrophic balance, and with friction there is Ekman balance. The effect of friction is to make the wind turn toward low pressure.

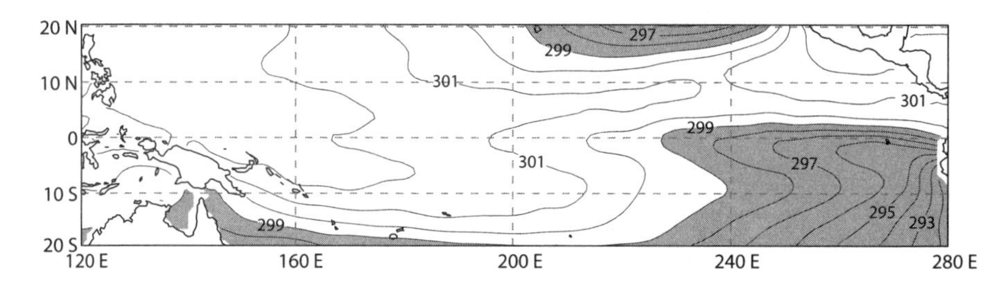

Figure 8.40. Climatological distribution of sea-surface temperature in the tropical Pacific Ocean. Note the cold water along the equator and throughout most of the eastern side of the ocean basin, and the warm water on the western side of the basin and stretching across the basin north of the equator.

The immediate effect of the surface wind stress is to push the ocean water in the same direction as the near-surface wind. The Coriolis acceleration then turns the current toward the right (relative to the wind) in the Northern Hemisphere, and toward the left (relative to the wind) in the Southern Hemisphere. This deviation has interesting and important consequences for the ocean circulation. In particular, the east-to-west trade winds along the equator bring about surface currents away from the equator in both hemispheres (i.e., toward the right of the wind north of the equator, and toward the left of the wind south of the equator), thus driving upwelling along the equator, which causes the surface waters to be colder along the equator than they are on either side. The equatorward winds along the west coasts of the continents drive equatorward currents, and also upwelling. Both the directions of the currents (from the poles) and the upwelling favor cold water, which is what is observed (see fig. 8.40).

Consider a surface wind blowing parallel to a coastline in the Northern Hemisphere, with the coastline on its left side, as in the California high in July. The surface current moves to the right of the wind, that is, away from the coast, driving coastal upwelling. The result is cool surface water. Such upwelling occurs near each of the subtropical highs, in both hemispheres. This is one reason why the subtropical highs tend to occur over cool water.

Pushing by the trade winds drives a westward current in the upper ocean, which causes warm water to pile up on the western side of each basin. The highly concentrated west-to-east return flow, very close to the equator and slightly below the surface, is called the *Equatorial Undercurrent*.

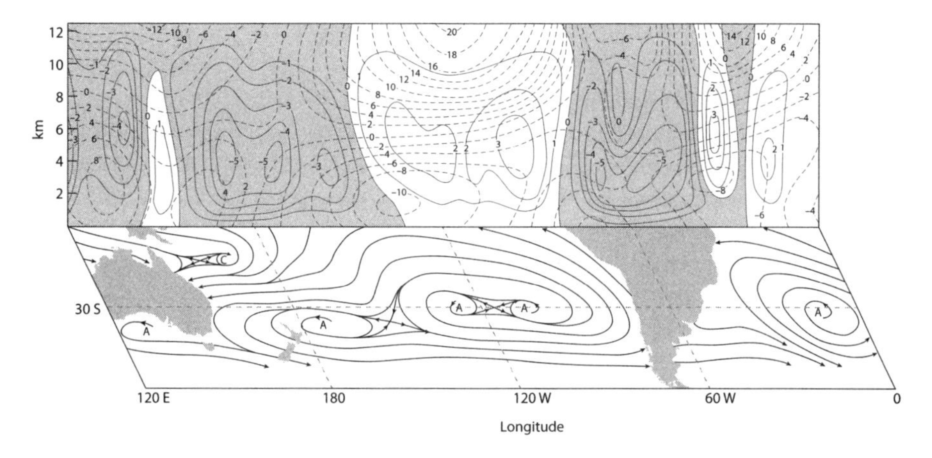

Figure 8.41. The Walker and Hadley circulations. The foreground shows the streamlines of the lower-tropospheric flow for January. The vertical cross section in the background shows the vertical motion (solid lines) and the zonal wind component at 5° S (dashed lines). From Philander (1990). Reprinted with permission from Elsevier.

The prevailing westerlies in middle latitudes induce an equatorward drift superimposed on a general eastward current in the upper ocean, and the prevailing easterlies in the tropics tend to cancel out this equatorward drift while driving the surface currents back towards the west. The circulation of the upper oceans thus takes the form of a pair of huge *gyres*, one in each hemisphere. The poleward currents, such as the Gulf Stream and Kuroshio, tend to be warmer than average at a given latitude, and the equatorward currents, such as the California Current and the Humboldt Current, tend to be cooler than average at a given latitude.

The *Walker circulation*, named by Bjerknes (1966), is an east-west overturning of the atmosphere above the tropical Pacific Ocean, with rising motion on the west side, over the so-called warm pool, and sinking motion on the east side. The Walker circulation can be viewed as a thermally excited stationary eddy. Although the Walker circulation is driven by the east-to-west SST gradient, it also helps maintain that gradient through mechanisms to be discussed later. For this reason, the Walker circulation is best understood as a coupled ocean-atmosphere phenomenon. It undergoes strong interannual variability. Figure 8.41 is a schematic illustration of the Walker circulation and its relation to the surface wind field in the Southern Hemisphere, taken from the book by Philander (1990). The equatorward flow just west of South America can be viewed as the inflow to the ITCZ (which is generally north of the equator in this region), and so it is in a sense a portion of the lower branch of the Hadley circulation.

The Hadley circulation is defined in terms of zonal averages, so a particle participating in the Hadley circulation through motions in the latitude-height plane cannot "escape" by moving to a different longitude. In contrast, a particle participating in the east-west Walker circulation can escape by moving to a different latitude; in fact, such meridional escapes are to be expected in view of the strong meridional motions associated with the Hadley circulation. For this reason, we should not think of the Walker circulation as a closed "racetrack." We can consider the Hadley and Walker circulations as closely linked. For example, a parcel may travel westward across the tropical Pacific in the lower branch of the Walker circulation, ascend to the tropopause over the pool, and then move both poleward and eastward away from the warm pool, possibly descending in the subtropical eastern Pacific. It can then join the trades and repeat its westward and equatorward journey through the boundary layer.

Bjerknes (1969) theorized that the cool, dry air of the trade winds is heated and moistened as it moves westward until it finally undergoes large-scale moist-adiabatic ascent over the warm pool. If there were no mass exchange with adjacent latitudes, a simple circulation would develop in which the flow would be easterly at low levels and westerly at upper levels. When meridional mass exchange is considered, this simple picture has to be altered, because absolute angular momentum is exported to adjacent latitudes. Under steady-state conditions, the flux divergence of angular momentum at the equator must be balanced by an easterly surface wind stress. Thus, surface easterlies on the equator are stronger than those imposed by the Walker circulation. The net result is that a thermally driven Walker cell is imposed on a background of easterly flow, the intensity of which depends on the strength of the angular momentum flux divergence.

Figure 8.42 shows the observed longitude-height cross section of the zonal wind, for January. Figure 8.43 shows that the 1000 hPa winds above the tropical Pacific (between 10° N and 10° S) have an easterly component in both solstitial seasons. Easterly flow occurs along and near the equator, west of about 90° W. In January, the easterly component is particularly strong above the central equatorial Pacific, and convergence is evident along the ITCZ near 8° N. In July, there is strong cross-equatorial flow in the eastern Pacific, and the convergence zone has moved to about 10° N. At the latitude of the ITCZ, the trade-wind flow extends to the east of Central America during both seasons.

Lindzen and Nigam (1987) used a linear model to show that SST gradients are capable of forcing low-level winds and convergence in the tropics. They assumed an Ekman balance, as illustrated in figure 8.39. Linearizing about a state of rest, they found a pressure field that qualitatively resembled the observations, although the wind speeds were unrealistically strong. Neelin et al. (1998) showed that the model used by Lindzen and Nigam is very similar to that of Gill (1980).

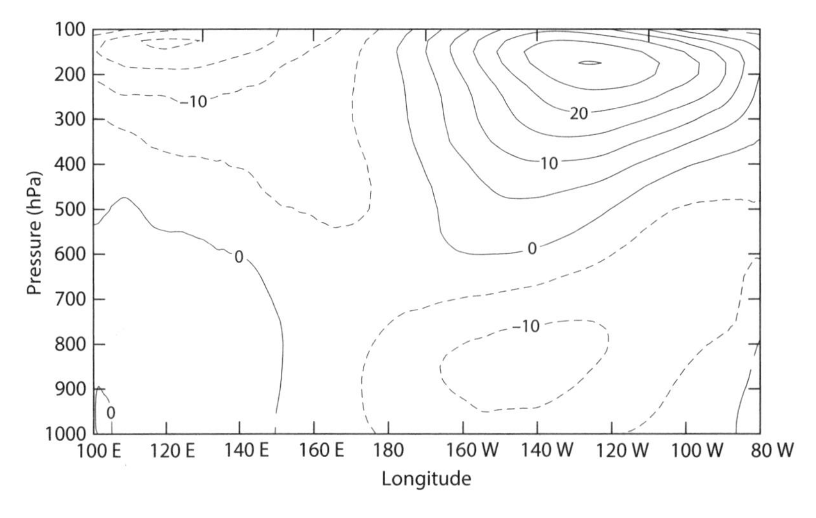

Figure 8.42. The longitude-height cross sections of the zonal wind along the equator, for January, as analyzed by ECMWF. The contour interval is 5 m s^{-1}.

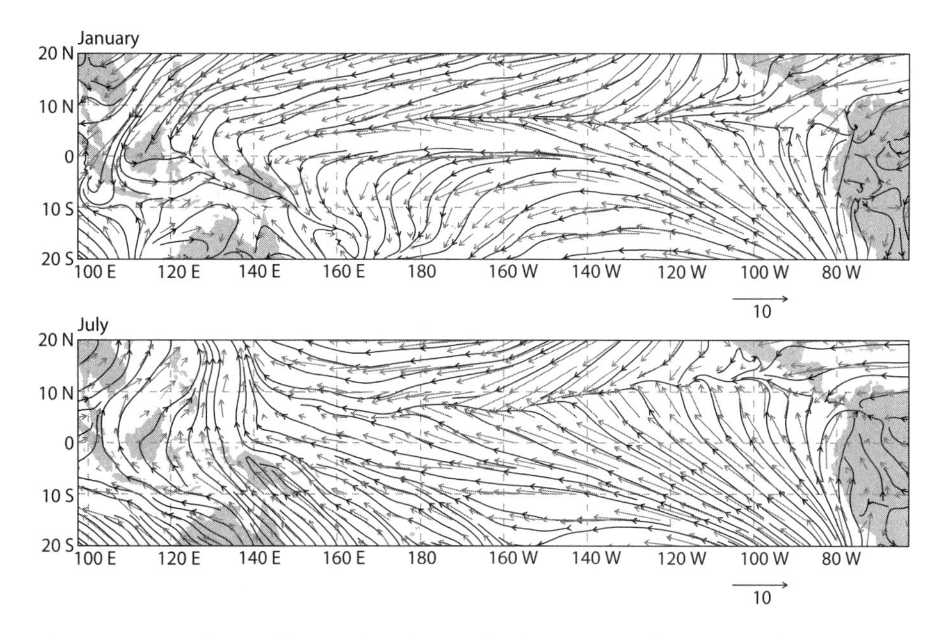

Figure 8.43. Streamlines and horizontal wind vectors for the tropical Pacific at 1000 hPa for January 1989 and July 1989. The units are m s^{-1}.

Newell et al. (1996; hereafter N96) compared water-vapor data from the Upper Atmosphere Research Satellite (UARS) with upper-air wind data from the ECMWF reanalysis dataset to deduce horizontal and vertical motions in the tropical atmosphere. Their results indicate regions of strong ascending motion over the western Pacific warm pool and the South Pacific convergence zone. The main regions of sinking motion, which are located off South America and extend westward to the date line just south of the equator, exhibit little seasonal movement. For comparison, figure 8.44 shows the vertical velocity fields at 300 hPa for January and July. The South Pacific convergence zone is clearly evident in the January data, with a large region of ascending motion that extends southeastward from 145° E to 160° W. Sinking motion straddles the equator and extends eastward from 160° E. During the northern summer, the intertropical convergence zone is well developed north of the equator. The general pattern is one in which ascending motion dominates over the tropical western Pacific, while sinking motion occurs over the tropical central and eastern Pacific. Easterlies extend across the equatorial Pacific from South America to 170° W. West of 160° E, the low-level winds are very weak along the equator, but easterlies span the Pacific at 5° S and 5° N.

Figure 8.45 completes the picture, showing the upper branch of the Walker circulation. West of the date line, the zonal winds over the equator are easterly. Upper-level westerly flow occurs to the east of the rising motion. In the Northern Hemisphere, weak westerly flow appears between 170° W and 140° W poleward of 15° N. In the Southern Hemisphere, a westerly component of the wind exists south of 5° S to the east of the date line. An interpretation is that the Walker circulation has migrated into the Southern Hemisphere. A reexamination of figure 8.44B indicates that sinking motion is confined mainly to the Southern Hemisphere and extends as far west as 165° E.

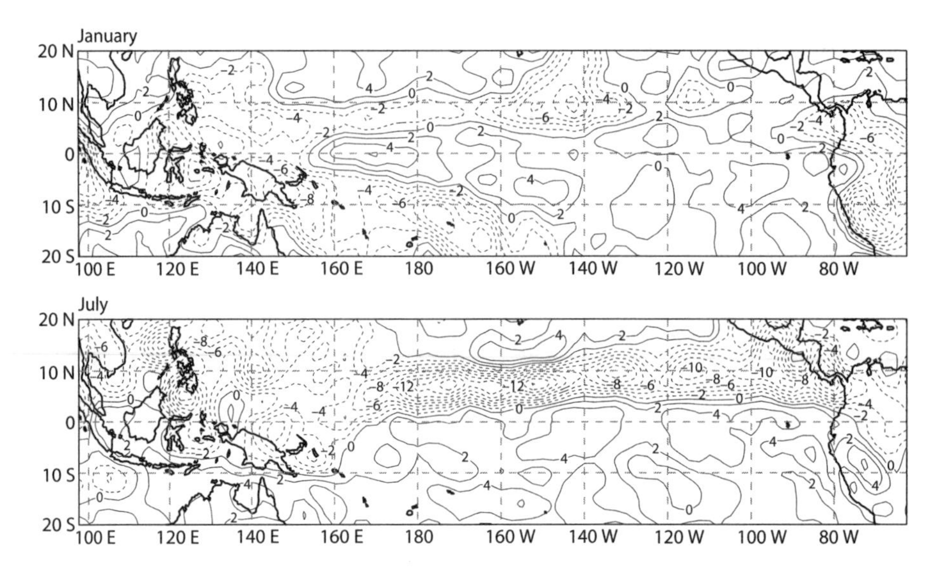

Figure 8.44. Contour plot of mean vertical velocity at 300 hPa (units are 10^{-2} Pa s^{-1}) from the ECMWF reanalysis dataset for January and July 1989. The contour interval is 2×10^{-2} Pa s^{-1}; negative contours are dashed.

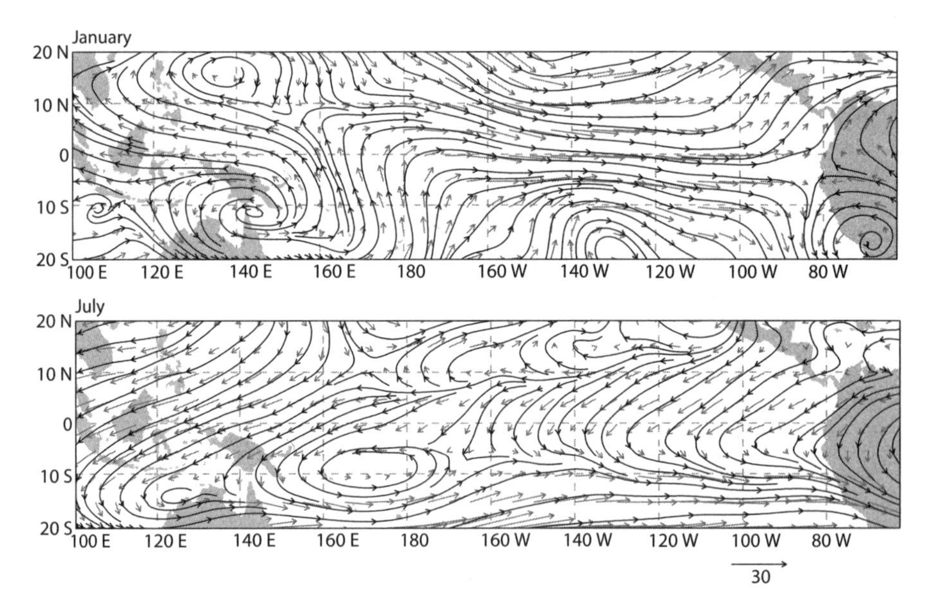

Figure 8.45. Streamlines and horizontal wind vectors for the tropical Pacific at 200 hPa for January and July.

The Walker circulation is closely tied to east-west SST gradients that are influenced by the Walker circulation itself, so it can be viewed as a coupled atmosphere-ocean phenomenon (Bjerknes, 1966, 1969). Figure 8.40 shows an SST maximum in the tropical western Pacific, and for this reason the region is known as the *tropical warm pool.* The "cold tongue," also seen in figure 8.40, is a band of relatively cold waters along the equator that stretches from South America westward to near 160° E. The distribution of tropical convection is influenced by the large-scale spatial pattern of the SST, rather

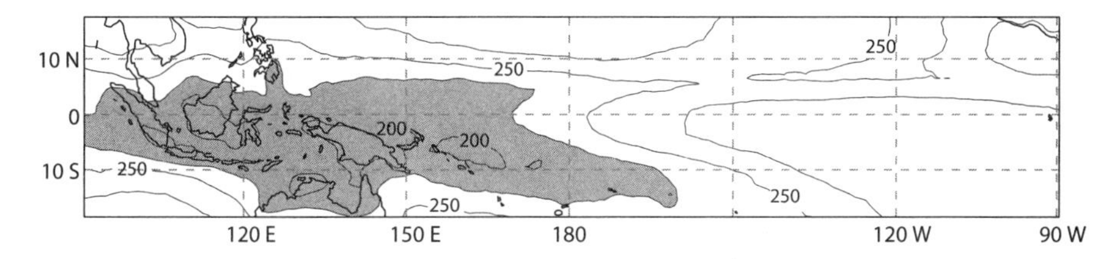

Figure 8.46. Tropical OLR for January averaged over 1985 to 1988. The contour interval is 25 W m^{-2}.

than the local SST. The tropical Pacific warm pool is a region of intense deep convection and heavy rainfall. In figure 8.46, regions in which the outgoing longwave radiation (OLR) is less than 225 W m^{-2} can be identified as areas of frequent convection (Webster 1994). The OLR threshold corresponds to a monthly-mean emission temperature of 250 K. Owing to longwave trapping by optically thick anvil clouds, which are produced by deep convection, the OLR is reduced, and threshold values of OLR can therefore be used as surrogates to infer the presence of convection. From the figure, we see that convection occurs throughout the warm pool and in the South Pacific convergence zone (SPCZ). In contrast, the OLR is generally larger than 275 W m^{-2} across the equatorial cold tongue, indicating that convection is infrequent there.

The high, cold, and sometimes bright clouds of the warm-pool region limit the radiative cooling of the atmosphere over the warm pool, but they also limit the solar warming of the ocean. Ramanathan and Collins (1991) hypothesized that cirrus clouds act as a thermostat to regulate tropical SST. They used Earth Radiation Budget Experiment (ERBE) data to deduce the interrelationships among shortwave and longwave cloud radiative forcings and radiative forcing of the clear atmosphere. They emphasized that the shortwave effects of clouds dominate the longwave effects in regulating SST. According to their hypothesis, as SST increases, the cloud albedo increases. Meanwhile, the atmosphere warms as a result of longwave cloud radiative effects, stronger latent-heat release by convection, and a stronger SST gradient over the tropical Pacific. This warming leads to an amplification of the large-scale flux convergence of moisture. The process continues until the reflectivity of the clouds increases sufficiently to cool the surface. A criticism of their study is that changes in the strength of the oceanic and atmospheric circulations were not included. Nevertheless, it is undoubtedly true that the blocking of shortwave radiation by deep cloud systems tends to limit the sea-surface temperature in the warm pool.

As seen in figure 8.47, in the eastern tropical Pacific, stratus clouds in the boundary layer intercept sunlight and strongly reduce the heat flux into the ocean below (e.g., Hartmann et al. 1992). In this way, the atmosphere helps maintain the cooler SSTs of the eastern Pacific. Stratus clouds form preferentially over cold water (Klein and Hartmann, 1993), so a positive feedback is at work here (Ma et al., 1996). Latent heat exchange between the ocean and atmosphere is influenced by the surface relative humidity and the surface winds. For fixed relative humidity and SST, the evaporative cooling of the ocean increases as the surface wind stress increases. The winds also influence the SST distribution by generating cold-water upwelling in the eastern Pacific and along the equator in the eastern and central Pacific.

The driving force behind the Walker circulation is the zonally varying heating that is balanced by zonally varying adiabatic heating/cooling due to sinking/rising motions. Over the warm waters of the western Pacific, latent heat release due to intense

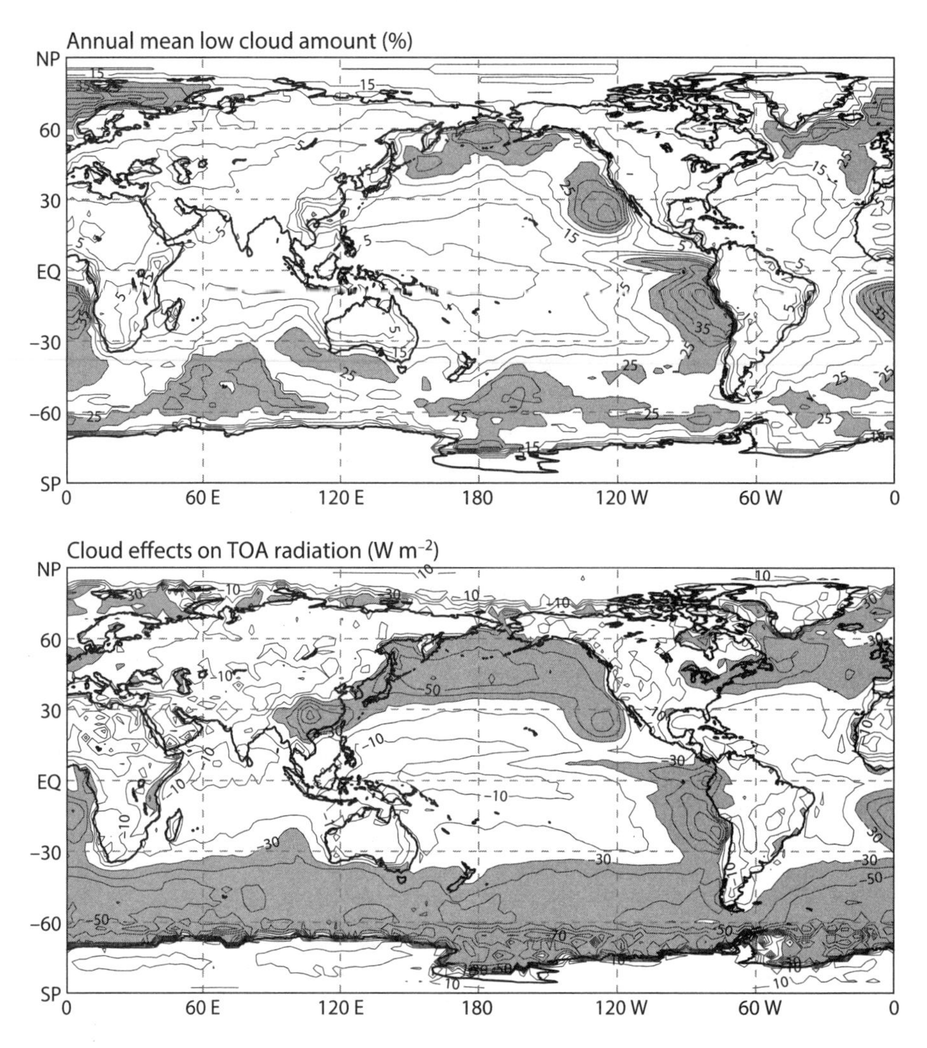

Figure 8.47. Observed annual-mean low-cloud amount (upper panel) and the net effects of clouds on the Earth's radiation budget (lower panel). The contour interval for low cloudiness is 5%; values greater than 25% are shaded. The contour interval for the lower panel is 10 W m^{-2}; values more negative than 30 W m^{-2} are shaded. Negative values in the lower panel indicate a cooling; that is, shortwave reflection dominates longwave trapping. A similar figure was published by Norris and Leovy (1994).

convection and radiative warming of the atmospheric column are balanced by the adiabatic cooling associated with rising motion (Webster 1987). Over the eastern tropical Pacific, where the SST is relatively cold, convection is infrequent, and so a balance between radiative cooling and subsidence exists.

Pierrehumbert (1995; hereafter P95) introduced an influential two-box model of the Hadley/Walker circulation. Figure 8.48 is a schematic of his "furnace/radiator-fin" model. The model includes separate energy budgets for its cold-pool and warm-pool "boxes." The SSTs of the cold pool and warm pool are assumed to be those that produce energy balance for each box of the model atmosphere and for the cold-pool ocean. Surface energy balance for the warm pool was not explicitly included in the model. A vertically and horizontally uniform lapse rate was assumed, and the

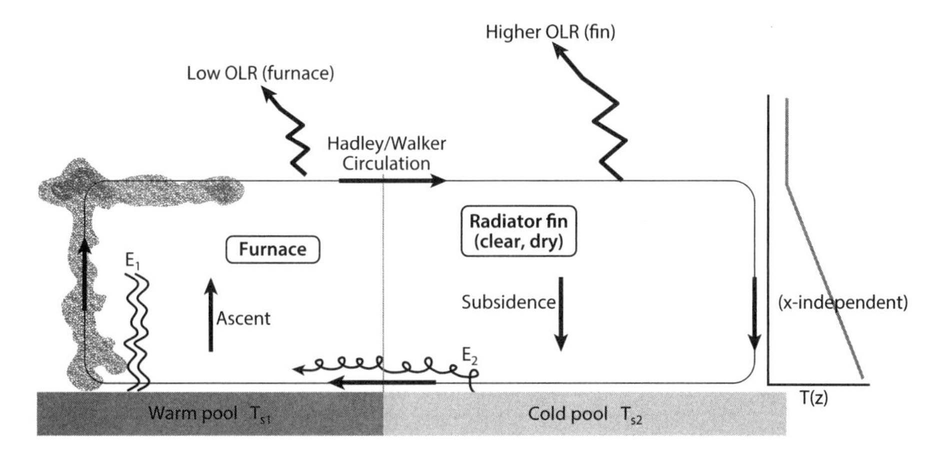

Figure 8.48. Pierrehumbert's (1995) schematic representation of the "furnace/radiator-fin" model of tropical circulation. The symbols E and TS represent the evaporation rate and SST, respectively. The subscript 1 denotes the warm pool or "furnace," and the subscript 2 denotes the cold pool or "radiator fin." © American Meteorological Association. Used with permission.

free-tropospheric temperature profile was assumed to be uniform across the tropics. The radiating temperature of the cold-pool free atmosphere was assumed to be the air temperature of the middle troposphere. The solution was obtained by first computing the net energy flux at the top of the warm-pool atmosphere for a given SST and relative humidity profile. The positive net radiative flux at top of the atmosphere over the warm pool was assumed to be balanced by horizontal energy transport to the cold pool. The cold-pool SST and radiating temperature were then computed by assuming that the net diabatic cooling must balance the energy imported from the warm pool.

The mass flux of the Hadley-Walker circulation was assumed to be that required to produce a balance between adiabatic warming by dry subsidence and the net radiative cooling of the cold-pool region. It can be shown that the horizontal heat transport by the warm-pool atmosphere is proportional to the diabatic cooling of the cold-pool atmosphere and to the ratio of cold-pool area and warm-pool area. This area ratio is a prescribed parameter of the model. P95 showed that for very small values of the cold-pool emissivity, the warm-pool SST increases without limit, because the cold pool cannot radiate enough energy to balance the energy absorbed in the warm pool. Because the warm pool controls the temperature profile, its equilibrium SST must decrease as the cold-pool radiating temperature decreases. As the cold-pool emissivity increases, the warm-pool cools off. The simulated cold-pool and warm-pool SSTs resemble the present-day climate for a range of conditions. The diagnosed mass flux is realistic.

A weakness of Pierrehumberts' model is that it fails to account for cloud-radiative effects. Miller (1997; hereafter M97) extended the model by including the radiative effects of low clouds in the cold-pool region. He constructed a three-box model with an updraft region, a warm-pool region, and a cold-pool region. Miller's model includes energy- and moisture-balance equations for the boundary layer and free troposphere, and a surface energy budget for each of three boxes. Miller assumed, following Bjerknes (1938), that the updraft region occupies a small area and that the lapse rate of the warm pool region is moist adiabatic. Miller implicitly included atmospheric dynamics by assuming a uniform free-tropospheric temperature sounding across both the warm- and cold-pool regions (Charney, 1963). Miller's main finding was that low clouds act as a thermostat that reduces the tropical SST. Although the low clouds act over the cold pool, their cooling effect extends into the warm pool.

The Madden-Julian Oscillation

The *Madden-Julian Oscillation* (MJO) can be defined as a broad region of humid air and vigorous precipitation that maintains itself as it drifts slowly eastward across the tropical Indian and western Pacific Oceans (Madden and Julian, 1971, 1972a). The MJO recurs with a period in the range of 30–60 days (see fig. 8.49) and involves many meteorological variables, including the zonal wind, surface pressure, temperature, and humidity. Observations show corresponding oscillations of various measures of cumulus activity (Murakami et al., 1986), including precipitation (Hartmann and Gross, 1988) and outgoing longwave radiation (e.g., Weickmann and Khalsa, 1990). The MJO is depicted schematically in figure 8.50, and the observed phase propagation is shown in figure 8.51. The MJO does not produce strong oscillations of convection over the Amazon basin or the Congo basin, even though an MJO signal is sometimes observed in the winds at those longitudes.

The MJO propagates eastward at about 5 m s^{-1} from the Indian Ocean to the date line, at which point it decouples from the convection and increases its phase speed to

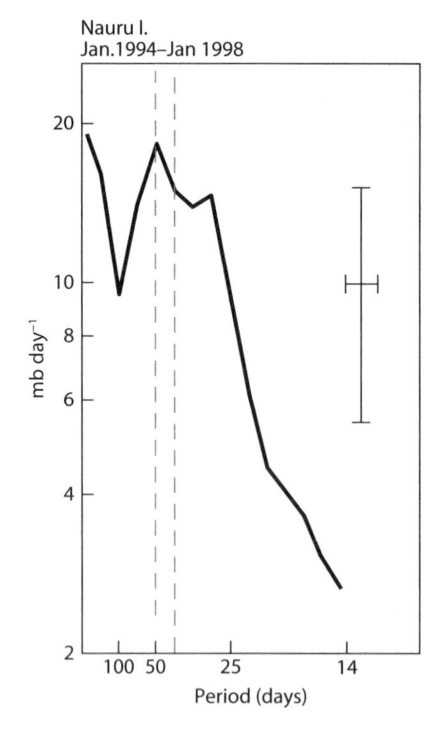

Figure 8.49. Variance spectrum for station pressures at Nauru Island, 0.4° S, 161.0° E. Ordinate (variance/frequency) is logarithmic, and abscissa (frequency) is linear. The 40–50 day range is indicated by the dashed vertical lines. Prior 95% confidence limits and the bandwidth of the analysis (0.008 day^{-1}) are indicated by the cross. From Madden and Julian (1994), based on Madden and Julian (1972a).

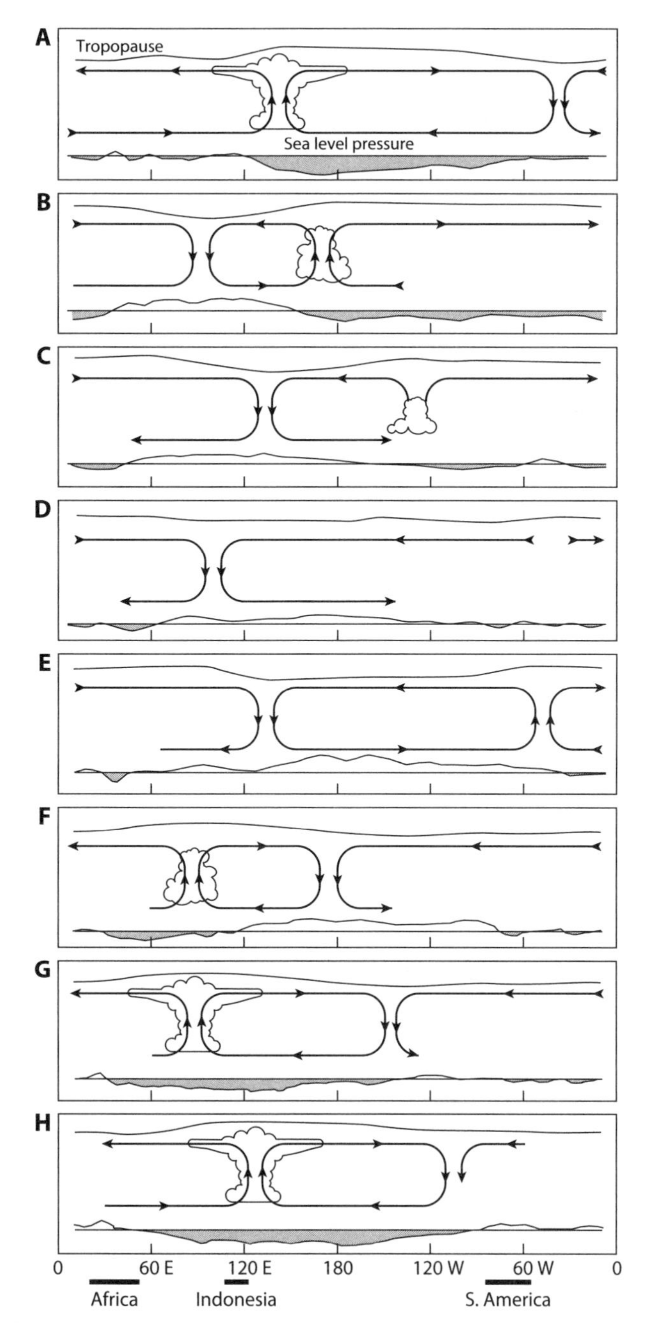

Figure 8.50. Schematic depiction of the time and space (zonal plan) variations of the disturbance associated with the 40–50 day oscillation. Dates are indicated symbolically by the letters at the left of each chart and correspond to dates associated with the oscillation in Canton's station pressure. The letter A refers to the time of low pressure at Canton, and E refers to the time of high pressure there. The other letters represent intermediate times. The mean pressure disturbance is plotted at the bottom of each chart, with negative anomalies shaded. The circulation cells are based on the mean zonal wind disturbance. Regions of enhanced large-scale convection are indicated schematically by the cumulus and cumulonimbus clouds. The relative tropopause height is indicated at the top of each chart. From Madden and Julian (1994), based on Madden and Julian (1972a). © American Meteorological Association. Used with permission.

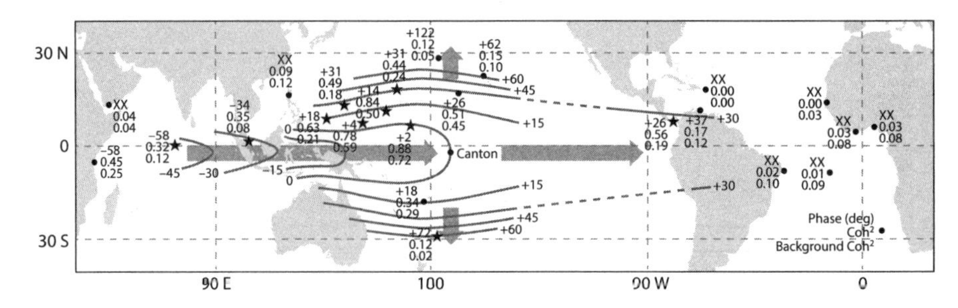

Figure 8.51. Mean phase angles (in degrees), coherence squares, and background coherence squares for approximately the 36–50 day range of cross spectra between surface pressures at all stations and those at Canton. The plotting model is given in the lower right-hand corner. Positive phase angle means Canton time series leads. Stars indicate stations where coherence squares exceed a smooth background at the 95% level. Mean coherence squares at Shemya (52.8° N, 174.1° E) and Campbell Island (52.6° S, 169.2 ° E; not shown) are 0.08 and 0.02, respectively. Both are below their average background coherence squares. Values at Dar es Salaam (0.8° S, 39.3° E) are from a cross spectrum with Nauru. The arrows indicate propagation direction. From Madden and Julian (1994), adapted from Madden and Julian (1972a). © American Meteorological Association. Used with permission.

about 12 m s^{-1}. Global circumnavigation can sometimes be seen in the upper-tropospheric winds but is harder to detect in other fields closer to the surface. Embedded within the broad convectively active region are fluctuations on smaller space- and timescales (Nakazawa, 1988; see fig. 8.52). In the convectively active phase, strong surface westerlies and high surface latent heat fluxes are observed.

As would be expected from the theoretical work of Matsuno (1966) and Gill (1980), the precipitation maximum of the MJO is accompanied by low-level winds that trace out twin cyclones on the west side of the precipitation maximum, and a zonally broader patch of easterly winds on the east side. The zonal wind field thus converges at low levels near the precipitation maximum, and diverges aloft.

The MJO has been the subject of intense research because of its intraseasonal timescale—which suggests that intraseasonal weather anomalies may be predictable—and also because of its apparent relationships with the Indian summer monsoon (Yasunari, 1979; Krishnamurti and Subrahmanyam, 1982), the likelihood of tropical Pacific storms (Gray, 1979), and the initiation of El Niño events (Lau and Chan, 1985). Since tropical convection can force Rossby waves that propagate into the extratropics, the MJO can also influence the weather in middle latitudes (e.g., Rueda, 1991).

Atmospheric *global circulation models* (GCMs) have difficulty in simulating the MJO. In addition, there is no consensus on the basic physical mechanisms that give rise to the MJO, although progress is being made. The problem of understanding the MJO can be separated into several linked parts. First, the steady motion generated by a moving heat source on the equatorial beta plane is well described by the model of Matsuno (1966) and Gill (1980), hereafter called the "MG model," and has been studied by Hendon and Salby (1994) and Schubert and Masarick (2006), among others. The relevance of the MG model to the MJO has been recognized for decades (e.g., Chao, 1987). The low-level wind pattern associated with the MJO closely resembles that shown in figure 8.25, for the case of heating centered on the equator. Given a realistic moving heat source, any GCM should be able to simulate a wind field similar to that of the observed MJO.

However, the failure of Matsuno's (1966) theory of equatorial waves to produce free modes resembling the MJO, despite the theory's success in predicting the other

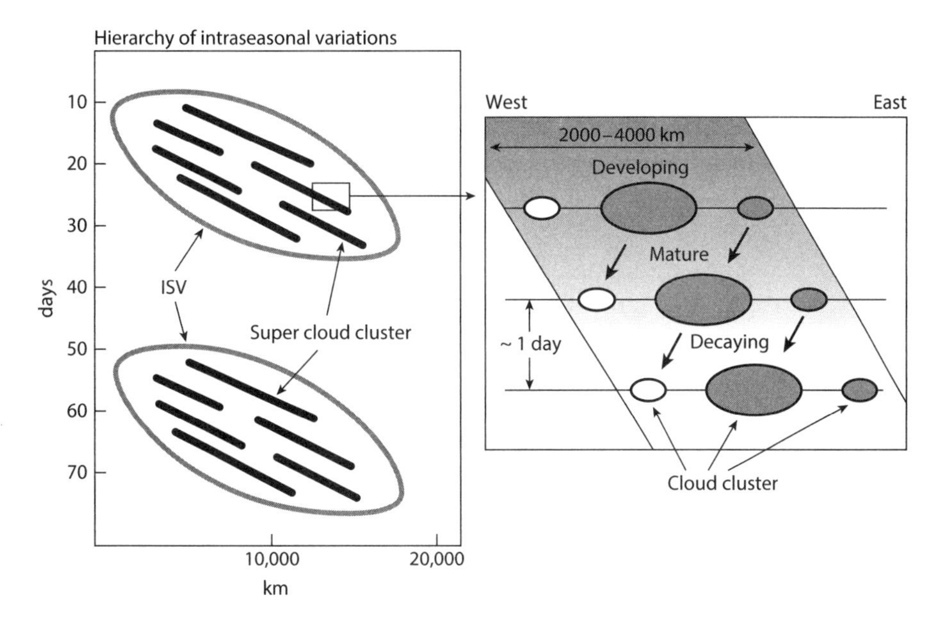

Figure 8.52. Schematic describing the details of the large-scale eastward-propagating cloud complexes (slanting ellipses marked ISV [intraseasonal variability] on the left-hand side). Slanting heavy lines represent super cloud clusters (SCC) within the larger complexes or ISV. The right-hand side illustrates the fine structure of the SCC with smaller westward-moving cloud clusters that develop, grow to maturity, and decay in a few days. From Nakazawa (1988).

observed equatorial waves (with the possible exception of the easterly waves mentioned earlier; Kiladis et al., 2009), implies that the MJO depends fundamentally on processes that were not included in Matsuno's model. It is now believed that moist processes are essential to the MJO (e.g., Raymond, 2001; Grabowski and Moncrief, 2004; Bony and Emanuel, 2005), and the term *moisture mode*, which was coined by Fuchs and Raymond (2007), is now widely used to describe the MJO (Sugiyama, 2009). The dry MG model cannot describe moisture modes.

Variations of longwave radiative heating and surface evaporation are believed be important for the MJO (e.g., Raymond, 2001; Bony and Emanuel, 2005; Andersen and Kuang, 2012; Arnold et al., 2013). GCMs should be able to simulate them, at least qualitatively, so they are probably not the missing ingredients that prevent many conventional GCMs from simulating the MJO.

The eastward propagation of the MJO is favored by moistening of the atmosphere on its east side and drying on its west side. Moisture advection is a key process in the MJO (Maloney, 2009; Maloney et al., 2010; Andersen and Kuang, 2012; Pritchard and Bretherton, 2014). Both meridional and vertical advection favor strong drying to the west of the humid, strongly precipitating core of the MJO; under such conditions, positive water vapor anomalies must move eastward or be destroyed. In modeling terms, the drying is due to resolved-scale advection rather than to parameterized processes. The precipitation rate is observed to decrease as advection dries the air on the west side of the MJO. Perhaps surprisingly, not all models simulate this process (Holloway et al., 2013). In an analysis of numerical experiments with an aqua-planet version of the SP-CAM, Andersen (2012) found that the eastward drift of the MJO speeds up if the subtropical reservoir of dry air is brought closer to the equator.

Wind-induced surface heat exchange (WISHE) theories of the MJO focus on the moistening on the east side (Emanuel, 1987a; Neelin and Yu, 1994; Yu and Neelin 1994). Neelin et al. (1987) and Emanuel hypothesized a feedback in which the maintenance and eastward propagation of convection depend on the existence of a "basic-state" easterly flow. The perturbation easterlies on the east side of the MJO reinforce the basic-state easterlies, favoring strong evaporation on the east side. Westward zonal advection can carry the evaporatively moistened air toward the convective disturbance (Sobel and Maloney, 2012, 2013). The observed specific humidity fluctuations are strongest near the 700 hPa level, however (Sherwood, 1999; Holloway and Neelin, 2009). The MJO needs a process that lifts the water vapor from the surface and moistens the air near the 700 hPa level. The *upward* transport of moisture on the east side of the MJO is due to processes that are parameterized in GCMs. Some GCMs that fail to simulate the MJO are unable to moisten the air in regions of strong precipitation (Thayer-Calder and Randall, 2009; Kim et al., 2009; Landu and Maloney, 2011; Mapes and Bacmeister, 2012; Hung et al., 2013; Kim et al., 2014).

As discussed by Thayer-Calder and Randall (2009), an essential requirement for successful simulation of the MJO is the ability to produce realistically deep layers of high relative humidity in regions of strong precipitation. Chikira (2014) presents an insightful analysis of the processes that lead to midtropospheric humidity changes in tropical convective systems. Recall from chapter 6 that the dry static energy and water vapor are governed by

$$\rho \frac{\partial \bar{s}}{\partial t} = -\rho \overline{\mathbf{V}} \cdot \nabla \bar{s} - \tilde{M} \frac{\partial \bar{s}}{\partial z} + \overline{Q_R} + \rho L \tilde{C} + D(s_c - \bar{s}), \tag{128}$$

$$\rho \frac{\partial \overline{q_v}}{\partial t} = -\rho \overline{\mathbf{V}} \cdot \nabla \overline{q_v} - \tilde{M} \frac{\partial \overline{q_v}}{\partial z} - \rho \tilde{C} + D\left[(q_v)_c - \overline{q_v}\right], \tag{129}$$

where $\tilde{M}$ is the environmental mass flux (positive upward), $\tilde{C}$ is the net environmental condensation rate, and D is the rate at which mass is detrained from the convective updrafts.

With the justification explained in chapter 3, we assume that in the tropical free atmosphere the dry static energy is independent of time and horizontal position. This assumption is used to justify the "weak temperature gradient approximation." Then, (128) becomes

$$0 \cong -\tilde{M} \frac{\partial \bar{s}}{\partial z} + \overline{Q_R} + \rho L \tilde{C} + D(s_c - \bar{s}), \tag{130}$$

which can be rearranged to

$$\tilde{M} \cong \left[\overline{Q_R} + \rho L \tilde{C} + D(s_c - \bar{s})\right] \left(\frac{\partial \bar{s}}{\partial z}\right)^{-1}. \tag{131}$$

According to (131), the environmental mass flux is whatever it takes to maintain a constant dry static energy. Assuming that $\partial \bar{s} / \partial z > 0$, the condition required for the environmental mass flux to be upward is $\overline{Q_R} + \rho L \tilde{C} + D(s_c - \bar{s}) > 0$. This constraint is relevant because an upward environmental mass flux will moisten the environment.

We can use (131) to eliminate $\tilde{M}$ in (129). The resulting equation does not contain an explicit vertical advection term. We obtain

$$\rho \frac{\partial \overline{q_v}}{\partial t} = -\rho \overline{\mathbf{V}} \cdot \nabla \overline{q_v} + \frac{\alpha}{L}\left[\overline{Q_R} + \rho L \tilde{C} + D(s_c - \bar{s})\right] - \rho \tilde{C} + D\left[(q_v)_c - \overline{q_v}\right], \tag{132}$$

where

$$\alpha \equiv -L\left(\frac{\partial \overline{q_v}}{\partial z}\right)\left(\frac{\partial \overline{s}}{\partial z}\right)^{-1}, \tag{133}$$

so that

$$1-\alpha = \left(\frac{\partial \overline{h}}{\partial z}\right)\left(\frac{\partial \overline{s}}{\partial z}\right)^{-1}. \tag{134}$$

From (133), we expect $\alpha > 0$. From (134), we see that that throughout most of the tropics α should be greater than 1 in the lower troposphere and close to zero in the upper troposphere. If the environment becomes saturated with a moist adiabatic lapse rate, then $\alpha = 1$.

By combining terms, we can rewrite (132) as

$$\rho\frac{\partial \overline{q_v}}{\partial t} = -\rho\overline{\mathbf{V}}\cdot\nabla\overline{q_v} + \frac{\alpha}{L}\overline{Q_R} - (1-\alpha)\rho\tilde{C} + D\left\{\left[(q_v)_c - \overline{q_v}\right] + \frac{\alpha}{L}(s_c - \overline{s})\right\}. \tag{135}$$

If detrainment occurs at levels where $s_c - \overline{s}$, then we can approximate (135) by

$$\boxed{\rho\frac{\partial \overline{q_v}}{\partial t} \cong -\rho\overline{\mathbf{V}}\cdot\nabla\overline{q_v} + \frac{\alpha}{L}\overline{Q_R} - (1-\alpha)\rho\tilde{C} + D\left[(q_v)_c - \overline{q_v}\right]}. \tag{136}$$

Inspection of (136) reveals the following:

- Radiative cooling dries, and radiative heating moistens. If water vapor and radiative heating are positively correlated, then radiation may be able to drive an instability, as suggested by Raymond (2001).
- For $\alpha > 1$, which is expected in the lower troposphere, evaporation into the environment (i.e., $\tilde{C} < 0$) dries the air. *This means that the evaporation of detrained cloud water and stratiform precipitation in the lower troposphere can actually have a drying effect.*
- Condensation in the environment (i.e., $\tilde{C} > 0$) leads to a net moistening when $\alpha > 1$. However, if the environment is saturated with a moist adiabatic lapse rate, then $\alpha \cong 1$ is expected, in which case environmental condensation has no effect on the water vapor.
- For $0 < \alpha < 1$, the evaporation into the environment of detrained cloud water and falling rain can drive an increase in moisture variance, consistent with the "stratiform instability" of Mapes (2000) and Kuang (2008). It appears that instability is most likely when $0 < \alpha < 1$.

Segue

In this chapter we described both observations and theories of some of the many eddies that fill the atmosphere. We began with Laplace's theory of the free and forced oscillations of a thin spherical atmosphere. There are many solutions, notably including inertia-gravity waves and Rossby waves. We discussed the vertical propagation of Rossby waves, using the QG framework. We then turned to Matsuno's theory of equatorially trapped waves, which was based on the shallow water equations but can be extended to a stratified atmosphere. Laplace's equations also describe the waves that Matsuno found, but with differences due to sphericity.

Most of the eddies that we discussed are "free modes," in the sense that they are not responses to external forcing. The monsoon is a major exception.

The effects of the eddies on the zonally averaged flow are the subject of the next chapter.

Problems

1. Show that the static stability increases strongly with height in an isothermal atmosphere.
2. Prove that an isothermal atmosphere has only one equivalent depth for free oscillations, given by

$$\hat{h} = \gamma H,$$

where $\gamma \equiv c_p / c_v$, and $H = RT/g$.

3. Show that for a resting isentropic basic state with no heating and no gravitational forcing, perturbations satisfy

$$\frac{\partial z'}{\partial t} + H_0 \nabla_p^2 \chi = 0$$

for all p, where $H_0 = R\overline{T}_0/g$, and subscript zero denotes a surface value. This equation looks very much like the continuity equation for shallow water.

4. Show that the free oscillation of the second class (FOSC) with the nondivergent approximation ($\chi = 0$) must satisfy

$$\nabla_p \cdot (f \nabla_p \psi) = g \nabla_p^2 z'.$$

This result implies that the oscillation is not in exact geostrophic balance.

5. Show that when there is a basic zonal current from the west with a constant angular velocity $\dot{\lambda}$, the apparent "phase speed" relative to the Earth's surface of the FOSC is given by

$$-\frac{\sigma}{s} = \dot{\lambda} - \frac{2(\Omega + \dot{\lambda})}{n(n+1)}.$$

6. Solberg (1936) showed that Laplace's theory has a special solution for which the period of oscillation is half a day and the zonal wavenumber is zero.
   a) Show that the solution is given by

$$\Theta_n = A \sin(\sqrt{\varepsilon_n} \mu) + B \cos(\sqrt{\varepsilon_n} \mu),$$

   where

$$\varepsilon_n = \frac{4\Omega^2 a^2}{g h_n}.$$

   b) Show that the two conditions

$$\varepsilon_n = \left(\frac{1}{2} n\pi\right)^2, \; n = 1, 2, \ldots \text{ and } A/B = \tan\left(\frac{n\pi}{2}\right)$$

   are required to ensure that the winds remain finite at the poles.

7. Matsuno derived the "meridional structure equation" that governs equatorially trapped waves (except for Kelvin waves):

$$\frac{d^2\hat{v}}{dy^2} + \left(\sigma^2 - k^2 + \frac{k}{\sigma} - y^2\right)\hat{v} = 0,$$

with the boundary conditions

$$\hat{v} \to 0 \text{ as } y \to \pm\infty.$$

The solution is

$$\hat{v}(y) = Ce^{-\frac{1}{2}y^2}H_n(y).$$

Prove that the dispersion equation is

$$\sigma^2 - k^2 + \frac{k}{\sigma} = 2n + 1 \text{ for } n = 0, 1, 2....$$

8. Consider linear quasi-geostrophic frictionally damped Rossby waves produced by shallow water flow over periodic mountains, as in the model of Charney and Eliassen. Assume that the "shallow water" has a uniform density ρ_0. Write an expression for the "form drag" that the mountains exert on the mean flow in the presence of resonant waves (i.e., $K^2 = K_S^2$).
9. As discussed in the text, Matsuno (1966) discovered equatorial Kelvin waves, which propagate eastward and have no meridional velocity component. It is also possible to have oceanic "coastal" Kelvin waves, such that the currents are everywhere parallel to the coastline. A coastal Kelvin wave propagates parallel to the coast.
   Consider a coastal Kelvin wave in the Northern Hemisphere. The wave is propagating along a north-south–oriented "west coast" with water to the west and a continent to the east. Does the wave propagate north or south? Include a sketch that shows the spatial structures of the height and current fields.

CHAPTER 9

What the Eddies Do

Interactions and Noninteractions of Gravity Waves with the Mean Flow

We have seen how eddies can affect the mean flow, through eddy flux divergences and energy conversions. Despite the presence of such terms in equations, however, it turns out that under surprisingly general conditions eddies do not affect the mean flow. Several related theorems demonstrate this finding. Reasonably enough, they are called *noninteraction theorems*. The earliest such ideas were published by Eliassen and Palm (1961), and the following discussion of this section is based on their paper. The same material is also discussed in more detail, and in somewhat more general form, in chapter 8 of Lindzen's (1990) book.

Consider the equation of zonal motion in the simplified form

$$\frac{\partial u}{\partial t} + u\frac{\partial u}{\partial x} + w\frac{\partial u}{\partial z} = -\frac{1}{\rho}\frac{\partial p}{\partial x}. \tag{1}$$

We have omitted the effects of rotation, sphericity, friction, and meridional motions. We will apply (1) to small-scale gravity waves that are forced by a mean flow over topography. We define a zonally uniform but vertically varying basic state, and eddies, by

$$\begin{aligned} u &= \overline{u(z)}^{\lambda} + u^{*}, \\ w &= w^{*}, \\ p &= \overline{p(z)}^{\lambda} + p^{*}, \\ \rho &= \overline{\rho(z)}^{\lambda} + \rho^{*}. \end{aligned} \tag{2}$$

Although we are using our notation for zonal means and departures from zonal means in (2), we are, in fact, considering only an isolated mountain or mountain range. We interpret the starred quantities (the eddies) as small-amplitude wavelike perturbations with zero means. The mean flow is assumed to respond to the momentum flux according to

$$\overline{\rho}^{\lambda}\frac{\partial \overline{u}^{\lambda}}{\partial t} \sim -\frac{\partial}{\partial z}\left(\overline{\rho}^{\lambda}\,\overline{w^{*}u^{*}}^{\lambda}\right). \tag{3}$$

We are interested in what determines the wave momentum flux divergence, $(\partial/\partial z)(\overline{\rho}^{\lambda}\overline{w^{*}u^{*}}^{\lambda})$.

Substituting (2) into (1) and linearizing, we obtain

$$\overline{\rho}^{\lambda}\frac{\partial u^{*}}{\partial t} = -\left(\overline{\rho}^{\lambda}\overline{u}^{\lambda}\frac{\partial u^{*}}{\partial x} + \overline{\rho}^{\lambda}w^{*}\frac{\partial \overline{u}^{\lambda}}{\partial z} + \frac{\partial p^{*}}{\partial x}\right). \tag{4}$$

We assume that the perturbations are steady, so that

$$\frac{\partial u^{*}}{\partial t} = 0. \tag{5}$$

This implies that the waves are neutral, that is, neither amplifying nor decaying, and also that they are stationary; that is, their phase speed is zero. The latter assumption is reasonable, for example, for mountain waves. Then, (4) reduces to

$$\begin{aligned} 0 &= \overline{\rho}^{\lambda}\overline{u}^{\lambda}\frac{\partial u^{*}}{\partial x} + \overline{\rho}^{\lambda}w^{*}\frac{\partial \overline{u}^{\lambda}}{\partial z} + \frac{\partial p^{*}}{\partial x} \\ &= \frac{\partial}{\partial x}(\overline{\rho}^{\lambda}\overline{u}^{\lambda}u^{*} + p^{*}) + \overline{\rho}^{\lambda}w^{*}\frac{\partial \overline{u}^{\lambda}}{\partial z}. \end{aligned} \tag{6}$$

This is the form of the steady-state equation of motion that we will use.

Next, we multiply (6) by $(\overline{\rho}^{\lambda}\overline{u}^{\lambda}u^{*} + p^{*})$ to obtain

$$0 = \frac{\partial}{\partial x}\left[\frac{(\overline{\rho}^{\lambda}\overline{u}^{\lambda}u^{*} + p^{*})^{2}}{2}\right] + (\overline{\rho}^{\lambda})^{2}\overline{u}^{\lambda}\frac{\partial \overline{u}^{\lambda}}{\partial z}w^{*}u^{*} + \overline{\rho}^{\lambda}\frac{\partial \overline{u}^{\lambda}}{\partial z}w^{*}p^{*}, \tag{7}$$

and we zonally average, which gives

$$\frac{\partial \overline{u}^{\lambda}}{\partial z}\left(\overline{\rho}^{\lambda}\overline{u}^{\lambda}\overline{w^{*}u^{*}}^{\lambda} + \overline{w^{*}p^{*}}^{\lambda}\right) = 0. \tag{8}$$

Equation (8) can be simplified to

$$\boxed{\overline{u}^{\lambda}\overline{\rho}^{\lambda}\overline{w^{*}u^{*}}^{\lambda} + \overline{w^{*}p^{*}}^{\lambda} = 0}, \tag{9}$$

provided that $\partial\overline{u}^{\lambda}/\partial z \neq 0$. Equation (9) shows that the wave momentum flux, $\overline{\rho}^{\lambda}\overline{w^{*}u^{*}}^{\lambda}$, and the wave energy flux, $\overline{w^{*}p^{*}}^{\lambda}$, are closely related. At a "critical" level, where $\overline{u}^{\lambda} = 0$, the wave energy flux must vanish; the only other possibility is that our assumptions, such as a steady state with no friction, do not apply at the critical level. For a wave forced by flow over a mountain, the energy flux is, of course, upward, but (9) shows that the upward propagation of the wave is blocked at the critical level. This means that the wave does not exist above the critical level. A more detailed analysis (Booker and Bretherton, 1967; Bretherton, 1969) shows that the degree of wave blocking at a critical level actually depends on the Richardson number, which measures the relative importance of buoyancy and shear.

Equation (9) also shows that a wave with an upward energy flux will produce a downward momentum flux in westerlies and an upward momentum flux in easterlies. In either case, the wave is driving the mean flow toward zero; that is, it is exerting a drag on the mean flow.

Let e_E be the total eddy energy per unit mass associated with the wave (the sum of the eddy kinetic, eddy internal, and eddy potential energies). It can be shown that e_E satisfies

$$\frac{\partial}{\partial x}(\overline{\rho}^{\lambda} e_E \overline{u}^{\lambda} + p^* u^*) + \frac{\partial}{\partial z}(w^* p^*) = -\overline{\rho}^{\lambda} u^* w^* \frac{\partial \overline{u}^{\lambda}}{\partial z}. \tag{10}$$

The right-hand side of (10) is a "gradient-production" term that represents conversion of the kinetic energy of the mean state into the total eddy energy, e_E. Equation (10) simply says that the production term on the right-hand side is balanced by the transport terms on the left-hand side. Integration over the domain gives

$$\boxed{\frac{\partial}{\partial z}\left(\overline{w^* p^*}^{\lambda}\right) = -\overline{\rho}^{\lambda}\, \overline{w^* u^*}^{\lambda} \frac{\partial \overline{u}^{\lambda}}{\partial z}}. \tag{11}$$

This means that the wave energy flux divergence balances conversion (via gradient production) to or from the kinetic energy of the mean flow.

By combining (9) and (11) we can show that

$$\boxed{\overline{u}^{\lambda} \frac{\partial}{\partial z}\left(\overline{\rho}^{\lambda}\, \overline{w^* u^*}^{\lambda}\right) = 0}. \tag{12}$$

Therefore, *when* $\overline{u}^{\lambda} \neq 0$, *the wave momentum flux* $\overline{\rho}^{\lambda}\overline{w^* u^*}^{\lambda}$ *is independent of height.* This result is very important because, as shown by (3), it implies that the wave momentum flux has no effect on $\overline{u}^{\lambda}(z)$ except at the critical level where $\overline{u}^{\lambda} = 0$. The wave momentum flux is absorbed at the critical level. From (3), it follows that $\overline{u}^{\lambda}$ will tend to change with time at the critical level, so $\overline{u}^{\lambda}$ will become different from zero. Therefore, the critical level will move.

If we allowed the phase speed, c, to be nonzero, we would find $\overline{u}^{\lambda} - c$ everywhere in place of $\overline{u}^{\lambda}$. The momentum would be absorbed at the critical level where $\overline{u}^{\lambda} = c$.

Since (12) tells us that $\int_{-\infty}^{\infty} \overline{\rho}^{\lambda}\overline{w^* u^*}^{\lambda}$ is independent of height (where $\overline{u}^{\lambda} \neq 0$), we see from (9) that the wave energy flux is just proportional to $\overline{u}^{\lambda}$. Alternatively, we can combine (9) and (12) to write

$$\overline{w^* p^*}^{\lambda} / \overline{u}^{\lambda} = \text{constant}. \tag{13}$$

The quantity $\overline{w^* p^*}^{\lambda} / \overline{u}^{\lambda}$ is called the *wave action.* Equation (9) can be read as "wave action plus wave momentum flux = zero."

The work of Eliassen and Palm, reviewed above, was published in 1961. The importance of their ideas for the global circulation was not widely appreciated until about 25 years later. Since the mid-1980s there has been much interest in the effects of gravity-wave momentum fluxes on the general circulation; because the waves act to decelerate the mean flow, these interactions are referred to as *gravity-wave drag* (McFarlane, 1987). Initially, most of the discussion centered on gravity waves forced by flow over topography, but after a few years the importance of gravity waves forced by convective storms was also recognized (e.g., Fovell et al., 1992).

Figure 9.1A shows the deceleration of the zonally averaged zonal wind induced by gravity-wave drag in a general circulation model, as reported by McFarlane (1987). Here the gravity-wave drag is parameterized using methods, which we will not discuss, based on the assumption that the waves are produced by flow over mountains. The plot shows the "tendency" of the zonally averaged zonal wind due to this orographic gravity-wave drag, for northern-winter conditions. The very strong response of the zonally averaged zonal wind is shown in figure 9.1B. For thermal wind balance to be maintained there have to be corresponding changes in the zonally averaged temperature; these are shown in figure 9.1C. The polar troposphere has warmed dramatically, to stay in balance with the weaker westerly jet. The responses to gravity-wave drag shown in figure 9.1A and B make the model results more realistic than without gravity-wave drag, suggesting that gravity-wave drag is an important process in nature.

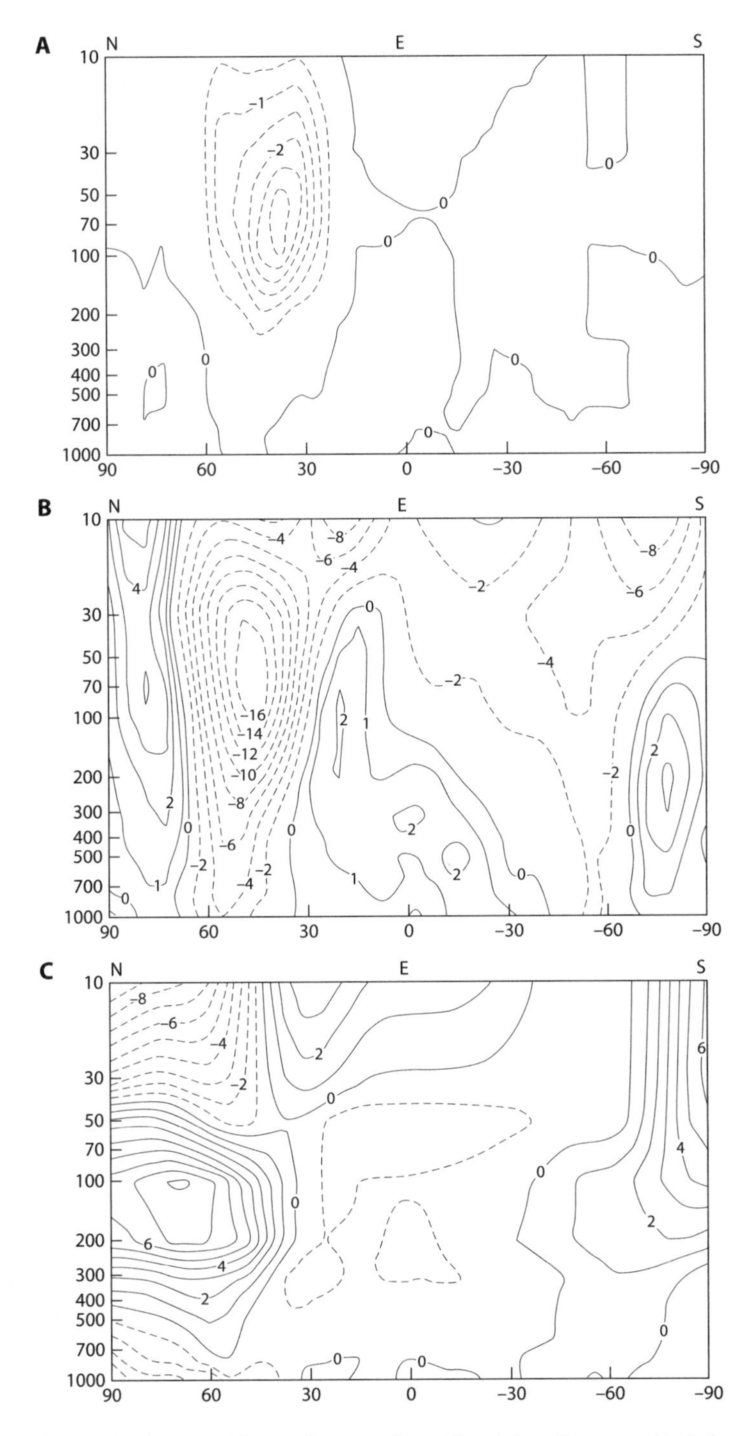

Figure 9.1. (A) The deceleration of the zonally averaged zonal flow, induced by orographically forced gravity waves, as simulated with a general circulation model. The units are m s^{-1} day^{-1}. (B) The actual change in the zonally averaged wind (in m s^{-1}) caused by the introduction of gravity-wave drag in a general circulation model, as inferred by comparison with a control run. (C) The actual change in the zonally averaged temperature (in K) caused by the introduction of gravity-wave drag in a general circulation model, as inferred by comparison with a control run. From McFarlane (1987). © American Meteorological Association. Used with permission.

Angular Momentum Transport by Rossby Waves

In chapter 8, we derived the dispersion equation for Rossby waves for the case of no mean flow. We now consider what happens when a mean zonal flow is present. For simplicity, we consider purely rotational horizontal motions in Cartesian coordinates.

The linearized vorticity equation with a constant zonal current is

$$\frac{\partial \zeta^*}{\partial t} + \overline{u}^\lambda \frac{\partial \zeta^*}{\partial x} + \beta v^* = 0. \tag{14}$$

For pure rotational flow, we can represent the solution in terms of a streamfunction, ψ^*:

$$v^* = \frac{\partial \psi^*}{\partial x}, u^* = -\frac{\partial \psi^*}{\partial y}, \zeta^* = \nabla^2 \psi^*. \tag{15}$$

We look for solutions of the form

$$\psi^* = A\cos(kx + ly - \sigma t), \tag{16}$$

where A is an arbitrary constant "amplitude" with dimensions of $m^2\ s^{-1}$, and, as usual, the zonal wavenumber, k, meridional wavenumber, l, and frequency, σ, are real numbers that are also constant in both space and time. Here we adopt the convention that the frequency is nonnegative, so that the signs of k and l determine the directions of zonal and meridional phase propagation, respectively. Substitution of (15) and (16) into (14) gives

$$-(-\sigma + \overline{u}^\lambda k)(k^2 + l^2) + \beta k = 0, \tag{17}$$

or

$$\sigma - \overline{u}^\lambda k = -\left(\frac{\beta k}{k^2 + l^2}\right). \tag{18}$$

The zonal phase speed, c, satisfies

$$\sigma = kc, \tag{19}$$

so (18) can also be written as

$$\boxed{c - \overline{u}^\lambda = -\left(\frac{\beta}{k^2 + l^2}\right) < 0}. \tag{20}$$

For $\overline{u}^\lambda = 0$, we have $c < 0$, which means westerly phase propagation, which was discussed in chapter 8. In any case, *the phase propagation is always westward relative to the mean zonal flow, because (20) guarantees that* $c - \overline{u}^\lambda < 0$. If $\overline{u}^\lambda$ is negative (easterlies), then c has to be even more negative.

The special case $c - \overline{u}^\lambda = 0$ occurs at *critical latitudes*. Equation (20) suggests that the total wavenumber $\sqrt{k^2 + l^2}$ becomes very large near a critical latitude, but this is only a suggestion (rather than a true deduction), because our assumed solution (16) uses zonal and meridional wavenumbers that are spatially and temporally constant.

We can derive the meridional flux of zonal momentum by the waves by using (16) in (15) to write

$$v^* = -Ak\sin(kx + ly - \sigma t), \text{ and } u^* = Al\sin(kx + ly - \sigma t), \tag{21}$$

so that

$$u^{*}v^{*} = -A^{2}kl\sin^{2}(kx + ly - \sigma t). \tag{22}$$

Averaging over longitude (which is equivalent to averaging over one zonal wavelength), we find that

$$\begin{aligned}\overline{u^{*}v^{*}}^{\lambda} &= -\frac{A^{2}l}{2\pi}\int_{0}^{2\pi}\sin^{2}(kx + ly - \sigma t)\,d(kx) \\ &= -\frac{A^{2}l}{2\pi}\left[\frac{kx}{2} - \frac{\sin(2kx)}{4}\right]_{0}^{2\pi} \\ &= -\frac{A^{2}kl}{2}.\end{aligned} \tag{23}$$

Conversely, the meridional component of the group velocity is

$$\begin{aligned}c_{gy} &= \frac{\partial\sigma}{\partial l} \\ &= \frac{2\beta kl}{(k^{2} + l^{2})^{2}}.\end{aligned} \tag{24}$$

The wave energy travels with speed c_{gy}. Comparing (23) and (24), we see that both involve the product kl, which is a measure of the "tilt" of the waves in the longitude-latitude plane. We can write

$$\boxed{\frac{\overline{u^{*}v^{*}}^{\lambda}}{c_{gy}} = -\frac{A^{2}(k^{2} + l^{2})^{2}}{4\beta} < 0}. \tag{25}$$

Equation (25) shows that the sign of the meridional flux of zonal momentum is always the opposite of the sign of the meridional flux of wave energy. If wave energy is going equatorward, then the momentum flux is toward the pole, and vice versa.

An important implication of this result is that *a latitude belt that is a source of Rossby-wave energy will tend to acquire westerly momentum*. The wave momentum flux is into the region of westerlies, so it is "up the gradient." Recall from chapter 7 that when the momentum transport is upgradient, the gradient-production term of the eddy kinetic energy equation converts the kinetic energy of the eddies into the kinetic energy of the mean flow. As discussed in chapter 7, this transport is observed, and it can be maintained over time only if the eddies have a source of kinetic energy *other than* the kinetic energy of the mean flow. That source can be baroclinic conversion from eddy available potential energy.

The wave energy flux is blocked when the wave encounters a critical latitude where $c - \overline{u}^{\lambda} = 0$. The wave momentum flux is drawn from the vicinity of the critical latitude. As a result, that latitude belt tends to acquire easterly momentum, which tends to maintain the easterlies and therefore favors the continuing existence of a critical latitude, although that latitude may move north or south.

Vertically Propagating Planetary Waves

Under what conditions do planetary waves transport energy and momentum, and how do they influence the zonally averaged flow? Recall from chapter 4 that within the framework of QG dynamics, the quasi-geostrophic pseudopotential vorticity

(QGPPV) controls the dynamics. We should therefore ask how eddies affect the QGPPV. Many of the ideas discussed in this section originated with Charney and Drazin (1961), Charney and Stern (1962), and Dickinson (1969).

From the QGPPV equation discussed in chapters 4 and 8, we can show that the zonally averaged QGPPV is governed by

$$\frac{\partial \overline{Z_{QG}}^{\lambda}}{\partial t} = -\frac{\partial}{\partial y}\left(\overline{v_g^* Z_{QG}^*}^{\lambda}\right). \tag{26}$$

Thus, apart from the effects of heating and friction, the zonally averaged QGPPV changes owing only to the convergence of the meridional eddy flux of QGPPV. We can show (see the problems at the end of this chapter) that this flux is related to the meridional eddy fluxes of both momentum and temperature, in the following interesting way:

$$\boxed{\begin{aligned} \overline{v_g^* Z_{QG}^*}^{\lambda} &= \overline{v_g^* \zeta_g^*}^{\lambda} - \frac{\partial}{\partial p}\left(\frac{Rf_0}{pS}\overline{v_g^* T^*}^{\lambda}\right) \\ &= -\frac{\partial}{\partial y}\left(\overline{u_g^* v_g^*}^{\lambda}\right) - \frac{\partial}{\partial p}\left(\frac{Rf_0}{pS}\overline{v_g^* T^*}^{\lambda}\right). \end{aligned}} \tag{27}$$

Equation (27) says that the meridional eddy flux of potential vorticity is related to the *convergence* of the meridional eddy flux of zonal momentum and to the *rate of change with height* of the meridional eddy sensible heat flux. When we form the convergence of the eddy potential vorticity flux, that is, $-\partial/\partial y(\overline{v_g^* Z_{QG}^*}^{\lambda})$, (27) will give us $\partial/\partial y[-\partial/\partial y(\overline{u_g^* v_g^*}^{\lambda})]$, which affects the meridional shear of $\overline{u}^{\lambda}$. We will also get a term proportional to $\partial/\partial p[-\partial/\partial y(\overline{v_g^* T^*}^{\lambda})]$, which affects the static stability.

The expression on the second line of (27) looks like the convergence of a flux vector. This is the quasi-geostrophic form of the so-called Eliassen-Palm flux vector, which we denote by $\mathbf{QGEPF} \equiv (0, QGEPF_{\varphi}, QGEPF_p)$, where

$$QGEPF_{\varphi} = -\overline{u_g^* v_g^*}^{\lambda}, \tag{28}$$

$$QGEPF_p = -\frac{Rf_0}{pS}\overline{v_g^* T^*}^{\lambda}. \tag{29}$$

When the QG pseudopotential vorticity flux vanishes, the Eliassen-Palm flux is nondivergent. Later in this chapter we will see that these ideas can be generalized considerably.

As discussed in chapter 4, the QG form of the thermodynamic energy equation is

$$\left(\frac{\partial}{\partial t} + \mathbf{V}_g \cdot \nabla\right)\frac{\partial \phi}{\partial p} + S\omega = 0. \tag{30}$$

Equation (30) can be written as

$$\left(\frac{\partial}{\partial t} + \mathbf{V}_g \cdot \nabla\right)\psi_z + \frac{N^2}{f_0} w = 0, \tag{31}$$

where w is defined by $-\omega/(\rho_{bs} g)$. Here $\psi_z \equiv \partial\psi/\partial z$, and z is the "log-p" coordinate defined in chapter 8. Linearization of (31) gives

$$\left(\frac{\partial}{\partial t} + \overline{u_g}^{\lambda}\frac{\partial}{\partial x}\right)\psi_z^* - v_g^*\frac{\partial \overline{u_g}^{\lambda}}{\partial z} + \frac{N^2}{f_0}w^* = 0. \tag{32}$$

Here we used the thermal wind equation.

Multiplying (32) by ψ_z^* gives a form of the *temperature variance equation*:

$$\left(\frac{\partial}{\partial t} + \overline{u}_g^{\lambda}\frac{\partial}{\partial x}\right)\left\{\frac{1}{2}(\psi_z^*)^2\right\} - v_g^*\psi_z^*\frac{\partial \overline{u}_g^{\lambda}}{\partial z} + \frac{N^2}{f_0}w^*\psi_z^* = 0. \tag{33}$$

Note the two gradient-production terms (refer to chapter 7). We take the zonal mean of (33), so that the $\overline{u}^{\lambda}(\partial/\partial x)$ term drops out, and we rearrange the result to isolate the meridional energy flux on the left-hand side:

$$\left(\overline{v^*\psi_z^*}^{\lambda}\right)\frac{\partial \overline{u}^{\lambda}}{\partial z} = \frac{\partial}{\partial t}\left[\frac{1}{2}\overline{(\psi_z^*)^2}^{\lambda}\right] + N^2\left(\frac{\overline{w^*\psi_z^*}^{\lambda}}{f_0}\right). \tag{34}$$

Note that $(\overline{w^*\psi_z^*}^{\lambda}/f_0)_0 > 0$ implies an upward temperature flux in either hemisphere, and similarly, $\overline{v_g^*\psi_z^*}^{\lambda} > 0$ implies a poleward temperature flux in either hemisphere.

Recall from chapter 7 that baroclinic eddies derive their kinetic energy by conversion from the potential energy of the mean state by means of an upward temperature flux that lowers the mean state's center of mass. Consider a baroclinically amplifying wave, for which $\partial/\partial t[1/2\,\overline{(\psi_z^*)^2}^{\lambda}] > 0$ and $(\overline{w^*\psi_z^*}^{\lambda}/f_0)_0 > 0$ so that the right-hand side of (34) is positive. Equation (34) shows that *a baroclinically amplifying eddy produces a poleward temperature flux (in either hemisphere) when* $\partial\overline{u}_g^{\lambda}/\partial z > 0$, *that is, when the temperature is decreasing toward the pole.* Such a temperature flux is downgradient, so the gradient-production term is positive.

Next, we consider a *neutral wave* of the form $e^{ik(x-ct)}$, for which $\partial/\partial t = -c(\partial/\partial x)$, where c is real. We multiply (32) by ψ^* and take the zonal mean to obtain

$$(\overline{u}_g^{\lambda} - c)\,\overline{v_g^*\psi_z^*}^{\lambda} = N^2\frac{\overline{w^*\psi^*}^{\lambda}}{f_0}. \tag{35}$$

Note that $\overline{w^*\psi^*}^{\lambda}/f_0 > 0$ means an upward propagation of wave energy in either hemisphere. Recall also that $\overline{u}_g^{\lambda} - c > 0$ is required for the wave to propagate. It follows that *an upward-propagating neutral wave transports energy poleward.* Such a wave might be excited, for example, by flow over mountains.

In summary, poleward energy transport is produced by either a baroclinically amplifying wave with $\partial\overline{u}_g^{\lambda}/\partial z > 0$ or a neutral wave propagating upward.

Recall from chapter 8 that the QG eddy PV equation is

$$\left(\frac{\partial}{\partial t} + \overline{u}_g^{\lambda}\frac{\partial}{\partial x}\right)Z_{QG}^* + v_g^*\frac{\partial}{\partial y}\overline{Z_{QG}}^{\lambda} = 0. \tag{36}$$

Application of (36) to a neutral wave gives

$$(\overline{u}_g^{\lambda} - c)\frac{\partial Z_{QG}^*}{\partial x} + v_g^*\frac{\partial}{\partial y}\overline{Z_{QG}}^{\lambda} = 0. \tag{37}$$

We multiply (37) by ψ^* and take the zonal mean to show that

$$\boxed{\overline{v_g^* Z_{QG}^*}^{\lambda} = 0 \text{ except where } \overline{u}_g^{\lambda} = c}; \tag{38}$$

that is, *the eddy QGPPV flux vanishes except at a critical line.* It follows from (26) that neutral waves do not affect $\overline{Z_{QG}}^{\lambda}$ except at a critical line. *This is a noninteraction theorem for quasi-geostrophic planetary waves,* analogous to the noninteraction theorem for gravity waves obtained by Eliassen and Palm (1961). Equation (38) implies that the QG Eliassen-Palm flux is nondivergent except at a critical line.

Referring to (27), we see that $\overline{v^* q_{QG}^*}^{\lambda} = 0$ means that

$$-\frac{\partial}{\partial y}\left(\overline{u_g^* v_g^*}^\lambda\right) + \frac{f_0^2}{\rho_{bs}}\frac{\partial}{\partial z}\left(\frac{\rho_{bs}}{N^2}\overline{v_g^* \psi_z^*}^\lambda\right) = 0. \tag{39}$$

We vertically integrate (39) through the depth of the atmosphere to obtain

$$-\int_0^{p_S} \frac{\partial}{\partial y}\left(\overline{u_g^* v_g^*}^\lambda\right)\frac{dp}{g} = f_0^2 \frac{\rho_{bs}}{N^2}\left(\overline{v_g^* \psi_z^*}^\lambda\right)_S \tag{40}$$

for neutral waves. The left-hand side of (40) represents the vertically integrated convergence of the meridional momentum flux, and the right-hand side represents the near-surface value of the eddy meridional energy flux. Recall our earlier conclusion that an upward-propagating neutral wave produces a poleward energy flux; that is, $\overline{v_g^* \psi_z^*}^\lambda > 0$. It follows from (40) that

$$\boxed{-\int_0^{p_S} \frac{\partial}{\partial y}\left(\overline{u_g^* v_g^*}^\lambda\right)\frac{dp}{g} > 0 \text{ for an upward-propagating neutral wave}}. \tag{41}$$

Equation (41) means that the vertically integrated meridional momentum flux convergence tends to accelerate the vertically integrated zonal-mean zonal wind. In other words, the eddy momentum flux tries to increase the speed of the jet. This finding is consistent with the discussion of Rossby-wave momentum transport given earlier in this chapter. Conversely, because the waves also transport temperature poleward, they tend to reduce the meridional temperature gradient and so (as implied by thermal wind balance) tend to reduce the strength of the jet. *The momentum flux and heat flux thus have opposing effects on the mean flow.* If these two opposing effects were to cancel, then eddies would have no net effect on the mean flow.

An upward propagating neutral wave in westerly shear tends to produce a downward momentum flux at the Earth's surface. To see why, let's consider the angular momentum equation, in Cartesian coordinates for simplicity:

$$\frac{\partial M}{\partial t} + \frac{\partial}{\partial x}(uM) + \frac{\partial}{\partial y}(uM) + \frac{1}{\rho_S}\frac{\partial}{\partial z}(\rho_S wM) = -\frac{\partial \phi}{\partial \lambda}. \tag{42}$$

We assumed no friction above the boundary layer. Taking the zonal mean of (42), we obtain

$$\frac{\partial \overline{M}^\lambda}{\partial t} + \frac{\partial}{\partial y}\left(\overline{v^* M^*}^\lambda\right) + \frac{1}{\rho_S}\frac{\partial}{\partial z}\left(\rho_S \overline{w^* M^*}^\lambda\right) = 0. \tag{43}$$

Here advection of $\overline{M}^\lambda$ by $\overline{v}^\lambda$ and $\overline{w}^\lambda$ is neglected; this can be justified for midlatitude winter. To the extent that $\overline{v}^\lambda$ is geostrophic, it vanishes anyway. Next, we assume that $\partial \overline{M}^\lambda / \partial t = 0$. This leads to

$$\frac{\partial}{\partial y}\left(\overline{v^* M^*}^\lambda\right) = -\frac{1}{\rho_S}\frac{\partial}{\partial z}\left(\rho_S \overline{w^* M^*}^\lambda\right). \tag{44}$$

Integrating (44) vertically with respect to mass, and using (41), we find that

$$\int_0^\infty \frac{\partial}{\partial z}\left(\rho_S \overline{w^* M^*}^\lambda\right) dz > 0. \tag{45}$$

We know that $\rho_S \overline{w^* M^*}^\lambda$ must vanish at great height, so (45) implies that

$$\boxed{\left(\rho_S \overline{w^* M^*}^\lambda\right)_S < 0 \text{ for an upward-propagating planetary wave}}. \tag{46}$$

Thus, in the presence of an upward-propagating planetary wave, surface friction and/or mountain torque must carry angular momentum into the Earth's surface. An alternative interpretation is that in a belt of westerlies where (44) is satisfied, surface friction and/or mountain torque will produce an upward-propagating planetary wave that, as discussed previously, will transport energy poleward.

Comparing (41) and (46), we see that the meridional momentum flux accelerates the westerlies, while the vertical momentum flux decelerates them. Angular momentum flows meridionally into the elevated jet and then downward into the Earth's surface. This is consistent with the observed angular momentum transport discussed in chapter 5.

The Transformed Eulerian Mean System

Previously, we discussed noninteraction theorems for pure gravity waves and for quasi-geostrophic eddies on a β-plane. It was discovered during the 1970s that noninteraction theorems can be derived for more general balanced flows. The following discussion is based on the work of Andrews et al. (1987).

The zonally averaged equations in spherical coordinates can be written as

$$\frac{\partial \overline{M}^{\lambda}}{\partial t} + \overline{v}^{\lambda}\frac{\partial \overline{M}^{\lambda}}{\partial \varphi} + \overline{w}^{\lambda}\frac{\partial \overline{M}^{\lambda}}{\partial z} - \overline{F}_x^{\lambda} a\cos\varphi = \frac{-1}{a\cos\varphi}\frac{\partial}{\partial \varphi}\left(\overline{v^{*}M^{*}}^{\lambda}\cos\varphi\right) - \frac{1}{\rho_S}\frac{\partial}{\partial z}\left(\rho_S \overline{w^{*}M^{*}}^{\lambda}\right), \tag{47}$$

$$\frac{\partial \overline{v}^{\lambda}}{\partial t} + \frac{1}{a}\overline{v}^{\lambda}\frac{\partial \overline{v}^{\lambda}}{\partial \varphi} + \overline{w}^{\lambda}\frac{\partial \overline{v}^{\lambda}}{\partial z} + \overline{u}^{\lambda}\left(f + \frac{\overline{u}^{\lambda}\tan\varphi}{a}\right) + \frac{1}{a}\frac{\partial \overline{\phi}^{\lambda}}{\partial \varphi} - \overline{F}_y^{\lambda} = \frac{-1}{a\cos\varphi}\frac{\partial}{\partial \varphi}\left[\overline{(v^{*})^{2}}^{\lambda}\cos\varphi\right] - \frac{1}{\rho_S}\frac{\partial}{\partial z}\left(\rho_S \overline{v^{*}M^{*}}^{\lambda}\right) - \frac{\overline{(u^{*})^{2}}^{\lambda}\tan\varphi}{a}, \tag{48}$$

$$\frac{\partial \overline{\theta}^{\lambda}}{\partial t} + \frac{\overline{v}^{\lambda}}{a}\frac{\partial \overline{\theta}^{\lambda}}{\partial \varphi} + \overline{w}^{\lambda}\frac{\partial \overline{\theta}^{\lambda}}{\partial z} - \overline{Q}^{\lambda} = \frac{-1}{a\cos\varphi}\frac{\partial}{\partial \varphi}\left(\overline{v^{*}\theta^{*}}^{\lambda}\cos\varphi\right) - \frac{1}{\rho_S}\frac{\partial}{\partial z}\left(\rho_S \overline{w^{*}\theta^{*}}^{\lambda}\right) \tag{49}$$

$$\frac{1}{a\cos\varphi}\frac{\partial}{\partial \varphi}\left(\rho_S \overline{v}^{\lambda}\cos\varphi\right) + \frac{\partial}{\partial z}\left(\rho_S \overline{w}^{\lambda}\right) = 0, \tag{50}$$

$$\frac{\partial \overline{\phi}^{\lambda}}{\partial z} - \frac{R\overline{\theta}^{\lambda}}{H}e^{-\frac{\kappa z}{H}} = 0. \tag{51}$$

Here $z \equiv -H\log(p/p_0)$ is the vertical coordinate, and $w \equiv Dz/Dt$. The scale height, H, is RT_0/g, where T_0 is a constant. In (49), Q represents a heating process. In (47)–(51), $\rho_{bs}(z) \equiv \rho_0 e^{-z/H}$, where ρ_0 is a constant. We have assumed for simplicity that the temperature is uniform with height, but this assumption is not really needed.

We now assume that the meridional momentum equation, (48), can be approximated by gradient wind balance; that is,

$$\overline{u}^{\lambda}\left(f + \overline{u}^{\lambda}\frac{\tan\varphi}{a}\right) + \frac{1}{a}\frac{\partial \overline{\phi}^{\lambda}}{\partial \varphi} \cong 0. \tag{52}$$

The smallness of $\overline{v}^{\lambda}$ outside the tropics, discussed in chapter 3, implies that the first three terms on the left-hand side of (48) are small. Equation (52) can then be obtained by neglecting friction and the various eddy terms on the right-hand side of (48). Equation (52) is essential to the following argument.

We *define* a *residual circulation* $(0, V, W)$ by

$$\boxed{V \equiv \overline{v}^{\lambda} - \frac{1}{\rho_S}\frac{\partial}{\partial z}\left(\frac{\rho_S \overline{v^* \theta^*}^{\lambda}}{\frac{\partial \overline{\theta}^{\lambda}}{\partial z}}\right)}, \tag{53}$$

$$\boxed{W \equiv \overline{w}^{\lambda} + \frac{1}{a\cos\varphi}\frac{\partial}{\partial\varphi}\left(\cos\varphi\frac{\overline{v^*\theta^*}^{\lambda}}{\frac{\partial \overline{\theta}^{\lambda}}{\partial z}}\right)}. \tag{54}$$

In the absence of eddies, $V = \overline{v}^{\lambda}$, which, again, is expected to be small outside the tropics, and $W = \overline{w}^{\lambda}$. Substitution shows that V and W satisfy a continuity equation analogous to (50). Use of (53) and (54) to eliminate $\overline{v}^{\lambda}$ and $\overline{w}^{\lambda}$ in favor of V and W allows us to rewrite (47) and (49) as

$$\boxed{\frac{\partial \overline{M}^{\lambda}}{\partial t} + \frac{V}{a}\frac{\partial \overline{M}^{\lambda}}{\partial \varphi} + W\frac{\partial \overline{M}^{\lambda}}{\partial z} - a\cos\varphi \overline{F_x}^{\lambda} = \frac{1}{\rho_S}(\nabla \cdot \mathbf{EPF})}, \tag{55}$$

and

$$\boxed{\frac{\partial \overline{\theta}^{\lambda}}{\partial t} + \frac{V}{a}\frac{\partial \overline{\theta}^{\lambda}}{\partial \varphi} + W\frac{\partial \overline{\theta}^{\lambda}}{\partial z} - \overline{Q}^{\lambda} = \frac{-1}{\rho_S}\frac{\partial}{\partial z}\left(\frac{\rho_S \overline{v^*\theta^*}^{\lambda}\frac{1}{a}\frac{\partial \overline{\theta}^{\lambda}}{\partial \varphi} + \rho_S \overline{w^*\theta^*}^{\lambda}\frac{\partial \overline{\theta}^{\lambda}}{\partial z}}{\frac{\partial \overline{\theta}^{\lambda}}{\partial z}}\right)}, \tag{56}$$

respectively, where

$$\mathbf{EPF} \equiv \left[0, (EPF)_{\varphi}, (EPF)_z\right] \tag{57}$$

is the *Eliassen-Palm flux*, whose components are

$$\boxed{(EPF)_{\varphi} \equiv \rho_S\left(\frac{\partial \overline{M}^{\lambda}}{\partial z}\cdot\frac{\overline{v^*\theta^*}^{\lambda}}{\frac{\partial \overline{\theta}^{\lambda}}{\partial z}} - \overline{v^* M^*}^{\lambda}\right)}, \tag{58}$$

and

$$\boxed{(EPF)_z \equiv \rho_S\left(-\frac{1}{a}\frac{\partial \overline{M}^{\lambda}}{\partial \varphi}\frac{\overline{v^*\theta^*}^{\lambda}}{\frac{\partial \overline{\theta}^{\lambda}}{\partial z}} - \overline{w^* M^*}^{\lambda}\right)}. \tag{59}$$

In (58), the $-\overline{v^* M^*}^{\lambda}$ term is usually dominant, and in (59) the $-\overline{v^*\theta^*}^{\lambda}$ term is usually dominant. When the **EPF** points upward, the meridional flux of potential temperature is in control. When the **EPF** points in the meridional direction, the meridional flux of zonal momentum is in control. From (55) we see that a positive **EPF** divergence tends to increase $\overline{M}^{\lambda}$. Compare (58) and (59) with the QG forms (28) and (29). Equations (55)–(56) are called the *transformed Eulerian mean* equations.

You should recognize the terms in the numerator inside $\partial/\partial z$ on the right-hand side of (56), the equation governing $\overline{\theta}^{\lambda}$, as the gradient-production terms of the equation for the eddy potential temperature variance, which was discussed in chapter 7.

The preceding derivation appears to be merely an algebraic shuffle. We wrote (53) and (54) without any explanation or motivation. What is the point of all this? The point is that for steady linear eddies with $F_x = F_y = 0$ and $Q = 0$, it can be shown that

$$\nabla \cdot (\mathbf{EPF}) = 0. \tag{60}$$

It turns out that the eddy forcing term of (56) is zero under the same conditions; that is,

$$\frac{\partial}{\partial z}\left(\frac{\rho_S \overline{v^*\theta^*}^\lambda \frac{1}{a}\frac{\partial \overline{\theta}^\lambda}{\partial \varphi} + \rho_S \overline{w^*\theta^*}^\lambda \frac{\partial \overline{\theta}^\lambda}{\partial z}}{\frac{\partial \overline{\theta}^\lambda}{\partial z}} \right) = 0. \tag{61}$$

Equation (61) follows essentially from our assumptions that $\overline{M}^\lambda$ does not change and that gradient-wind balance is maintained.

For the case of steady, linear eddies, in the absence of friction and heating, our system of equations reduces to

$$\begin{aligned}
\frac{\partial \overline{M}^\lambda}{\partial t} + \frac{V}{a}\frac{\partial \overline{M}^\lambda}{\partial \varphi} + W\frac{\partial \overline{M}^\lambda}{\partial z} &= 0, \\
\overline{u}^\lambda\left(f + \overline{u}^\lambda \frac{\tan\varphi}{a}\right) + \frac{1}{a}\frac{\partial \overline{\phi}^\lambda}{\partial \varphi} &= 0, \\
\frac{\partial \overline{\theta}^\lambda}{\partial t} + \frac{V}{a}\frac{\partial \overline{\theta}^\lambda}{\partial \varphi} + W\frac{\partial \overline{\theta}^\lambda}{\partial z} &= 0, \\
\frac{1}{a\cos\varphi}\frac{\partial}{\partial \varphi}(\rho_S V \cos\varphi) + \frac{\partial}{\partial z}(\rho_S W) &= 0, \\
\frac{\partial \overline{\phi}^\lambda}{\partial z} - \frac{R\overline{\theta}^\lambda}{H} e^{\frac{-\kappa z}{H}} &= 0.
\end{aligned} \tag{62}$$

This system has the following steady-state solution:

$$\begin{aligned}
\frac{\partial \overline{M}^\lambda}{\partial t} &= 0, M \text{ in gradient wind balance,} \\
V &= 0, W = 0, \\
\frac{\partial \overline{\theta}^\lambda}{\partial t} &= 0, \overline{\theta}^\lambda \text{ specified from the past history or radiative-convective equilibrium.}
\end{aligned} \tag{63}$$

From the definitions of V and W, we can solve for the mean meridional circulation implied by $V = 0$ and $W = 0$:

$$\rho_S \overline{v}^\lambda = \frac{\partial}{\partial z}\left(\rho_S \frac{\overline{v^*\theta^*}^\lambda}{\frac{\partial \overline{\theta}^\lambda}{\partial z}} \right), \tag{64}$$

$$\overline{w}^\lambda = -\frac{1}{a\cos\varphi}\frac{\partial}{\partial \varphi}\left(\cos\varphi \frac{\overline{v^*\theta^*}^\lambda}{\frac{\partial \overline{\theta}^\lambda}{\partial z}} \right). \tag{65}$$

To aid in interpreting these results, suppose we had a solution with no eddies at all. "No eddies" certainly qualify as "steady linear eddies," so the preceding argument applies, and from (64) and (65) we conclude that the MMC vanishes in the absence of eddies (and heating).

Now, we add steady linear eddies, so that $\nabla \cdot (\mathbf{EPF})$ continues to be zero. Exactly the same $\overline{M}^\lambda$ and $\overline{\theta}^\lambda$ will satisfy the equations! Of course, v and w will be different, that is, the MMC will be different, because it will have to be whatever it takes to ensure that $V = W = 0$, that is, to satisfy (64) and (65). *This MMC is eddy-driven*. The system produces the MMC to prevent the eddies from disrupting the gradient wind balance. Perhaps a better way to say this is that the processes that act to maintain thermal-wind balance (i.e., geostrophic and hydrostatic adjustment) accomplish this feat by using the wave-induced MMC as a tool.

The interpretation of this amazing result is that if we try to modify $\overline{M}^{\lambda}$ and $\overline{\theta}^{\lambda}$ by applying steady, linear eddy forcing such that $\nabla \cdot (\mathbf{EPF}) = 0$, we will be disappointed! The only result will be that the MMC will change in such a way that V and W will continue to be zero. In effect, the eddies will induce an MMC that exactly cancels the direct effects of the eddies on $\overline{M}^{\lambda}$ and $\overline{\theta}^{\lambda}$.

When the eddies are unsteady, the residual circulation is different from zero, and $\overline{M}^{\lambda}$ and $\overline{\theta}^{\lambda}$ are modified by the combined effects of the eddies and/or the eddy-induced MMC. The effects of the eddies and the MMC still tend to cancel, but the cancellation is incomplete.

Edmon et al. (1980) used the QG form of the noninteraction theorem to analyze the data of Oort and Rasmussen (1971). Figure 9.2 shows the Eliassen-Palm fluxes

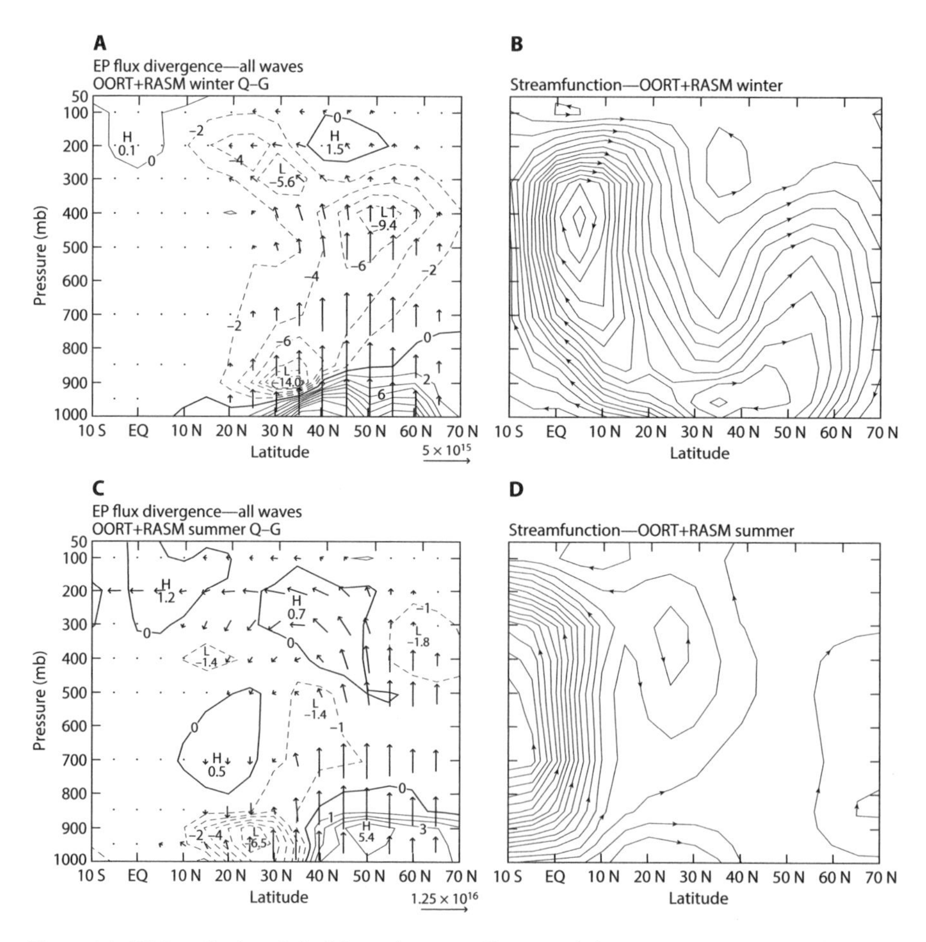

Figure 9.2. (A) Contribution of all eddies to the seasonally averaged Eliassen-Palm cross sections for winter. The contour interval is 2×10^{15} m^3. The horizontal arrow scale (in units of m^3) is indicated at bottom right. (B) The streamfunction of the winter residual circulation. The contour interval is 7.5×10^{16} m^2 s Pa. (C) Same as A, but for summer, and with a contour interval of 1×10^{15} m^3. (D) Same as B but for summer. From Edmon et al. (1980). © American Meteorological Association. Used with permission.

(arrows) and their divergences (contours), as well as the streamfunctions of the residual circulations. First, we consider the northern winter results, shown in the upper panels. In middle latitudes, near the surface, the arrows point strongly upward, indicating an intense poleward potential temperature flux. Near the tropopause, the arrows curve over and become horizontal, pointing toward the tropics, indicating a strong poleward eddy momentum flux. Keep in mind that $\nabla \cdot (\mathbf{EPF}) > 0$ means $\partial \overline{M}^{\lambda}/\partial t > 0$; that is, a positive EPF divergence favors westerly acceleration. The negative divergence (i.e., convergence) near 200 hPa at about 30° N indicates that the net effect of the eddies is to decelerate the jet. In fact, the westerlies are decelerated throughout middle latitudes, except near the surface. Note that this EPF convergence results mainly from the decrease with height of the upward component of the flux; that is, it is due mainly to the meridional temperature flux.

The residual circulation for winter (panel B) looks suspiciously like a giant Hadley cell extending from the tropics to the poles. It is not a coincidence that this is reminiscent of the mean meridional circulation as seen in isentropic coordinates, which was discussed in chapter 5. In summer, the northern edge of a Hadley cell apparently extends into the Southern Hemisphere. Clearly, the residual circulation can be regarded as partially driven by tropical heating.

The results for northern summer are quite similar, except that the action is generally weaker and shifted poleward. Further analysis shows that the contributions of the transient eddies dominate those of the stationary eddies, in both seasons.

Figure 9.2 shows seasonal means. In the remainder of this chapter, we discuss three fairly dramatic examples of time-dependent interactions between the eddies and the zonally averaged circulation, and one example of a seasonal-mean interaction.

Blocking

Blocking (Berggren et al., 1949; Rex, 1950a, 1950b) is a low-frequency, midlatitude phenomenon characterized by a nearly stationary anticyclone that persists for at least several days and sometimes up to several weeks. The anticyclone splits the westerly jet and steers cyclonic disturbances around itself, mainly on the poleward side. The flow in the vicinity of a blocking high is strongly meridional and is an example of what the older literature calls a "low-index" regime (Rossby, 1939); a high-index regime is strongly zonal. Blocking highs tend to fluctuate in intensity, weakening briefly and then reintensifying. They remain nearly stationary in an average sense, although they may wander around a little over their lifetime.

Blocks tend to occur preferentially in the Northern Hemisphere, especially in the eastern North Atlantic and the eastern North Pacific in winter, and northern Asia in summer. Southern Hemisphere blocking does occur, most commonly near New Zealand. Blocking can occur in both summer and winter but is more common in winter. Wintertime blocking events sometimes appear to be associated with stratospheric sudden warmings (Quiroz, 1986), which are discussed later in this chapter. The formation of such a block can be associated with upward propagation of a Rossby wave, which then interacts with the stratospheric zonal flow to produce the sudden stratospheric warming (e.g., Martius et al., 2009; Woolings et al., 2010).

Blocks strongly influence weather patterns for one or more weeks at a time, sometimes in association with heat waves and/or persistent droughts (e.g., Green, 1977). If the formation and dissipation of blocks can be foreseen, then it may be possible to predict some weather patterns weeks in advance. Until recently, weather-prediction models were not very successful in forecasting blocks, and the same models produced blocks less often than observed when run in climate simulation mode. Within the last few years, this situation has improved dramatically, apparently as a result of increased model resolution.

What causes the formation of a blocking anticyclone? A simple and widely accepted dynamical theory of blocking is still lacking. As discussed by Colucci (1985), blocks sometimes begin when particularly intense cyclones advect air with low potential vorticity (PV) from the subtropics (where low PV is the norm) into middle latitudes (where low PV is an anomaly). An example is shown in figure 9.3, which is taken from Shutts (1986). At the same time the air with low PV is advected poleward, a mass of high-PV air is advected into position on the equatorward side of the low-PV anomaly. The block thus has a dipole structure in terms of PV. Near the longitude of the block, the PV decreases poleward, whereas it normally increases poleward. The blocking anticyclone could be called a "cut-off high," and the cyclone a "cut-off low." Figure 9.4 shows the corresponding sea-level pressure and 500 hPa–height fields. The dipole structure is clearly evident in the latter. Note that the westerlies have split into a strong branch to the north of the high, and a weaker branch to the south of the low.

How is it possible for blocking highs to persist as well-defined, isolated "objects" even while they are embedded in the turmoil of the midlatitude winter circulation? The destruction of the low seems natural; the persistence of the high demands an explanation. Hoskins et al. (1985) speculated that lows are disrupted by convection, whereas highs can survive because convection is suppressed there. Observations suggest that the smaller-scale transient eddies that are steered around a blocking high actually help maintain the high (e.g., Hansen and Chen, 1982; Hoskins et al., 1983; Egger et al., 1986; Shutts, 1986; Dole, 1986; Mullen, 1987; Nakamura et al., 1997; Chang et al., 2002; Luo and Chen, 2006; Ren et al., 2009). Blocks may be examples of up-scale energy transport, which, as discussed in chapter 10, is expected on theoretical grounds in two-dimensional turbulence. It has been suggested that blocking anticyclones in the Earth's atmosphere are dynamical cousins of the Great Red Spot that has persisted in the atmosphere of Jupiter for at least several hundred Earth years.

The nearly stationary character of blocks suggests that they are somehow anchored to fixed geographic features such as topography, although Hu et al. (2008) reported a simulation of blocks with an aqua-planet model that uses zonally symmetric boundary conditions. A theory that involves topographic anchoring was proposed by Charney and DeVore (1979; hereafter CDV), who suggested that in the presence of topography the large-scale circulation can adopt either of two equilibrium states, one of which corresponds to blocking, and the other to a more zonal flow. The basic idea is that for some configurations of the mean flow stationary waves can exist that are resonantly forced by the topography. The waves then feed back on the mean flow. Under certain conditions, the resonant waves alter the mean flow in a way that facilitates the formation of resonant waves; that is, the waves create a mean flow that favors their own continuing existence. However, if the waves are not present, the resulting mean flow is not conducive to wave formation. Hence, the system can exist in either of two possible configurations, referred to as *multiple equilibria*. The possible existence of multiple equilibria has been a key reason for continuing interest in the CDV theory. Because the theory involves the interactions of the waves with the mean flow, it is inherently nonlinear.

A simplified version of the CDV theory is as follows. The topography is assumed to vary sinusoidally in longitude; that is,

$$h(x) \equiv h_T \cos(K_m x). \tag{66}$$

The motion is described by a streamfunction, ψ, which is assumed to be of the form

$$\psi(x,y,t) = -U(t)y + A(t)\cos(K_m x) + B(t)\sin(K_m x). \tag{67}$$

We can consider (67) as a highly truncated functional expansion (see fig. 9.5). Recall that the streamfunction is defined by the relations $u \equiv -\partial\psi/\partial y$ and $v \equiv -\partial\psi/\partial x$; using these formulas, we see that the zonal component of the flow described by (67) is simply $U(t)$; that is, it depends only on time. The A-part of the wave is in phase with the topography, in the sense that for $A > 0$ the maximum of the streamfunction (corresponding to ridging behavior) occurs over the mountain, whereas for $B > 0$ the

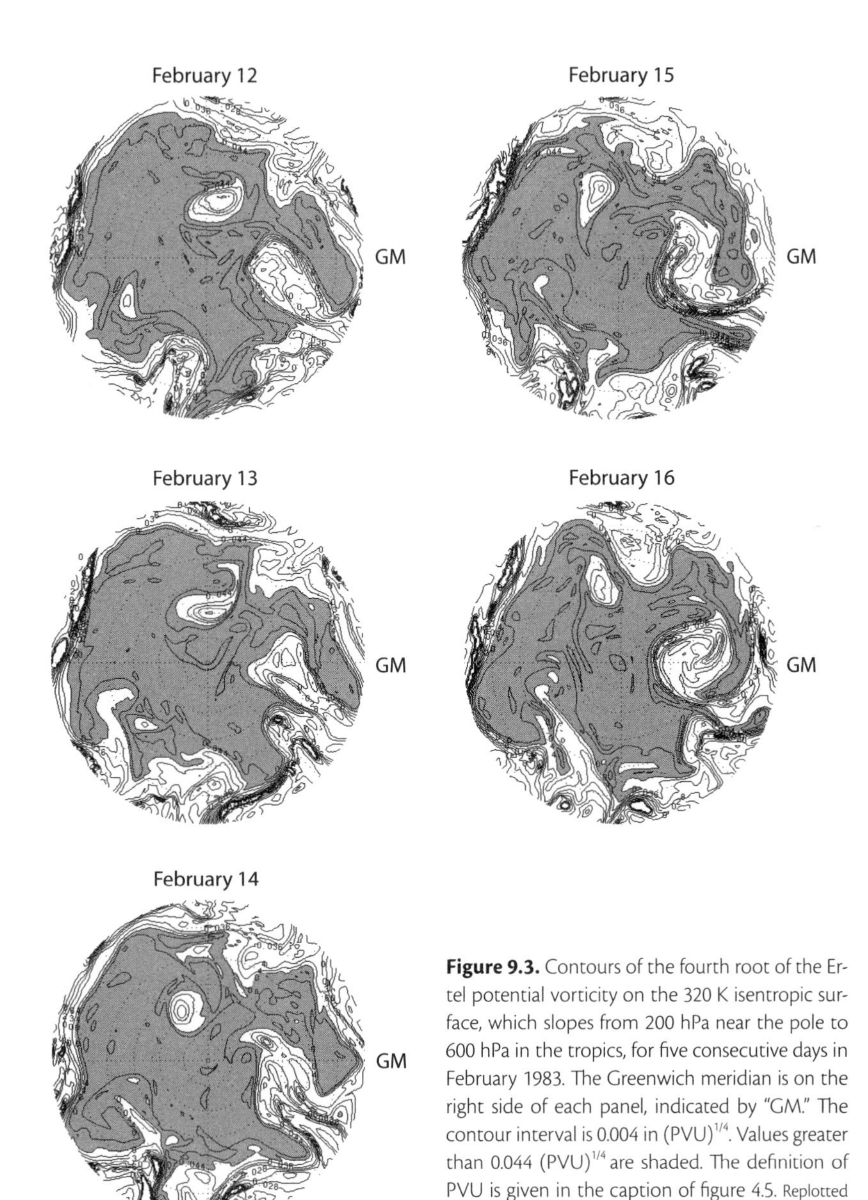

Figure 9.3. Contours of the fourth root of the Ertel potential vorticity on the 320 K isentropic surface, which slopes from 200 hPa near the pole to 600 hPa in the tropics, for five consecutive days in February 1983. The Greenwich meridian is on the right side of each panel, indicated by "GM." The contour interval is 0.004 in $(\text{PVU})^{1/4}$. Values greater than 0.044 $(\text{PVU})^{1/4}$ are shaded. The definition of PVU is given in the caption of figure 4.5. Replotted based on a figure in Shutts (1986).

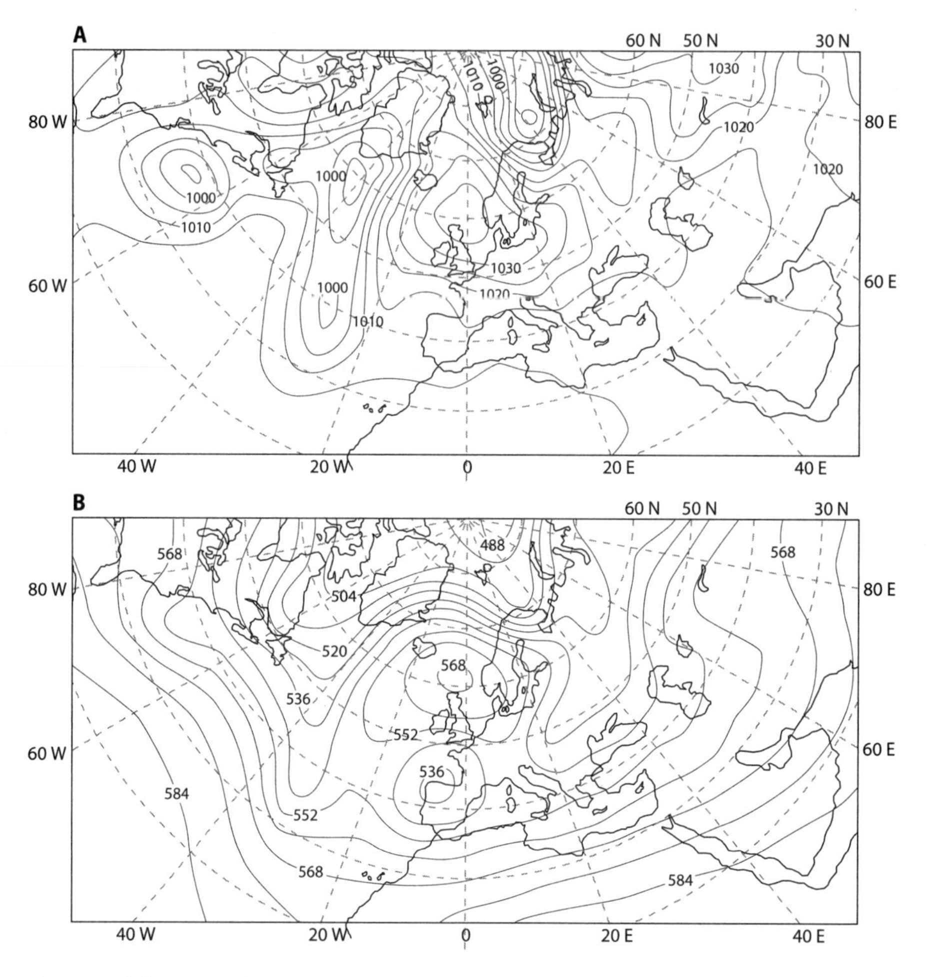

Figure 9.4. (A) Mean sea-level pressure field for February 15, 1983, 12Z. The contour interval is 5 hPa. (B) Height of the 500 hPa surface for February 15, 1983, 12Z. The contour interval is 80 m. From Shutts (1986). Reprinted with permission from Elsevier.

B-part of the wave represents a trough downstream of the mountain. The wavelike meridional component of the flow varies with both x and t:

$$v(x,t) = K_m\{-A(t)\sin(K_m x) + B(t)\cos(K_m x)\}. \tag{68}$$

By substituting (67) into a nonlinear vorticity equation describing a Rossby-wave flow with friction, CDV showed that the wave motion satisfies

$$\frac{1}{K_m}\frac{dA}{dt} + \left(\frac{v}{K_m}\right)A + \left(U - \frac{\beta}{K_m^2}\right)B = 0, \tag{69}$$

$$\frac{1}{K_m}\frac{dB}{dt} - \left(U - \frac{\beta}{K_m^2}\right)A + \left(\frac{v}{K_m}\right)B + \left(\frac{f_0 h_T}{K_m^2 H}\right)U = 0, \tag{70}$$

while the zonal flow obeys

$$\frac{dU}{dt} = \left(\frac{f_0 h_T K_m}{4H}\right)B - v(U - U^*). \tag{71}$$

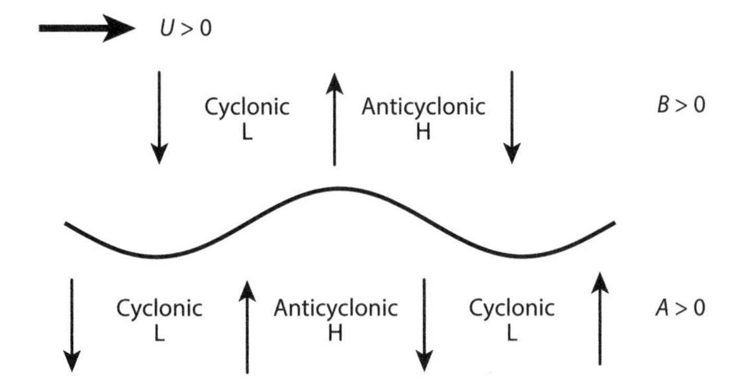

Figure 9.5. The relationships between the *A*- and *B*-components of the meridional wind and the topography in the model of Charney and Devore. The topography is indicated by the wavy line in the center. The arrows indicate the direction of the v-wind component, with "up" corresponding to "southerly." The arrows above the topography are for the *B*-component of the wave, and the arrows below are for the *A*-component, assuming that *A* and *B* are positive. The indicated regions are cyclonic or anticyclonic for the case of $f > 0$ (the Northern Hemisphere).

In (69)–(70), the terms involving v represent friction. In (71), the B-term represents the orographic form drag (or "mountain torque") that the mountain exerts on the mean flow when the wave is oriented with the trough over the mountain, and the U^*-term represents a "momentum forcing" that maintains the mean flow against friction. Note that (69) and (70) "blow up" if the wavenumber of the topography, K_m, is equal to zero. This simply means that there is no wave solution in the absence of topography; that is, the wave is topographically forced. Also note that (69) and (70) are nonlinear, because they involve the products of A and B with U. This nonlinearity represents the wave-mean flow interactions.

CDV considered equilibrium (steady-state) solutions of (69)–(71). These equilibria can be found by setting the time-rate-of-change terms to zero, solving the resulting linear system (69)–(70) for A and B as functions of U, and selecting the appropriate value of U by requiring that the steady-state version of (71) also be satisfied. Figure 9.6 shows an example, in which the straight line represents the (B, U) pairs that satisfy (71), while the curve represents the (B, U) pairs that satisfy (69)–(70). There are three equilibria, but it can be shown that the middle one is unstable. The stable equilibrium with large U has a small wave amplitude, and the stable equilibrium with small U has a large wave amplitude. This is understandable, because the wave term of (71) represents a drag on U due to mountain torque. The solution with large wave amplitude and a weak zonal flow is interpreted as representing blocking. CDV suggested that if the model is subjected to stochastic forcing, representing the random fluctuations of the weather—perhaps manifested through fluctuations of U^*, then the system can undergo occasional transitions back and forth between the neighborhood of the "blocked" equilibrium and the neighborhood of the "unblocked" equilibrium.

Charney and Straus (1980) extended the CDV theory to the baroclinic case, and it has been studied by many other authors. The CDV theory has been heavily criticized (e.g., Tung and Rosenthal, 1985), in part because its extreme idealizations limit the possible behaviors of the model, suggesting that the small number of discrete equilibria (i.e., two) is an artifact. Nevertheless, the theory continues to be cited frequently as an important background concept for the interpretation of blocking and

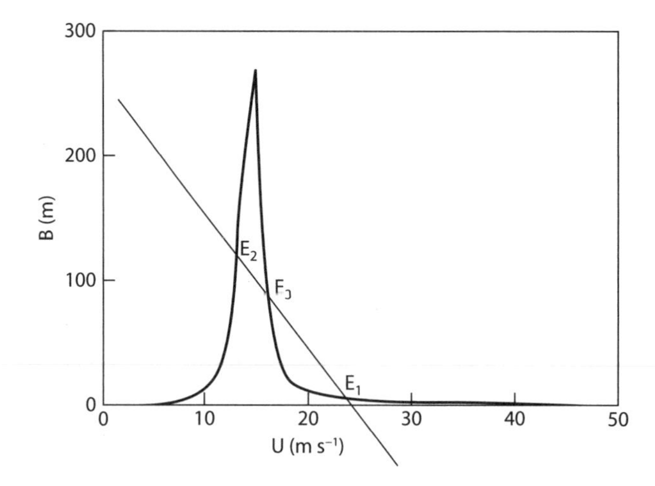

Figure 9.6. Equilibrium solutions of (57)–(58), found by setting the time-rate-of-change terms to zero, solving the resulting linear system for A and B as functions of U, and selecting the appropriate value of U by requiring that (115) also be satisfied. The straight line represents the (B, U) pairs that satisfy (71), while the peaked curve represents the (B, U) pairs that satisfy (69)–(70). Equilibrium solutions occur where the line and the curve intersect. The figure shows that there are three solutions, but it turns out that the middle one is unstable. From Speranza (1986). Reprinted with permission from Elsevier.

also in research on the possible existence of multiple discrete weather "regimes" (e.g., O'Kane et al., 2013).

McWilliams (1980) suggested that *modons* could be considered as idealized models of blocks. Modons are exact solutions of the nonlinear vorticity equation (Flierl, 1978). They have dipole structures, in which a high (a negative vorticity center) is paired with a neighboring low (a positive vorticity center). Modons must have finite amplitudes to exist—there is no such thing as a "linear" modon. An interesting and block-like property of modons is that they are resistant to disruption by perturbations. In addition, a modon "translates" relative to the mean flow, and under special conditions it is possible to set up a modon that is stationary relative to the Earth, in the presence of background westerlies.

As mentioned earlier, we do not yet have a simple, widely accepted theory of blocking. The following issues, among others, have been identified in connection with blocking:

- What causes a blocking anticyclone to form?
- What determines the preferred geographic locations of blocking activity?
- How are blocking highs maintained against the noisy background flow?
- Why are blocks nearly stationary even though they are embedded in strong westerly currents?
- What causes a block to break down?
- Why are persistent, quasi-stationary anticyclones observed but not persistent, quasi-stationary cyclones?

The preceding discussion gave at least partial answers to some of these questions.

Stratospheric Sudden Warmings

Stratospheric sudden warmings (SSWs) were discovered by Richard Scherhag (1952, 1960), and early reviews were published by Schoeberl (1978) and Holton (1980). In an SSW, polar stratospheric temperatures increase by tens of kelvins in just a few days. As would be expected from thermal-wind balance, the usually strong polar stratospheric westerlies (see chapter 3) dramatically weaken and sometimes even give way to easterlies. As shown in figure 9.7, SSWs occur in two different ways. Sometimes the polar vortex simply shifts away from the pole, but in other cases it splits into two parts (Charlton and Polvani, 2007). The polar vortex is reestablished when a positive vorticity anomaly moves back into position over the pole.

Major SSWs occur a bit more often than once every other year in the Northern Hemisphere, usually in January or February. They occur much less often in the Southern Hemisphere, although there was a spectacular south polar event in September 2002 (e.g., Newman and Nash, 2005; Simmons et al., 2005).

As first proposed by Matsuno (1970, 1971; see also Matsuno and Nakamura, 1979), SSWs are triggered by the rapid growth of quasi-stationary tropospheric planetary waves that propagate upward into the stratosphere. Recall from chapter 8 that planetary waves can propagate from the troposphere into the stratosphere in winter but not in summer. The rapid intensification of wave activity is often associated with the formation of a blocking high in the troposphere (e.g., Barriopedro and Calvo, 2014).

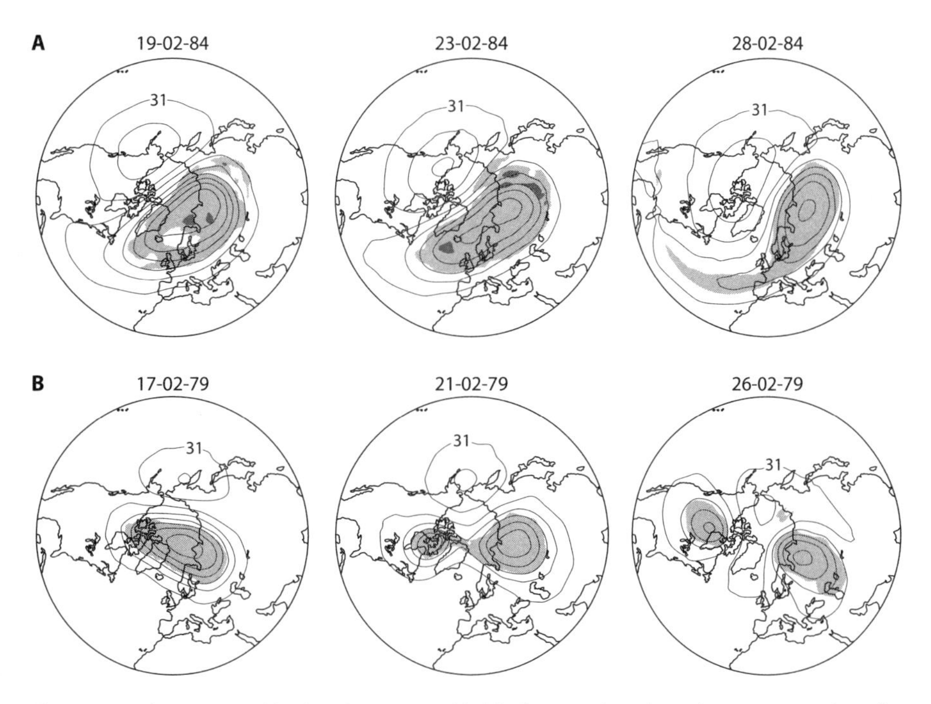

Figure 9.7. Polar stereographic plot of geopotential height (contours) on the 10 hPa pressure surface. The contour interval is 0.4 km, and shading indicates potential vorticity greater than 4.0 PVUs. The definition of PVU is given in the caption of figure 4.5. (A) A vortex displacement–type warming that occurred in February 1984. (B) A vortex splitting–type warming that occurred in February 1979. From Charlton and Polvani (2007).

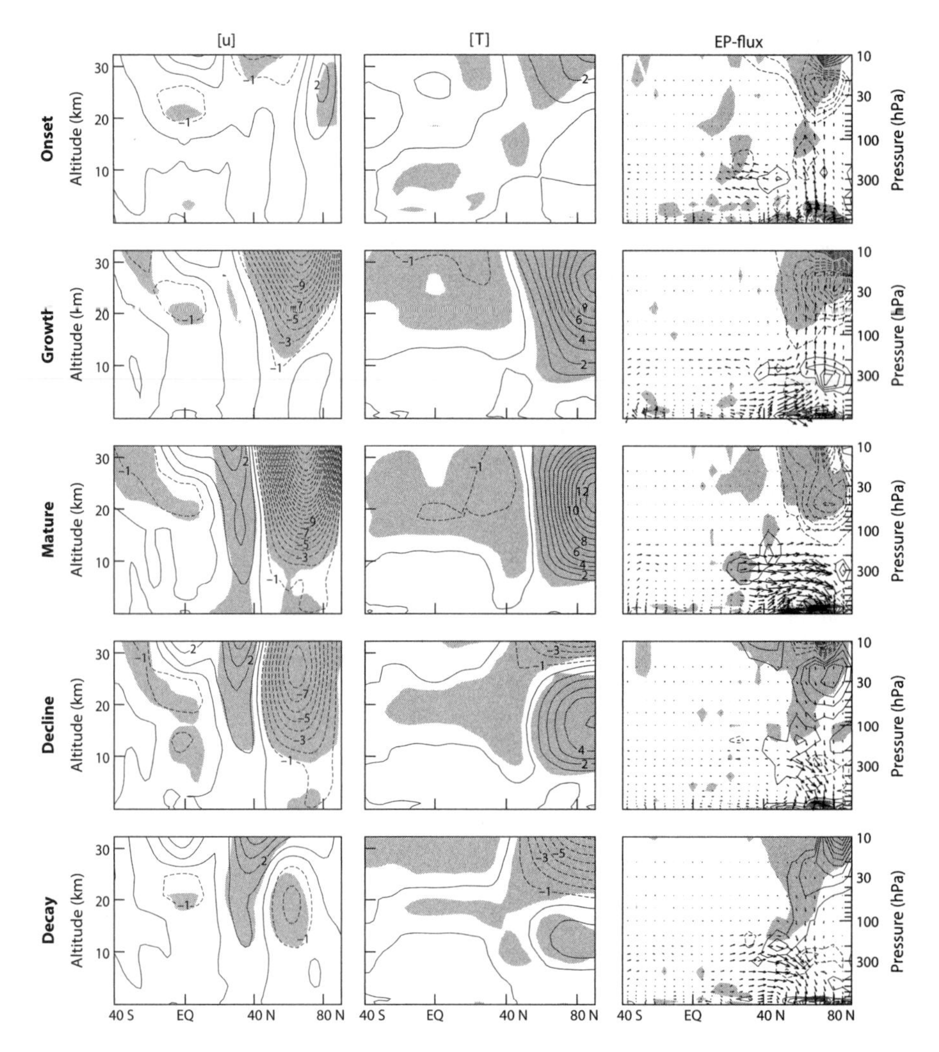

Fig. 9.8. (left) Anomalous zonal-mean zonal wind, (middle) zonal-mean temperature, and (right) EP flux with its divergence during the composite life cycle of SSWs. Negative contours are shown as dashes. Zero contours are given as solid lines. The wind (temperature) contour interval is 1 m s^{-1} (1 K). The EP flux divergence (divided by $\rho a \cos\varphi$) is contoured in the right column at every 0.25 m s^{-1} day^{-1}. The vector lengths in the right column are referenced with respect to the top figure in the column. Gray shading indicates areas with a 95% confidence level (based on t-statistics). Adapted from Limpasuvan et al. (2004). Varavut Limpasuvan of Coastal Carolina University kindly provided a black-and-white version of this figure. © American Meteorological Association. Used with permission.

SSWs are far from "steady," so noninteraction theorems do not apply. Figure 9.8 shows the life cycle of a composite SSW, constructed from observations averaged over many SSWs. The polar stratosphere is observed to warm up, while the lower latitudes cool off. As the polar warming progresses, the stratospheric meridional temperature gradient actually reverses, causing the usually strong westerly polar vortex to give way to polar easterlies! When easterlies form above, further wave propagation is blocked. The deceleration of the polar westerlies is observed to coincide with a strong convergence of the Eliassen-Palm vectors. The vectors are nearly vertical in the stratosphere, indicating a dominance of the meridional energy flux. As the SSW matures, the warm anomaly and the easterlies migrate downward toward the troposphere. It has recently been recognized that SSWs are often followed by persistent changes in tropospheric weather (Baldwin and Dunkerton, 2001; Thompson et al., 2005).

The Quasi-Biennial Oscillation

As discussed by Baldwin et al. (2001), the zonal winds of the tropical stratosphere undergo an amazing *quasi-biennial oscillation* (QBO), with a period of about 26 months. A "ring" of air in the tropical stratosphere, extending over all longitudes, is observed to reverse the direction of its zonal motion roughly every two years, like a giant zonally oriented Ferris wheel, as shown in figure 9.9.

The QBO was discovered by Richard Reed in about 1960 (see the early review by Reed, 1966). The early evidence was not sufficient to establish the existence of a true oscillation in a statistically significant fashion, but additional decades of data have made it clear that a quasi-periodic oscillation does in fact exist. The oscillation is strong: the winds shift from about 20 m s^{-1} westerly to 20 m s^{-1} easterly, and back again. They are observed to propagate down from the middle stratosphere to near the tropopause. Corresponding oscillations are seen in other stratospheric fields, and much more weakly in the troposphere.

Lindzen and Holton (1968) and Holton and Lindzen (1972) proposed that the oscillation is due to the interactions of upward-propagating Kelvin and Yanai waves with the mean flow. More recently, it has been suggested that eastward- and westward-propagating gravity waves play an important role, and the Kelvin and Yanai waves are now being deemphasized. The various waves are observed to produce energy propagation upward into the stratosphere. The source of wave energy must therefore be in the troposphere and is believed to be associated with latent heating. Each type of wave can be blocked by a critical level where $\overline{u}^{\lambda} - c = 0$. *Eastward-propagating waves, such as Kelvin waves and eastward-propagating inertia-gravity* (*EIG*) *waves, can have critical levels in westerlies. Westward-propagating waves, such as Rossby waves, Yanai waves, and westward-propagating inertia-gravity* (*WIG*) *waves, can have critical levels in easterlies* (see fig. 9.10).

As shown in chapter 5, the zonally averaged angular momentum equation can be written as

$$\frac{\partial}{\partial t}(\overline{\rho_\theta M}^{\lambda}) + \frac{1}{a\cos\varphi}\frac{\partial}{\partial\varphi}(\overline{\rho_\theta v M}^{\lambda}\cos\varphi) + \frac{\partial}{\partial\theta}\left(\overline{\rho_\theta \dot{\theta} M}^{\lambda} + \overline{F}_u^{\lambda} a\cos\varphi - \overline{p\frac{\partial z}{\partial\lambda}}^{\lambda}\right) = 0. \quad (72)$$

It is easy to show that

$$\overline{p\frac{\partial z}{\partial\lambda}}^{\lambda} = \overline{p^{*}\frac{\partial z^{*}}{\partial\lambda}}^{\lambda}, \quad (73)$$

from which it follows that

$$\boxed{\text{upward flux of zonal momentum due to the wave} = -\overline{p^{*}\frac{\partial z^{*}}{\partial\lambda}}^{\lambda}}. \quad (74)$$

At this point it is useful to recall the form of the mechanical energy equation in isentropic coordinates, which was stated in a homework problem at the end of chapter 4 and is repeated here for convenience:

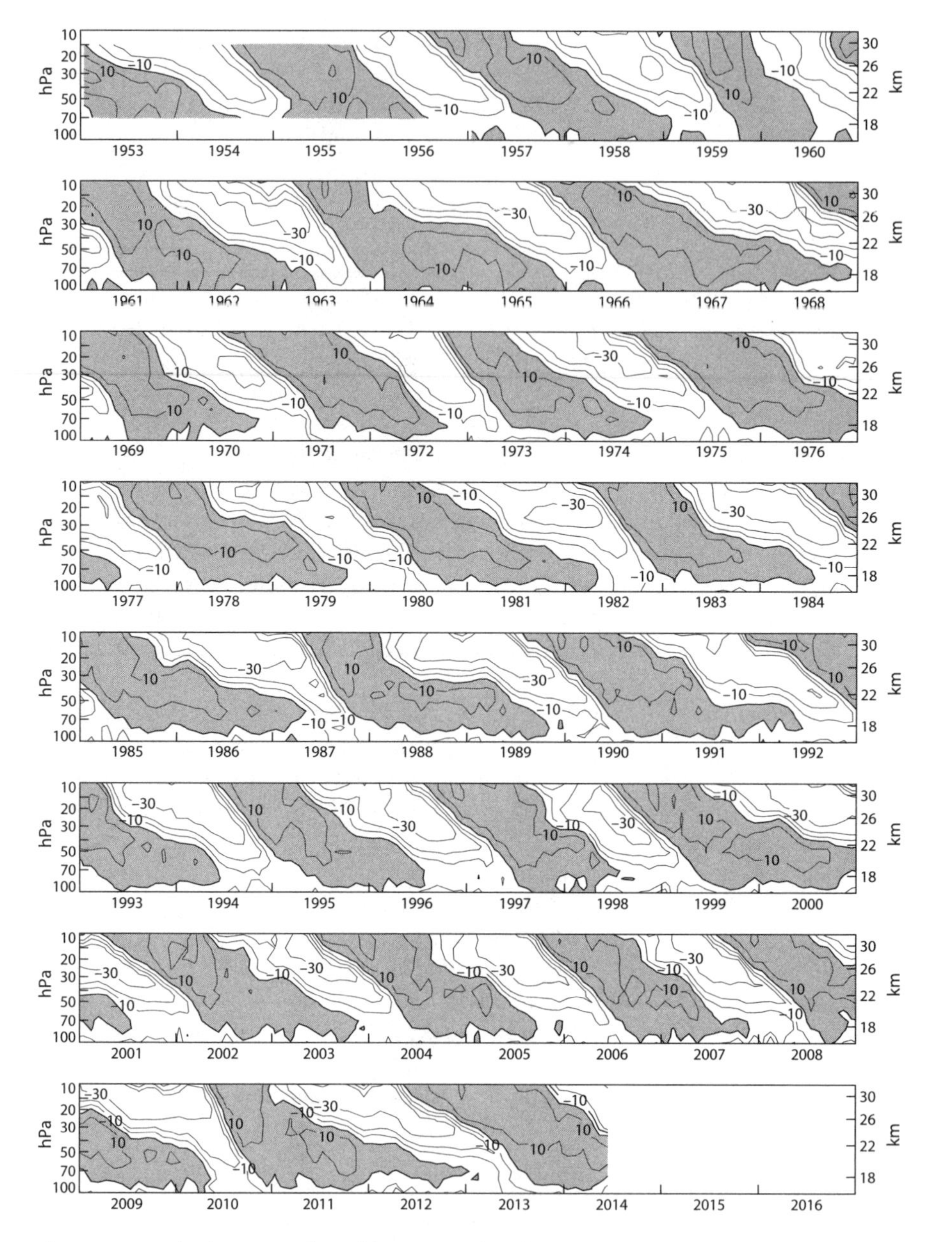

Figure 9.9. Time-height section of monthly-mean zonal winds at equatorial stations: Canton Island, 3° S/172° W (Jan 1953–Aug 1967); Gan/Maldive Islands, 1° S/73° E (Sep 1967–Dec 1975); and Singapore, 1° N/104° E (since Jan 1976). The contour interval is 10 m s^{-1}, and westerlies are shaded. Used with permission of Markus Kunze. Marquardt (1998) created an earlier version of the figure by updating one published by Naujokat (1986). From http://www.geo.fu-berlin.de/en/met/ag/strat/produkte/qbo/.

$$\left[\frac{\partial}{\partial t}(\rho_\theta K)\right]_\theta + \nabla_\theta \cdot [\rho_\theta \mathbf{V}(K+\phi)] + \frac{\partial}{\partial \theta}\left[\rho_\theta \dot{\theta}(K+\phi) - z\left(\frac{\partial p}{\partial t}\right)_\theta + \mathbf{V}\cdot\mathbf{F}_\mathrm{V}\right] = -\rho_\theta \omega\alpha - \rho_\theta \delta. \tag{75}$$

The term $\partial/\partial\theta[-z(\partial p/\partial t)_\theta]$ on the left-hand side of (75) represents the vertical transport of energy via "pressure work." The upward flux of wave energy is therefore given

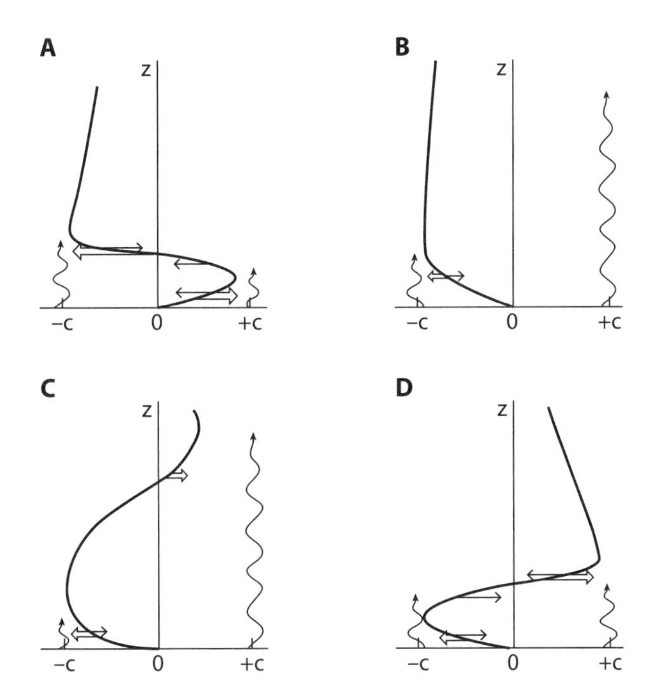

Figure 9.10. In each panel, the vertical axis is height, and the horizontal axis is wind speed and/or phase speed. The curves show the zonal wind as a function of height. The labels +c and −c denote eastward- and westward-phase speeds, respectively. From Plumb (1984). Used with kind permission from Springer Science and Business Media.

by $-\overline{z^*(\partial p^*/\partial t)_\theta}^\lambda$. Recall that $\partial/\partial t = (-c/a\cos\varphi)\,\partial/\partial\lambda$ for a neutral wave propagating zonally with phase velocity c. Then, for such a wave,

$$-\overline{z^*\left(\frac{\partial p^*}{\partial t}\right)_\theta}^\lambda = \frac{c}{a\cos\varphi}\overline{z^*\left(\frac{\partial p^*}{\partial \lambda}\right)_\theta}^\lambda = -\frac{c}{a\cos\varphi}\overline{p^*\left(\frac{\partial z^*}{\partial \lambda}\right)_\theta}^\lambda, \tag{76}$$

so that

$$\boxed{\text{upward flux of wave energy} = -\frac{c}{a\cos\varphi}\overline{p^*\left(\frac{\partial z^*}{\partial \lambda}\right)_\theta}^\lambda}. \tag{77}$$

Comparing (77) with (74), we conclude that for a neutral wave

$$\boxed{\text{upward wave energy flux} = \frac{c}{a\cos\varphi}\text{ times the upward wave angular momentum flux}}. \tag{78}$$

Thus, for eastward-propagating neutral waves (with $c > 0$) the momentum flux and the energy flux have the same sign, while for westward-propagating neutral waves (with $c < 0$) the momentum and energy fluxes have opposite signs.

In particular, upward-propagating Kelvin waves (i.e., Kelvin waves that produce an upward energy flux) transport westerly momentum upward from the level where they are generated and deposit it aloft. They produce a westerly acceleration when they encounter a critical level at the base of a layer of westerlies, thus causing the westerlies to descend with time, as observed in the QBO. Eastward-propagating inertia-gravity waves will produce a similar effect.

Upward-propagating Yanai waves (i.e., Yanai waves that produce an upward energy flux) transport easterly momentum upward from the level where they are generated, and deposit it aloft. They produce an easterly acceleration when they encounter a critical level at the base of a layer of easterlies. Westward-propagating inertia-gravity waves will produce a similar effect, and recent work suggests that they are in fact important for the QBO.

Plumb and McEwan (1978) produced a remarkable laboratory simulation of the QBO in an annular tank full of stratified salt water. In such an experiment, "eastward " and "westward-" propagating inertia-gravity waves are artificially excited by an oscillating diaphragm at the bottom of the tank. At a given level, the direction of the mean flow reverses periodically with time, and these reversals propagate downward. The oscillations are caused by wave-mean flow interactions. A video of this laboratory experiment can be found on the Internet.

For many years, GCMs failed to simulate the QBO. Cariolle et al. (1993) produced a somewhat promising simulation, and Takahashi and Shiobara (1995) successfully simulated a QBO in a simplified GCM. Finally, M. Takahashi (1996) produced fairly successful simulations of the QBO using a full GCM that included both the troposphere and stratosphere. During 1997 the ECMWF model began to produce a QBO when run in climate-simulation mode, that is, without data assimilation. This improvement in the model's performance was associated with an increase in the vertical resolution. It appears that high vertical resolution and possibly also weak damping are needed for a successful numerical simulation of the QBO.

There is some evidence for phenomena similar to the QBO in the atmospheres of Jupiter (Orton et al., 1991; 1994; Leovy et al., 1991; Friedson, 1999; Flasar et al., 2004) and Saturn (Fouchet et al., 2008).

The Eliassen-Palm Theorem in Isentropic Coordinates

The Eliassen-Palm theorem is simpler, more general, and easier to interpret when expressed using isentropic coordinates. The following analysis is based on that of Andrews (1983), Tung (1986), and Andrews et al. (1987). The tendency term of the angular momentum equation can be broken into two parts:

$$\frac{\partial}{\partial t}\left(\overline{\rho_\theta M}^\lambda\right) = \frac{\partial}{\partial t}\left(\overline{\rho_\theta}^\lambda \overline{M}^\lambda\right) + \frac{\partial}{\partial t}\left(\overline{\rho_\theta^* M^*}^\lambda\right). \tag{79}$$

Using (79) in (72), we obtain

$$\begin{aligned} &\frac{\partial}{\partial t}\left(\overline{\rho_\theta}^\lambda \overline{M}^\lambda\right) + \frac{1}{a\cos\varphi}\frac{\partial}{\partial\varphi}\left(\overline{\rho_\theta v M}^\lambda \cos\varphi\right) = -\frac{\partial}{\partial t}\left(\overline{\rho_\theta^* M^*}^\lambda\right) \\ &+\frac{\partial}{\partial\theta}\left(\overline{p^*\frac{\partial z^*}{\partial\lambda}}^\lambda - \overline{\rho_\theta M\dot{\theta}}^\lambda + \overline{F_u}^\lambda a\cos\varphi\right). \end{aligned} \tag{80}$$

Here the "eddy part" of the time-rate-of-change term has been moved to the right-hand-side of the equals sign, for reasons that will be discussed shortly. We want to derive an "advective form" of (80), so we subtract $\overline{M}^\lambda$ times the zonally averaged continuity equation from (80) to obtain

$$\begin{aligned} &\overline{\rho_\theta}^\lambda \frac{\partial \overline{M}^\lambda}{\partial t} + \frac{1}{a\cos\varphi}\left\{\frac{\partial}{\partial\varphi}\left(\overline{\rho_\theta v M}^\lambda \cos\varphi\right)\right\} - \frac{\overline{M}^\lambda}{a\cos\varphi}\frac{\partial}{\partial\varphi}\left(\overline{\rho_\theta v}^\lambda \cos\varphi\right) \\ &= -\frac{\partial}{\partial t}\left(\overline{\rho_\theta^* M^*}^\lambda\right) + \overline{M}^\lambda \frac{\partial}{\partial\theta}\left(\overline{\rho_\theta\dot{\theta}}^\lambda\right) + \frac{\partial}{\partial\theta}\left(\overline{p^*\frac{\partial z^*}{\partial\lambda}}^\lambda - \overline{\rho_\theta M\dot{\theta}}^\lambda + \overline{F_u}^\lambda a\cos\varphi\right). \end{aligned} \tag{81}$$

We cannot yet combine the meridional and vertical derivative terms of (81) to obtain an advective form.

To obtain the desired advective form (a few steps further on), we introduce a *mass-weighted zonal mean*, defined by

$$\hat{A} \equiv \frac{\overline{\rho_\theta A}^\lambda}{\overline{\rho_\theta}^\lambda}, \tag{82}$$

where A is an arbitrary scalar. We note that with this definition the "hat" quantities are independent of longitude. Using (82), we can write

$$\begin{aligned} \rho_\theta A &= \overline{\rho_\theta A}^\lambda + (\rho_\theta A)^* \\ &= \overline{\rho_\theta}^\lambda \hat{A} + (\rho_\theta A)^*, \end{aligned} \tag{83}$$

and

$$\begin{aligned} \overline{\rho_\theta A B}^\lambda &= \overline{\rho_\theta A}^\lambda \overline{B}^\lambda + \overline{(\rho_\theta A)^* B^*}^\lambda \\ &= \overline{\rho_\theta}^\lambda \hat{A} \overline{B}^\lambda + \overline{(\rho_\theta A)^* B^*}^\lambda, \end{aligned} \tag{84}$$

where B is a second arbitrary variable. As special cases of (84), we can write

$$\begin{aligned} \overline{\rho_\theta v B}^\lambda &= \overline{\rho_\theta v}^\lambda \overline{B}^\lambda + \overline{(\rho_\theta v)^* B^*}^\lambda \\ &= \overline{\rho_\theta}^\lambda \hat{v} \overline{B}^\lambda + \overline{(\rho_\theta v)^* B^*}^\lambda, \end{aligned} \tag{85}$$

$$\begin{aligned} \overline{\rho_\theta \dot{\theta} B}^\lambda &= \overline{\rho_\theta \dot{\theta}}^\lambda \overline{B}^\lambda + \overline{(\rho_\theta \dot{\theta})^* B^*}^\lambda \\ &= \overline{\rho_\theta}^\lambda \hat{\dot{\theta}} \overline{B}^\lambda + \overline{(\rho_\theta \dot{\theta})^* B^*}^\lambda. \end{aligned} \tag{86}$$

Using (85) and (86), we can rewrite the zonally averaged continuity and angular momentum equations as

$$\frac{\partial \overline{\rho_\theta}^\lambda}{\partial t} + \frac{1}{a\cos\varphi}\frac{\partial}{\partial \varphi}\left(\overline{\rho_\theta}^\lambda \hat{v}\cos\varphi\right) = -\frac{\partial}{\partial \theta}\left(\overline{\rho_\theta}^\lambda \hat{\dot{\theta}}\right) \tag{87}$$

and

$$\begin{aligned} &\overline{\rho_\theta}^\lambda \frac{\partial \overline{M}^\lambda}{\partial t} + \frac{1}{a\cos\varphi}\frac{\partial}{\partial\varphi}\left\{\left[\overline{\rho_\theta}^\lambda \hat{v}\overline{M}^\lambda + \overline{(\rho_\theta v)^* M^*}^\lambda\right]\cos\varphi\right\} - \frac{\overline{M}^\lambda}{a\cos\varphi}\frac{\partial}{\partial\varphi}\left(\overline{\rho_\theta}^\lambda \hat{v}\cos\varphi\right) \\ &= -\left(\overline{\rho_\theta^* M^*}^\lambda\right) + \overline{M}^\lambda \frac{\partial}{\partial\theta}\left(\overline{\rho_\theta}^\lambda \hat{\dot{\theta}}\right) + \frac{\partial}{\partial\theta}\left[\overline{p^* \frac{\partial z^*}{\partial\lambda}}^\lambda - \overline{\rho_\theta}^\lambda \hat{\dot{\theta}}\overline{M}^\lambda - \overline{(\rho_\theta\dot{\theta})^* M^*}^\lambda + \overline{F_u}^\lambda a\cos\varphi\right], \end{aligned} \tag{88}$$

respectively. We can now combine the meridional and vertical derivatives in (88) to obtain the desired advective form. We also divide by $\overline{\rho_\theta}^\lambda$, simplify, and rearrange, to obtain

$$\begin{aligned} &\frac{\partial \hat{M}}{\partial t} + \frac{\hat{v}}{a}\frac{\partial \overline{M}^\lambda}{\partial\varphi} + \hat{\dot{\theta}}\frac{\partial \overline{M}^\lambda}{\partial\theta} = -\frac{1}{\overline{\rho_\theta}^\lambda}\frac{\partial}{\partial t}\left(\overline{\rho_\theta^* M^*}^\lambda\right) \\ &-\frac{1}{\overline{\rho_\theta}^\lambda\cos\varphi}\frac{\partial}{\partial\varphi}\left[\overline{(\rho_\theta v)^* M^*}^\lambda\cos\varphi\right] + \frac{1}{\overline{\rho_\theta}^\lambda}\frac{\partial}{\partial\theta}\left[\overline{p^*\frac{\partial z^*}{\partial\lambda}}^\lambda - \overline{(\rho_0\dot{\theta})^* M^*}^\lambda + \overline{F_u}^\lambda a\cos\varphi\right]. \end{aligned} \tag{89}$$

Here all the eddy terms (and friction) have been collected on the right-hand side, and the remaining terms have been collected on the left-hand side.

Now, we define the isentropic Eliassen-Palm flux vector as

$$\mathbf{IEPF} \equiv \left(0, IEPF_\varphi, IEPF_\theta\right),$$

where

$$\boxed{IEPF_{\varphi} \equiv -\overline{(\rho_{\theta}v)^{*}M^{*}}^{\lambda}}, \text{ and } \boxed{IEPF_{\theta} \equiv \overline{p^{*}\frac{\partial z^{*}}{\partial \lambda}}^{\lambda} - \overline{(\rho_{0}\dot{\theta})^{*}M^{*}}^{\lambda}}. \tag{90}$$

The meridional component is the negative of the eddy angular momentum flux. The vertical component is the negative of the "total" vertical eddy angular momentum flux, owing to the combination of isentropic form drag and the vertical mass flux associated with heating. The divergence of the isentropic Eliassen-Palm flux is given by

$$\nabla \cdot \mathbf{IEPF} = \frac{-1}{a\cos\varphi}\frac{\partial}{\partial\varphi}\left[\overline{(\rho_{\theta}v)^{*}M^{*}}^{\lambda}\cos\varphi\right] + \frac{\partial}{\partial\theta}\left[\overline{p^{*}\frac{\partial z^{*}}{\partial\lambda}}^{\lambda} - \overline{(\rho_{0}\dot{\theta})^{*}M^{*}}^{\lambda}\right], \tag{91}$$

where it is understood that the meridional derivative is taken along an isentropic surface. With these definitions, we can write (89) as

$$\boxed{\frac{\partial\hat{M}}{\partial t} + \frac{\hat{v}}{a}\frac{\partial\overline{M}^{\lambda}}{\partial\varphi} + \hat{\dot{\theta}}\frac{\partial\overline{M}^{\lambda}}{\partial\theta} = \frac{1}{\overline{\rho_{\theta}}^{\lambda}}\left[-\frac{\partial}{\partial t}\left(\overline{\rho_{\theta}^{*}M^{*}}^{\lambda}\right) + \nabla\cdot\mathbf{IEPF}\right] + \frac{1}{\overline{\rho_{\theta}}^{\lambda}}\frac{\partial}{\partial\theta}\left(\overline{F_{u}}^{\lambda}a\cos\varphi\right)}. \tag{92}$$

This derivation does *not* rely on an assumption of gradient wind balance or any other form of balance. It is therefore more general than the results presented earlier.

We now consider a steady state (or time average) with no heating. Then, the continuity equation (87) reduces to

$$\frac{\partial}{\partial\varphi}\left(\overline{\rho_{\theta}v}^{\lambda}\cos\varphi\right) = 0, \text{ for steady flow without heating.} \tag{93}$$

Since $\cos\varphi = 0$ at both poles, we conclude that

$$\overline{\rho_{\theta}v}^{\lambda} = 0 \text{ for all } \varphi, \text{ for steady flow without heating,} \tag{94}$$

from which it follows that

$$\hat{v} = 0 \text{ for all } \varphi, \text{ for steady flow without heating.} \tag{95}$$

In other words, $\hat{v} \neq 0$ in a time average is due to diabatic processes, as was discussed in chapter 5.

Equation (95) tells us that for steady flow with no heating the meridional advection term of (92) vanishes. Since the flow is steady, the tendency term of (92) is also zero. When friction is also negligible, it follows from (92) that the isentropic Eliassen-Palm flux is nondivergent:

$$\nabla_{\theta}\cdot\mathbf{IEPF} = 0 \text{ for steady flow without heating or friction.} \tag{96}$$

In other words, for steady flow in the absence of heating and friction, the zonally averaged meridional transport of angular momentum is due only to the eddies and is balanced by the form drag on isentropic surfaces. This beautifully simple result is very nearly exact. It is a statement of the Eliassen-Palm theorem.

With the isentropic system, there is no need to define a "residual" circulation, because *the true zonally averaged circulation as seen in isentropic coordinates* is *the residual circulation.* This circulation vanishes for a steady state (or time average) with no heating, even when friction is present. A time-averaged mean meridional circulation in isentropic coordinates is possible only when there is heating.

In chapter 4 we derived the potential vorticity equation in the form

$$\frac{\partial}{\partial t}(\rho_{0}Z) + \nabla_{\theta}\cdot\left[\rho_{0}\mathbf{V}_{h}Z - \mathbf{k}\times\left(\dot{\theta}\frac{\partial\mathbf{V}_{h}}{\partial\theta} + \frac{1}{\rho_{\theta}}\frac{\partial\mathbf{F}_{V}}{\partial\theta}\right)\right] = 0. \tag{97}$$

Applying zonal and time averages to (97), we find that

$$\frac{\partial}{\partial\varphi}\left\{\left[\overline{\rho_\theta v Z}^{\lambda.t} - \left(\overline{\dot{\theta}\frac{\partial u}{\partial\theta}}^{\lambda,t} + \frac{1}{\rho_\theta}\frac{\partial\overline{F_u}^{\lambda,t}}{\partial\theta}\right)\right]\cos\varphi\right\} = 0. \tag{98}$$

This result was derived by Haynes and McIntyre (1987; their eq. 3.4). Equation (98) says that the quantity in square brackets is independent of latitude. It is zero at both poles, because of the factor $\cos\varphi$. It must therefore be zero at every latitude, which implies that

$$\overline{\rho_\theta v Z}^{\lambda.t} = \left(\overline{\dot{\theta}\frac{\partial u}{\partial\theta}}^{\lambda,t} + \frac{1}{\rho_\theta}\frac{\partial\overline{F_u}^{\lambda,t}}{\partial\theta}\right) = 0 \text{ for all } \varphi. \tag{99}$$

Equation (99) can also be obtained directly from the temporally and zonally averaged zonal momentum equation. We can further conclude that

$$\boxed{\overline{\rho_\theta v Z}^{\lambda.t} = 0 \text{ for all } \varphi \text{ except where there is heating or friction.}} \tag{100}$$

Compare (100) with (38), which was derived for the quasi-geostrophic case. An interpretation of (100) is that the time-averaged meridional eddy flux of potential vorticity can be different from zero only where there is heating and/or friction. The conditions under which the eddy PV flux vanishes are the same as those under which the isentropic Eliassen-Palm flux is nondivergent. Again, this result is consistent with the quasi-geostrophic case discussed earlier.

The Brewer-Dobson Circulation

The Brewer-Dobson circulation is a slow poleward drift of the air in the winter stratosphere. It was named by Newell (1963) and is discussed in the recent review by Butchart (2014). It is our final, rather sedate example of an interaction of the eddies with the zonally averaged flow.

Figure 9.11 shows the isentropic streamfunction for the mass circulation, plotted from the surface up to 800 K, which is in the upper stratosphere. The figure shows rising motion in the equatorial lower stratosphere, which is made possible by radiative heating there. The upward-moving current diverges toward the winter pole, where it converges to feed the slow subsidence made possible by radiative cooling.

The middle two panels of the figure show the streamfunction of the isentropic angular momentum transport. In the stratosphere, the contours of the angular momentum streamfunction resemble those of the mass streamfunction. Angular momentum is carried poleward to about 60° of latitude in the winter hemisphere, where the meridional flux sharply converges. The bottom two panels of the figure show that this convergence is compensated for by a strong downward flow of angular momentum due to isentropic form drag.

In effect, we have told the story backward. The real starting point is the downward momentum flux due to isentropic form drag in high latitudes of the winter hemisphere stratosphere. The tropospheric continuation of this downward momentum transport was discussed in chapter 5. The meridional mass flow of the Brewer-Dobson circulation is "induced" by the wave drag; it is required that angular momentum be carried poleward to compensate for the downward angular momentum transport by the waves.

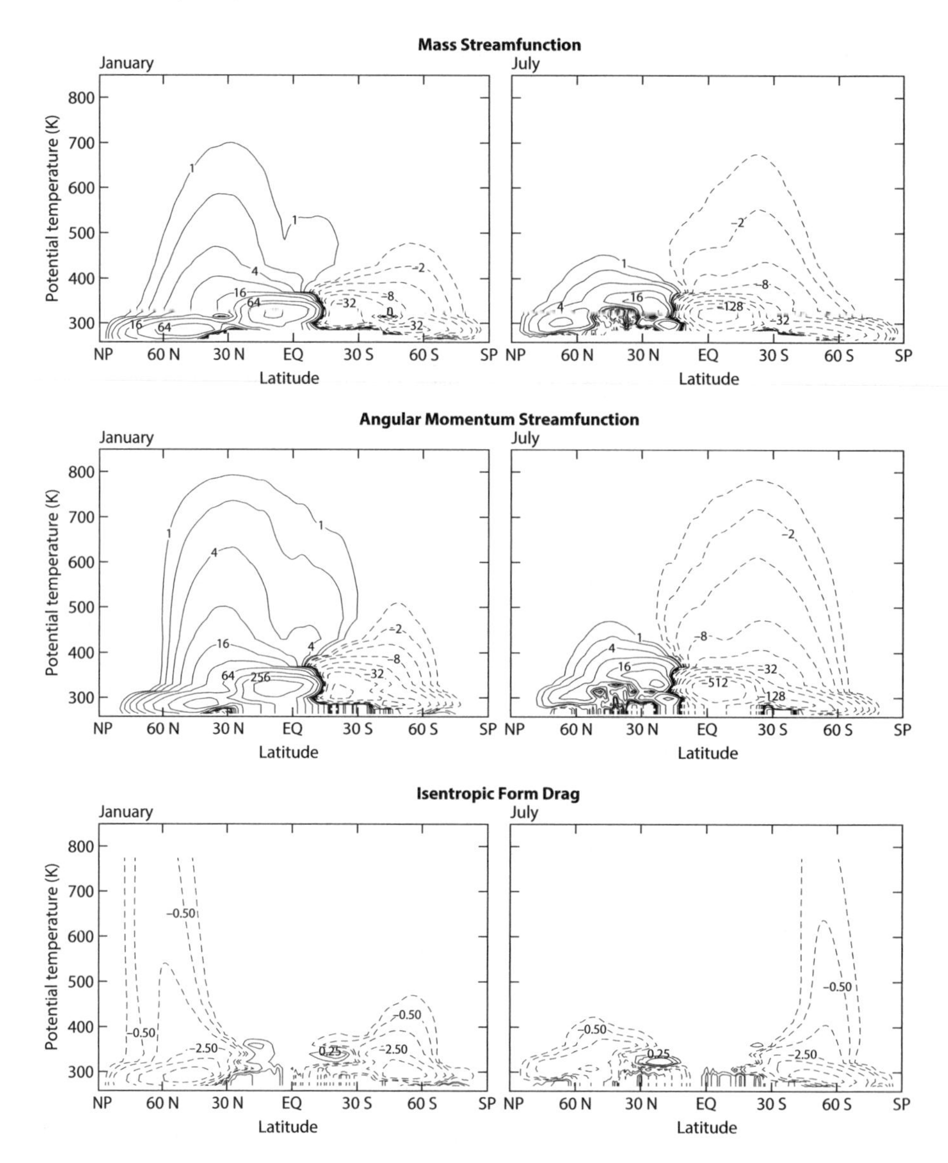

Figure 9.11. Top two panels: The isentropic mass streamfunction, plotted to the 800 K level, for January and July. The contours are logarithmically spaced, in units of 10^9 kg s^{-1}. Center panels: The streamfunction of the total angular momentum transport. The contours are logarithmically spaced, in units of hadleys. Bottom panels: The downward momentum flux due to isentropic form drag. The contour interval is 10^5 N m^{-1}.

Segue

As discussed in chapter 8, atmospheric eddies exist on a wide range of space- and timescales. This chapter examined the interactions of the eddies with the zonally averaged flow.

The eddies interact with one another, too. Chapter 10 considers the global circulation as a type of large-scale turbulence, with a continuous range of scale interactions.

Problems

1. Prove that

$$\overline{v_g^* Z_{QG}^*}^{\lambda} = -\frac{\partial}{\partial y}\left(\overline{u_g^* v_g^*}^{\lambda}\right) - \frac{\partial}{\partial p}\left(\frac{Rf_0}{pS}\overline{v_g^* T^*}^{\lambda}\right).$$

2. Prove that for nondivergent horizontal motion on the sphere the vorticity flux and the angular momentum flux divergence are related by

$$\overline{v^* \zeta^*}^{\lambda} \cos\varphi = -\frac{1}{a\cos\varphi}\frac{\partial}{\partial\varphi}\left(\overline{v^* M^*}^{\lambda}\cos\varphi\right).$$

3. Solve the one-dimensional steady-state version of the Charney-Devore model, as given by

$$\left(\frac{v}{K_m}\right)A + \left(U - \frac{\beta}{K_m^2}\right)B = 0,$$

$$\left(U - \frac{\beta}{K_m^2}\right)A + \left(\frac{v}{K_m}\right)B + \left(\frac{f_0 h_T}{K_m^2 H}\right)U = 0,$$

$$\left(\frac{f_0 h_T K_m}{4H}\right)B - v\left(U - U^*\right) = 0.$$

   To do this you will have to choose appropriate values for the various parameters of the model. Explain your choices.

4. Show that

$$\psi_z^* = \frac{g}{f_0}\frac{T^*}{T_S}.$$

CHAPTER 10

A Fluid Dynamical Commotion

Turbulence Is Made of Vortices

What is turbulence? It is not easy to define. This chapter gives the answer in several parts.

The first piece of the answer is that turbulence contains many interacting vortices of different sizes, created through a chain of shearing instabilities. To understand this concept, we investigate the dynamics of small-scale vortices in three dimensions. The relevant momentum equation is

$$\frac{\partial \mathbf{V}}{\partial t} = \mathbf{V} \times \boldsymbol{\omega} - \nabla\left(\frac{\mathbf{V} \cdot \mathbf{V}}{2}\right) - \nabla\left(\frac{\delta p}{\rho_0}\right) + g\frac{\delta \theta}{\theta_0}\mathbf{k} + \upsilon \nabla^2 \mathbf{V}. \tag{1}$$

Here $\mathbf{V}$ is the three-dimensional wind vector; $\boldsymbol{\omega}$ is the three-dimensional vorticity vector; ρ_0 and θ_0 are reference values of the density and potential temperature, respectively; $\delta p \equiv p - p_0$; $\delta\theta \equiv \theta - \theta_0$; $\mathbf{k}$ is a unit vector pointing upward; and υ is the (molecular) kinematic viscosity. The term $g(\delta\theta/\theta_0)\mathbf{k}$ represents the effects of buoyancy. In writing (1), we used the Boussinesq approximation, for simplicity, and we neglected the effects of rotation. The corresponding Boussinesq form of the continuity equation is

$$\nabla \cdot \mathbf{V} = 0. \tag{2}$$

Taking $\nabla \cdot$ (1) and using (2), we obtain a diagnostic equation for δp:

$$\nabla^2\left(\frac{\delta p}{\rho_0}\right) = \nabla \cdot \left[\mathbf{V} \times \boldsymbol{\omega} - \nabla\left(\frac{\mathbf{V} \cdot \mathbf{V}}{2}\right) + g\frac{\delta \theta}{\theta_0}\mathbf{k} + \upsilon \nabla^2 \mathbf{V}\right]. \tag{3}$$

Equation (3) shows that δp is entirely determined by the wind and temperature fields; it plays only a passive role. For this reason, it is useful to eliminate δp by taking $\nabla \times$ (1) and using (2). The result is

$$\boxed{\frac{\partial \boldsymbol{\omega}}{\partial t} = -(\mathbf{V} \cdot \nabla)\,\boldsymbol{\omega} + (\boldsymbol{\omega} \cdot \nabla)\mathbf{V} + \nabla \times \left(g\frac{\delta \theta}{\theta_0}\mathbf{k}\right) + \upsilon \nabla^2 \boldsymbol{\omega}}. \tag{4}$$

This is the three-dimensional vector vorticity equation. The buoyancy term of (4) acts on the part of the vorticity vector that lies in the horizontal plane.

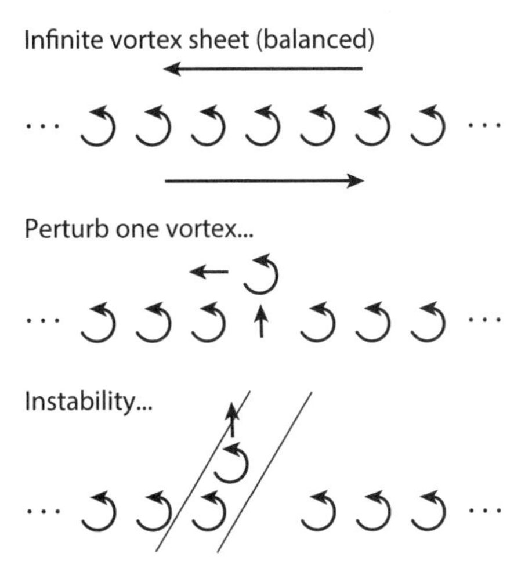

Figure 10.1. The basic mechanism of shearing instability. If a balanced vortex sheet is perturbed, the perturbation is amplified by vorticity advection.

When v is sufficiently small (more precisely, when the Reynolds number, ZL^2/v, is sufficiently large, where Z is a vorticity scale, and L is the diameter of the largest vortices), the flow described by (4) becomes turbulent, owing to shearing instability. The mechanism of shearing instability is illustrated in figure 10.1. The upper panel of the figure shows a balanced vortex sheet that is considered to extend to infinity in both directions. The sheet is in balance because each vortex is advected up by its neighbor to the left, and down by its neighbor to the right, so that no net vertical motion occurs. If one vortex is perturbed upward, however, as in the middle panel, it is carried to the left by the combined effects of the vortices left behind. After being displaced to the left, the vortex experiences a net upward advection, away from the sheet. Thus, the initial upward perturbation is amplified, and so the balanced sheet is unstable. The tilted lines in the bottom panel show a new shear zone, which is also unstable. After some time, the flow becomes highly disordered. This is the mechanism that leads to turbulence. It can work in either two or three dimensions.

The preceding simplified discussion of shearing instability refers only to vorticity advection, that is, to the $-(\mathbf{V} \cdot \nabla)\boldsymbol{\omega}$ term of (4), which is of course nonlinear. *Vorticity advection is the essential process that gives rise to shearing instability, and it is the essential process that gives rise to turbulence.*

The buoyancy term of (4) can make a strong contribution to vorticity production, and so it may be asked whether it represents a second mechanism for the production of turbulence. The answer is "not really." *The buoyancy term of* (4) *is linear; as discussed later, turbulence itself is intrinsically nonlinear.* Buoyancy acts as an indirect source of turbulence, but not as a direct source. Buoyancy generates highly organized coherent structures that contain regions of strong shear, such as the vortex rings associated with thermals. The shear in these buoyancy-driven structures sets the stage for shearing instability. In other words, buoyancy creates conditions in which turbulence can develop through shearing instability, but buoyancy does not actually produce the turbulence itself.

The $(\boldsymbol{\omega} \cdot \nabla)\mathbf{V}$ term of (4), which comes from the advection term of (1), represents both stretching and twisting; it vanishes in two-dimensional flows and is not essential

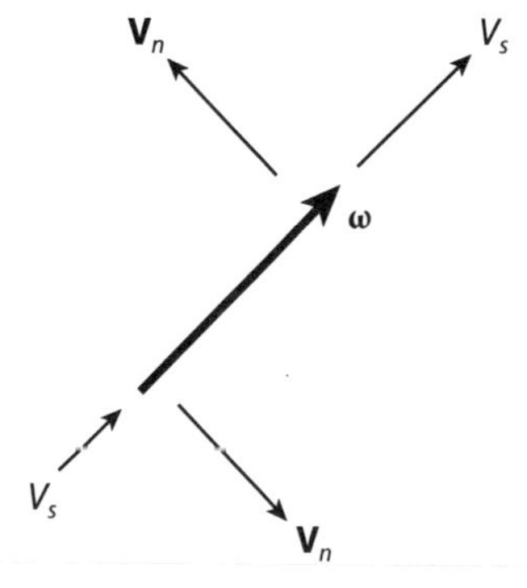

Figure 10.2. The components of the velocity normal and tangent to the vorticity vector, and the stretching and twisting processes associated with these velocity components. The thick arrow represents the vorticity vector, and the thinner arrows represent various velocity vectors or velocity components.

for the generation of turbulence. It does, however, very powerfully influence three-dimensional turbulence. The term can be written as

$$(\boldsymbol{\omega} \cdot \nabla)\mathbf{V} = |\boldsymbol{\omega}|\frac{\partial \mathbf{V}}{\partial s}, \tag{5}$$

where s is a curvilinear coordinate that points in the direction of the vorticity vector (see fig. 10.2). Note that the vector velocity appears. We can divide $|\boldsymbol{\omega}|(\partial \mathbf{V}/\partial s)$ into two parts:

- *Stretching*, $\mathbf{e}_s|\boldsymbol{\omega}|(\partial V_s/\partial s)$, where V_s is the component of $\mathbf{V}$ in the direction of $\boldsymbol{\omega}$, and $\mathbf{e}_s$ is a unit vector parallel to $\boldsymbol{\omega}$ (see fig. 10.2). The stretching term is of course a vector quantity, and it always points in the same direction as the vorticity vector. Positive stretching $\partial V_s/\partial s > 0$ causes $\partial \boldsymbol{\omega}/\partial t > 0$. There is a tendency for the vorticity field and $\partial V_s/\partial s$ to be positively correlated, because viscosity causes convergence into regions of positive vorticity, and divergence from regions of negative vorticity. This will be proven below. A consequence is that stretching causes $|\boldsymbol{\omega}|$ to increase in an average sense. As demonstrated below, we can also say that the stretching term causes the mean of the squared vorticity to increase; the square of the magnitude of the vorticity vector is called the *enstrophy* (or, alternatively, half of that).
- *Twisting*, $|\boldsymbol{\omega}|(\partial \mathbf{V}_n/\partial s)$, where $\mathbf{V}_n$ is the (vector) part of $\mathbf{V}$ that is perpendicular to $\boldsymbol{\omega}$. The twisting term of the vorticity equation is always perpendicular to the vorticity vector. Twisting changes the direction of $\boldsymbol{\omega}$, but not its magnitude.

Integrated over a volume, the enstrophy is a measure of the strength of the vorticity, just as the kinetic energy is a measure of the strength of the wind. To derive an equation for the time change of the enstrophy, we form the dot product of (4) with the vorticity vector, $\boldsymbol{\omega}$:

$$\begin{aligned}\frac{\partial}{\partial t}\left(\frac{|\boldsymbol{\omega}|^2}{2}\right) = -(\mathbf{V}\cdot\nabla)\left(\frac{|\boldsymbol{\omega}|^2}{2}\right) + |\boldsymbol{\omega}|^2\frac{\partial V_s}{\partial s} + \boldsymbol{\omega}\cdot\left[\nabla\times\left(g\frac{\delta\theta}{\theta_0}\mathbf{k}\right)\right] \\ + \nu\{\nabla\cdot[\boldsymbol{\omega}\cdot(\nabla\boldsymbol{\omega})] - [(\nabla\boldsymbol{\omega})\cdot\nabla]\cdot\boldsymbol{\omega}\}.\end{aligned} \tag{6}$$

Here we have exposed the effects of vortex stretching on the enstrophy, using

$$\boldsymbol{\omega} \cdot [(\boldsymbol{\omega} \cdot \nabla)\mathbf{V}] = |\boldsymbol{\omega}|^2 \frac{\partial V_s}{\partial s}, \tag{7}$$

which you should prove as an exercise. The tilting term contributes nothing in (6), because the direction of the tilting term is always perpendicular to the direction of $\boldsymbol{\omega}$. By analogy with our analysis of the friction terms of the kinetic energy equation in chapter 4, we separated the effects of viscosity on the enstrophy into two parts:

$$\begin{aligned} \boldsymbol{\omega} \cdot [v\nabla^2 \boldsymbol{\omega}] &= v\boldsymbol{\omega} \cdot [\nabla \cdot (\nabla \boldsymbol{\omega})] \\ &= v\{\nabla \cdot [\boldsymbol{\omega} \cdot (\nabla \boldsymbol{\omega})] - [(\nabla \boldsymbol{\omega}) \cdot \nabla] \cdot \boldsymbol{\omega}\}. \end{aligned} \tag{8}$$

The term $-v[(\nabla \boldsymbol{\omega}) \cdot \nabla] \cdot \boldsymbol{\omega}$ represents *enstrophy dissipation*, and it is always an enstrophy sink.

Now, we use the continuity equation to rewrite (6) in flux form, drop the buoyancy term, and combine the two flux-divergence terms:

$$\frac{\partial}{\partial t}\left(\frac{1}{2}|\boldsymbol{\omega}|^2\right) = \nabla \cdot \left\{-\mathbf{V}\frac{1}{2}|\boldsymbol{\omega}|^2 + v[\boldsymbol{\omega} \cdot (\nabla \boldsymbol{\omega})]\right\} + |\boldsymbol{\omega}|^2 \frac{\partial s_s}{\partial s} - v[(\nabla \boldsymbol{\omega}) \cdot \nabla] \cdot \boldsymbol{\omega}. \tag{9}$$

In a steady state or time average, the left-hand side of (9) drops out, and when we spatially average over the whole domain, the flux-divergence terms on the right-hand side vanish as well, so that we are left with

$$\int_M \left[|\boldsymbol{\omega}|^2 \frac{\partial v_s}{\partial s} - v[(\nabla \boldsymbol{\omega}) \cdot \nabla] \cdot \boldsymbol{\omega}\right] dM = 0. \tag{10}$$

Because the viscous term of (10) is a sink of enstrophy, we conclude that, *on the average, the stretching term has to be a source of enstrophy*; that is,

$$\boxed{\int_M |\omega|^2 \frac{\partial V_s}{\partial s} dM > 0}. \tag{11}$$

Similar arguments were used in chapter 4 in connection with the energy and entropy budgets of the global atmosphere.

Recall that the rate of production of vorticity by vortex stretching is $|\boldsymbol{\omega}|(\partial V_s/\partial s)$. Equation (11) implies that in a statistically steady turbulence there is a tendency for $\partial V_s/\partial s$ to be positive where $|\boldsymbol{\omega}|$ is larger than average, and for $\partial V_s/\partial s$ to be negative where $|\boldsymbol{\omega}|$ is smaller than average. This means that the rich get richer—strong vortices are stretched and made even stronger.

We conclude that the net effect of $(\boldsymbol{\omega} \cdot \nabla)\mathbf{V}$ is to increase $|\boldsymbol{\omega}|$ in an average sense. The advective terms of the kinetic energy equation do not change the domain-averaged kinetic energy, however. *A process that increases the enstrophy without changing the kinetic energy tends to shift the spectrum of kinetic energy toward shorter scales.* To see why, note that the ratio of the kinetic energy to the enstrophy has units of a length squared:

$$\frac{\frac{1}{2}|\mathbf{V}|^2}{|\boldsymbol{\omega}|^2} \sim L^2. \tag{12}$$

An interpretation is that L is the diameter of the most energetic vortices. The energy of circulations with lots of vorticity per unit kinetic energy is concentrated in relatively small vortices.

The term *inertial process* refers to momentum advection and also includes the effects of the Earth's rotation, although we are neglecting rotation here. An inertial process increases $|\boldsymbol{\omega}|^2$ through vortex stretching, but it does not change $\frac{1}{2}|\mathbf{V}|^2$. It therefore tends to decrease L. This systematic migration of the kinetic energy from larger to smaller scales under inertial processes is called a kinetic energy *cascade*. The terminology evokes a waterfall in which a single stream falls off a cliff and repeatedly splits on the way down as the water strikes rocks or other obstacles. The kinetic energy cascade is like a "flow" of kinetic energy from larger scales to smaller scales. Because viscosity acts most effectively on small scales, *vortex stretching promotes kinetic energy dissipation*. If viscosity could somehow be eliminated, kinetic energy would accumulate (over time) on small scales, leading to a very noisy wind field. Viscosity prevents this accumulation by removing (dissipating) the noise.

Later, we will use the term *anticascade* to describe energy transfers from small scales to larger scales.

Nonlinearity and Scale Interactions

Scale interactions are intrinsically nonlinear; that is, they can arise only from the nonlinear terms of the equations, such as the stretching term. To demonstrate this from a mathematical perspective, we consider the following simple example. Suppose that two modes are given by

$$A(x) = \hat{A}e^{ikx} \text{ and } B(x) = \hat{B}e^{ilx}, \tag{13}$$

respectively. Here the wavenumbers of modes A and B are denoted by k and l, respectively. If we combine A and B linearly, for example, if we form

$$\alpha A + \beta B, \tag{14}$$

where α and β are spatially constant coefficients, then no "new" waves are generated; k and l continue to be the only wavenumbers present. In contrast, if we multiply A and B together, which is a nonlinear operation, then we generate the new wavenumber $k + l$:

$$AB = \hat{A}\hat{B}e^{i(k+l)x}. \tag{15}$$

Other nonlinear operations such as division and exponentiation will also generate new wavenumbers.

Two-Dimensional Turbulence

In two dimensions, the stretching/twisting term of (4) is zero, because $\partial/\partial s$ is zero. It follows that vorticity and enstrophy are both conserved under inertial processes in two-dimensional turbulence. Of course, kinetic energy is also conserved under inertial processes. Since both kinetic energy and enstrophy are conserved under inertial processes in two-dimensional flows, the length scale L is also conserved. The implication is that kinetic energy does not cascade in frictionless two-dimensional flows.

When the effects of viscosity are included in two-dimensional turbulence, they act most effectively on the smallest scales, which are the scales on which the enstrophy is concentrated. Friction therefore removes or "dissipates" enstrophy quite effectively

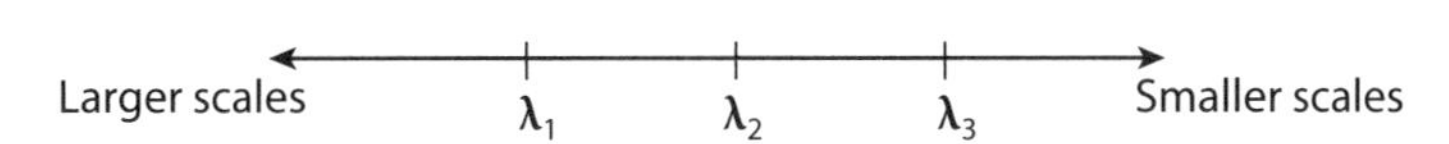

Figure 10.3. Diagram used in the explanation of Fjortoft's (1953) analysis of the exchanges of energy and enstrophy among differing scales in two-dimensional motion.

at the small-scale end of the spectrum. For enstrophy dissipation to continue, additional small-scale enstrophy must be supplied by nonlinear transfer of enstrophy from larger scales to the smaller scales. It is as if the enstrophy dissipation "pulls" enstrophy from larger to smaller scales, through the nonlinear terms. This is an enstrophy cascade. Note that enstrophy dissipation also tends to increase L, over and above the increase (discussed below) that comes through inertial processes. The scale of the most energetic vortices increases with time. This is a kinetic energy "anticascade," which will be discussed shortly.

We conclude that in three-dimensional flow both kinetic energy and enstrophy cascade and are dissipated, while in a two-dimensional flow enstrophy cascades and is dissipated, but kinetic energy is nearly conserved and migrates toward larger scales.

The exchanges of energy and enstrophy among scales in two-dimensional turbulence were studied by Fjortoft (1953), who obtained some very fundamental and famous results, which can be simply summarized as follows. Consider three equally spaced wave wavenumbers, as shown in figure 10.3. By "equally spaced" we mean that

$$\begin{aligned} \lambda_2 - \lambda_1 &= \lambda_3 - \lambda_2 \\ &\equiv \Delta\lambda. \end{aligned} \tag{16}$$

The enstrophy, E, is

$$E = E_1 + E_2 + E_3, \tag{17}$$

and the kinetic energy is

$$K = K_1 + K_2 + K_3. \tag{18}$$

It can be shown that

$$E_n = \lambda_n^2 K_n, \tag{19}$$

where λ_n is a wavenumber, and the subscript n denotes a particular Fourier component.

Consider an *inertial process* such that kinetic energy and enstrophy are redistributed in a two-dimensional frictionless flow; that is,

$$K_n \rightarrow K_n + \delta K_n, \tag{20}$$

$$E_n \rightarrow E_n + \delta E_n. \tag{21}$$

Because kinetic energy and enstrophy are both conserved under two-dimensional inertial processes, we have

$$\sum \delta K_n = 0, \tag{22}$$

$$\sum \delta E_n = 0. \tag{23}$$

From (22) we see that

$$\delta K_1 + \delta K_3 = -\delta K_2. \tag{24}$$

We note from (19) that

$$\delta E_n = \lambda_n^2 \delta K_n. \tag{25}$$

From (23) and (25), we obtain

$$\begin{aligned} \lambda_1^2 \delta K_1 + \lambda_3^2 \delta K_3 &= -\lambda_2^2 \delta K_2 \\ &= \lambda_2^2 (\delta K_1 + \delta K_3). \end{aligned} \tag{26}$$

Collecting terms in (26), we find that

$$\frac{\delta K_3}{\delta K_1} = \frac{\lambda_2^2 - \lambda_1^2}{\lambda_3^2 - \lambda_2^2}. \tag{27}$$

Using (16), we can simplify (27) to

$$\frac{\delta K_3}{\delta K_1} = \frac{\lambda_2 + \lambda_1}{\lambda_3 + \lambda_2} < 1. \tag{28}$$

This is the first result.

Equation (28) shows that the energy transferred to higher wavenumbers (δK_3) is less than the energy transferred to lower wavenumbers (δK_1). This conclusion rests on both (22) and (23), that is, on both kinetic energy conservation and enstrophy conservation. The implication is that kinetic energy actually "migrates" from higher wavenumbers to lower wavenumbers, that is, from smaller scales to larger scales. This process is sometimes called an *anticascade*.

We now perform a similar analysis for the enstrophy. As a first step, we note from (25) and (28) that

$$\begin{aligned} \frac{\delta E_3}{\delta E_1} &= \frac{\lambda_3^2}{\lambda_1^2}\left(\frac{\lambda_2 + \lambda_1}{\lambda_3 + \lambda_2}\right) \\ &= \frac{(\lambda_2 + \Delta\lambda)^2 (\lambda_2 - \frac{1}{2}\Delta\lambda)}{(\lambda_2 - \Delta\lambda)^2 (\lambda_2 + \frac{1}{2}\Delta\lambda)}. \end{aligned} \tag{29}$$

To show that $\delta E_3 / \delta E_1$ is greater than 1, we demonstrate that it can be written as $a \cdot b \cdot c$, where a, b, and c are each greater than 1. We can choose

$$a = \frac{\lambda_2 + \Delta\lambda}{\lambda_2 + \frac{1}{2}\Delta\lambda} > 1, \quad b = \frac{\lambda_2 - \frac{1}{2}\Delta\lambda}{\lambda_2 - \Delta\lambda} > 1, \quad c = \frac{\lambda_2 + \Delta\lambda}{\lambda_2 - \Delta\lambda} > 1. \tag{30}$$

The conclusion is that enstrophy does cascade to higher wavenumbers in two-dimensional turbulence. In the presence of viscosity, such a cascade ultimately leads to enstrophy dissipation.

Quasi-Two-Dimensional Turbulence

Large-scale motions are quasi-two-dimensional, so it is reasonable to suspect that an enstrophy cascade occurs for them just as it does for purely two-dimensional motion. We consider the potential vorticity equation derived earlier:

$$\frac{\partial(\rho_\theta q)}{\partial t} + \nabla_\theta \cdot (\rho_\theta \mathbf{V} q) = \nabla_\theta \cdot \left(\rho_\theta \dot{\theta} \frac{\partial \mathbf{V}}{\partial \theta} + \mathbf{F}\right), \tag{31}$$

and the continuity equation in isentropic coordinates

$$\frac{\partial \rho_\theta}{\partial t} + \nabla_\theta \cdot (\rho_\theta \mathbf{V}) + \frac{\partial(\rho_\theta \dot{\theta})}{\partial \theta} = 0. \tag{32}$$

Here ρ_θ is the pseudodensity. By combining (31) and (32), we obtain an advective form of the potential vorticity equation:

$$\rho_\theta \frac{\partial q}{\partial t} + \rho_\theta \mathbf{V} \cdot \nabla_\theta q = q \frac{\partial(\rho_\theta \dot{\theta})}{\partial \theta} + \nabla_\theta \cdot \left(\rho_\theta \dot{\theta} \frac{\partial \mathbf{V}}{\partial \theta} + \mathbf{F} \right). \tag{33}$$

According to (33), in the absence of heating and friction the potential vorticity is conserved following a particle; we recall that in the absence of heating the particle moves along an isentropic surface.

In a region where latent heating increases with height to a maximum in the middle troposphere and then decreases with height from the middle troposphere to the tropopause, a positive potential vorticity anomaly is generated in the lower troposphere, and a negative anomaly is generated in the upper troposphere.

Multiplying (33) by q, we get

$$\rho_\theta \frac{\partial}{\partial t}\left(\frac{q^2}{2}\right) + \rho_\theta \mathbf{V} \cdot \nabla_\theta \left(\frac{q^2}{2}\right) = q^2 \frac{\partial(\rho_\theta \dot{\theta})}{\partial \theta} + q \nabla_\theta \cdot \left(\rho_\theta \dot{\theta} \frac{\partial \mathbf{V}}{\partial \theta} + \mathbf{F} \right). \tag{34}$$

The quantity q^2 is called the *potential enstrophy*. Equation (34) shows that potential enstrophy tends to be generated locally where the heating rate increases with height, and destroyed where the heating rate decreases with height. For example, in a region where latent heating increases with height to a maximum in the middle troposphere and then decreases with height from the middle troposphere to the tropopause, potential enstrophy is generated in the lower troposphere and destroyed in the upper troposphere.

Using (32), we can convert (34) back to flux form:

$$\frac{\partial}{\partial t}\left(\rho_\theta \frac{q^2}{2}\right) + \nabla_\theta \cdot \left(\rho_\theta \mathbf{V} \frac{q^2}{2}\right) = \frac{q^2}{2} \frac{\partial(\rho_\theta \dot{\theta})}{\partial \theta} + q \nabla_\theta \cdot \left(\rho_\theta \dot{\theta} \frac{\partial \mathbf{V}}{\partial \theta} + \mathbf{F} \right). \tag{35}$$

Equation (35) implies that

$$\overline{\rho_\theta q^2}^{\theta} = \text{constant in the absence of heating and friction.} \tag{36}$$

Here $\overline{(\ )}^{\theta}$ denotes an average over an isentropic surface.

First, we consider a nondivergent flow (i.e., nondivergent on an isentropic surface) with no heating. For that very special case, we can write (32) as

$$\frac{D\rho_\theta}{Dt} = 0, \tag{37}$$

which says that ρ_θ is constant following the motion. Since

$$\rho_\theta q^2 = \frac{\eta^2}{\rho_\theta}, \tag{38}$$

where η is the absolute vorticity, (36) reduces to

$$\overline{\eta^2}^{\theta} = \text{constant for nondivergent flow with no heating or friction.} \tag{39}$$

It follows that, in this limiting case,

$$L^2 \equiv \frac{\overline{K}^{\theta}}{\overline{\eta^2}^{\theta}} = \text{constant}. \tag{40}$$

This result is closely analogous to the conclusion we reached earlier for the case of two-dimensional turbulence.

For a more general divergent flow, we can write

$$\overline{\eta^2}^{\theta} < \overline{\left\{\frac{(\rho_\theta)_{\max}}{\rho_\theta}\right\}\eta^2}^{\theta}, \tag{41}$$

where $(\rho_\theta)_{\max}$ is an appropriately chosen constant *upper bound* on ρ_θ. We can consider $(\rho_\theta)_{\max}$ to be a constant. Then, we can rewrite (41) as

$$\overline{\eta^2}^{\theta} < (\rho_\theta)_{\max}\overline{\left(\frac{\eta^2}{\rho_\theta}\right)}^{\theta} = (\rho_\theta)_{\max}\overline{(\rho_\theta q^2)}^{\theta} = \text{constant}. \tag{42}$$

According to (42), $\overline{\eta^2}^{\theta}$ has an upper bound. Then, from (40) we see that L^2 has a *lower* bound. Thus, even for divergent motion the kinetic energy cannot cascade into arbitrarily small scales; it tends to stay in the larger scales. This is a partial explanation for the "smoothness" of the global circulation.

Charney (1971) showed that even though real atmospheric motions are not two-dimensional, the constraint of quasi-geostrophy causes large-scale geostrophic turbulence to behave much like idealized two-dimensional turbulence, so that potential enstrophy cascades to smaller scales and is dissipated, while kinetic energy is conserved (is not dissipated) and anticascades to larger scales.

When kinetic energy is conserved while enstrophy decreases due to dissipation, the effect is that L must increase. (Our earlier conclusion that L remains constant was based on the assumption that both kinetic energy and enstrophy are invariant.) Because the total amount of "room" available to eddies is fixed (the planet is not getting any bigger), the only way for L to increase is for the "number of eddies" to decrease; for this reason, "eddy mergers" tend to occur in two-dimensional or geostrophic turbulence.

In an idealized numerical simulation, McWilliams et al. (1994) showed that through vortex mergers pure two-dimensional turbulence gradually organizes itself into a single large cyclone-anticyclone pair (see fig. 10.4).

It has been hypothesized that the anticyclonic Great Red Spot on Jupiter, and similar phenomena observed elsewhere on Jupiter and the other giant gaseous planets of the outer solar system, may be the end products of a kinetic energy anticascade in quasi-two-dimensional turbulence.

Dimensional Analysis of the Kinetic Energy Spectrum

We now turn to an analysis of the distribution of kinetic energy with scale, that is, the kinetic energy "spectrum." We consider a three-dimensional flow. Let ε be the rate of kinetic energy dissipation per unit mass, and υ the kinematic viscosity. These quantities have physical dimensions (see appendix B) as follows:

$$\begin{aligned}\varepsilon &\sim L^2T^{-3},\\ \upsilon &\sim L^2T^{-1}.\end{aligned} \tag{43}$$

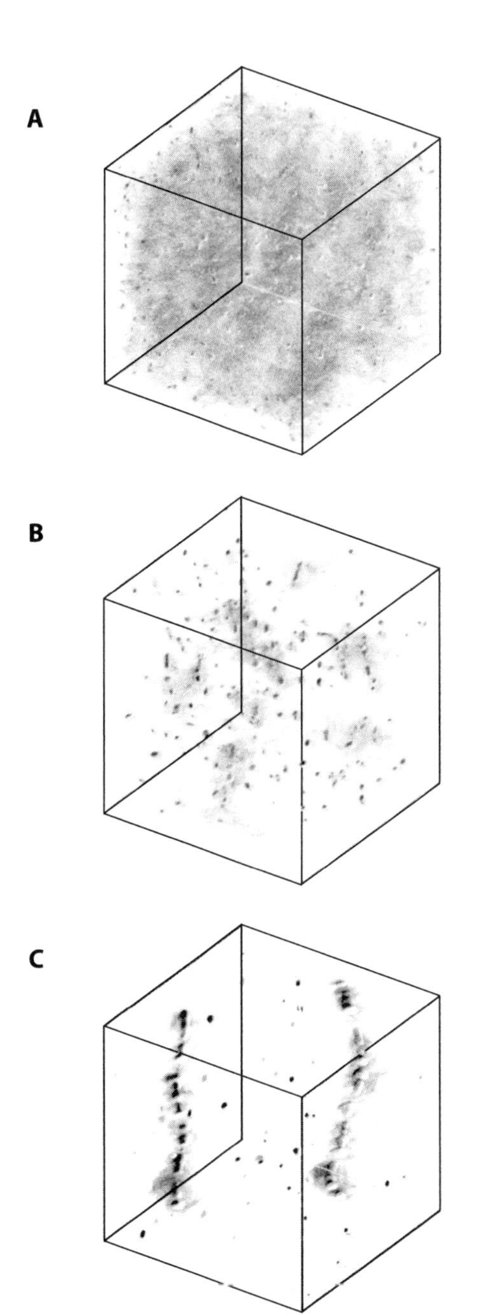

Figure 10.4. A time sequence of the vorticity distribution in an idealized numerical simulation of quasi-two-dimensional turbulence, by McWilliams et al. (1994). In panel A the vorticity is highly disorganized. Through a process of enstrophy dissipation and vortex mergers, the circulation gradually organizes itself into a single pair of vortices—one cyclonic and one anticyclonic, as shown in panel C. A color version of the figure can be found in McWilliams et al. (1994). Thanks to Profs. James McWilliams of UCLA and Jeffrey Weiss of the University of Colorado for assistance with this figure.

Kolmogorov (1941) hypothesized that for locally homogeneous and isotropic three-dimensional turbulence, the turbulence statistics are determined by ε and v; according to his hypothesis, if ε and v are known, no other information is needed. Note that ε is a property of the flow, while v is a property of the fluid.

The *viscous subrange* consists of scales for which both ε and v are important. Kolmogorov further hypothesized that there exists an *inertial subrange*—a range of scales—within which energy is neither generated nor dissipated but just "passes through," like a little town on a major highway where nobody stops except perhaps to buy gas. The inertial subrange consists of homogeneous isotropic eddies that are larger than λ_k. Together, the viscous subrange and the *inertial subrange* make up the homogeneous isotropic component of the turbulence. The smallest scale in the inertial subrange, which is the same as the largest scale in the viscous subrange, can be estimated by forming a length from v and ε:

$$L_K = \left(\frac{v^3}{\varepsilon}\right)^{\frac{1}{4}}. \tag{44}$$

This quantity is called the *Kolmogorov microscale*. According to (44), if ε increases for a given v (which means for a given kind of fluid), then L_K has to become smaller, but weakly. For the Earth's atmosphere, a typical value of L_K is in the range 10^{-2} to 10^{-3} m.

Kolmogorov hypothesized that for the scales of motion that lie within the inertial subrange, the turbulence statistics are determined, as functions of wavenumber, by a single dimensional parameter that characterizes the particular flow in question: the dissipation rate, ε. Note that ε represents dissipation that occurs on scales outside (shorter than those of) the inertial subrange.

In the boundary layer of the Earth's atmosphere, where the turbulent eddies have scales on the order of a few kilometers or less, kinetic energy is typically generated at lower wavenumbers (on the scale of the largest turbulent eddies), migrates to higher wavenumbers by way of the inertial subrange, and is finally dissipated on the smallest scales. This process is illustrated in figure 10.5. We want to find $K(k)$, called the *spectrum of kinetic energy*, in the inertial subrange. This is the squared modulus of the Fourier transform of the velocity and has units of energy per unit mass per unit wavenumber. Dimensionally,

$$K(k) \sim L^3T^{-2}. \tag{45}$$

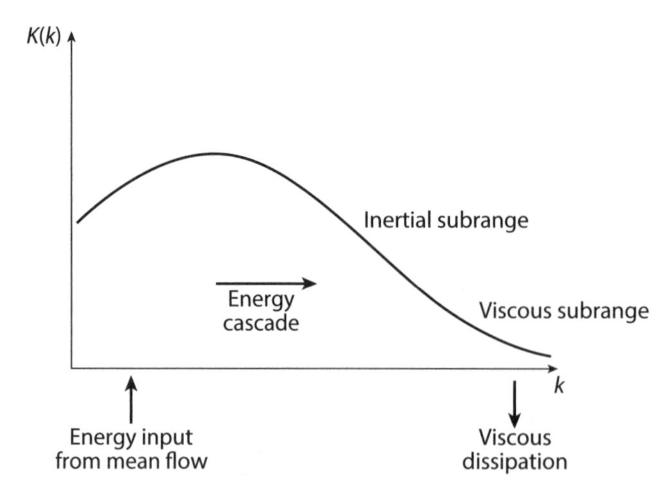

Figure 10.5. Sketch illustrating, in a highly idealized way, the flow of kinetic energy through wavenumber space, from a low-wavenumber region with an energy source, to a high-wavenumber region in which dissipation removes kinetic energy. This diagram applies to the case of three-dimensional turbulence, as found, for example, in the boundary layer.

Because the turbulence statistics in the inertial subrange depend only on ε, $K(k)$ can depend only on ε and k. Assuming that

$$K(k) \sim \alpha \varepsilon^m k^n, \tag{46}$$

where $\alpha \cong 1.5$ is an empirically determined nondimensional constant, we find that $m = 2/3$, and $n = -5/3$. In other words,

$$K(k) \sim \alpha \varepsilon^{2/3} k^{-5/3}. \tag{47}$$

As discussed, for example, by Lesieur (1995, p. 178), equation (47) is remarkably well supported by measurements of small-scale turbulence in the laboratory, the atmosphere, and the oceans.

As discussed earlier, kinetic energy does not cascade in two-dimensional turbulence, whereas enstrophy does. Suppose that in two-dimensional turbulence there exists an inertial subrange, presumably on rather large scales, in which enstrophy is neither generated nor dissipated, and that within this inertial subrange the turbulence statistics are determined by the rate of enstrophy dissipation. Using methods similar to those used to derive (47), we can show that the kinetic energy spectrum follows k^{-3}. This conclusion is supported by numerical simulations of two-dimensional turbulence. Note that k^{-3} decreases more steeply with increasing k than does $k^{-5/3}$. This means that there is less kinetic energy at high wave numbers in two-dimensional turbulence than in three-dimensional turbulence. This finding is consistent with our earlier conclusion that kinetic energy dissipation is weak in two-dimensional turbulence and strong in three-dimensional turbulence.

An analysis very similar to that used to derive (47) can be used to determine the kinetic energy spectrum for scales longer than those at which kinetic energy is generated. We assume that in an inertial subrange that lies up-scale from the kinetic energy source, the kinetic energy spectrum depends only on the strength of the source and the wavenumber. The conclusion is that up-scale from the energy source, the kinetic energy spectrum follows $k^{-5/3}$. This is true for either two- or three-dimensional turbulence.

In figure 10.6, which is a more general and more refined version of figure 10.5, four distinct inertial subranges are indicated. The "low" wavenumber k_B denotes the scale at which baroclinic instability acts as a kinetic energy source. The "high" wavenumber k_C denotes the scale at which convection acts as a kinetic energy source. For $k > k_C$,

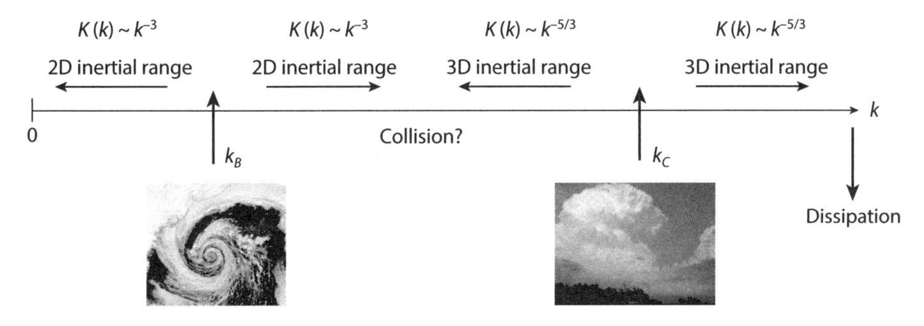

Figure 10.6. Cartoon illustrating the kinetic energy spectrum in the Earth's atmosphere, as implied by dimensional analysis. Baroclinic instability adds kinetic energy at wavenumber k_B, and convection adds kinetic energy at wavenumber k_C. See the text for details.

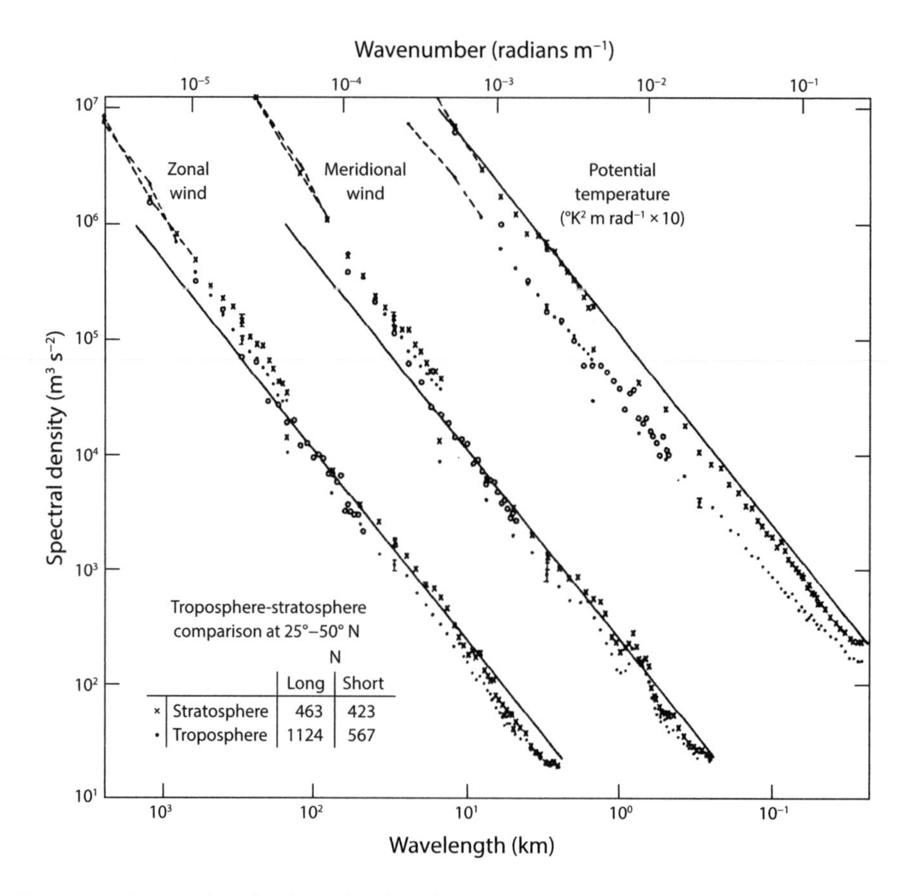

Figure 10.7. Spectra of zonal and meridional winds and potential temperature near the tropopause, obtained from aircraft data over many parts of the world. The meridional wind spectrum is displaced one decade to the right, and the potential temperature spectrum two decades. This is one of the most frequently reproduced figures in atmospheric science. From Lilly (1998), after Nastrom and Gage (1985). © American Meteorological Association. Used with permission.

the inertial range is three-dimensional, with $K(k) \sim k^{-5/3}$ and an energy cascade to smaller scales, ultimately feeding kinetic energy dissipation. For k slightly less than k_C, the inertial range is three-dimensional, with $K(k) \sim k^{-5/3}$ and energy flowing up-scale. For k slightly greater than k_B, the inertial subrange is two-dimensional, with $K(k) \sim k^{-3}$ and energy flowing down-scale. For k less than k_B, energy flows up-scale, away from the baroclinic energy source. Clearly, the shape of the spectrum has to change somewhere between k_B and k_C, where the word "Collision?" appears in the figure.

Lilly (1998) discussed observations that appear to show such a change in the spectral slope, as seen in figure 10.7. The spectrum follows $k^{-5/3}$ for scales less than about 100 km, and k^{-3} for scales greater than about 100 km. There is a "kink" in the spectrum close to 100 km. A second kink is visible at a scale of several thousand kilometers. The correspondence between figures 10.6 and 10.7 should be clear.

There is another complication. Rhines (1975) pointed out that sufficiently large vortices will feel the β-effect, which exerts a "restoring force" opposing large meridional excursions by the particles making up the vortex. This means that the β-effect tends to resist meridional widening of vortices beyond some limit. Rhines suggested that eddies will in fact begin to behave as Rossby-wave packets for scales large enough so that the characteristic eddy velocity, U, is comparable to the phase speed of a Rossby wave. Dimensional analysis suggests that this behavior occurs for

$$k \sim k_\beta \equiv \sqrt{\frac{\beta}{U}}. \tag{48}$$

This scale is sometimes called the *Rhines length* or the *Rhines barrier*. For $k < k_\beta$, further meridional broadening is resisted by β, *but longitudinal broadening can continue*. The eddies therefore become elongated in the zonal direction and ultimately give rise to alternating zonal jets of width k_β^{-1}, like those seen on Jupiter. In recent years a number of numerical modeling studies have tended to support this idea (e.g., Huang and Robinson, 1998).

In summary, vorticity and enstrophy are conserved in two-dimensional flow but not in three-dimensional flow. Kinetic energy is conserved under inertial processes in both two- and three-dimensional flows. Because both energy and enstrophy are conserved in two-dimensional flows, a two-dimensional motion field "has fewer options" than does a three-dimensional one. Because kinetic energy does not cascade in two-dimensional flow, the motion remains smooth and is dominated by "large" eddies.

Observations of the Kinetic Energy Spectrum

Boer and Shepherd (1983) analyzed observations to examine the spectra of kinetic energy, enstrophy, and available potential energy, and also the exchanges of kinetic energy among various scales. Following a suggestion of Baer (1972), they used the two-dimensional index associated with the spherical harmonics as a measure of scale, much as Blackmon did in the work described in chapter 8. The vertically integrated spectra obtained by Boer and Shepherd are shown in figure 10.8, for kinetic energy and enstrophy only. The slope of the kinetic energy spectrum is plotted as a function of height in figure 10.9. A k^{-3} behavior is evident, particularly at the upper levels. Boer and Shepherd evaluated the exchanges of kinetic energy among the various scales, as shown in figure 10.10; Chen and Wiin-Nielsen (1978) reported similar computations. The smaller scales generally experience a kinetic energy cascade toward even smaller scales, as would be expected for three-dimensional turbulence, but the larger scales experience an inverse cascade, as would be expected for two-dimensional turbulence.

It has also been suggested that the formation and maintenance of a blocking high represents an example of up-scale energy transfer. As mentioned in chapter 9, the formation of a block is associated with the advection of subtropical potential vorticity into middle latitudes. The "agent" that carries out this advection is a rapidly developing cyclone, such as over the Gulf Stream (Hoskins et al., 1985). Moreover, it is the interaction of the block with small-scale eddies that allows the block to maintain itself for an extended period (Shutts, 1986).

Dissipating Enstrophy but Not Kinetic Energy

Sadourny and Basdevant (1985) suggested an interesting approach to representing the effects of quasi-two-dimensional, geostrophic turbulence in the momentum equation. The issue is that in such a flow enstrophy is dissipated but kinetic energy is

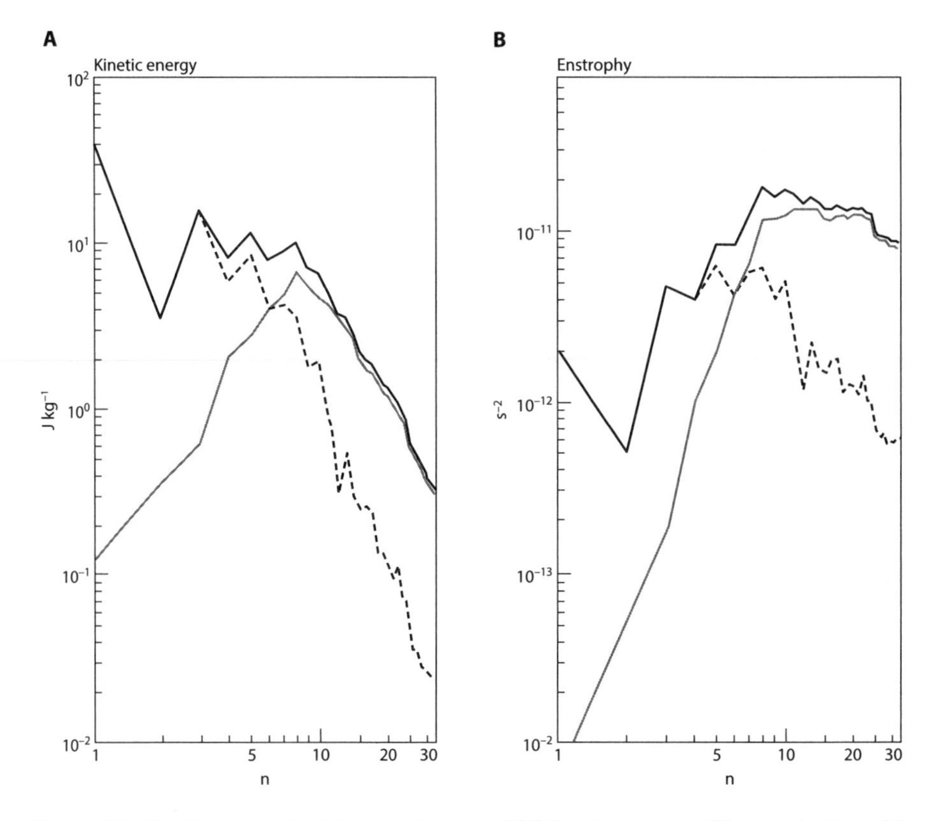

Figure 10.8. The observed vertically integrated spectra of (A) kinetic energy and (B) enstrophy. The solid lines denote total, the dashed stationary, and the gray transient. Note that both axes are logarithmic. From Boer and Shepherd (1983). © American Meteorological Association. Used with permission.

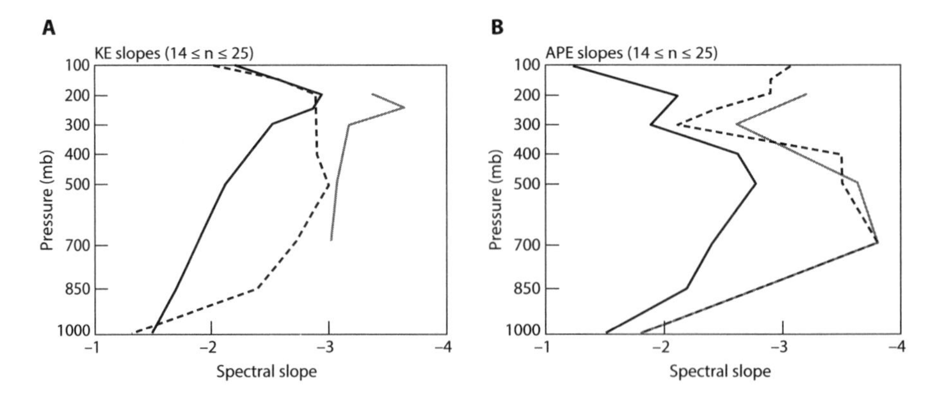

Figure 10.9. (A) Slope of line fitted to the kinetic-energy spectrum, for two-dimensional indexes in the range 14 to 25. The gray line shows the results of Baer (1972), and the dashed line those of Chen and Wiin-Nielsen (1978). From Boer and Shepherd (1983). © American Meteorological Association. Used with permission.

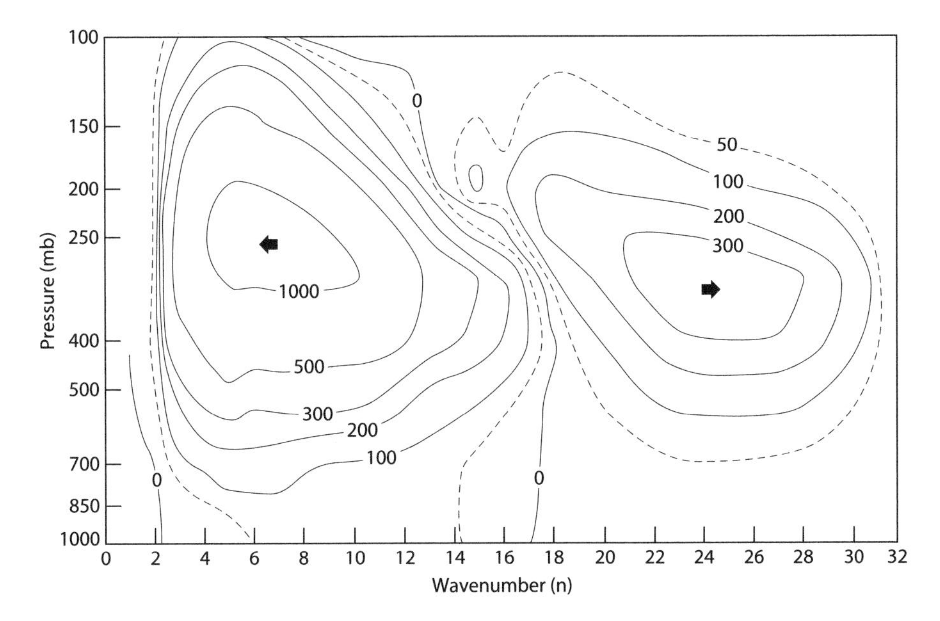

Figure 10.10. Observed nonlinear kinetic energy exchanges as a function of height. The units are 10^{-3} W m^{-2}. From Boer and Shepherd (1983). © American Meteorological Association. Used with permission.

not. How can a "frictional" term be formulated in the momentum equation that has this property?

To examine the idea of Sadourny and Basdevant, we start from the equation of motion in "invariant" form, using isentropic coordinates:

$$\frac{\partial \mathbf{V}}{\partial t} + q\mathbf{k} \times (\rho_\theta \mathbf{V}) + \nabla\left(\frac{1}{2}\mathbf{V} \cdot \mathbf{V} + s\right) = 0. \tag{49}$$

Here q is the potential vorticity, as before. To include friction in such a way that it does not affect the kinetic energy we introduce a parameter D, as follows:

$$\frac{\partial \mathbf{V}}{\partial t} + (q - D)\mathbf{k} \times (\rho_\theta \mathbf{V}) + \nabla\left(\frac{1}{2}\mathbf{V} \cdot \mathbf{V} + s\right) = 0. \tag{50}$$

We can interpret $(q - D)$ as a modified potential vorticity. When we take the dot product of (50) with $\mathbf{V}$ to form the kinetic energy equation, the term involving D drops out, regardless of the form of D. This means that D does not contribute to the tendency of the kinetic energy.

We want to choose the form of D in such a way that potential enstrophy is dissipated. The first step is to construct the potential vorticity equation, by taking the curl of (50) and using continuity:

$$\frac{Dq}{Dt} = \frac{1}{\rho_\theta} \nabla \cdot (D\rho_\theta \mathbf{V}). \tag{51}$$

The right-hand side of (51) vanishes, as it should, for $D \equiv 0$. We let the potential enstrophy averaged over an entire isentropic surface (globally) be given by

$$Z(\theta) = \frac{1}{S}\int_S\int \frac{q^2}{2}\rho_\theta dS. \tag{52}$$

Then, (51) implies that

$$\frac{dZ(\theta)}{dt} = -\int_S\int (D\rho_\theta \mathbf{V}\cdot\nabla q)\, dS. \tag{53}$$

To guarantee dissipation of $Z(\theta)$, we choose

$$D = \tau \mathbf{V}\cdot\nabla q, \tag{54}$$

with $\tau \geq 0$. By substituting (54) into (53), we find that

$$\frac{dZ(\theta)}{dt} = -\tau\int_S\int \rho_\theta(\mathbf{V}\cdot\nabla q)^2 dS. \tag{55}$$

Equation (55) guarantees that $Z(\theta)$ decreases with time for $\tau > 0$. We can recover potential vorticity conservation and potential enstrophy conservation by setting $\tau = 0$. From (54), we see that

$$q - D = q - \tau(\mathbf{V}\cdot\nabla q) \equiv q_{\text{anticipated}}. \tag{56}$$

Here $q_{\text{anticipated}}$ can be interpreted as the value of q that we expect or "anticipate" by looking upstream to see what value of q is being advected toward us. Equation (56) is equivalent to

$$\frac{q_{\text{anticipated}} - q}{\tau} = -\mathbf{V}\cdot\nabla q. \tag{57}$$

For this reason, the technique is referred to as the *anticipated potential vorticity method* to parameterize the effects of geostrophic turbulence in the momentum equation. Sadourny and Basdevant (1985) showed that this approach gives realistic kinetic energy and enstrophy spectra in a numerical model.

The point of this discussion is that it is possible to conceive of specific processes that dissipate potential enstrophy without dissipating kinetic energy. The particular approach suggested by Sadourny and Basdevant is merely used here as an example, although it has recently been applied by Ringler et al. (2011).

The Global Circulation as a Blender

To the extent that the eddies of the global circulation act like turbulence, it is natural to ask whether they "mix" things, and if so, which things. A variable that is mixed by turbulence is a conserved variable; that is, it remains unchanged over time following fluid particles. Linear momentum is not very well conserved; it is subject to a variety of nonconservative effects, including the Coriolis acceleration and pressure gradients. The angular momentum, $M \equiv a\cos\varphi(u + \Omega a\cos\varphi)$, is somewhat more conservative, because it is not affected by the Coriolis acceleration; a little thought shows, however, that uniform angular momentum is impossible unless the uniform value is zero, because a finite angular momentum at the pole would imply infinite zonal wind and vorticity there.

We are thus led to look for evidence that eddies mix the Ertel potential vorticity, which is well conserved following fluid particles. In the absence of heating, particles stay on their isentropic surfaces, so we would expect the large-scale turbulence to

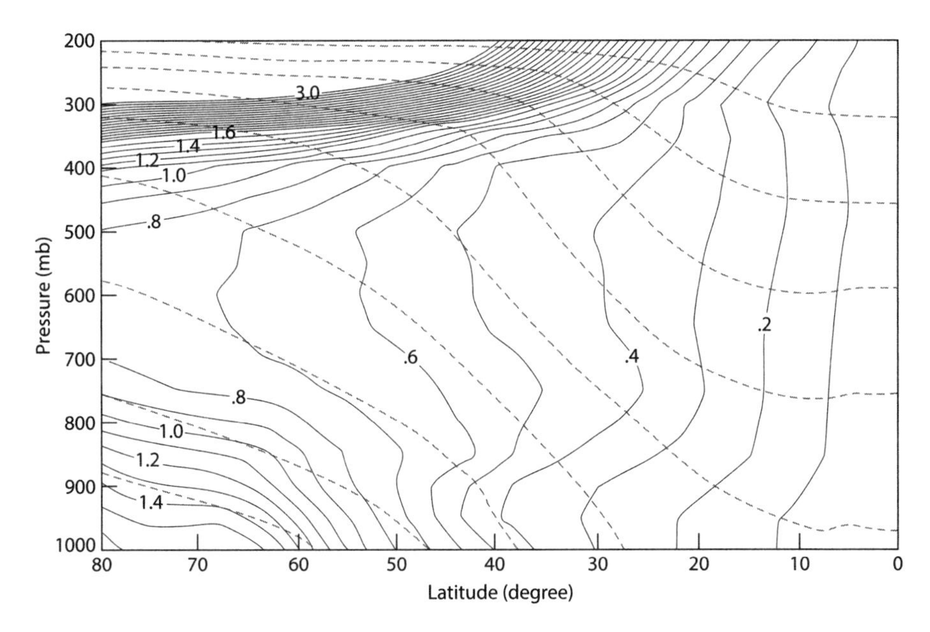

Figure 10.11. Potential vorticity (solid lines) and potential temperature (dashed lines) plotted as functions of pressure and latitude for the Northern Hemisphere. From Sun and Lindzen (1994). © American Meteorological Association. Used with permission.

homogenize potential vorticity (and other conservative variables) along isentropic surfaces. When we discussed the zonally averaged PV in chapter 4, we remarked that the PV is quite homogenous in the troposphere, compared with the stratosphere. Figure 10.11, from Sun and Lindzen (1994), shows that in the midlatitude troposphere there is some tendency for zonally averaged PV contours to be parallel to zonally averaged potential temperature profiles.

The Limits of Deterministic Weather Prediction

When we make a weather forecast, we are solving an *initial-value problem*. The current state of the atmosphere is the "initial condition." The governing equations are integrated forward in time to predict the future state of the atmosphere.

Is it possible in principle to make a perfect (or arbitrarily accurate) forecast? Broadly speaking, there are three sources of forecast error:

- Properties of the atmosphere itself:
  - The uncertainty principle of quantum mechanics (not important for this problem);
  - Nonlinearity and instability (very important). As discussed below, these lead to sensitive dependence on initial conditions.
- The observing system:
  - Imperfect measurements of the initial conditions (e.g., imperfect thermometers);
  - Imperfect spatial coverage of the initial conditions;
  - Mistakes.

- The model:
  - Wrong equations;
  - Imperfect resolution.

We focus on the first and most fundamental source of error, that is, properties of the atmosphere itself that limit predictability.

Lorenz (1963) discussed the predictability of a "deterministic nonperiodic flow." A system is said to be *deterministic* if its future evolution is completely determined by a set of rules. Models of the atmosphere (e.g., the primitive equations) are examples of sets of rules. The atmosphere is therefore a deterministic system. The behavior of the atmosphere is obviously nonperiodic in time; its previous history is not repeated. The predictability of periodic flows is a rather boring subject. If the behavior of the atmosphere were periodic in time, the weather would certainly be predictable!

How does nonperiodic behavior arise? The forcing of the atmosphere by the seasonal and diurnal cycles is at least approximately periodic. For linear systems, periodic forcing always leads to a periodic response. For nonlinear systems, however, periodic forcing (including no forcing at all as a special case) can lead to a nonperiodic response. *Nonperiodic behavior arises from nonlinearity.*

Lorenz (1963) studied an idealized set of nonlinear convection equations and found that for some values of the parameters all the steady-state and periodic solutions are unstable. The model exhibits nonperiodic solutions; again, a periodic solution is, by definition, predictable. The equations of Lorenz's toy model are remarkably simple:

$$\begin{aligned} \dot{X} &= -\sigma X + \sigma Y, \\ \dot{Y} &= -XZ + rX - Y, \\ \dot{Z} &= XY - bZ. \end{aligned} \tag{58}$$

Here

$$\sigma = 10,\ b = 8/3,\ \text{and}\ r = 24.74 \tag{59}$$

are parameters that are specified before the model is run. The numerical values given in (59) are particular choices (not unique ones) that lead to nonperiodic behavior. A solution of (58) is shown in figure 10.12. The state of the model is plotted in a phase space. Most of the time the solution is near one of two "attractors; that is, at a randomly chosen point in the integration the probability of finding the solution near one of the attractors is very high. Occasionally, the solution wanders from one attractor to the other. Partly because of the appearance of this plot, the solution is sometimes called the "butterfly attractor."

This example illustrates the important point that *even a simple nonlinear system can be unpredictable*. Lack of predictability and complex behavior are not necessarily due to complexity in the definition of the system itself.

The following discussion explains why there is a finite limit to deterministic predictability. Two slightly different states of the atmosphere diverge from each other with time because the atmosphere is unstable (see the sketch in fig. 10.13). This divergence leads to sensitive dependence on the initial conditions, on a scale-by-scale basis. Systems that exhibit sensitive dependence on initial conditions are called *chaotic*. The sensitive dependence of the state of the atmosphere on its past history

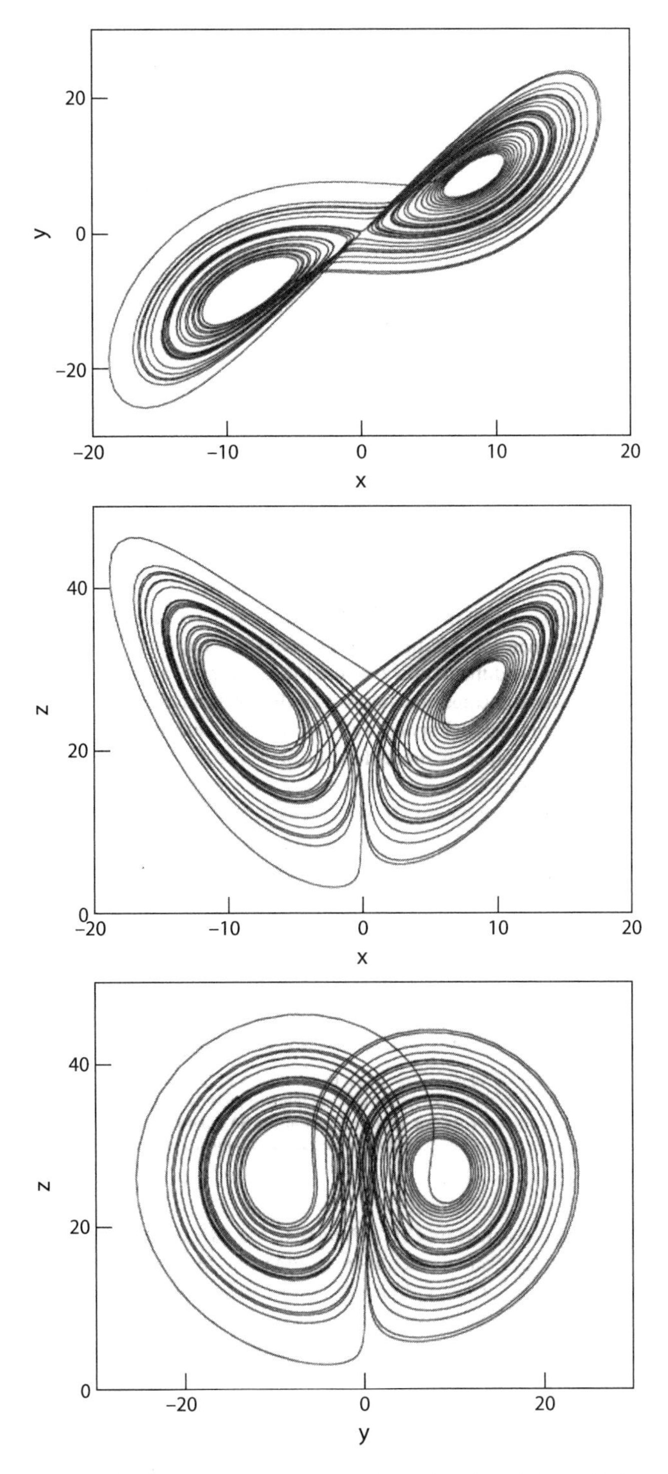

Figure 10.12. The Butterfly Attractor of Lorenz (1963), obtained as the solution of (74). From Drazin (1992). Reprinted with the permission of Cambridge University Press.

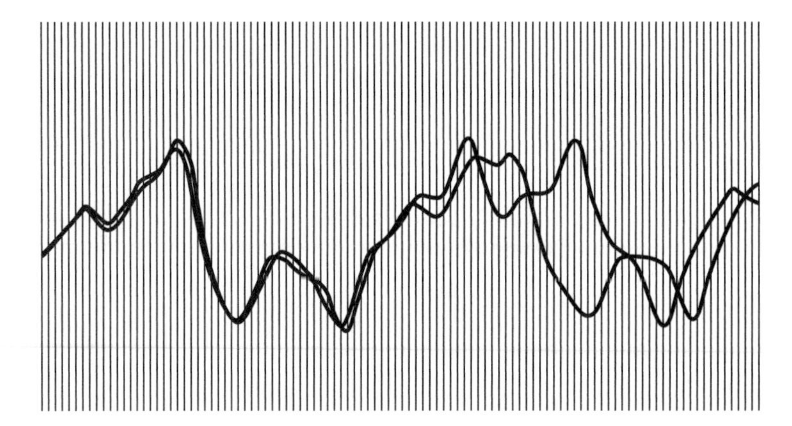

Figure 10.13. The divergence of two weather patterns. Lorenz noticed that two numerical solutions that started from nearly the same point grew increasingly farther apart until all resemblance disappeared. From Gleick (1987). Used with permission from James Gleick.

suggests that the flap of a butterfly's wings in China could noticeably change the weather in North America a few days later—a second reason for calling figure 10.12 the Butterfly Attractor.

There are many kinds of instability, acting on virtually all spatial scales. Small-scale shearing instabilities act on scales of meters or less. Buoyant instabilities, including cumulus instability, occur primarily on scales of a few hundred meters to a few kilometers. Baroclinic instability occurs on scales of thousands of kilometers.

Although Lorenz discovered the importance of sensitive dependence on initial conditions, he was not the first to do so; Poincaré (1912) recognized the phenomenon and even discussed the fact that it makes long-term weather forecasting impossible. James Clerk Maxwell was also aware, during the nineteenth century, that as a result of instabilities, deterministic physical laws do not necessarily permit deterministic predictions (Harman, 1998; 206–8). Lorenz was the first to realize that the phenomenon of sensitive dependence on initial conditions permits complex unpredictable behavior to occur even in very simple systems. He also emphasized the importance of nonlinearity, in addition to instability.

Recall that small-scale eddies produce fluxes that modify larger scales. In this way, errors on small scales can produce errors on larger scales through nonlinear processes, as discussed in chapter 7. Recall that scales cannot interact in linear systems.

We conclude that it is the combination of instability and nonlinearity that limits our ability to make skillful forecasts of the largest scales of motion. This concept is illustrated in figure 10.14. Both instability and nonlinearity are properties of the atmosphere itself that lead to an intrinsic limit on deterministic predictability; we cannot eliminate or circumvent them by improving our models or our observing systems. We can say that sensitive dependence on initial conditions imposes a "predictability time" or "predictability limit," that is, a limit beyond which the weather is unpredictable in principle. This limitation is not intrinsic to numerical weather prediction or any other specific forecast technique. It applies to all methods.

Errors on smaller scales double (grow proportionately) faster than errors on larger scales, simply because the intrinsic timescales of smaller-scale circulations are

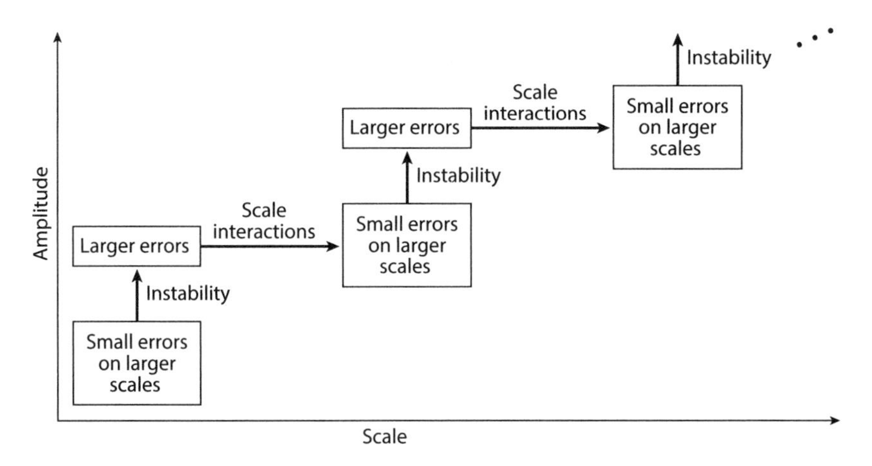

Figure 10.14. The road to chaos. Sketch illustrating the roles of instability in leading to error growth, and of nonlinearity in leading to the spreading of error from small scales to larger scales.

shorter. For example, the intrinsic timescale of a buoyant thermal in the boundary layer might be on the order of 20 minutes; that of a thunderstorm circulation, one hour; that of a baroclinic eddy, two or three days; and that of planetary wavenumber 1, one to two weeks. Because the predictability time is related to the eddy turnover time, the predictability limit is a function of scale—larger scales are generally more predictable than smaller scales.

Eliminating errors on smaller spatial scales by adding more observations increases the range of skillful forecasts by a time increment approximately equal to the predictability time of the newly resolved smaller scales. Pushing the initial error to increasingly smaller spatial scales is therefore a strategy for improving forecasts that yields diminishing returns (see fig. 10.15).

The following is an example of this concept. Suppose that we forecast the temperature at a particular house and that the forecast error grows exponentially with the length of the forecast. We can write

$$T_{\text{forecast}}(t) = T_{\text{true}}(t) + E_0 e^{\lambda t}, \tag{60}$$

where E_0 is the initial error, t is the length of the forecast, and λ is the exponential growth rate. If $E_0 = 0$ (no initial error), then the error remains zero for all time. We can reduce the error at a given forecast range, say five days, by reducing the initial error. In other words, by reducing the initial error, we can postpone the time at which the error attains a certain "reference" value, E_{ref}. For example, suppose that

$$E_{\text{ref}} = E_0 e^{\lambda t} = \frac{E_0}{2} e^{\lambda(t+\Delta t)}. \tag{61}$$

With this formula, cutting the initial error in half postpones the time when the error grows to E_{ref} by an amount Δt, which is a measure of the gain in forecast skill, and satisfies

$$\Delta t = \frac{\ln(2)}{\lambda}. \tag{62}$$

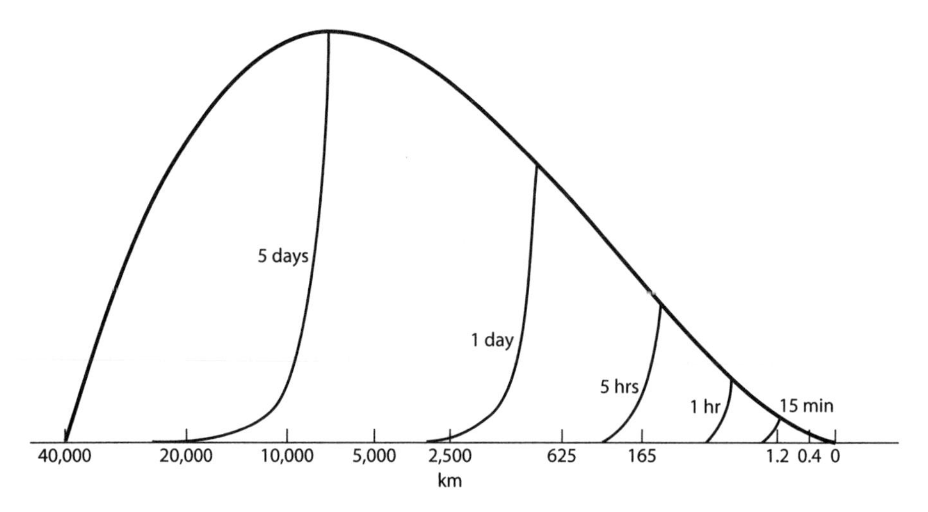

Figure 10.15. A simplified depiction of the energy spectrum $E(k)$ (upper curve), and the error-energy spectra (lower curves) at 15 minutes, 1 hour, 5 hours, 1 day, and 5 days, as interpolated from the results of a numerical study. The lower curves coincide with the upper curve to the right of their intersections with the upper curve. Areas are proportional to energy. From Lorenz (1969).

A large value of Δt is "good," because it means that the error grows more slowly; that is, a large Δt requires a small error growth rate, λ.

The growth rate is expected to be larger (faster) for smaller spatial scales, and smaller (slower) for larger spatial scales. Qualitatively, we can write

$$\lambda = \frac{V}{d}, \tag{63}$$

where V is an "eddy velocity scale," and d is the linear spatial scale (e.g., the radius or "size") of the eddy. We can interpret $1/\lambda$ as the time required to travel distance d at speed V, so it is roughly the time required for an air parcel to make one trip around the eddy. We will call this the "eddy turnover time." Substituting into our formula for Δt gives us

$$\Delta t = \frac{d \ln(2)}{V}. \tag{64}$$

Thus, the gain in forecast skill is less for small values of d, for a fixed value of V.

The reason that most of the error resides on small spatial scales is that they are inadequately sampled. To reduce the error on the small scales requires an increase in the density of the observing network. In two dimensions, quadrupling the number of observations cuts the average distance between stations in half. This means that d, that is, the linear spatial scale of the smallest eddies that are "resolved" by the observing network, is reduced by a factor of 2.

As an example, suppose that we improve an initial observing network by quadrupling the number of points. This increases the cost of the network by a factor of 4 and reduces d by a factor of 2. We gain forecast skill by an amount $(\Delta t)_1$. Then, we quadruple the number of points again, increasing the cost of the network by another factor of 4. The forecast *improvement* is measured by $(\Delta t)_2 = 1/2(\Delta t)_1$ that is, *it is only half as large as* $(\Delta t)_1$. Further costly improvements to the network will result in increasingly smaller improvements in forecast skill.

As will be discussed shortly, estimates show that small errors on the smallest spatial scales can grow in both amplitude and scale to significantly contaminate the largest scales (comparable to the radius of the Earth) in about two to three weeks. Some aspects of atmospheric behavior may nevertheless be predictable on longer timescales, particularly if they are forced by slowly changing external influences. An obvious example is the seasonal cycle. Another example is the statistical character of the weather anomalies associated with long-lasting sea-surface temperature anomalies, such as those due to El Niño. This point will be discussed further, later in this chapter.

At this point, we can offer a second piece of the definition of turbulence: *a turbulent circulation is not predictable on the timescale of interest*. For example, midlatitude winter storms can be considered as turbulent eddies on seasonal timescales, but they behave as highly predictable, orderly circulations in terms of a one-day forecast. With this definition, turbulence is in the eye of the beholder.

As an interesting analogy, the orbits of the planets of our solar system are well known to be highly predictable. Nevertheless, the solar system is known to be chaotic on longer timescales (e.g., Baytgin and Laughlin, 2008; Laskar, 1994; Laskar and Gastineau, 2009). If the timescale of interest is a few million years or less, the motions of the planets are analogous to laminar flow. If the timescale of interest is hundreds of millions of years, the motions of the planets are analogous to turbulent flow.

Quantifying the Limits of Predictability

The following three approaches to determining the limits of predictability were discussed by Lorenz (1969).

The Dynamical Approach

In the dynamical approach, two or more model solutions are produced, starting from similar but not quite identical initial conditions. This procedure is similar to what Lorenz did by accident when he discovered sensitive dependence on initial conditions. The model is used in place of the atmosphere; no real data are used. Problems with this approach are (1) truncation error, (2) imperfect equations, and (3) lack of information about very small scales. Studies of this type suggest that the doubling time for small errors with spatial scales of a few hundred kilometers is about five days, which implies that the limit of predictability is about two weeks.

One of the earliest examples of the dynamical approach is the study of Charney et al. (1966), who used several atmospheric global circulation models (GCMs) to study the growth of small perturbations. Figure 10.16 shows some of their results, obtained with an early version of the UCLA GCM. The root-mean-square temperature error grows in both hemispheres, but more rapidly in the winter hemisphere, where the circulation is more unstable. The error does not continue to grow indefinitely; it stops growing at the point where the "forecast" is no better than a guess. The error is said to have *saturated*. A suitable guess might consist of a state of the system chosen at random from a very long record of such states, analogous to pulling a weather map at random out of a huge cabinet full of many such maps.

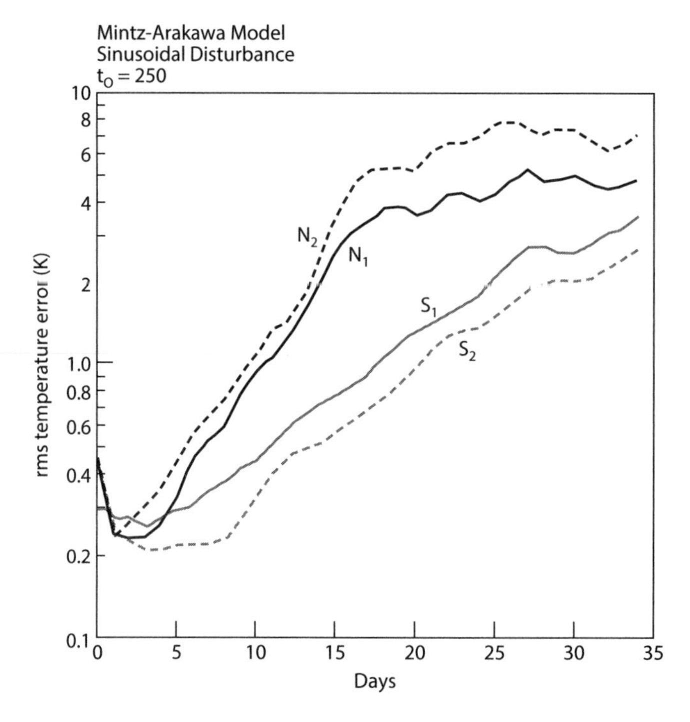

Figure 10.16. Root-mean-square temperature error in January simulations performed with the two-level Mintz-Arakawa model. "N" and "S" denote the Northern and Southern Hemispheres, respectively. The subscripts 1 and 2 denote the two model levels. From Charney et al. (1966). © American Meteorological Association. Used with permission.

A later example of the dynamical approach is discussed by Shukla (1981, 1985). The results shown in figure 10.17 are based on computations with a GCM, as reported by Shukla (1985). The model is used to perform multiple simulations, differing only in very small perturbations of the initial conditions. The two panels on the left side of the figure show the growth and saturation of errors in the sea-level pressure for winter and summer. The winter errors grow more rapidly than the summer ones. The errors are large in middle latitudes, where baroclinic instability is active, especially in the winter. The errors are much smaller in the tropics.

Recall from chapter 3, however, that the tropical sea-level pressure normally does not vary much. This means that a small error in the tropics can be important. To take this factor into account, the two panels on the right show the errors normalized by the temporal standard deviation of the sea-level pressure. From this perspective, we see that the tropical errors actually grow more rapidly than those of middle latitudes and saturate at about the same (normalized) values. At all latitudes the saturation values are close to 1. This means that the errors stop growing when they become as large as the temporal standard deviation based on day-to-day variability.

Figure 10.18 is taken from Shukla (1981). For numerical experiments on the growth of errors, similar to those discussed above, the figure shows how error growth varies with zonal wavenumber. In each panel, the solid curve shows the growth of error in the numerical model, and the dashed curve shows the corresponding growth of error when the "forecast" used is simply persistence. We can say that the model forecast is no longer skillful when it is no better than a forecast based on persistence.

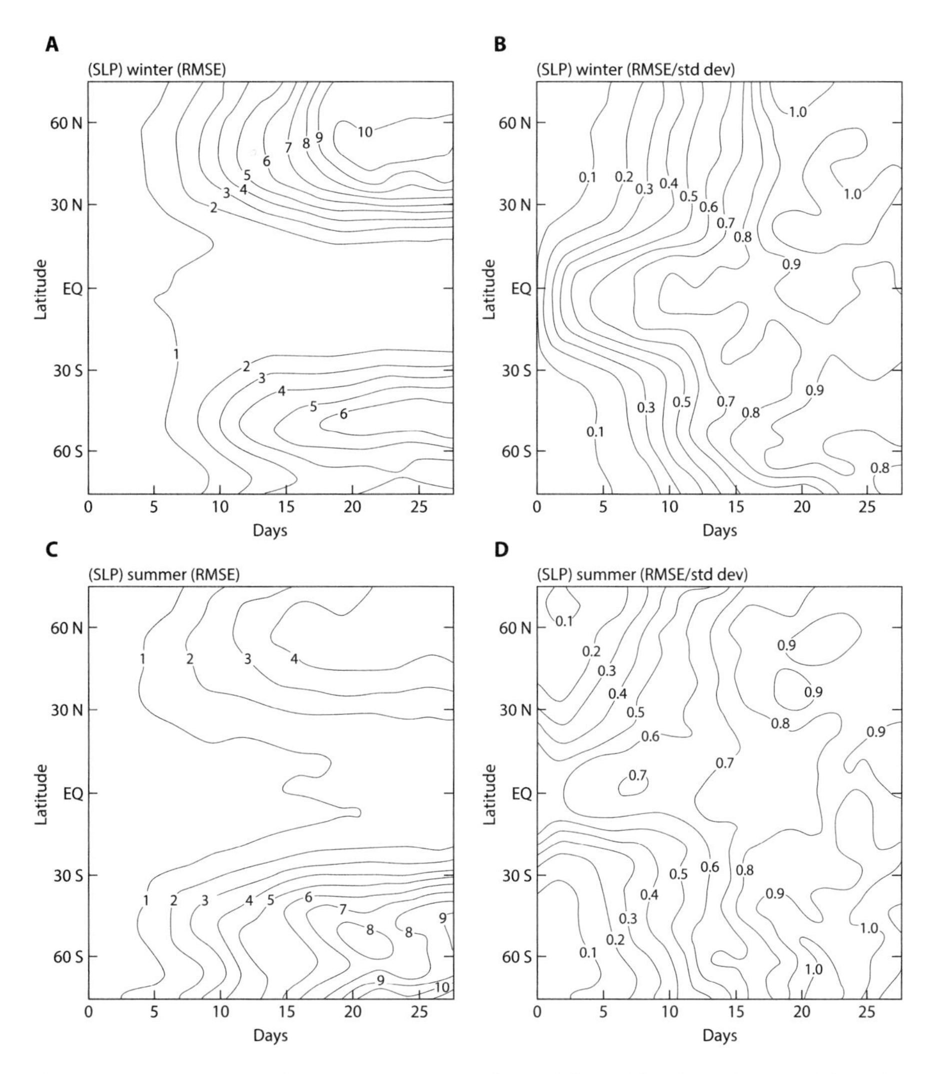

Figure 10.17. Zonal average of root-mean-square error (RMSE; left panels) and ratio (RMSE/std dev; right panels) of root mean square and standard deviation of daily values for sea-level pressure. (A) RMSE and (B) RMSE/std dev for six pairs of control and perturbation runs during winter; (C) RMSE and (D) RMSE/std dev for three pairs of runs during summer. From Shukla (1985). Reprinted with permission from Elsevier.

After 30 days, the numerical model still has some skill at the lower wavenumbers, which correspond to the largest spatial scales. For higher wavenumbers, the model's skill disappears more quickly.

The Empirical Approach

In the empirical approach, the atmosphere is used in place of a model; the atmosphere itself is used to predict the atmosphere. Lorenz (1969) examined the observational records of the 200, 500, and 850 hPa heights, for the Northern Hemisphere only. He searched for pairs of similar states, or "analogs," occurring within one month

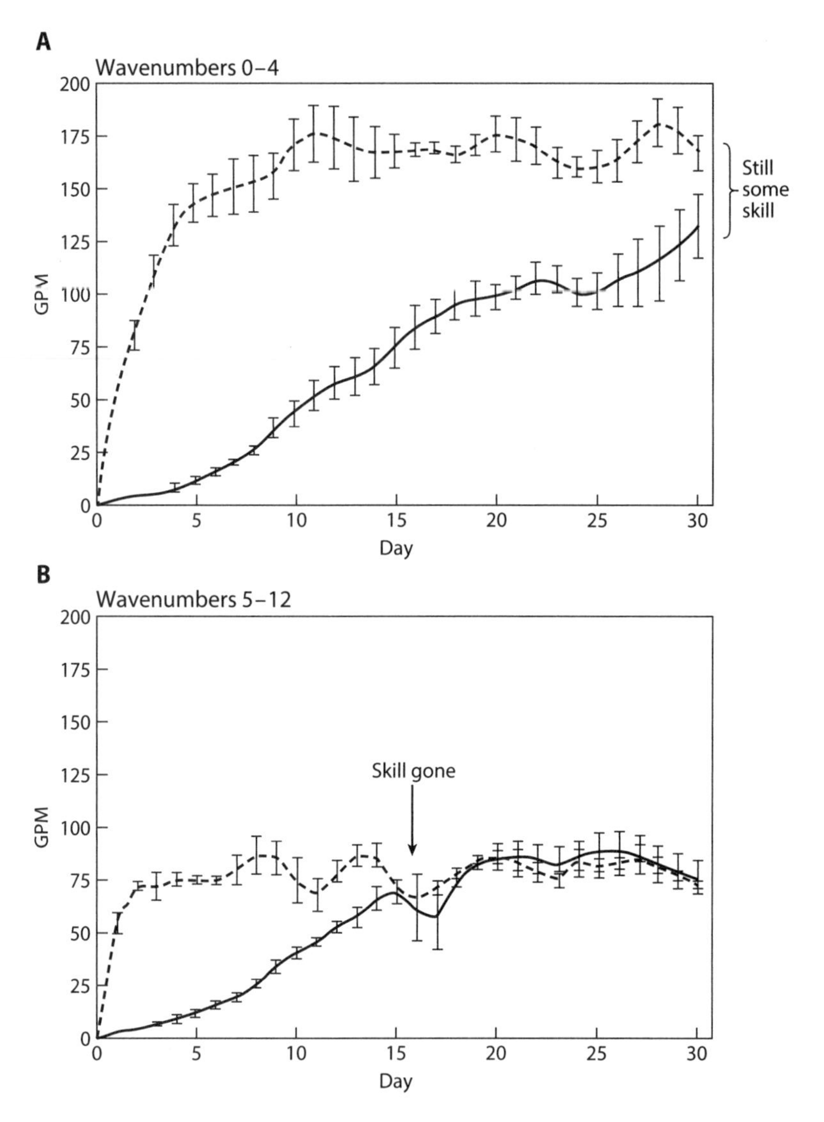

Figure 10.18. Root-mean-square error (solid line) averaged for six pairs of control and perturbation runs and averaged for latitude belt 40°–60° N for 500 mb height for (A) wavenumbers 0–4 and (B) wave numbers 5–12. Dashed line is the persistence error averaged for the three control runs. Vertical bars denote the standard deviation of the error values. From Shukla (1981). © American Meteorological Association. Used with permission.

of the same day of the year. He chose 30 December 1963 and 13 January 1965 as the best-available analogs within a five-year record.

The results of the empirical approach show that the doubling time for errors on the smallest resolved scales is less than eight days. They are thus reasonably consistent with the results of the dynamical approach.

The following are among the problems with the empirical approach: (1) it is hard to find "good" analogs, in that the smallest error is about half the average error; (2) it is not possible to experiment with the initial error, because it is given by nature; and (3) the data cannot be used to study the growth of errors on very small scales, simply because such scales are not adequately observed.

The Dynamical-Empirical Approach

The dynamical-empirical approach is the most difficult of the three to understand. The basic idea, as first conceived by Lorenz (1969), is to derive an equation for the time change of "error kinetic energy" (using a model) and then to Fourier transform the error energy equation, so that the spectrum of the kinetic energy appears. The key step is to stipulate this kinetic energy spectrum from observations, down to very small scales (~40 m). The resulting semiempirical equation is then used to draw conclusions about error growth. The dynamical-empirical approach reveals that errors in the smallest scales amplify most quickly and soon dominate. Again, the dynamical-empirical approach suggests that the limit of deterministic predictability is about two to three weeks.

A modern example of the dynamical-empirical approach was described by Lorenz (1982). Suppose that we make a large number of one-day and two-day forecasts, as shown in figure 10.19. Let $z_{i,1}$ be the one-day forecast for z on day i, and let $z_{i,2}$ be the two-day forecast for z on day i (the same day). Let $E_{1,2}$ be the root-mean-square (rms) difference between the one-day and two-day forecasts *for the same day*, averaged over the globe and over all verification days:

$$E_{1,2}^2 = N^{-1}S^{-1}\sum_{i=1}^{N}\int_S (z_{i,1} - z_{i,2})^2\,dS. \tag{65}$$

Here the integral is over the area, S, and the sum is over the verification days, which are distinguished by subscript i. We can interpret $E_{1,2}$ as the average or typical growth of the forecast error between the first and second days of the forecasts. If all forecasts were perfect, $E_{1,2}$ would be zero. If the one-day forecasts were perfect but the two-day forecasts were not, $E_{1,2}$ would be positive, and so on.

More generally, we can compute the rms error growth between the j-day forecast and the k-day forecast, still verifying on the same day (i.e., day i), from

$$E_{j,k}^2 = N^{-1}S^{-1}\sum_{i=1}^{N}\int_S (z_{i,j} - z_{i,k})^2\,dS. \tag{66}$$

We assume without loss of generality that $k \geq j$. It should be clear that $E_{j,k}$ compares j-day forecasts with k-day forecasts. If $E_{j,k} > 0$, the implication is that k-day forecasts are less skillful, on the average, than j-day forecasts. This is, of course, expected. Note that $E_{0,k}$ compares "zero-day forecasts," which are actually analyses rather than forecasts, with k-day forecasts. If the k-day forecasts were perfect, $E_{0,k}$ would be zero. Because of the intrinsic limit of predictability and the inevitable small errors of the initial conditions, however, even a perfect model will give $E_{0,k} > 0$, for $k > 0$. Also, $E_{0,k}$ increases initially as k increases, but then it "saturates" when the k-day forecast becomes no better than a guess, as shown in figure 10.20. Even if the model's simulated climate is perfect, $E_{0,k}$ will be different from zero because of the limit of deterministic predictability.

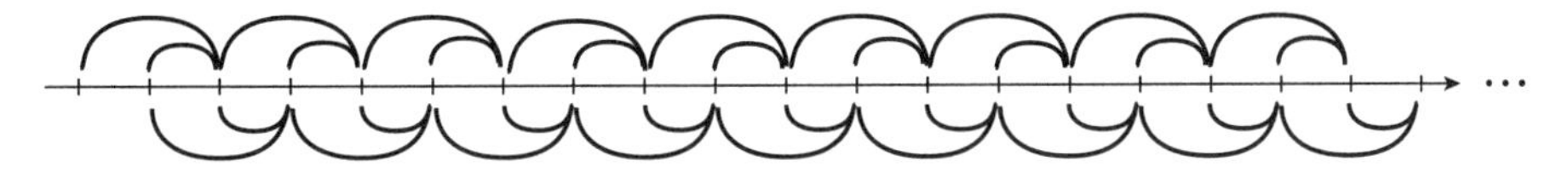

Figure 10.19. A long sequence of one- and two-day forecasts.

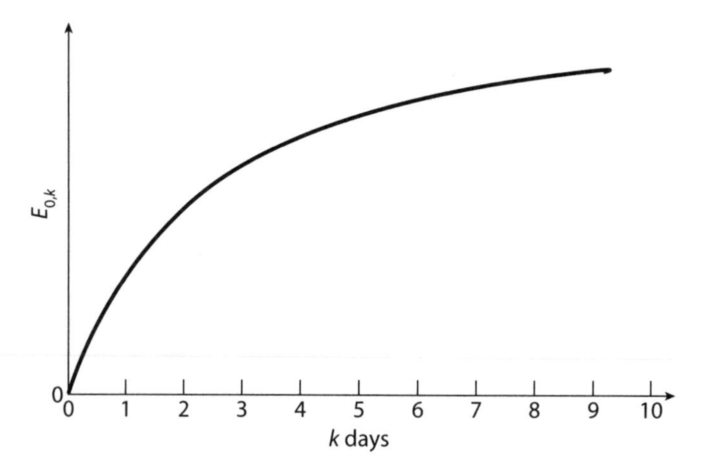

Figure 10.20. The growth of $E_{0,k}$ with time. Rapid initial growth is followed by saturation.

As k becomes large we expect $E_{j,k}$ to saturate, no matter what the value of j is. For $k \to \infty$, we might expect that $E_{j,k}$ "should be" independent of j. It is not, at least not for "small" j (<20). The reason is that the model's climate is different from the real climate. Since the model is started from real data, a j-day forecast looks like the real world when j is small. As j increases, the model goes to its own climate, and so the forecast increasingly departs from the ensemble of real-world states. However, $E_{j,k}$ becomes independent of j when both j and k are large, because then the model's climate is compared with itself.

A special case, illustrated in figure 10.21, compares $(k-1)$-day forecasts with k-day forecasts for the same day. For example, for $k=10$, $E_{9,10}$ compares 9-day forecasts with 10-day forecasts for the same day. Of course, 9-day forecasts and 10-day forecasts are both rather bad, and so, generally speaking, they will be quite different from each

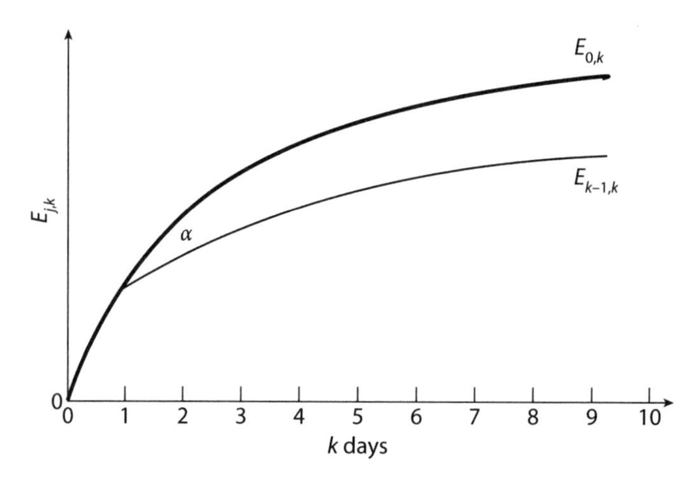

Figure 10.21. The growth of $E_{0,k}$ and $E_{k-1,k}$ with time. The thin line starts at $k=1$ because $E_{k-1,k}$ is not defined for $k<1$. The angle α measures how much the model can be improved by reducing the systematic error.

other, even though they are supposed to represent the weather on the same day. For this reason, as $k \to \infty$, $E_{k-1,k}$ is not small. Instead, as $k \to \infty$, we expect $E_{k-1,k}$ to approach a constant, which is *a measure of the variability of* z *in the model's climate*. The same is true for any other $E_{j,k}$, when $k - j$ is a constant.

For a perfect model, we would have

$$\lim_{k \to \infty} E_{0,k} = \lim_{k \to \infty} E_{k-1,k} \tag{67}$$

because the model's climate would be identical with the true climate. For an imperfect model, however, we expect

$$\lim_{k \to \infty} E_{0,k} > \lim_{k \to \infty} E_{k-1,k} \tag{68}$$

because the model's climate differs from the true climate. This is a key point.

To explore these ideas with real data, Lorenz chose 1 Dec 1980–10 March 1981 (100 days) as the verification dates. Because he worked with 100 verification days, $N = 100$ in (66). He used an archive of forecasts and analyses performed with the ECMWF model. Since ECMWF routinely makes 10-day forecasts, Lorenz had 100 one-day forecasts, 100 two-day forecasts, and so on, out to 100 ten-day forecasts. He plotted $E_{j,k}$ as shown in figure 10.22. Each point represents an average over 100 pairs of forecasts.

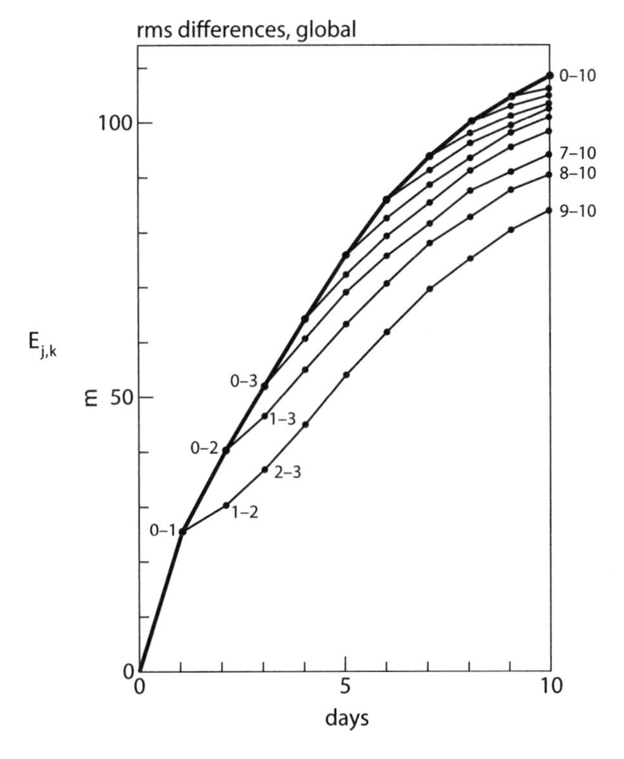

Figure 10.22. Global root-mean-square 500 mb height differences, in meters, between j-day and k-day forecasts made by the ECMWF operational model for the same day, for $j < k$, plotted against k. Values of (j, k) are shown beside some of the points. The heavy curve connects values of $E_{0,k}$. The thin curves connect values of $E_{j,k}$ for constant $k - j$. From Lorenz (1982).

The "thin curves" in figure 10.22 appear to run nearly parallel to one another, in the sense that $dE/dk = f(E)$. In other words, for any given value of E, the thin curves all have about the same slope. This means that the rate of error growth is strongly influenced by the size of the error; of course we know that this cannot be the only factor involved, but it can be the dominant one.

Lorenz wanted to estimate the growth rate of very small errors in forecasts performed with a *perfect model*. To do this, he essentially fit a curve to the error growth rates in the ECMWF forecasts. He hypothesized that *for a perfect model*

$$\frac{dE}{dk} = aE - bE^2. \tag{69}$$

Here $E \equiv E_{j,k}$, where $k - j = \text{constant}$, and so $dE/dk \cong (E_{j+1,k+1} - E_{j,k})$ per day. The second index minus the first index is the same for both E's; these are the thin curves. Equation (69) is consistent with the hypothesis that the rate of error growth is determined by the size of the error. *For small E, the exponential growth rate is a.* For larger E, saturation occurs, so that

$$E_{\text{sat}} = \frac{a}{b}. \tag{70}$$

This means that (a/b) is the maximum error for the perfect model. Note from (69) that if the initial error is zero, there will be no error growth, as expected from a "perfect model."

We want to deduce a, the growth rate of small errors, from the data, even though the data do not contain truly small errors. We can do this by fitting the function $f(E) = aE - bE^2$ against dE/dk, as evaluated from the data (see figure 10.23). In performing this curve fit, we find the "best" values of a and b. This calculation can be interpreted as an example of the dynamical-empirical approach to determining the growth rate of small errors, because both a model and data were used.

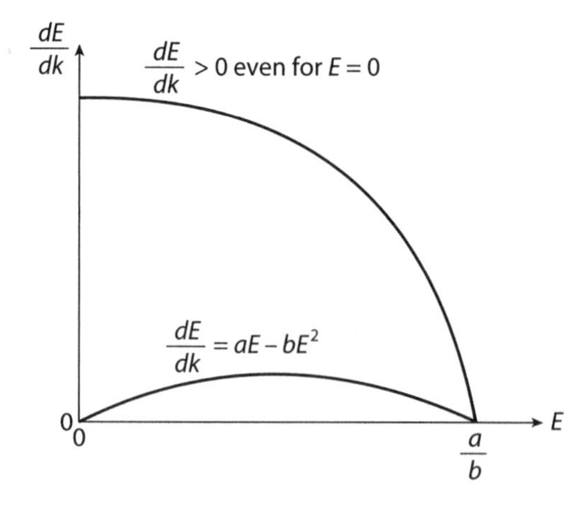

Figure 10.23. Diagram illustrating the expected variation of the error growth rate with the magnitude of the error. The lower curve shows the expected behavior for a perfect model, and the upper curve shows the expected behavior for a real, imperfect model.

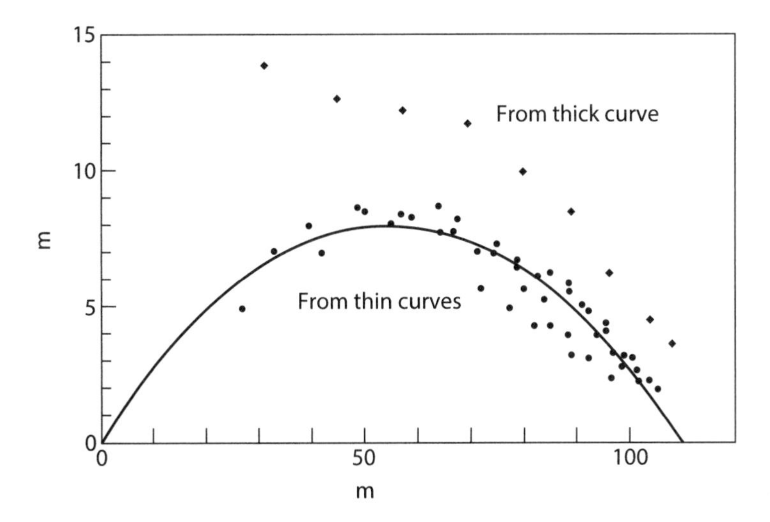

Figure 10.24. Increases in global root-mean-square 500 mb height differences, $(E_{j+1,k+1} - E_{j,k})$, plotted against average height differences, $(E_{j+1,k+1} + E_{j,k})/2$, in meters, for each one-day segment of each thin curve in figure 10.22 (dots), and increases, $(E_{0,k+1} - E_{0,k})$, plotted against average differences, $(E_{0,k+1} + E_{0,k})/2$, for each one-day segment of the heavy curve in figure 10.22 (boxes). The parabola of best fit to the large dots is shown. From Lorenz (1982).

Figure 10.24 shows the same data as in figure 10.22, plotted in a different way. The dots correspond to the thin curves in figure 10.22 and so represent the "perfect model." The squares correspond to the bold curve in figure 10.22 and so represent the "imperfect model." The conclusion of this exercise is that the doubling time for small errors on the smallest scales resolved by the model was 2.4 days.

In another publication, Lorenz discussed a later set of more skillful forecasts made with an improved model and started from better observed initial conditions. He found that the heavy curve in figure 10.22 was shifted downward, but so were the thin curves. The gap between them was cut in half. By itself, the downward shift of the heavy curve suggests that the newer model's simulated climate was more realistic than that of the older model. Alternatively, it could mean that the analyses are more realistic. The downward shifts of the thin lines suggest that the newer model's weather is less active than that of the old model.

Ensemble Forecasting

Since 1992, major weather prediction centers, including ECMWF and the U.S. National Centers for Environmental Prediction, have run forecast ensembles as a key element of their operational systems. A *forecast ensemble* consists of multiple forecasts based on slightly different initial conditions. The approach is useful because the ensemble members diverge with time, for the reasons explained by Lorenz. For any given location, the degree of the spread is an indication of the reliability of the forecast; a larger spread means less confidence. Figure 10.25 is an example of an ensemble of forecasts performed at the National Centers for Environmental Prediction.

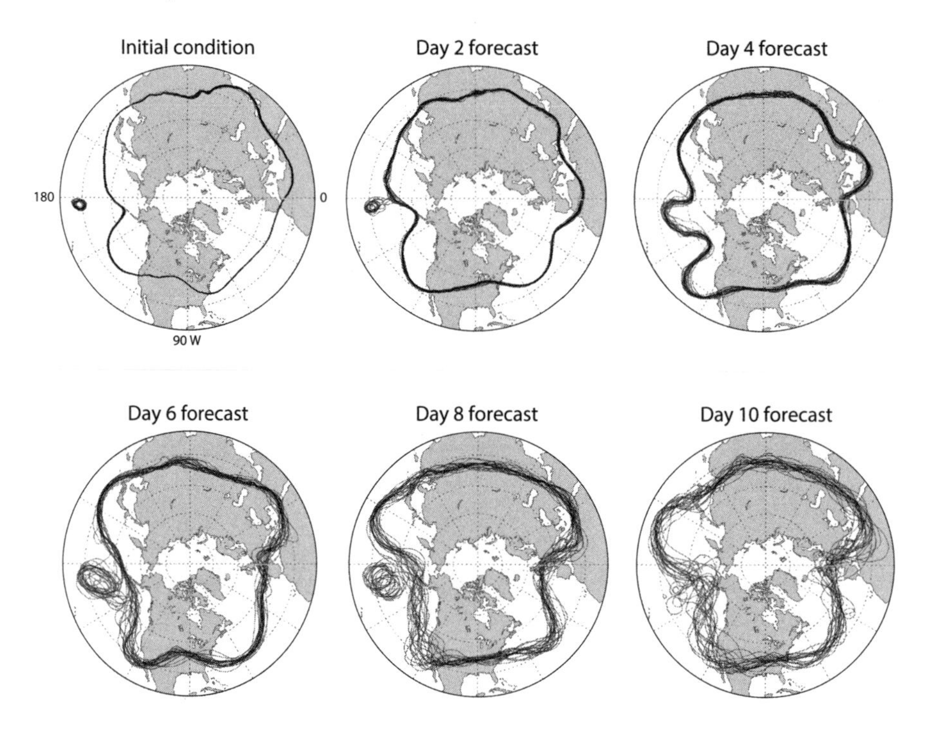

Figure 10.25. Actual forecasts of the shapes of two selected height contours for a pressure surface in the middle troposphere, looking down on the North Pole. North America is at the bottom center of each panel. The forecast times shown are the initial conditions (top left) and then, to the right across the upper row and then left to right across the bottom row, every 2 days after that, out to 10 days. The two heavy curves show the climatological positions of the height contours, which change very little over 10 days. In each panel, the thick curve shows the predicted position of the contours in a forecast made with a high-resolution version of the forecast model. The thin curves show the positions of the contours as predicted in an ensemble of forecasts (with a lower-resolution version of the same model) from initial conditions that differ very slightly from those of the control run. For obvious reasons, these are called "spaghetti diagrams." This figure was made available by Louis Uccellini and Dennis Grumm of the National Centers for Environmental Prediction.

The Response of the Atmosphere to Changes in Sea-Surface Temperature

Over the past few decades it has become clear that the statistics of the weather respond in predictable ways to changes in sea-surface temperature and other surface boundary conditions. Figure 10.26 shows an early example of research on this topic, taken from Lau (1985). The figure shows temporal variations of monthly indices of the observed sea-surface temperature (top) and the simulated zonal wind at 200 and 950 hPa, precipitation, east-west sea-level pressure gradient across the South Pacific, 200 hPa height, and an index of what is called the "Pacific–North American pattern," as obtained in two 15-year model simulations performed with an atmospheric GCM. The two runs are shown in the left- and right-hand panels, respectively. The smooth curves superimposed on the various time series were obtained using a kind of running mean.

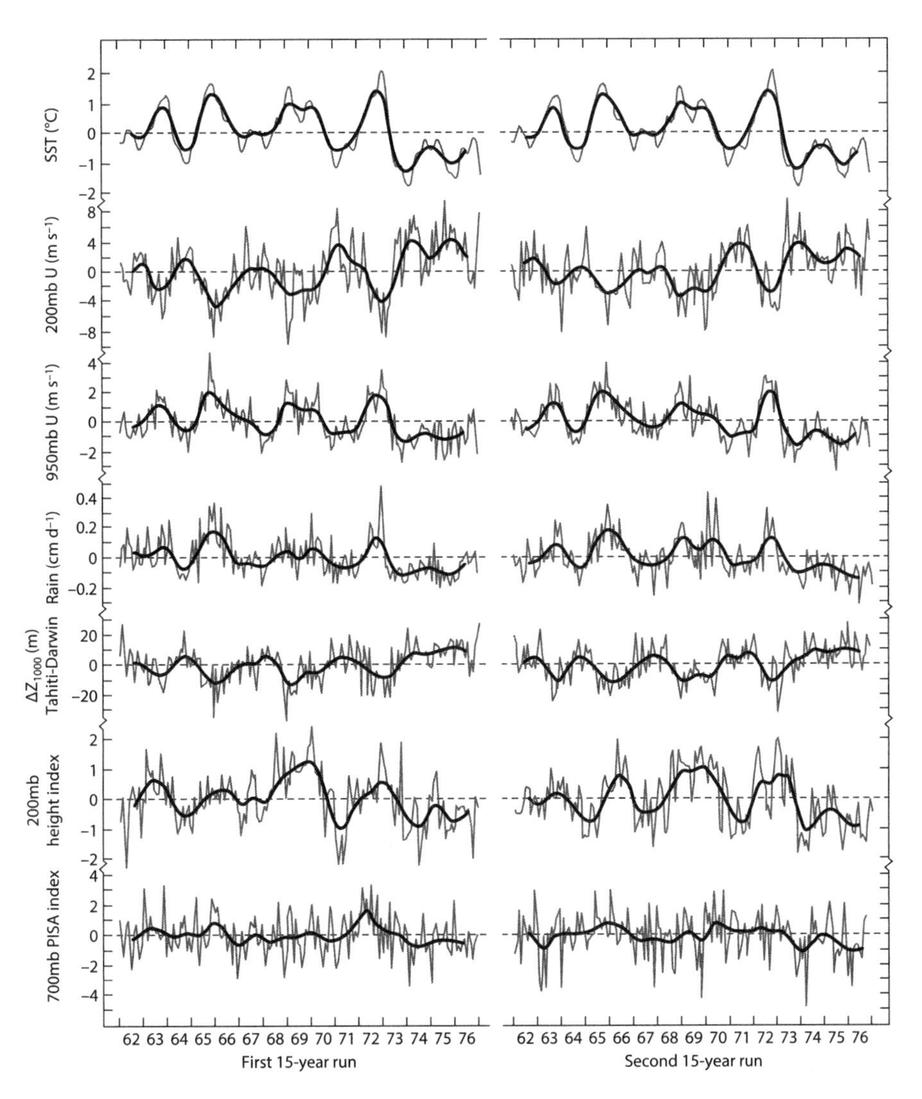

Figure 10.26. Temporal variations of monthly indexes of sea-surface temperature, zonal wind at 200 and 950 hPa, precipitation, east-west sea-level pressure gradient across the South Pacific, 200 hPa height, and the 700 hPa Pacific–North American pattern for the first (left half of figure) and second (right half) 15-year model runs. The smooth curves superimposed on these time series were obtained using a running mean. From Lau (1985). © American Meteorological Association. Used with permission.

The imposed sea-surface temperature changes were exactly the same in the two runs, which differed only in perturbations of their initial conditions. The main point here is that the two runs gave very similar results, showing that the indicated statistics were controlled by the year-to-year variations of the sea-surface temperature and were predictable far beyond the limit of deterministic predictability for individual weather events.

Lau's study was groundbreaking and computationally expensive at the time, but similar calculations have been routine since the 1990s, in the context of the Atmospheric Model Intercomparison Project (AMIP) (Gates, 1992). Today, AMIP simulations are a standard way of evaluating the performance of GCMs.

Climate Prediction

Weather can be defined as the instantaneous distribution of the atmospheric state variables, and *climate* as the long-term statistical properties of the same variables. The ocean, land surface, cryosphere, and biosphere influence the climate. Together with the atmosphere, they form the climate system. All parts of the climate system interact, on sufficiently long timescales.

If long-range weather prediction is impossible, how can climate prediction be contemplated at all? If all memory of the initial conditions is "forgotten," what is the point of solving an initial-value problem? Two factors have the potential to make seasonal forecasting and/or climate change prediction possible. First, the system has components with very long memories, including especially the ocean. At present, however, the observations needed to fully initialize these components of the system are lacking.

Second, the system responds in systematic and statistically predictable ways to changes in external forcing. This means that it is possible to make a prediction without solving an initial-value problem! A good example is the seasonal cycle. If systematic differences in weather between summer and winter are predicted at a given location, is that prediction based on solution of an initial-value problem? If the Sun's energy output decreased significantly, would the prediction that a general cooling of the Earth's climate would ensue be based on solution of an initial-value problem? These examples illustrate that predictions can be based on knowledge of an external forcing (e.g., the diurnal and seasonal cycles, or a change in solar output) that changes in a predictable way.

"The Simplest Possible GCM" (Lorenz, 1984, 1990) is described by the following equations:

$$
\begin{aligned}
\dot{X} &= -Y^2 - Z^2 - aX + aF, \\
\dot{Y} &= XY - bXZ - Y + G, \\
\dot{Z} &= bXY + XZ - Z.
\end{aligned}
\tag{71}
$$

The symbols have the following interpretations:

- X represents the strength of the westerlies;
- Y and Z are the sine and cosine components of a planetary wave train;
- F represents the meridional heating contrast (a "forcing");
- G represents land-sea contrast (another "forcing"); and
- a and b are parameters whose values are set in advance.

For $F = G = 0$, the model has the trivial steady-state solution $X = Y = Z = 0$. For $G = 0$, the steady-state solution $X = F$ is found with $Y = Z = 0$. This eddy-free solution can be unstable, however. Instability can lead to the growth of Y and Z. We interpret "large" $F(\sim 8)$ as "winter," and "small" $F(\sim 6)$ as "summer." Compare (71) with (58).

For fixed F, there can be one, two, or three steady-state solutions, depending on the value of G, as shown in figure 10.27. Here $F = 2$, so this is "super summer." The steady-state solutions shown in figure 10.27 are found most easily by fixing X and solving for G, Y, and Z.

Figure 10.28 demonstrates that the model exhibits sensitive dependence on initial conditions. The three solutions shown are for the same values of the external

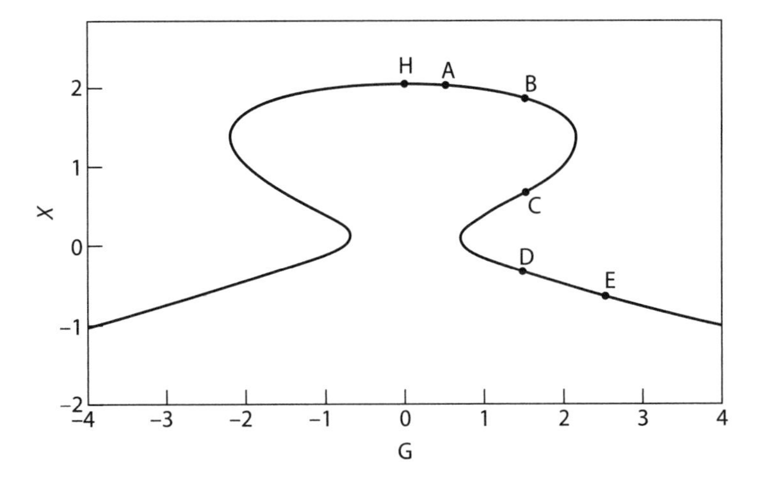

Figure 10.27. Steady-state solutions of (71). The vertical axis is X, and the horizontal axis is G. The figure is constructed for $a = 0.25$, $b = 4.0$, and $F = 2.0$. From Lorenz (1984).

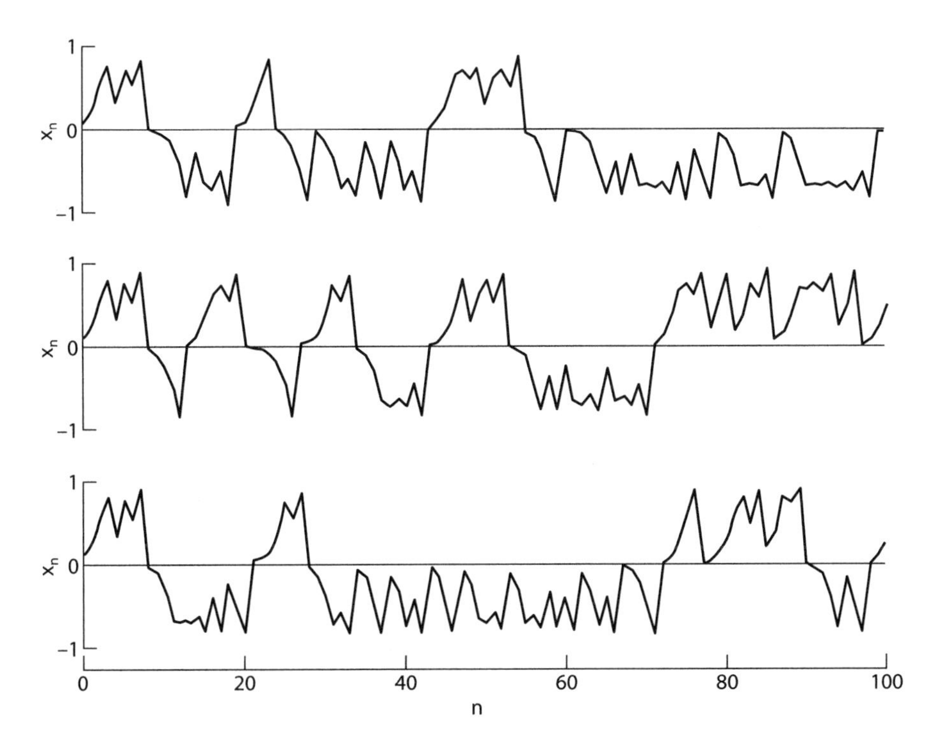

Figure 10.28. Solutions of (82) extending for 100 time steps, starting from initial values of 0.0999 (upper), 0.1000 (middle), and 0.1001 (lower). The straight-line segments joining consecutive points are solely for the purpose of making the chronological order easier to see. This figure illustrates sensitive dependence on initial conditions. From Lorenz (1976). Reprinted with permission from Elsevier.

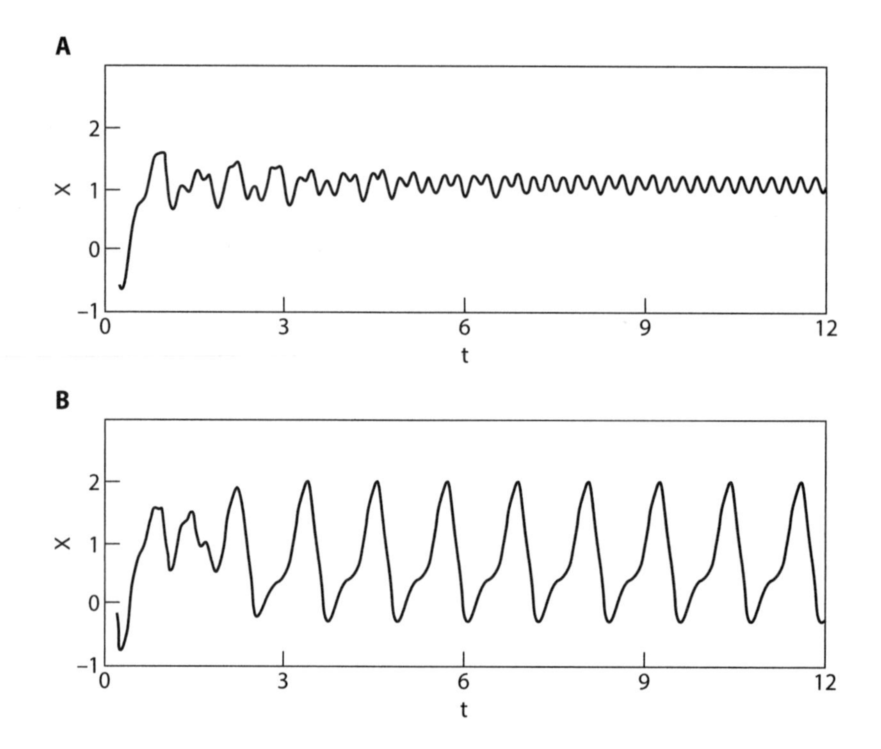

Figure 10.29. (A) The variations of X (dimensionless) with t (months) in a numerical solution of (82), with $a = 0.25$, $b = 4.0$, $F = 6.0$, and $G = 1.0$ (summer conditions). The initial state is (2.4, 1.0, 0). (B) The same as panel A, except that the initial state is (2.5, 1.0, 0). From Lorenz (1990).

parameters but with slightly different initial conditions. The solutions clearly diverge after some time.

Figure 10.29 shows two "summer" solutions from different initial conditions. Note that these two summers are both fairly regular in appearance but nevertheless look quite different from each other, indicating that the model is capable of producing two "kinds" of summers: active summers (like the one in the lower panel) and inactive summers (like the one in the upper panel).

Similarly, figure 10.30 shows two "winter" solutions from different initial conditions. The two winters are highly irregular but look much the same, suggesting that all model winters are essentially equivalent; the model makes only one "kind" of winter.

Figure 10.31 shows results from a six-year run in which two different kinds of summers occur. The interpretation is very simple and interesting. Winters are chaotic. The model locks into either an active summer or an inactive summer, based on the "initial conditions" at the end of the winter, which are essentially random. When winter returns, all information about the previous summer is obliterated by nonlinear scrambling. The dice are rolled again at the beginning of the next summer.

We define σ as the (nondimensional) standard deviation of X within the period July through September. Active summers have large σ, and inactive summers have small σ. We can calculate one value of σ for each summer or, in other words, one for each year. Figure 10.32 shows the variations of σ (dimensionless) with time in years,

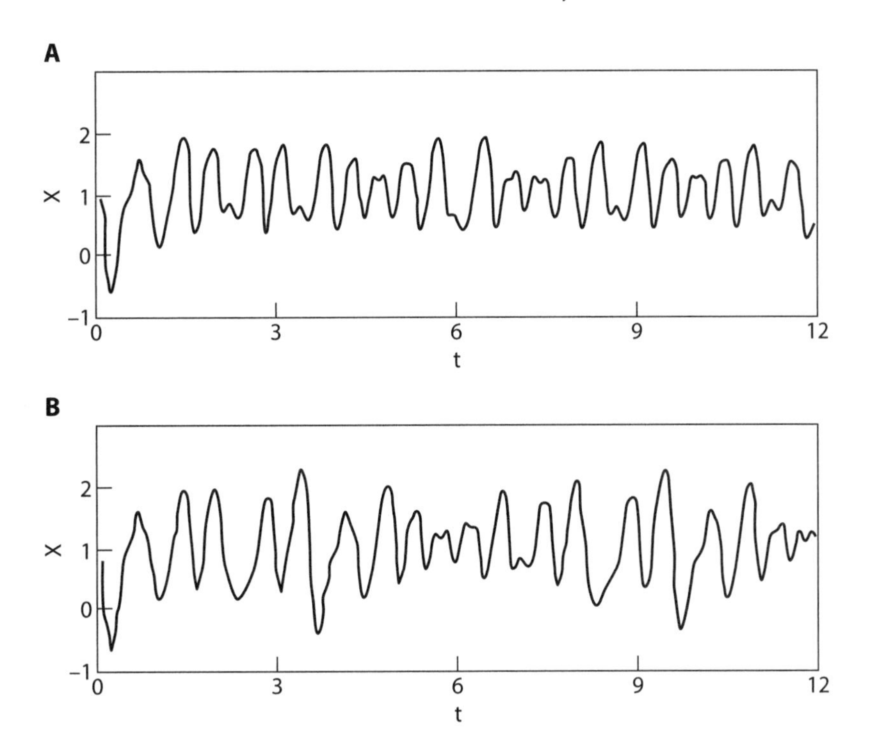

Figure 10.30. (A) The same as figure 10.29, except that $F = 8.0$ (winter conditions). (B) The same as panel A, except that the initial state is (2.5, 1.0, 0). From Lorenz (1990).

in a 100-year numerical solution of (71), for the conditions of figure 10.31. Active and inactive summers alternate, at irregular intervals.

Pushing the Attractors Around

Palmer (1993, 1999) has argued that the climate system can be viewed as occupying a collection of attractors. For example, one attractor might represent an El Niño state, and another a La Niña state. The climate itself is a set of statistics that tells what the attractors look like and how frequently each is visited. Palmer argues that slowly varying external forcings can alter the frequency with which each attractor is visited and that this is how climate change manifests itself. For example, some climate states may have frequent and/or persistent El Niños, while others may have very few El Niños.

To illustrate these ideas, Palmer used a modified version of the model given by (58):

$$\begin{aligned} \dot{X} &= -\sigma X + \sigma Y + f_0 \cos\theta, \\ \dot{Y} &= -XZ + rX - Y + f_0 \sin\theta, \\ \dot{Z} &= XY - bZ. \end{aligned} \tag{72}$$

Here f_0 is a forcing that tries to push X and Y in the direction of the angle θ in the (X,Y) plane. If we put $f_0 = 0$, then (72) reduces to (58). Figure 10.33 shows how

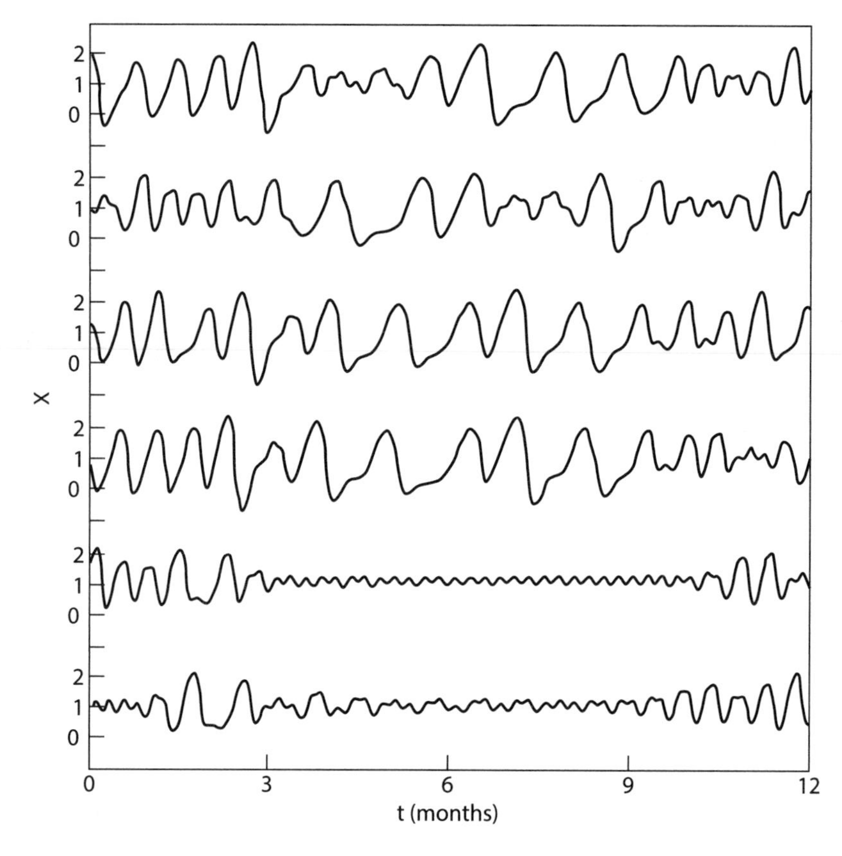

Figure 10.31. The variations of X (dimensionless) with t (months) in a six-year numerical solution of (71), with $A = 0.25$, $b = 4.0$, $F = 7 + 2\cos(2\pi t/\tau)$, and $G = 1.0$, where $\tau = 12$ months. Each row begins on January 1, and except for the first, each row is a continuation of the previous one. From Lorenz (1990).

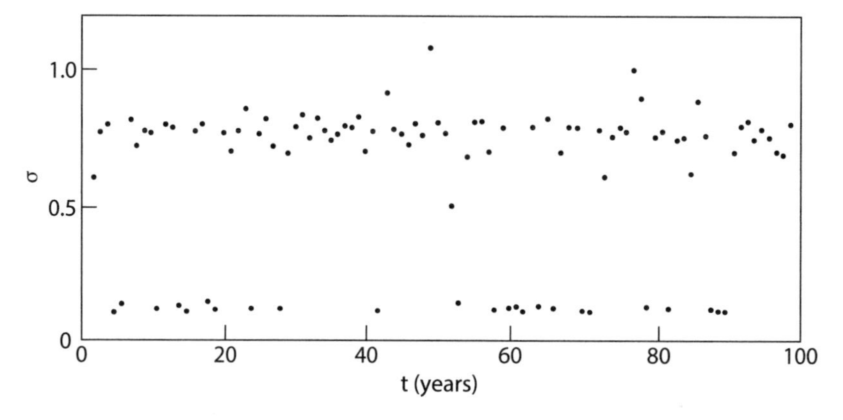

Figure 10.32. The variations of σ (dimensionless) with t (years) in a 100-year numerical solution of (71), for the conditions of figure 10.31, where σ is the standard deviation of X within the summer season, that is, July through September. From Lorenz (1990).

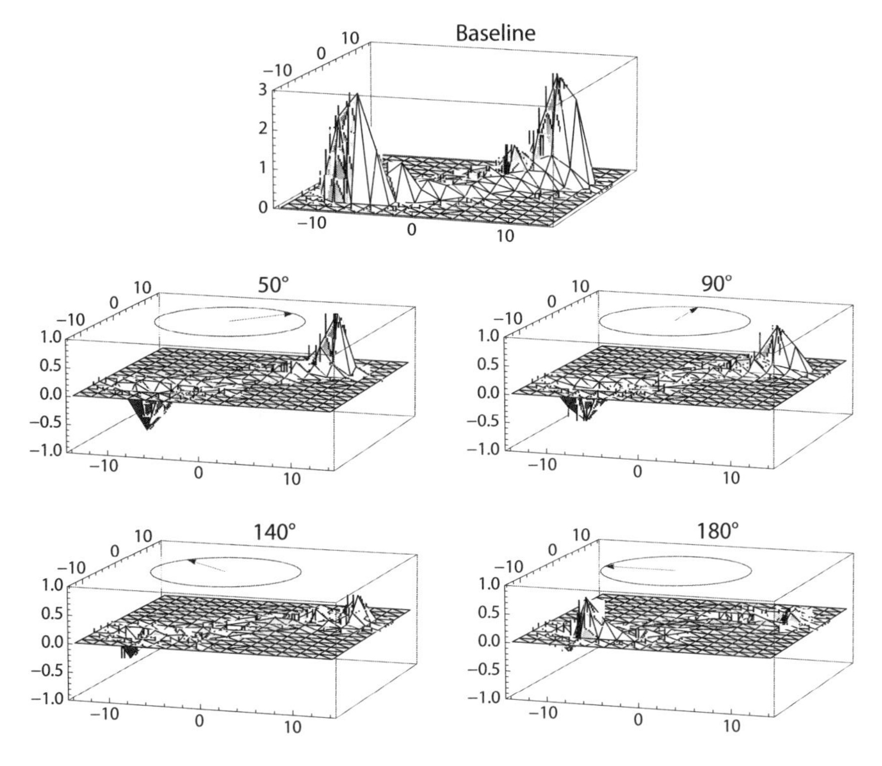

Figure 10.33. The effects of various imposed forcings on the PDF of the butterfly-model state vector with running time mean, in the (X, Y) plane based on Palmer (1999). From Palmer (1999).

different choices of θ affect the *probability density function* (PDF) of the solution in the (X,Y) plane. The maxima of the PDF are the attractors of the model. As θ changes, the locations of the maxima of the PDFs do not change much. This means that the attractors of the model are insensitive to θ; however, the maxima become stronger or weaker as θ changes. This means that varying the forcing can cause some attractors to be visited more often and others to be visited less often. Also, there is some tendency for the solution to be pushed in the direction in which the forcing acts.

Segue

Here is a terse definition of turbulence: A circulation is turbulent to the extent that its predictability time is shorter than the timescale of interest. The fundamental cause of turbulence is shearing instability. Vortex stretching leads to a kinetic energy cascade in three-dimensional turbulence. In contrast, enstrophy cascades in two-dimensional turbulence, while kinetic energy anticascades. The Rhines scale limits the meridional growth of the turbulent vortices, which become elongated into pairs of zonal jets.

Deterministic weather prediction is impossible beyond about two weeks. It may be possible to predict some *statistics* of the weather for a month or a season ahead, because of the external forcing due to persistent boundary anomalies. This forecasting appears to be most feasible in the tropics, because of the favorable signal-to-noise ratio. Longer-term climate anomalies are also predictable provided they are driven by an external forcing that is itself predictable, and provided the response to this forcing is large enough to detect against the climatic noise due to natural variability and competing external forcings.

The ocean circulation is also chaotic but has a much longer predictability time than the atmosphere; therefore, to the extent that the sea-surface temperature is predictable, and to the extent that the statistics of the weather are influenced by the sea-surface temperature, the statistics of the weather are predictable beyond the predictability limit of the atmosphere itself.

Even a chaotic system responds in a statistically predictable way to sufficiently strong external forcing; thus, summers are predictably warm and winters are predictably cold. The climate responds in a predictable way to a sufficiently strong external forcing, provided, of course, the forcing itself is predictable. Chapter 11 deals with the ongoing and future response of the global circulation to an external forcing that is gradually becoming stronger.

Problems

1. Show that in two-dimensional flow the variance of a passive scalar, per unit wavenumber, varies as k^{-1}.
2. Show that in two-dimensional flow the geopotential variance, per unit wavenumber, varies as k^{-5}.
3. Consider Lorenz's butterfly model:

$$\begin{aligned}\dot{X} &= -\sigma(X - Y),\\ \dot{Y} &= -XZ + rX - Y,\\ \dot{Z} &= XY - bZ.\end{aligned}$$

   Here σ, b, and r are constants. Assume $r \geq 1$, $b \geq 1$, and $\sigma \geq 1$. Find the stationary solutions for X, Y, and Z, and discuss the stability of these solutions as functions of σ, b, and r. One of the stationary solutions is particularly simple (you can see it by inspection of the equations). Discuss how the stability of this simplest equilibrium depends on the values of σ, b, and r.
4. Find the steady-state solutions of Lorenz's (1984) "Simplest Possible GCM," and analyze their stability.
5. Program the "Simplest Possible GCM," as discussed by Lorenz (1984). The equations of the model are

$$\begin{aligned}\dot{X} &= -Y^2 - Z^2 - aX + aF,\\ \dot{Y} &= XY - bXZ - Y + G,\\ \dot{Z} &= bXY + XZ - Z.\end{aligned}$$

   Use the fourth-order Runge-Kutta time-differencing scheme, which is discussed in many numerical analysis books, with a time step of $\Delta t = 1/30$. Following Lorenz, we interpret a time unit as corresponding to four hours, so that six time units correspond to one day. The time step is thus 4/30 hours.
   We use two slightly different versions of the model. The first, which we think of as the "real world," uses $a = 0.25$, $b = 4$, $F = 8$, and $G = 1$. The second, which we

think of as the "model," is identical with the first, except that $b = 4.01$. The code is set up in "real-world" mode.

The "standard" initial conditions are $X = 2.5, Y = 1.0, Z = 0$.

a) Demonstrate that two identical runs of the "real world" give identical results.
b) Do a test to show that the "real world" exhibits sensitive dependence on initial conditions.
c) Perform a 100-day simulation of the "real world" and save the results for X, Y, and Z once per simulated day.

Starting from *each of the first 90 days* of this "real world" weather record, make a 10-day forecast by running the "model." Save the results from each of these forecasts, once per simulated day.

Produce a plot similar to figure 10.22, with at least the curves $E_{0,k}$ and $E_{k-1,k}$. (You may want to do more than just these two.) Here use X as the variable you study, just as Lorenz (1982) used the 500 hPa height.

d) Still using X as your variable, plot dE/dk versus E, as in figure 10.24, for both the "perfect model" and the "imperfect model." Estimate the value of a, the error e-folding rate.

6. Consider an idealized state in which a two-dimensional atmosphere is at rest with respect to the rotating Earth. Imagine that starting from this state of rest, the absolute vorticity is mixed throughout the atmosphere so that it becomes uniform. Plot the resulting meridional distributions of the zonal wind and the absolute angular momentum per unit mass. Compare the angular momentum of the uniform-vorticity state with the angular momentum of the resting state.

CHAPTER 11

The Future of the Circulation

Our understanding of the global circulation of the atmosphere is very incomplete but rapidly filling in. The last several decades have seen tremendous advances. During the past 20 years or so, global circulation research has become a subdiscipline within the much broader field of climate research. At the same time, the scientific foundations of global circulation research have become deeper and more intellectually challenging.

Today we possess overwhelming evidence that the Earth's climate is rapidly changing owing to the accumulation of anthropogenic greenhouse gases in the atmosphere (IPCC, 2013; hereafter the IPCC report). The carbon dioxide concentration of the atmosphere has increased by about 40% since the middle of the nineteenth century. In the latter part of the twentieth century, there was a sharp decrease in stratospheric ozone due to anthropogenic increases in long-lived chlorine and bromine compounds (Douglass et al., 2014). Humans have also increased the aerosol load of the atmosphere through various processes including combustion, and have altered the albedo, roughness, and other properties of the land surface through agriculture, city building, and so on.

As a result of these human-induced changes, the global circulation of the atmosphere will be significantly different in the coming decades and beyond. Some changes have already become apparent in observations. It should be possible to learn more about how the circulation works by studying its response to the ongoing anthropogenic perturbations. This brief closing chapter summarizes some of the current trends and expectations for the future.

The 1500-page bulk of the recent IPCC report contains an assessment of the most recent research on this very hot topic. It is impossible to cover all the interesting aspects of the changing circulation in this chapter, so only a few topics have been selected. Interested readers should consult the IPCC report for a comprehensive discussion.

As discussed by Levitus et al. (2000, 2009, 2012) and Loeb et al. (2012), in recent decades the oceans have been accumulating thermal energy at a rate consistent with the satellite-derived imbalance of about 0.5 W m^{-2} in the Earth's radiation budget. This increase has led to a general warming of the Earth's surface and troposphere, on the order of 0.8 K since the mid-nineteenth century. The near-surface warming is strongest in the winter and greater at night than during the day (e.g., Vinnikov et al., 2002;

Vose et al., 2005). The Arctic surface air temperature has warmed about twice as much as the global mean (Screen et al., 2012), with the largest increase in winter. The Arctic warming has been accompanied by a substantial reduction in late-summer Arctic sea-ice extent and a year-round decrease in sea-ice thickness (Stroeve et al., 2014).

The troposphere above the surface is also warming (Karl et al., 2006). In fact, the tropical upper troposphere is expected to warm faster than the surface (e.g., Lorenz and DeWeaver, 2007) because, as discussed in chapter 6, the tropical lapse rate is constrained by convection to be close to the saturated moist adiabatic lapse rate, which decreases at warmer temperatures. A decrease of the lapse rate means an increase in the (dry) static stability, which inhibits baroclinic eddies. In addition, a decrease in the lapse rate is a negative feedback that tends to limit changes in the Earth's temperature, because infrared emission to space is relatively efficient from the upper troposphere (e.g., Bony et al., 2006).

Meanwhile, the stratosphere is observed to be cooling about twice as fast as the surface is warming (e.g., Thompson et al., 2012). An increase in stratospheric carbon dioxide (CO_2) promotes cooling, because infrared photons emitted from the stratosphere can easily escape to space. A decrease in stratospheric ozone concentrations also promotes cooling, because ozone warms the air by efficiently absorbing solar ultraviolet radiation.

Both cooling of the stratosphere and warming of the troposphere favor an increase in tropopause height (Hoskins, 2003; Lu et al., 2009), and a variety of observations show that the height of the tropopause has indeed been increasing (Highwood and Hoskins, 1998; Highwood et al., 2000; Randel et al., 2000; Seidel et al., 2001; Gettelman et al., 2009; Seidel and Randel, 2006; Schmidt et al., 2008; Austin and Reichler, 2008; Son et al., 2009). Tropical tropopause temperatures are decreasing (Wang et al., 2012). Lu et al. (2009) present evidence that the increase in tropopause height has been due more to the cooling of the stratosphere than to the warming of the troposphere.

Thus, the picture is complicated: the tropospheric meridional temperature gradient has decreased (especially in the Northern Hemisphere), the tropospheric static stability has increased, and the tropopause height has increased. Each of these factors can be expected to influence both baroclinic eddy activity and the width of the Hadley circulation, which in turn are linked, because the poleward limit of the Hadley regime more or less coincides with the equatorward limit of the baroclinic eddy regime. Observations show that there has been a widening of the Hadley cells (Seidel and Randel, 2007; Hu and Fu, 2007; Lu et al., 2007, 2009; Birner, 2010), which includes a poleward shift of the subtropical subsidence that inhibits precipitation. Observations also show a poleward shift of the westerly jet streams (e.g., Strong and Davis, 2007; Seidel et al., 2008; Archer and Caldeira, 2008; Barton and Ellis, 2009; Fu and Lin, 2011) and the baroclinic eddies (Cornes and Jones, 2011), at least in the Northern Hemisphere.

As mentioned in chapter 2, for typical current surface temperatures the saturation vapor pressure increases at the rate of 7% per kelvin (Held and Soden, 2006). The slope will become even larger in a future, warmer climate. Because the actual vapor pressure near the surface is closely coupled to the saturation vapor pressure at the sea surface, a warmer atmosphere will contain much more water vapor. Observations show an upward trend in the atmosphere's water vapor content over the past few decades (Trenberth et al., 2005; Wentz et al., 2007; Jin et al., 2007).

However, as also mentioned in chapter 2, the globally averaged rate of precipitation is strongly linked to the rate at which the atmosphere cools radiatively. Increases in temperature, water vapor, and CO_2 concentrations favor stronger emission of infrared radiation, that is, more intense radiative cooling. The precipitation rate is also constrained by the supply of energy available to evaporate water from the surface. Increased downward emission of infrared radiation to the Earth's surface can encourage faster evaporation and is therefore consistent with stronger precipitation.

For these reasons, the hydrologic cycle is expected to run faster in a warmer climate, and observations suggest that this is already occurring. A particularly interesting line of evidence is based on observations of sea-surface salinity. Salinity is increasing in the subtropics where precipitation is weak, and decreasing in the tropics where precipitation is strong (e.g., Durack and Wijffels, 2010; Durack et al., 2012). These changes are consistent with more evaporation in the subtropics and more precipitation in the tropics. According to Durack et al. (2012), the speed of the hydrologic cycle is increasing at the very rapid rate of 8% per kelvin of surface warming, which is about equal to the rate of increase of the surface saturation vapor pressure, as mentioned above.

In contrast, most models predict that the speed of the hydrologic cycle will increase by a much smaller 2% or 3% per kelvin (e.g., Vecchi and Soden, 2007), determined by the rate at which atmospheric radiative cooling increases (see fig. 11.1). It

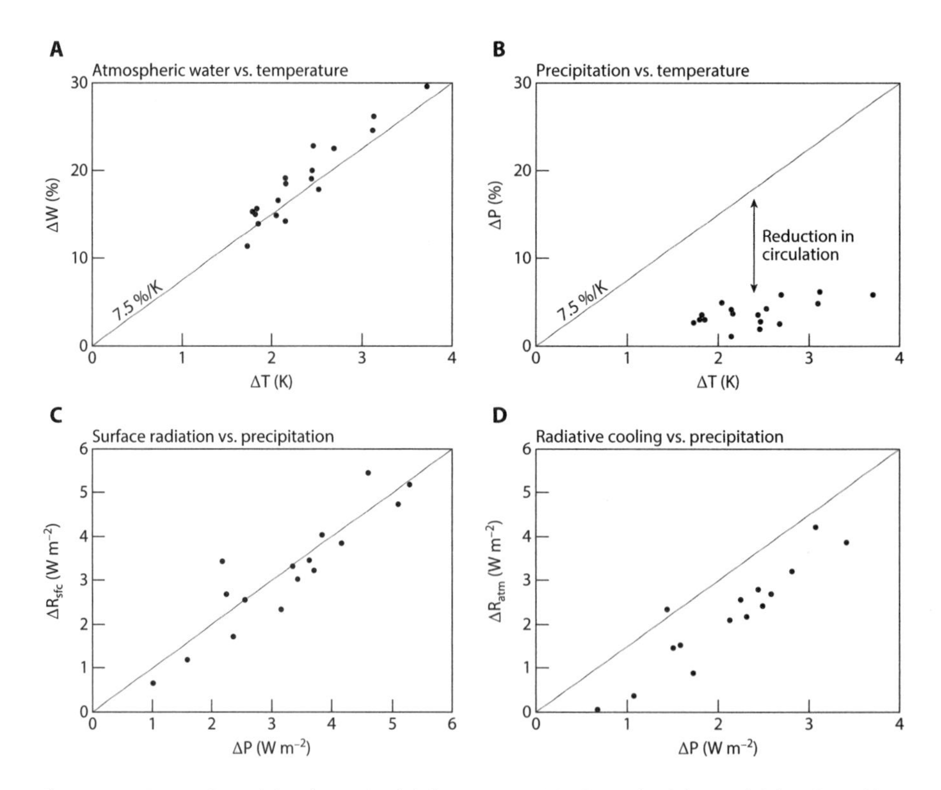

Figure 11.1. Scatterplots of the change in global-mean quantities for each of the models listed in table 1 of Vecchi and Soden (2007) for which the necessary data were available in the archive. Results are shown for (A) temperature versus column-integrated water vapor (%); (B) temperature versus precipitation (%); (C) precipitation (W m^{-2}) versus net downward radiation at the surface; and (D) precipitation (W m^{-2}) versus net radiative cooling of the atmosphere. The differences are computed by subtracting the decadal means from the first 10 years and last 10 years of the twenty-first century, as simulated by climate models. Panels A and B reproduce figs. 2.a and 2.b of Held and Soden (2006). From Vecchi and Soden (2007).

has been suggested that if the precipitation rate increases much more slowly than the atmospheric water vapor content, then weaker Hadley and Walker circulations will be sufficient to meet the needs of the hydrologic cycle (Held and Soden, 2006). This hypothesis is based on the idea that the precipitation rate is given by the atmospheric water content times a measure of the strength of the circulation. The conclusion is that the atmospheric circulation will run more slowly in a warmer, more humid climate (Vecchi and Soden, 2007; Tokinaga et al., 2012; Hsu and Li, 2012; Hsu et al., 2013; Kitoh et al., 2013). There is evidence that the global monsoon circulations have been weakening (Zhou et al., 2008; Annamalai et al., 2013). Conversely, l'Heureux et al. (2013) report a recent strengthening of the Walker circulation (see also England et al., 2014). There is also some evidence that the annular modes will be different in a warmer world (Thompson et al., 2000; Previdi and Liepert, 2007).

As discussed in chapter 8, the physical processes that give rise to the Madden-Julian oscillation are still controversial, and many of today's global atmospheric models have difficulty simulating the MJO. Climate-change simulations with models that are able to simulate the present-day MJO with some fidelity suggest that the MJO will be much more active in the warmer, more humid world of the future (e.g., Schubert et al., 2013; Arnold et al., 2014). If true, this will have momentous consequences for the people who live in the tropical Eastern Hemisphere.

Fifty years ago, the study of the global circulation of the atmosphere was a relatively obscure academic pursuit. Today it is a much more mature discipline with societally relevant practical applications ranging from weather prediction to climate simulation. In a twist of fate, we find that the global circulation of the atmosphere is rapidly evolving before our eyes, just at the moment when our science has become capable of understanding and predicting such changes.

APPENDIX A

Vectors, Vector Calculus, and Coordinate Systems

Physical Laws and Coordinate Systems

For the present discussion, we define a *coordinate system* as a system for describing positions in space. Coordinate systems are human inventions and therefore are not part of physics, although they can be used in a discussion of physics. Any physical law should be expressible in a form that is invariant with respect to our choice of coordinate systems; we certainly do not expect the laws of physics to change when we switch from spherical coordinates to Cartesian coordinates! It follows that *we should be able to express physical laws without making reference to any coordinate system.* Nevertheless, it is useful to understand how physical laws can be expressed in different coordinate systems and, in particular, how various quantities "transform" as we change from one coordinate system to another.

Scalars, Vectors, and Tensors

A *tensor* is a quantity that is defined without reference to any particular coordinate system. A tensor is simply "out there" and has a meaning that is the same whether we happen to be working in spherical coordinates, or Cartesian coordinates, or whatever. Tensors are therefore just what we need to formulate physical laws.

The number of directions associated with a tensor is called the *rank* of the tensor. In principle, the rank can be arbitrarily large, but tensors with ranks higher than 2 are rarely found in atmospheric science. The simplest kind of tensor, called a *tensor of rank 0*, is a *scalar*, which is represented by a single number—essentially a magnitude with no direction. An example of a scalar is temperature. Not all quantities that are represented by a single number are scalars, because not all of them are defined without reference to any particular coordinate system. An example of a (single) number that is *not* a scalar is the longitudinal component of the wind, which is defined with respect to a particular coordinate system, that is, spherical coordinates.

A scalar is expressed in exactly the same way regardless of what coordinate system may be in use to describe nonscalars in a problem. For example, if someone tells you the temperature in Fort Collins, you don't have to ask whether they are using spherical coordinates or some other coordinate system, because it makes no difference at all.

Vectors are *tensors of rank 1*; a vector can be represented by a magnitude and one direction. An example is the wind vector. In atmospheric science, vectors are normally either three dimensional or two dimensional, but in principle they have any number of dimensions. A scalar can be considered to be a vector in a one-dimensional space.

A vector can be expressed in a particular coordinate system by an ordered list of numbers, called the *components* of the vector. The components have meaning only with respect to the particular coordinate system. More or less by definition, the number of components needed to describe a vector is equal to the number of dimensions in which the vector is "embedded."

We can define *unit vectors* that point in each of the coordinate directions, and then we can write the vector as the vector sum of each of the unit vectors times the component associated with the unit vector. In general, the directions in which the unit vectors point depend on position.

Unit vectors are always nondimensional; note that here we are using the word *dimension* to refer to physical quantities, such as length, time, and mass. Because the unit vectors are nondimensional, *all components of a vector must have the same dimensions as the vector itself.*

Spatial coordinates may or may not have the dimensions of length. In the familiar Cartesian coordinate system, each of the three coordinates (x, y, z) has dimensions of length. In spherical coordinates, (λ, φ, r), where λ is longitude, φ is latitude, and r is distance from the origin, the first two coordinates are nondimensional angles, while the third has units of length.

When we change from one coordinate system to another, an arbitrary vector $\mathbf{V}$ transforms according to

$$\mathbf{V}' = \{\mathbf{M}\}\mathbf{V}. \tag{1}$$

Here $\mathbf{V}$ is the representation of the vector in the first coordinate system (i.e., $\mathbf{V}$ is the list of the components of the vector in the first coordinate system), $\mathbf{V}'$ is the representation of the vector in the second coordinate system, and $\{\mathbf{M}\}$ is a *rotation matrix*. The rotation matrix used to transform a vector from one coordinate system to another is a property of the two coordinate systems in question; it is the same for all vectors.

The transformation rule (1) is actually part of the definition of a vector; that is, a vector must, by definition, transform from one coordinate system to another via a rule of the form (1). It follows that not all ordered lists of numbers are vectors. The list

(mass of the moon, distance from Fort Collins to Denver)

is not a vector.

Let $\mathbf{V}$ be a vector representing the three-dimensional velocity of a particle in the atmosphere. The Cartesian and spherical representations of $\mathbf{V}$ are

$$\mathbf{V} = \dot{x}\mathbf{i} + \dot{y}\mathbf{j} + \dot{z}\mathbf{k}, \tag{2}$$

$$\mathbf{V} = \dot{\lambda} r \cos\varphi \mathbf{e}_\lambda + r\dot{\varphi}\mathbf{e}_\varphi + \dot{r}\mathbf{e}_r. \tag{3}$$

Here a "dot" denotes a Lagrangian time derivative, that is, a time derivative following a moving particle; $\mathbf{i}$, $\mathbf{j}$, and $\mathbf{k}$ are unit vectors in the Cartesian coordinate system; and $\mathbf{e}_\lambda$, $\mathbf{e}_\varphi$, and $\mathbf{e}_r$ are unit vectors in the spherical coordinate system. Equations (2) and

(3) both describe the same vector, **V**; that is, the meaning of **V** is independent of the coordinate system that is chosen to represent it.

A tensor of rank 2 that is important in atmospheric science is the flux of momentum. The momentum flux, also called a *stress*, and equivalent to a force per unit area, has a magnitude and "two directions." One of the directions is associated with the force vector itself, and the other is associated with the vector normal to the unit area in question. The momentum flux tensor can be written as $\rho \mathbf{V} \otimes \mathbf{V}$, where ρ is the density of the air, **V** is the wind vector, and $\otimes$ is the *outer* or *dyadic* product that accepts two vectors as input and delivers a rank-2 tensor as output.

Like a vector, a tensor of rank 2 can be expressed in a particular coordinate system; that is, the "components" of the tensor can be defined with respect to a particular coordinate system. The components of a tensor of rank 2 can be arranged in the form of a two-dimensional matrix, in contrast with the components of a vector, which form an ordered one-dimensional list. When we change from one coordinate system to another, a tensor of rank 2 transforms according to

$$\mathsf{T}' = \{\mathbf{M}\}\mathsf{T}\{\mathbf{M}\}^{-1}, \tag{4}$$

where T is the representation of a rank-2 tensor in the first coordinate system, T' is the representation of the same tensor in the second coordinate system, $\{\mathbf{M}\}$ is the matrix introduced in (1), and $\{\mathbf{M}\}^{-1}$ is its inverse. Note that we use sans serif Myriad bold type to represent a tensor.

Differential Operators

Several familiar differential operators can be defined without reference to any coordinate system. These operators are more fundamental than, for example, $\partial/\partial x$, where x is a particular spatial coordinate. The following are the coordinate-independent operators needed most often for atmospheric science (as well as for most other branches of physics):

$$\text{the gradient, denoted by } \nabla A, \text{ where } A \text{ is an arbitrary scalar;} \tag{5}$$

$$\text{the divergence, denoted by } \nabla \cdot \mathbf{Q}, \text{ where } \mathbf{Q} \text{ is an arbitrary vector; and} \tag{6}$$

$$\text{the curl, denoted by } \nabla \times \mathbf{Q}, \tag{7}$$

$$\text{the Laplacian, given by } \nabla^2 A \equiv \nabla \cdot (\nabla A). \tag{8}$$

Note that the gradient and curl are vectors, while the divergence is a scalar. The gradient operator accepts scalars as "input," while the divergence and curl operators consume vectors.

In discussions of two-dimensional motion, it is often convenient to introduce an additional operator called the *Jacobian*, denoted by

$$\begin{aligned} J(\alpha,\beta) &\equiv \mathbf{k} \cdot (\nabla\alpha \times \nabla\beta) \\ &= \mathbf{k} \cdot \nabla \times (\alpha\nabla\beta) \\ &= -\mathbf{k} \cdot \nabla \times (\beta\nabla\alpha). \end{aligned} \tag{9}$$

Here the gradient operators are understood to produce vectors in the two-dimensional space, α and β are arbitrary scalars, and **k** is a unit vector perpendicular to the two-dimensional surface. The second and third lines of (9) can be derived with the use of vector identities given later in this appendix.

A definition of the gradient operator that does not make reference to any coordinate system is

$$\nabla A \equiv \lim_{S\to 0}\left[\frac{1}{V}\oint_S \mathbf{n} A dS\right], \tag{10}$$

where S is the surface bounding a volume V, and $\mathbf{n}$ is the outward normal on S. Here the terms *volume* and *bounding surface* are used in the following generalized sense. In a three-dimensional space, "volume" is literally a volume, and "bounding surface" is literally a surface. In a two-dimensional space, "volume" means an area, and "bounding surface" means the curve bounding the area. In a one-dimensional space, "volume" means a curve, and "bounding surface" means the endpoints of the curve. The limit in (10) is one in which the volume and the area of its bounding surface shrink to zero.

As an example, consider a Cartesian coordinate system (x, y) on a plane, with unit vectors $\mathbf{i}$ and $\mathbf{j}$ in the x and y directions, respectively. Consider a "box" of width Δx and height Δy, as shown in figure A.1. We can write

$$\begin{aligned}\nabla A \equiv \lim_{(\Delta x,\Delta y)\to 0}&\left\{\frac{1}{\Delta x \Delta y}\left[A\left(x_0+\frac{\Delta x}{2}, y_0\right)\Delta y\mathbf{i} + A\left(x_0, y_0+\frac{\Delta y}{2}\right)\Delta x\mathbf{j}\right.\right.\\ &\left.\left.- A\left(x_0-\frac{\Delta x}{2}, y_0\right)\Delta y\mathbf{i} - A\left(x_0, y_0-\frac{\Delta y}{2}\right)\Delta x\mathbf{j}\right]\right\}\\ &= \frac{\partial A}{\partial x}\mathbf{i} + \frac{\partial A}{\partial y}\mathbf{j}.\end{aligned} \tag{11}$$

This is the answer that we expect.

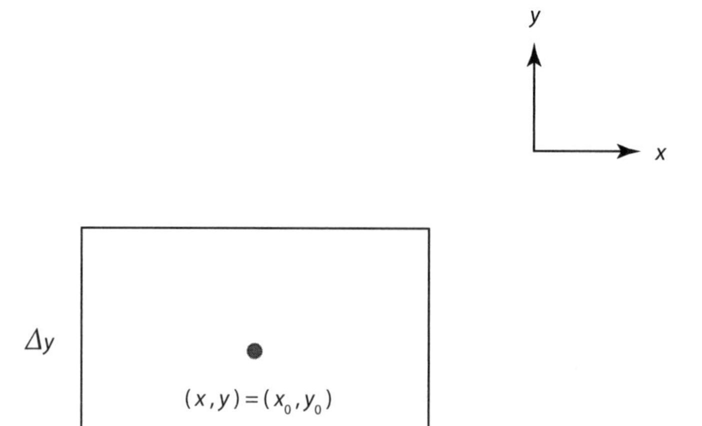

Figure A.1. Diagram illustrating a rectangular box in a planar two-dimensional space, with center at (x_0, y_0), width Δx, and height Δy.

Definitions of the divergence and curl operators that do not make reference to any coordinate system are

$$\nabla \cdot \mathbf{Q} \equiv \lim_{S \to 0} \left[\frac{1}{V} \oint_S \mathbf{n} \cdot \mathbf{Q} dS \right], \tag{12}$$

$$\nabla \times \mathbf{Q} \equiv \lim_{S \to 0} \left[\frac{1}{V} \oint_S \mathbf{n} \times \mathbf{Q} dS \right]. \tag{13}$$

It is possible to work through exercises similar to (11) for these operators, too. You might want to try it yourself, to see if you understand.

Finally, the Jacobian on a two-dimensional surface can be defined by

$$J(A,B) = \lim_{A \to 0} \left[\oint_C A \nabla B \cdot \mathbf{t} dl \right], \tag{14}$$

where $\mathbf{t}$ is a unit vector that is tangent to the bounding curve C.

Vector Identities

Many useful identities relate the divergence, curl, and gradient operators. Most of the following identities can be found in any mathematics reference manual, e.g., Beyer (1984). As before, we let α and β be arbitrary scalars; we let $\mathbf{V}$, $\mathbf{A}$, $\mathbf{B}$, and $\mathbf{C}$ be arbitrary vectors; and we let $\mathbf{T}$ be an arbitrary tensor of rank 2. Then,

$$\nabla \times (\nabla \alpha) = 0, \tag{15}$$

$$\nabla \cdot (\nabla \times \mathbf{V}) = 0, \tag{16}$$

$$\mathbf{A} \times \mathbf{B} = -\mathbf{B} \times \mathbf{A}, \tag{17}$$

$$\nabla \cdot (\alpha \mathbf{V}) = \alpha (\nabla \cdot \mathbf{V}) + \mathbf{V} \cdot \nabla \alpha, \tag{18}$$

$$\nabla \cdot (\mathbf{A} \times \mathbf{B}) = (\nabla \times \mathbf{A}) \cdot \mathbf{B} - (\nabla \times \mathbf{B}) \cdot \mathbf{A}, \tag{19}$$

$$\nabla \times (\alpha \mathbf{V}) = \nabla \alpha \times \mathbf{V} + \alpha (\nabla \times \mathbf{V}), \tag{20}$$

$$\mathbf{A} \cdot (\mathbf{B} \times \mathbf{C}) = (\mathbf{A} \times \mathbf{B}) \cdot \mathbf{C} = \mathbf{B} \cdot (\mathbf{C} \times \mathbf{A}), \tag{21}$$

$$\mathbf{A} \times (\mathbf{B} \times \mathbf{C}) = \mathbf{B}(\mathbf{C} \cdot \mathbf{A}) - \mathbf{C}(\mathbf{A} \cdot \mathbf{B}), \tag{22}$$

$$\nabla \times (\mathbf{A} \times \mathbf{B}) = \mathbf{A}(\nabla \cdot \mathbf{B}) - \mathbf{B}(\nabla \cdot \mathbf{A}) - (\mathbf{A} \cdot \nabla)\mathbf{B} + (\mathbf{B} \cdot \nabla)\mathbf{A}, \tag{23}$$

$$\nabla (\mathbf{A} \cdot \mathbf{B}) = (\mathbf{A} \cdot \nabla)\mathbf{B} + (\mathbf{B} \cdot \nabla)\mathbf{A} + \mathbf{A} \times (\nabla \times \mathbf{B}) + \mathbf{B} \times (\nabla \times \mathbf{A}), \tag{24}$$

$$J(\alpha,\beta) \equiv \mathbf{k} \cdot (\nabla \alpha \times \nabla \beta) = \mathbf{k} \cdot \nabla \times (\alpha \nabla \beta) = -\mathbf{k} \cdot \nabla \times (\beta \nabla \alpha) = -\mathbf{k} \cdot (\nabla \beta \times \nabla \alpha), \tag{25}$$

$$\nabla^2 \mathbf{V} \equiv (\nabla \cdot \nabla) \mathbf{V} = \nabla (\nabla \cdot \mathbf{V}) - \nabla \times (\nabla \times \mathbf{V}), \tag{26}$$

$$\nabla \cdot (\mathbf{A} \otimes \mathbf{B}) = (\mathbf{A} \cdot \nabla)\mathbf{B} + (\mathbf{B} \cdot \nabla)\mathbf{A}, \tag{27}$$

$$\nabla \cdot (\alpha \mathbf{T}) = (\nabla \alpha) \cdot \mathbf{T} + \alpha (\nabla \cdot \mathbf{T}). \tag{28}$$

In (27), $\mathbf{A} \otimes \mathbf{B}$ denotes the outer product of two vectors, which yields a tensor of rank 2.

A useful result that is a special case of (23) is

$$\mathbf{e}_r \cdot \nabla \times (\mathbf{e}_r \times \mathbf{V}) = \nabla \cdot \mathbf{V}, \tag{29}$$

where $\mathbf{e}_r$ is the unit vector pointing upward, and $\mathbf{V}$ is the *horizontal* velocity vector. In words, the curl of $\mathbf{e}_r \times \mathbf{V}$ is equal to the divergence of $\mathbf{V}$. Similarly, a useful special case of (19) is

$$\nabla \cdot (\mathbf{e}_r \times \mathbf{V}) = -(\nabla \times \mathbf{V}) \cdot \mathbf{e}_r. \tag{30}$$

This means that the divergence of $\mathbf{e}_r \times \mathbf{V}$ is equal to minus the curl of $\mathbf{V}$.

A special case of (24) is

$$\frac{1}{2}\nabla(\mathbf{V} \cdot \mathbf{V}) = (\mathbf{V} \cdot \nabla)\mathbf{V} + \mathbf{V} \times (\nabla \times \mathbf{V}). \tag{31}$$

This identity is used to write the advection terms of the momentum equation in alternative forms.

Identity (26) says that the Laplacian *of a vector* is the gradient of the divergence of the vector, minus the curl of the curl of the vector. This relation can be used, for example, in a parameterization of momentum diffusion.

Spherical Coordinates

For obvious reasons, spherical coordinates are of special importance in geophysics. The unit vectors in spherical coordinates are denoted by $\mathbf{e}_\lambda$ pointing toward the east, $\mathbf{e}_\varphi$ pointing toward the north, and $\mathbf{e}_r$ pointing outward from the origin (in geophysics, outward from the center of the Earth).

The gradient, divergence, and curl operators can be expressed in spherical coordinates as follows:

$$\nabla A = \left(\frac{1}{r\cos\varphi}\frac{\partial A}{\partial \lambda}, \frac{1}{r}\frac{\partial A}{\partial \varphi}, \frac{\partial A}{\partial r}\right), \tag{32}$$

$$\nabla \cdot \mathbf{V} = \frac{1}{r\cos\varphi}\frac{\partial V_\lambda}{\partial \lambda} + \frac{1}{r\cos\varphi}\frac{\partial}{\partial \varphi}(V_\varphi \cos\varphi) + \frac{1}{r^2}\frac{\partial}{\partial r}(V_r r^2), \tag{33}$$

$$\nabla \times \mathbf{V} = \left\{\frac{1}{r}\left[\frac{\partial V_r}{\partial \varphi} - \frac{\partial}{\partial r}(rV_\varphi)\right], \frac{1}{r}\frac{\partial}{\partial r}(rV_\lambda) - \frac{1}{r\cos\varphi}\frac{\partial V_r}{\partial \lambda}, \frac{1}{r\cos\varphi}\left[\frac{\partial V_\varphi}{\partial \lambda} - \frac{\partial}{\partial \varphi}(V_\lambda \cos\varphi)\right]\right\}, \tag{34}$$

$$\nabla^2 A = \frac{1}{r^2\cos\varphi}\left[\frac{\partial}{\partial \lambda}\left(\frac{1}{\cos\varphi}\frac{\partial A}{\partial \lambda}\right) + \frac{\partial}{\partial \varphi}\left(r^2\cos\varphi\frac{\partial A}{\partial r}\right)\right], \tag{35}$$

$$J(A,B) = \frac{1}{a^2\cos\varphi}\left(\frac{\partial A}{\partial \lambda}\frac{\partial B}{\partial \varphi} - \frac{\partial B}{\partial \lambda}\frac{\partial A}{\partial \varphi}\right). \tag{36}$$

Here A is an arbitrary scalar, and $\mathbf{V}$ is an arbitrary vector. If $\mathbf{V}$ is separated into a horizontal vector and a vertical vector, as in $\mathbf{V} = \mathbf{V}_h + V_r\mathbf{e}_r$, then (34) can be written as

$$\nabla \times (\mathbf{V}_h + V_r\mathbf{e}_r) = (\nabla_r \times \mathbf{V}_h) + \mathbf{e}_r \times \left[\frac{1}{r}\frac{\partial}{\partial r}(r\mathbf{V}_h) - \nabla_r V_r\right]. \tag{37}$$

As an example of the application of (34), the vertical component of the vorticity is

$$\zeta = \frac{1}{r\cos\varphi}\left[\frac{\partial V_\varphi}{\partial \lambda} - \frac{\partial}{\partial \varphi}(V_\lambda \cos\varphi)\right]. \tag{38}$$

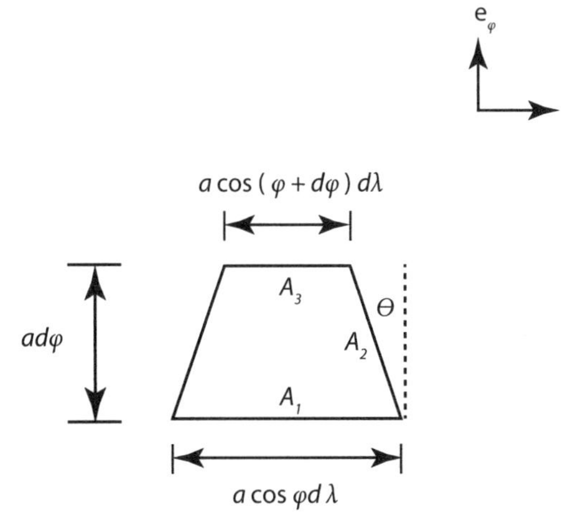

Figure A.2. A patch of the sphere, with longitudinal width $a\cos\varphi d\lambda$, and latitudinal height $ad\varphi$.

For the case of pure solid-body rotation of the atmosphere about the Earth's axis of rotation, we have

$$V_\lambda = \dot{\lambda} r \cos\varphi \text{ and } V_\varphi = 0, \tag{39}$$

where $\dot{\lambda}$ is independent of φ. Substitution gives

$$\begin{aligned} \zeta &= \frac{-1}{r\cos\varphi}\frac{\partial}{\partial\varphi}(\dot{\lambda} r \cos^2\varphi) \\ &= 2\dot{\lambda}\sin\varphi. \end{aligned} \tag{40}$$

As a second, trickier example, consider how the two-dimensional version of (32) can be derived from (10). Figure A.2 illustrates the problem. Here r has been replaced by a, the radius of the Earth. The angle θ depicted in the figure arises from the gradual rotation of $\mathbf{e}_\lambda$ and $\mathbf{e}_\varphi$, the unit vectors associated with the spherical coordinates, as the longitude changes; the directions of $\mathbf{e}_\lambda$ and $\mathbf{e}_\varphi$ in the center of the area element, where ∇A is defined, are different from their respective directions on either east-west wall of the area element. Inspection of figure A.2 shows that θ satisfies

$$\begin{aligned} \tan\theta &= \frac{-\frac{1}{2}\left[a\cos(\varphi + d\varphi) - a\cos\varphi\right]d\lambda}{ad\varphi} \\ &\to -\frac{1}{2}\left(\frac{\partial}{\partial\varphi}\cos\varphi\right)d\lambda \\ &= \frac{1}{2}\sin\varphi d\lambda \\ &\cong \sin\theta. \end{aligned} \tag{41}$$

The angle θ is of "differential" or infinitesimal size. Nevertheless, it is needed in the derivation of (32). The line integral in (10) can be expressed as

$$\frac{1}{\text{Area}}\oint A\mathbf{n}dl = \frac{1}{a^2\cos\varphi d\lambda d\varphi}\left[-\mathbf{e}_\varphi A_1 a\cos\varphi d\lambda + \mathbf{e}_\lambda A_2 ad\varphi\cos\theta + \mathbf{e}_\varphi A_2 ad\varphi\sin\theta \right.$$
$$\left. + \mathbf{e}_\varphi A_3\cos(\varphi + d\varphi)\,d\lambda - \mathbf{e}_\lambda A_4 ad\varphi\cos\theta + \mathbf{e}_\varphi A_4 ad\varphi\sin\theta\right]$$
$$= \mathbf{e}_\lambda\frac{(A_2 - A_4)\cos\theta}{a\cos\varphi d\lambda} \tag{42}$$
$$+ \mathbf{e}_\varphi\left\{\frac{[A_3 a\cos(\varphi + d\varphi) - A_1 a\cos\varphi]d\lambda + (A_2 + A_4)\sin\theta ad\varphi}{a^2\cos\varphi d\lambda d\varphi}\right\}.$$

Note that the angle θ has entered here. We put $\cos\theta \to 1$ and $\sin\theta \to \frac{1}{2}\sin\varphi d\lambda$, to obtain

$$\frac{1}{\text{Area}}\oint A\mathbf{n}dl = \mathbf{e}_\lambda\frac{(A_2 - A_4)}{a\cos\varphi d\lambda} + \mathbf{e}_\varphi\left\{\left[\frac{A_3 a\cos(\varphi + d\varphi) - A_1 a\cos\varphi}{a\cos\varphi d\varphi}\right] + \left(\frac{A_2 + A_4}{2}\right)\frac{\sin\varphi}{a\cos\varphi}\right\}$$
$$\to \mathbf{e}_\lambda\frac{1}{a\cos\varphi}\frac{\partial A}{\partial\lambda} + \mathbf{e}_\varphi\left[\frac{1}{a\cos\varphi}\frac{\partial}{\partial\varphi}(A\cos\varphi) + \frac{A\sin\varphi}{a\cos\varphi}\right] \tag{43}$$
$$= \mathbf{e}_\lambda\frac{1}{a\cos\varphi}\frac{\partial A}{\partial\lambda} + \mathbf{e}_\varphi\frac{1}{a}\frac{\partial A}{\partial\varphi},$$

which agrees with the two-dimensional version of (32).

Similar (but more straightforward) derivations can be given for (33)–(36).

We can prove the following about the unit vectors in spherical coordinates:

$$\nabla\cdot\mathbf{e}_\lambda = 0,$$
$$\nabla\cdot\mathbf{e}_\varphi = -\frac{\tan\varphi}{r}, \tag{44}$$
$$\nabla\cdot\mathbf{e}_r = \frac{2}{r},$$

$$\nabla\times\mathbf{e}_\lambda = \frac{\mathbf{e}_\varphi}{r} + \frac{\tan\varphi}{r}\mathbf{e}_r,$$
$$\nabla\times\mathbf{e}_\varphi = -\frac{\mathbf{e}_\lambda}{r}, \tag{45}$$
$$\nabla\times\mathbf{e}_r = 0.$$

Finally, the following relations are useful when working with the momentum equation in spherical coordinates:

$$(\mathbf{V}_h\cdot\nabla)\mathbf{e}_\lambda = -\frac{u\sin\varphi}{r}\mathbf{e}_\varphi + \frac{u\cos\varphi}{r}\mathbf{e}_r,$$
$$(\mathbf{V}_h\cdot\nabla)\mathbf{e}_\varphi = \frac{u\sin\varphi}{r}\mathbf{e}_\lambda + \frac{v\sin\varphi}{r}\mathbf{e}_r, \tag{46}$$
$$(\mathbf{V}_h\cdot\nabla)\mathbf{e}_r = \frac{-\mathbf{V}_h}{r}.$$

Here $\mathbf{V}_h = u\mathbf{e}_\lambda + v\mathbf{e}_\varphi$ is the horizontal wind vector.

Conclusions

This brief overview is intended mainly as a refresher for students who learned these concepts once upon a time but may have not thought about them for a while. The bibliography provides much more information. Resources can also be found on the Web.

APPENDIX B

Dimensional Analysis, Scale Analysis, and Similarity Theories

Dimensions and Units

So far as we know, nature can be described using the four "dimensions" or "primary quantities" of length, time, mass, and electric charge. Here the word "dimension" is used to refer to aspects of nature that are independent in the sense that they are not interconvertible. Length cannot be rescaled somehow as a mass. Time cannot be rescaled as an electric charge.

All physical quantities can be measured as combinations of these four dimensions. For example, velocity is length divided by time, and energy can be expressed as mass times length squared divided by time squared.

Temperature is an anomaly; there has been a debate as to whether it should be included as a primary quantity independent of mass, length, and time (Huntley, 1967). It is a statistic of molecular motions that can be defined in terms of energy per unit mass. In atmospheric science, temperature is usually treated as a fifth primary quantity.

Units are different from dimensions. The various primary quantities are measured using units, which can be defined in very arbitrary ways. For example, length can be measured using meters, feet, furlongs, or the size of King Henry VIII's foot. Today, scientists almost always use the metric or the international system of units (SI, Système International d'Unités).

It is possible to define natural or fundamental units (e.g., Barrow, 2002; Wilczek, 2005, 2006a, 2006b), but these are not convenient for use in atmospheric science.

The starting point for the following discussion is that physical principles must be independent of the choice of units. For example, Newton's law $F = ma$—that is, force equals mass times acceleration—must predict the same physical phenomena whether English units or SI units are used.

Some definitions are necessary:

1. *Physical quantity*: a conceptual property of a physical system that can be expressed numerically in terms of one or more standards. Example: the radius of the Earth.
2. *Primary quantities:* a set of quantities (hereafter called qs) chosen arbitrarily for the description of a problem, subject to the constraint that the units of measurement chosen for the quantities can be assigned independently. Primary

quantities are sometimes called *fundamental quantities*. Examples: length, time, mass.

3. *Dimension*: the relationship of a derived physical quantity to whatever primary quantities have been selected. Example: velocity = length / time.
4. *Standard*: an arbitrary reference measure adopted for purposes of communication. Example: meter.
5. *Unit*: an arbitrary fraction or multiple of a standard, used to avoid inconveniently large (or small) numbers. Example: kilometer.
6. *Extraneous standard*: a standard that is irrelevant for a particular problem. Example: the length of King Henry VIII's foot, which is irrelevant except when he visits a shoe store.
7. *Extraneous unit*: a unit based on an extraneous standard. Example:1 mile = 5280 ft.
8. *Dimensionless quantity*: a quantity that is expressed in units derivable from the problem (*not* in extraneous units). There are no *intrinsically* dimensionless quantities. A quantity is dimensionless or not only with respect to a particular problem. Example: Rossby number.
9. *Dimensional analysis*: the process of removing extraneous information from a problem by forming dimensionless groups.
10. *Nondimensionalization*: conversion of a system of dimensional equations to a system that contains only nondimensional quantities.
10. *Scale analysis, sometimes called scaling*: using *chosen* numerical values for the dimensional parameters to compare the orders of magnitude of various terms of a system of nondimensional equations. A scale analysis can be performed only when the governing equations are known.
11. *Similar systems*: those for which the dimensionless quantities have identical values, even though the dimensional quantities may be very different. Example: wind tunnel.
12. *Similarity theory*: a theory based on the hypothesis that functional relationships exist among the nondimensional parameters describing a physical system. The functions themselves must be determined empirically. Similarity theories can be useful when the desired functions cannot be derived from the governing equations.

Consistent Use of Dimensional and Nondimensional Quantities in Equations

It makes no sense to add or subtract quantities that have different dimensions. A length cannot be added to a mass. Thus, all terms of an equation must have the same dimensions. This is called the requirement of *dimensional homogeneity*.

An exponent must be nondimensional, because it is the number (pure number) of times that something is multiplied by itself.

Similarly, the arguments of (inputs to) mathematical functions must be nondimensional. Examples include exponentials, logarithms, and trigonometric functions. You can take the sine of 2, but you cannot take the sine of 2 meters.

It is not unusual to see dimensional quantities as the arguments of logarithms, even though the expression makes no sense. For example, you might see

$$\frac{1}{\theta}\frac{D\theta}{Dt} = \frac{D}{Dt}(\ln\theta),$$

where θ is the potential temperature. The correct (but longer) statement is

$$\frac{1}{\theta}\frac{D\theta}{Dt} = \frac{D}{Dt}\left[\ln\left(\frac{\theta}{\theta_{ref}}\right)\right],$$

where θ_{ref} is a constant reference value. It is necessary to include θ_{ref} for the equation to make mathematical sense.

The Buckingham Pi Theorem

The fundamental theorem of dimensional analysis is due to Buckingham and is stated here without proof:

The Buckingham Pi Theorem

If the equation

$$\phi(q_1, q_2 \ldots, q_n) = 0 \tag{1}$$

is the only relationship among the q_i, and if it holds for any arbitrary choice of the units in which $q_1, q_2 \ldots, q_n$ are measured, then (1) can be written in the form

$$\phi(\pi_1, \pi_2 \ldots, \pi_m) = 0, \tag{2}$$

where $\pi_1, \pi_2, \ldots, \pi_m$, are independent dimensionless products of the q's.

Further, if k is the minimum number of primary quantities necessary to express the dimensions of the qs, then

$$m = n - k. \tag{3}$$

Since $k > 0$, (3) implies that $m < n$. According to (3), the number of dimensionless products is the number of dimensional parameters minus the number of primary quantities.

Another way of writing (2) is

$$\phi'(\pi_1, \pi_2, \ldots, \pi_m; 1, \ldots, 1) = 0, \tag{4}$$

where the number of 1s appearing in the argument list is k. Clearly the 1s carry no information about the functional relationship among the π's, so that we can just omit them, as was done in (2). In (4), the 1s clearly represent "extraneous" information, which entered the problem through extraneous units of the q's. By dropping the extraneous information, we don't change the problem; we just simplify it to its essentials.

The q's can be chosen by inspection of the governing equations (if known) or by inspection of the physical system.

The dimensions of the q's can be determined in terms of primary quantities that can be chosen arbitrarily, provided their units can be assigned independently. It is necessary to choose enough primary quantities to ensure that nondimensional combinations can be formed in all cases.

The following is an example of dimensional analysis. We consider thermal convection of a shallow fluid in a laboratory tank, with no mean flow or rotation. The linearized governing equations are

$$\left(\frac{\partial}{\partial t} - \upsilon\nabla^2\right)u' = -\frac{\partial}{\partial x}\left(\frac{p'}{\rho_0}\right),$$

$$\left(\frac{\partial}{\partial t} - \upsilon\nabla^2\right)v' = -\frac{\partial}{\partial y}\left(\frac{p'}{\rho_0}\right),$$

$$\left(\frac{\partial}{\partial t} - \upsilon\nabla^2\right)w' = -\frac{\partial}{\partial z}\left(\frac{p'}{\rho_0}\right) + g\alpha T',$$

$$\frac{\partial u'}{\partial x} + \frac{\partial v'}{\partial y} + \frac{\partial w'}{\partial z} = 0$$

$$\left(\frac{\partial}{\partial t} - \kappa\nabla^2\right)T' = -w'\Gamma.$$

Here x, y, and z are the spatial coordinates; u, v, and w are the corresponding components of the velocity vector; T is temperature; p is pressure; ρ_0 is a constant reference-state density; g is the acceleration of gravity; α is a thermal expansion coefficient; υ is the molecular viscosity; κ is the molecular conductivity; and $\Gamma \equiv \partial\overline{T}/\partial z$ is the rate at which temperature increases with height in the mean state. An overbar denotes the mean state, and a prime denotes the departure from the mean state.

By inspection of the equations, we see that the six dimensional parameters of the problem are

$$g, \Gamma, h, \upsilon, \kappa, \alpha.$$

Here g is gravity, Γ is the lapse rate, h is the depth of the fluid, υ is the molecular viscosity, κ is the molecular thermal conductivity, and α is a parameter that measures the amount of thermal expansion per unit temperature change. These six parameters are determined by the design of the laboratory experiment, such as the depth of the tank and the choice of convecting fluid (water, air, oil, etc.). We don't include the dependent variables u, v, w, p, or T in the list, because they are part of the solution and so depend on the parameters listed. We don't include ρ_0 in the list because it appears only in the combination p'/ρ_0, and so we just think of p'/ρ_0 as one of the dependent variables.

As primary quantities, we choose length (L), time (T), and temperature (Θ). There are three primary quantities and six dimensional parameters, so the π-theorem tells us that we should be able to eliminate $6-3=3$ pieces of extraneous information. The dimensions of the q's are shown in table B.1. As our pertinent (nonextraneous) unit of length, we choose h. Forming products, we systematically eliminate the "lengths" from our set of quantities, as shown in table B.2. As our unit of time, we use $h^2\upsilon^{-1}$. (We could just as well take $h^2\kappa^{-1}$.) Again forming products, we obtain the results shown in table B.3. Obviously, to complete the procedure, we simply form the product $\Gamma\alpha h$. The final results are shown in table B.4. All together, we then have three 1s in the "Quantity" column of table B.4. This means that three pieces of extraneous information have been eliminated, as promised by the π-theorem. But we have the *three* nondimensional combinations

$$gh^3\upsilon^{-2} \equiv x_1, \quad (5)$$

Table B.1. The Dimensions of the Dimensional Quantities Used in the Nondimensionalization of the Equations Governing Thermal Advection

	Dimensions		
Quantity	L	T	Θ
g	1	−2	0
α	0	0	−1
Γ	−1	0	1
h	1	0	0
υ	2	−1	0
κ	2	−1	0

Table B.2. The Data of Table A2.1 with the Dimension of Length Eliminated

	Dimensions		
Quantity	L	T	Θ
gh^{-1}	0	−2	0
α	0	0	−1
Γh	0	0	1
1	0	0	0
υh^{-2}	0	−1	0
κh^{-2}	0	−1	0

Table B.3. The Data of Table A2.2 with the Dimension of Time Eliminated

	Dimensions		
Quantity	L	T	Θ
$gh^{3}\upsilon^{-2}$	0	0	0
α	0	0	−1
Γh	0	0	1
1	0	0	0
1	0	0	0
$\kappa\upsilon^{-2}$	0	0	0

Table B.4. The Data of Table A2.3 with All Dimensional Quantities Eliminated

	Dimensions		
Quantity	L	T	Θ
$gh^{3}\upsilon^{-2}$	0	0	0
$\alpha\Gamma h$	0	0	0
1	0	0	0
1	0	0	0
1	0	0	0
$\kappa\upsilon^{-1}$	0	0	0

$$\kappa v^{-1} \equiv Pr, \tag{6}$$

and

$$\Gamma \alpha h \equiv x_2 . \tag{7}$$

Notice that

$$Ra = Pr\, x_1 x_2 \tag{8}$$

so that we can, alternatively, regard Pr, Ra, and x_1 (say) as our three combinations.

Chandrasekhar (1961) showed that only *two* dimensionless combinations matter for the convection problem, namely, Pr and Ra. So then, why have we found three? In the governing equations, g and α appear only in the combination $g\alpha$, so they don't have to be *separately* included in our list of q's. Because this reduces the number of q's by one, without reducing the number of primary quantities, the π-theorem tells us that the number of dimensionless combinations will also be reduced by one. The lesson is that we should not enter dimensional quantities separately if they appear in the equations only in some combination.

Scale Analysis

In atmospheric science and oceanography, scale analysis is very widely used to justify various approximations that can aid in the solution of the governing equations. Scale analysis always starts from a set of governing equations.

The following is an important example. Quasi-geostrophic theory was derived by Charney (1948) using scale analysis. We now perform a simple scale analysis of the equation of motion, in the spirit of Charney's work. We can write the equation of motion in simplified form as

$$\frac{D\mathbf{V}}{Dt} + f\mathbf{k} \times \mathbf{V} = -\nabla_p \phi . \tag{9}$$

Here $\mathbf{V}$ is the horizontal wind vector. We have omitted friction, for simplicity. The three terms included in (9) suffice to describe the evolution of the large-scale horizontal wind throughout most of the atmosphere.

We can nondimensionalize (9) through a straightforward, almost mechanical procedure, as follows. We let U be a velocity scale and write

$$\mathbf{V} = U\hat{\mathbf{V}},$$

where the carat notation denotes a nondimensional variable. Similarly, we let U be a length scale. We can then construct a timescale as $T = L/U$ and write

$$t = T\hat{t} = (L/U)\hat{t}, \text{ and}$$
$$\frac{D}{Dt} = \frac{1}{T}\frac{D}{D\hat{t}} = \frac{U}{L}\frac{D}{D\hat{t}} .$$

Finally, we let $f = f_0 f$, where f_0 is a suitably *chosen* representative (dimensional) value of the Coriolis parameter. Note that to nondimensionalize (9) we do not need scales for mass or temperature.

Making the various substitutions into (9), we can rewrite it as

$$\frac{U^2}{L}\frac{D\hat{\mathbf{V}}}{D\hat{t}} + f_0\hat{f}\mathbf{k}\times U\hat{\mathbf{V}} = -\nabla_p\phi.$$

Dividing through by f_0U, we find that

$$Ro\frac{D\hat{\mathbf{V}}}{D\hat{t}} + \hat{f}\mathbf{k}\times\hat{\mathbf{V}} = -\frac{\nabla_p\phi}{f_0U}, \tag{10}$$

where $Ro \equiv U/Lf_0$ is the (nondimensional) Rossby number. Equation (10) is the nondimensional form of (9).

In converting the dimensional equation (9) to the nondimensional form (10), we did not change anything. The problem to be solved remains the same. Nondimensionalization removes extraneous information, however, and that simplifies things. Nondimensionalization also reveals nondimensional parameters of physical importance, such as the Rossby number. If we nondimensionalized the equations governing Rayleigh convection, the Rayleigh and Prandtl numbers would emerge in much the same way that the Rossby number emerged previously.

As mentioned earlier, two systems are said to be "similar" if their nondimensional parameters have the same values, even though their dimensional parameters are quite different. The most familiar example is a wind tunnel, in which small models are used to investigate the aerodynamic properties of much larger, full-scale aircraft. Laboratory analogs of the atmosphere also can be constructed. The simplified and incomplete scale analysis presented above suggests that to be a useful analogue of the atmosphere, the laboratory system must be designed to have the same Rossby number as the atmosphere. A more complete scale analysis would reveal the importance of several additional nondimensional parameters.

Nondimensionalization is the first step of a scale analysis. The second step is to choose the numerical values of the scales *so that the nondimensional dependent variables are of order* 1 *for the problem of interest*. For example, to investigate large-scale midlatitude motions in the Earth's atmosphere we would choose U to be 10 m s^{-1}, L to be 10^6 m, and f_0 to be 10^{-4} s^{-1}, because these are about the right size for midlatitude large-scale motions.

Using these scales, we find that $Ro = 0.1$. This means that the (leading) acceleration term of (10) is an order of magnitude smaller than the Coriolis term. For the equation to be satisfied, the Coriolis term has to be balanced by the only remaining term, which represents the pressure-gradient force. Then, (10) reduces to

$$\hat{f}\mathbf{k}\times\hat{\mathbf{V}} = -\frac{\nabla_p\phi}{f_0U},$$

which is an expression of geostrophic balance. Since $\nabla_p \sim 1/L$, we can conclude that the horizontal variations of ϕ on pressure surfaces, denoted by $\delta\phi$, satisfy

$$\delta\phi \sim f_0UL.$$

The scale analysis thus allowed us to deduce the order of magnitude of $\delta\phi$.

The conclusions drawn depended on our *choices* for the numerical values of the dimensional scales U, L, and f_0. If we were interested in a different meteorological phenomenon, such as individual turbulent eddies in the boundary layer, we would make different choices for the scales and reach different conclusions.

Similarity Theories

Scale analysis makes use of the equations that describe a physical system. Unfortunately, there are many cases in which those equations are not known.

For example, the wind experiences a drag force as it moves near the Earth's surface. We can hypothesize that there exists (i.e., it is possible to find) a "formula" that tells us how to compute the drag given the wind speed, the lapse rate of temperature, the roughness of the ground (or ocean), and perhaps several other dimensional quantities. We don't know how to derive the formula from the equation of motion and the other basic physical principles of our science, but we believe that the formula exists.

Similarity theories aim to find such formulas by means of hypotheses, which could be described as inspired guesses. We start by listing dimensional parameters that appear in the basic governing equations, including the boundary conditions. In general, these parameters would include such factors as the spatial coordinates and time, as well as the Earth's rotation rate and the acceleration of gravity. The list of dimensional parameters can be very long, especially considering that it could in principle include detailed information about such things as the Earth's topography.

To make the list shorter, as a first step we can hypothesize, often with good reason, that some of the dimensional quantities are irrelevant to the problem at hand. For example, the height of Mount Everest is irrelevant to the drag that the air experiences as it flows over Kansas City. As a less ludicrous example, we might omit the density of water vapor from the list of dimensional parameters to be used in our search for the drag formula.

Having settled on a reasonably short list of dimensional parameters, we then nondimensionalize, and the π-theorem tells us that this process will yield an even shorter list of nondimensional parameters. We then assert that there are functional relationships among the nondimensional combinations of interest. If the number of parameters is small enough, then we will have a reasonable hope of finding the functional relationships empirically. This is exactly what was done in the development of the famous Monin-Obukhov similarity theory, which provides very useful empirical formulas for determining the surface fluxes of momentum and sensible heat in terms of the mean-wind and temperature profiles near the Earth's surface.

As a second example, related to the discussion of the preceding section, we can imagine two laboratory convection experiments in which h v, κ, $g\alpha$, and Γ are all different. We would like to find a formula that relates the upward heat flux in the convection tank to the nondimensional parameters of the problem, namely, Pr and Ra. We don't know how to derive such a formula from first principles. A similarity hypothesis might be that if the nondimensional parameters Pr and Ra are the same in the two experiments, then the (nondimensional) heat flux will also be the same.

This turns out to be true. If the suitable nondimensionalized heat fluxes from many "similar" experiments are plotted in nondimensional form, all the data fall neatly onto families of curves. For example, for a given value of Pr, we can plot the experimentally determined nondimensional heat flux against Ra. For each value of Pr, the data fall onto orderly curves. The similarity theory does not actually give us the shapes of the curves. The shapes have to be determined empirically, but at least the similarity theory tells us what to look for. If we plotted the same data in dimensional form, a "scattered" set of points would result; no order would be apparent.

A famous example of similarity analysis was provided by G. I. Taylor (1950a, 1950b), who analyzed nuclear blasts. Using his (formidable) intuition, he chose the

Table B.5. The Dimensional Quantities Used by G. I. Taylor in His Analysis of Nuclear Explosions

Symbol	Definition	Representative value or first guess
R	Radius of wavefront	10^2 m
t	Time	10^{-2} s
p_0	Ambient pressure	10^5 Pa
ρ_0	Ambient density	1 kg m^{-3}
E	Energy released	10^{14} J

q's for the problem that are shown in table B.5. Using the five dimensional parameters in the table, which can be described using the three primary quantities length, time, and mass, Taylor used dimensional analysis to identify the following two nondimensional parameters:

$$\frac{\rho_0 R^5}{Et^2}, \text{ and } \frac{p_0 R^3}{E}. \tag{11}$$

He could then assert that

$$\frac{\rho_0 R^5}{Et^2} = f\left(\frac{p_0 R^3}{E}\right), \tag{12}$$

where f is a function to be determined empirically. But actually, he did better than that.

With the numerical values given in the third column of table B.5, including a "first guess" at the value of E, Taylor estimated that

$$\frac{\rho_0 R^5}{Et^2} = 1,$$

$$\frac{\rho_0 R^3}{E} = 10^{-3}.$$

Because the second parameter is much less than 1, Taylor concluded that it was physically irrelevant. The second parameter involves the ambient pressure, so this amounts to the (plausible) assumption that the relatively puny ambient pressure is inconsequential in comparison with the nuclear fireball, with its huge internal pressure. This similarity hypothesis led Taylor to conclude that (12) can be replaced by

$$\frac{\rho_0 R^5}{Et^2} = A = \text{constant},$$

where A is expected to be close to 1. This implies that

$$R^5 = \left(\frac{AE}{\rho_0}\right)t^2. \tag{13}$$

The factor AE/ρ_0 is expected to be independent of time, because A is a constant, and both ρ_0 and E are independent of time. Taylor was able to confirm that R^5 increased in proportion to t^2 using published (unclassified) magazine photos of the explosion. He then estimated E by assuming that $A = 1$. His estimate was accurate, embarrassing the government, which had not declassified the amount of energy released in the explosion.

Summary

Dimensional reasoning is very common in atmospheric science and engineering. It is used in several different ways.

Mass, length, time, and temperature suffice to describe the physical quantities used in most atmospheric science work. The Buckingham pi theorem states that a physical problem can be described most concisely if it is expressed using only nondimensional combinations. Dimensional analysis can be used to identify such combinations.

Scale analysis goes further, by using chosen dimensional scales—that is, numerical values expressed in units—to determine the dominant terms of the nondimensionalized equations that describe a physical system. The scales are chosen so that the nondimensional unknowns are of order 1 for the problem of interest. Scale analysis always starts from the known governing equations.

Finally, similarity theories are used to find empirical relationships among the nondimensional parameters that characterize a physical system. Similarity theories are used when the formulas that are sought cannot be derived from first principles.

Similar systems are those for which the relevant nondimensional parameters take the same values. The concept of similar systems is relevant to both scale analysis and similarity theories.

Figure B.1 summarizes the logical relationships between dimensional analysis, scale analysis, and similarity theories.

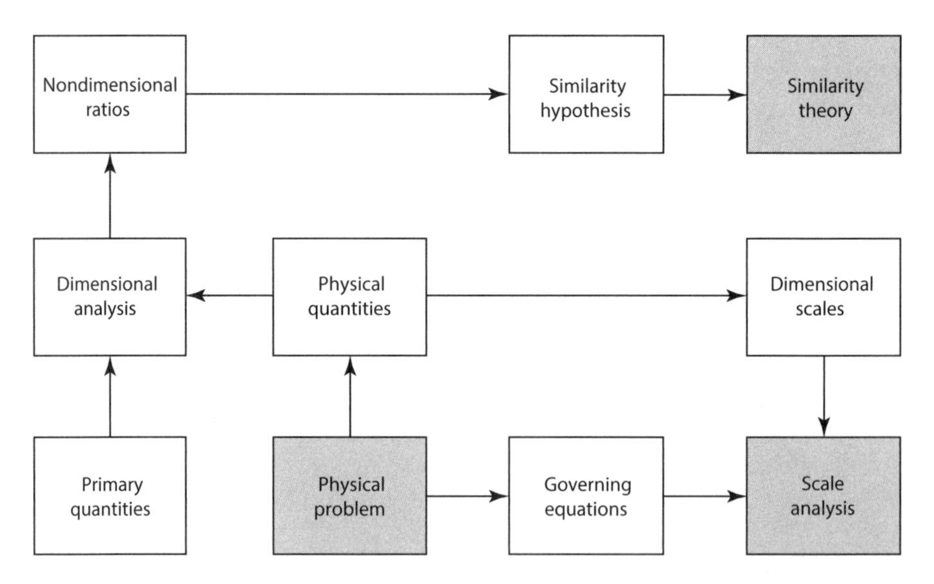

Figure B.1. An attempt to summarize the logical connections among the various topics discussed in this essay. Everything flows from the physical problem, and the endpoints are similarity theory and scale analysis.

APPENDIX C

Why Is the Dissipation Rate Positive?

As shown in shown in chapter 4, the rate of kinetic energy dissipation, per unit mass, is

$$\delta \equiv -\alpha(\mathbf{F} \cdot \nabla) \cdot \mathbf{V}. \tag{1}$$

We will prove that

$$\delta \geq 0. \tag{2}$$

For simplicity, we consider a Cartesian coordinate system that is applied in some small volume of the atmosphere. We name the coordinates (x, y, z) and call the corresponding velocity components (u, v, w). We can write the stress tensor as a matrix:

$$\mathbf{F} = \begin{bmatrix} 0 & F_{v,x} & F_{w,x} \\ F_{u,y} & 0 & F_{w,y} \\ F_{u,z} & F_{v,z} & 0 \end{bmatrix}. \tag{3}$$

As an example, $F_{u,y}$ is the flux of u-momentum in the y-direction. The diagonal elements of the matrix are set to zero because they would represent "normal stresses" (e.g., the pressure), and dissipation arises only from shearing stresses, which "deform" the fluid elements in the same way that dough is deformed when it is kneaded. It can be shown that the stress tensor has to be symmetric about its diagonal; that is, $F_{v,x} = F_{u,y}$, $F_{w,x} = F_{u,z}$, and $F_{w,y} = F_{v,z}$. If this were not true, the stresses would exert a finite torque on an infinitesimal air particle. Invoking this symmetry, we can rewrite (3) as

$$\mathbf{F} = \begin{bmatrix} 0 & F_{u,y} & F_{u,z} \\ F_{u,y} & 0 & F_{v,z} \\ F_{u,z} & F_{v,z} & 0 \end{bmatrix}, \tag{4}$$

and we can rewrite the divergence operator as a row vector; that is,

$$\nabla \cdot = \begin{bmatrix} \frac{\partial}{\partial x} & \frac{\partial}{\partial y} & \frac{\partial}{\partial z} \end{bmatrix}. \tag{5}$$

We find that

$$\nabla \cdot \mathbf{F} = \left(\frac{\partial F_{v,x}}{\partial y} + \frac{\partial F_{w,x}}{\partial z} \right) \mathbf{i} + \left(\frac{\partial F_{u,y}}{\partial x} + \frac{\partial F_{w,y}}{\partial z} \right) \mathbf{j} + \left(\frac{\partial F_{u,z}}{\partial x} + \frac{\partial F_{v,z}}{\partial y} \right) \mathbf{k}, \tag{6}$$

where **i**, **j**, and **k** are the unit vectors in the x-, y- and z-directions, respectively. The application of (5) to (4) to obtain (6) illustrates how and why the divergence of a tensor is a vector. Similarly,

$$\mathbf{F}\cdot\mathbf{V} = (F_{v,x}v + F_{w,x}w)\,\mathbf{i} + (F_{u,y}u + F_{w,y}w)\,\mathbf{j} + (F_{u,z}u + F_{v,z}v)\,\mathbf{k} \qquad (7)$$

is the (vector) flux of kinetic energy due to work done by friction on the air "next door." For example, $F_{u,z}u$ is the component of the energy exchange vector that acts in the z-direction (hence, multiplied by **k**) owing to the work done as u-momentum is transferred in the z-direction by friction.

The dissipation rate, $-\alpha(\mathbf{F}\cdot\mathbf{V})\cdot\mathbf{V}$, can be constructed as follows. We start by writing

$$\begin{aligned}\mathbf{F}\cdot\nabla &= \begin{bmatrix} 0 & F_{v,x} & F_{w,x} \\ F_{u,y} & 0 & F_{w,y} \\ F_{u,z} & F_{v,z} & 0 \end{bmatrix} \begin{bmatrix} \frac{\partial}{\partial x} \\ \frac{\partial}{\partial y} \\ \frac{\partial}{\partial z} \end{bmatrix} \\ &= \left[\left(F_{v,x}\frac{\partial}{\partial y} + F_{w,x}\frac{\partial}{\partial z}\right)\mathbf{i} \;\; \left(F_{u,y}\frac{\partial}{\partial x} + F_{w,y}\frac{\partial}{\partial z}\right)\mathbf{j} \;\; \left(F_{u,z}\frac{\partial}{\partial x} + F_{v,z}\frac{\partial}{\partial y}\right)\mathbf{k}\right].\end{aligned} \qquad (8)$$

We call $\mathbf{F}\cdot\nabla$ an "operator" because it can form dot products to differentiate vectors. Forming the dot product with **V**, we obtain

$$\begin{aligned}(\mathbf{F}\cdot\nabla)\cdot\mathbf{V} &= \left(F_{v,x}\frac{\partial}{\partial y} + F_{w,x}\frac{\partial}{\partial z}\right)u + \left(F_{u,y}\frac{\partial}{\partial x} + F_{w,y}\frac{\partial}{\partial z}\right)v + \left(F_{u,z}\frac{\partial}{\partial x} + F_{v,z}\frac{\partial}{\partial y}\right)w \\ &= F_{u,y}\left(\frac{\partial v}{\partial x} + \frac{\partial u}{\partial y}\right) + F_{u,z}\left(\frac{\partial w}{\partial x} + \frac{\partial u}{\partial z}\right) + F_{v,z}\left(\frac{\partial w}{\partial y} + \frac{\partial v}{\partial z}\right).\end{aligned} \qquad (9)$$

We obtained the second line of (9) by using the symmetry of the stress tensor, as discussed previously.

To finish our proof that the dissipation rate is nonnegative, we must formulate the stresses in terms of the wind components. Air is an example of a *Newtonian fluid*, for which the stress is proportional to the spatial derivatives of the motion (called the *strain*), as follows:

$$F_{v,x} = F_{u,y} = -\mu\left(\frac{\partial v}{\partial x} + \frac{\partial u}{\partial y}\right), \qquad (10)$$

$$F_{w,x} = F_{u,z} = -\mu\left(\frac{\partial w}{\partial x} + \frac{\partial u}{\partial z}\right), \qquad (11)$$

$$F_{w,y} = F_{v,z} = -\mu\left(\frac{\partial w}{\partial y} + \frac{\partial v}{\partial z}\right). \qquad (12)$$

Here μ is the positive, nearly constant molecular viscosity, which is a "material property" rather than a property of the motion field. Equations (10)–(12) are called *stress-strain relationships*. As an example, (11) says that an upward increase of u (a "strain") tends to favor a negative (downward) value of the stress $F_{u,z}$. The flow of momentum is from "fast" to "slow" and thus tends to homogenize the momentum over time. The jargon is that such a flux is "downgradient." Substituting (10)–(12) into the second line of (9), we find that

$$(\mathbf{F} \cdot \nabla) \cdot \mathbf{V} = -\mu \left[\left(\frac{\partial v}{\partial x} + \frac{\partial u}{\partial y} \right)^2 + \left(\frac{\partial w}{\partial x} + \frac{\partial u}{\partial z} \right)^2 + \left(\frac{\partial w}{\partial y} + \frac{\partial v}{\partial z} \right)^2 \right], \tag{13}$$

which establishes (2). It also shows that dissipation occurs only when shear is present, as in flows that have vorticity; the dissipation rate is zero in a uniform or purely divergent (i.e., irrotational) flow.

APPENDIX D

Vertical Coordinate Transformations

We consider two vertical coordinates, denoted by z and ζ, respectively. Although the z symbol suggests height, no such implication is intended here; z and ζ can be any variables at all, so long as they vary monotonically with height.

Suppose that we have a rule telling how to compute ζ for a given value of z, and vice versa. For example, we might define $\zeta \equiv z - z_S(x, y, t)$, where $z_S(x, y, t)$ is the distribution of z along the Earth's surface.

We consider the variation of an arbitrary dependent variable A with the independent variable x, as sketched in figure D.1. Our goal is to relate $(\partial A/\partial x)_\zeta$ to $(\partial A/\partial x)_z$. With reference to figure D.1, we can write

$$\begin{aligned} \frac{A_3 - A_1}{x_3 - x_1} &= \frac{A_2 - A_1}{x_3 - x_1} + \left(\frac{A_3 - A_2}{x_3 - x_1}\right) \\ &= \frac{A_2 - A_1}{x_2 - x_1} + \left(\frac{A_3 - A_2}{\zeta_3 - \zeta_2}\right)\left(\frac{\zeta_3 - \zeta_2}{x_3 - x_1}\right) \\ &= \frac{A_2 - A_1}{x_2 - x_1} + \left(\frac{A_3 - A_2}{\zeta_3 - \zeta_2}\right)\left(\frac{\zeta_1 - \zeta_2}{x_2 - x_1}\right) \\ &= \frac{A_2 - A_1}{x_2 - x_1} - \left(\frac{A_3 - A_2}{\zeta_3 - \zeta_2}\right)\left(\frac{\zeta_2 - \zeta_1}{x_2 - x_1}\right). \end{aligned} \tag{1}$$

Taking the limit as the increments become small, we obtain

$$\left(\frac{\partial A}{\partial x}\right)_\zeta = \left(\frac{\partial A}{\partial x}\right)_z - \left(\frac{\partial A}{\partial \zeta}\right)_x \left(\frac{\partial \zeta}{\partial x}\right)_z. \tag{2}$$

Similarly, the horizontal gradient satisfies

$$\nabla_\zeta A = \nabla_z A - \left(\frac{\partial A}{\partial \zeta}\right)_x (\nabla \zeta)_z. \tag{3}$$

Naturally, the preceding derivation works in exactly the same way if the independent variable is time rather than a horizontal coordinate. Analogous identities apply with other operators. For example, for an arbitrary horizontal vector $\mathbf{H}$, we can write

$$\nabla_r \times \mathbf{H} = \nabla_\theta \times \mathbf{H} + \frac{\partial \mathbf{H}}{\partial r} \times \nabla_\theta r. \tag{4}$$

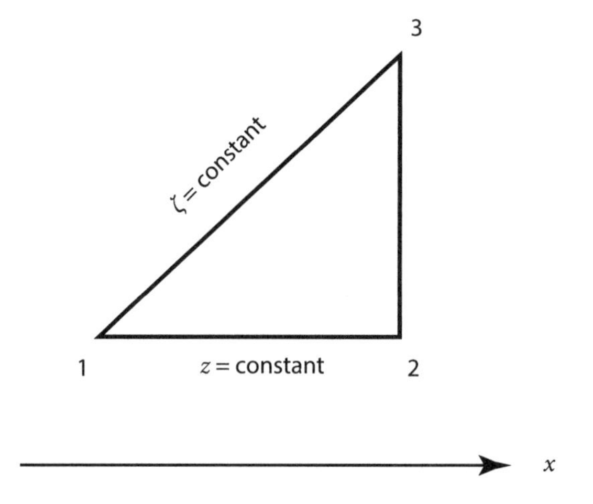

Figure D.1. Sketch used to derive the rule relating derivatives on surfaces of constant z to those on surfaces of constant ζ.

In the example suggested earlier, with $\zeta \equiv z - z_S(x, y, t)$, (3) reduces to

$$\nabla_\zeta A = \nabla_z A + \frac{\partial A}{\partial \zeta}\nabla(z_S). \tag{5}$$

As a second example, we let ζ be the pressure, and A the geopotential, denoted by ϕ. Then, (3) becomes

$$\begin{aligned}\nabla_p \phi &= \nabla_z \phi - \frac{\partial \phi}{\partial p}\nabla_z p \\ &= -\frac{\partial \phi}{\partial p}\nabla_z p \\ &\cong \alpha \nabla_z p.\end{aligned} \tag{6}$$

The third line follows in the hydrostatic limit, where $\alpha = RT/p$ is the specific volume. As a third example, again using hydrostatics, we can write

$$\begin{aligned}\nabla_p \phi &= \nabla_\theta \phi - \frac{\partial \phi}{\partial p}\nabla_\theta p \\ &\cong \nabla_\theta \phi + \alpha \nabla_\theta p \\ &= \nabla_\theta \phi + \frac{RT}{p}\nabla_\theta p.\end{aligned} \tag{7}$$

Using

$$T = \theta\left(\frac{p}{p_0}\right)^{R/c_p}, \tag{8}$$

we can show that

$$\frac{\nabla_\theta p}{p} = \frac{c_p}{R}\frac{\nabla_\theta T}{T}. \tag{9}$$

Use of (9) in (7) gives us

$$\nabla_p \phi = \nabla_\theta (c_p T + \phi). \tag{10}$$

APPENDIX E

The Moist Adiabatic Lapse Rate

As discussed in chapter 4, the moist static energy,

$$h \equiv c_p T + gz + Lq_v, \tag{1}$$

is approximately conserved under moist adiabatic processes. The saturation moist static energy is given by

$$h_{sat} \equiv c_p T + gz + Lq_{sat}(T,p), \tag{2}$$

where $q_{sat}(T,p)$ is the saturation mixing ratio, which depends, as indicated, on temperature and pressure. Differentiating (2) with respect to height gives us

$$\frac{\partial h_{sat}}{\partial z} = c_p \frac{\partial T}{\partial z} + g + L\left[\left(\frac{\partial q_{sat}}{\partial T}\right)_p \frac{\partial T}{\partial z} + \left(\frac{\partial q_{sat}}{\partial p}\right)_T \frac{\partial \overline{p}}{\partial z}\right], \tag{3}$$

where

$$\gamma \equiv \frac{L}{c_p}\left(\frac{\partial q_{sat}}{\partial T}\right)_p \tag{4}$$

is evaluated using the mean-state temperature and pressure. The nondimensional parameter γ is positive and of order 1. Using the hydrostatic equation, we can rearrange (3) to

$$\frac{\partial h_{sat}}{\partial z} = c_p \left\{(1+\gamma)\frac{\partial T}{\partial z} + \frac{g}{c_p}\left[1 - L\left(\frac{\partial q_{sat}}{\partial p}\right)_T \overline{\rho}\right]\right\}. \tag{5}$$

The saturation moist static energy is conserved by a saturated parcel in the absence of mixing and radiative heating. Setting $\partial h_{sat}/\partial z = 0$ in (5) therefore leads us to an expression for the moist adiabatic lapse rate:

$$\begin{aligned} \Gamma_m &\equiv -\left(\frac{\partial T}{\partial z}\right)_{\text{moist adiabatic}} \\ &= \frac{(g/c_p)}{(1+\gamma)}\left[1 - L\left(\frac{\partial q_{sat}}{\partial p}\right)_T \rho\right]. \end{aligned} \tag{6}$$

To go further, we need formulas for $(\partial q_{sat}/\partial p)_T$ and γ.

The saturation mixing ratio is defined by

$$q_{sat}(T,p) \equiv \frac{(\rho_v)_{sat}}{\rho_d}, \tag{7}$$

where $(\rho_v)_{sat}$ is the saturation vapor density, and ρ_d is the density of dry air. Both vapor and dry air obey the equation of state, with the same temperature but different gas constants:

$$p_d = \rho_d R_d T \text{ and } e = \rho_v R_v T. \tag{8}$$

Here e is the vapor pressure, and $R_v \cong 462 \text{ J kg}^{-1} \text{ K}^{-1}$. When we take the ratio of the two equations in (8), the temperatures cancel, and we find that

$$\frac{e}{\rho_d} = \frac{\rho_v R_v}{\rho_d R_d}, \tag{9}$$

which is equivalent to

$$\begin{aligned} q &\equiv \frac{\rho_v}{\rho_d} \\ &= \frac{\varepsilon e}{p_d}, \end{aligned} \tag{10}$$

where

$$\varepsilon \equiv \frac{R_d}{R_v} \cong 0.622. \tag{11}$$

Throughout the troposphere, $p_d \gg e$, so we can approximate (10) by

$$q \cong \frac{\varepsilon e}{p}, \tag{12}$$

and similarly

$$q_{sat}(T,p) \cong \frac{\varepsilon e_{sat}(T)}{p}. \tag{13}$$

From (13), we see that

$$\left(\frac{\partial q_{sat}}{\partial p}\right)_T \cong -\frac{q_{sat}(T,p)}{p} < 0, \tag{14}$$

and

$$\begin{aligned} \gamma &\equiv \frac{L}{c_p}\left(\frac{\partial q_{sat}}{\partial T}\right)_p \\ &= \frac{L}{c_p}\frac{\varepsilon}{p}\frac{d}{dT}e_{sat}(T). \end{aligned} \tag{15}$$

The Clausius-Clapeyron equation is approximately given by

$$\frac{d}{dT}e_{sat}(T) \cong \frac{L e_{sat}(T)}{R_v T^2}. \tag{16}$$

Substitution into (15) gives

$$\begin{aligned} \gamma &\equiv \frac{L}{c_p}\left(\frac{\partial q_{sat}}{\partial T}\right)_p \\ &= \frac{L^2 q_{sat}}{c_p R_v T^2}. \end{aligned} \tag{17}$$

Then, we can rewrite (6) as

$$\boxed{\Gamma_m = \Gamma_d \left[\frac{1 + \dfrac{Lq_{sat}}{R_d T}}{1 + \dfrac{L^2 q_{sat}}{c_p R_v T^2}} \right]}, \tag{18}$$

where

$$\begin{aligned} \Gamma_d &= -\left(\frac{\partial T}{\partial z}\right)_{\text{dry adiabatic}} \\ &= \frac{g}{c_p} \end{aligned} \tag{19}$$

is the dry adiabatic lapse rate. The denominator of (18) is larger than the numerator, so $\Gamma_m < \Gamma_d$, although $\Gamma_m \rightarrow \Gamma_d$ at cold temperatures. For example, with a pressure of 1000 mb and a temperature of 288 K, we find that $\Gamma_m = 4.67$ K km^{-1}. As the temperature increases, the moist adiabatic lapse rate decreases.

APPENDIX F

Eddy Kinetic Energy and Zonal Kinetic Energy

The derivations of the eddy kinetic energy, zonal kinetic energy, and zonally averaged total kinetic energy equations follow methods similar to those used to derive the conservation equation for the potential energy variance and so will be omitted here for brevity.

We define the eddy kinetic energy per unit mass by

$$KE \equiv \frac{1}{2}\left[\overline{(u^*)^2}^{\lambda} + \overline{(v^*)^2}^{\lambda}\right], \tag{1}$$

and the zonal kinetic energy by

$$KZ \equiv \frac{1}{2}\left[(\overline{u}^{\lambda})^2 + (\overline{v}^{\lambda})^2\right]. \tag{2}$$

It follows that

$$\overline{K}^{\lambda} = KZ + KE. \tag{3}$$

All three quantities in (3) are independent of longitude.

To derive equations that govern KE and KZ, we start from the zonal and meridional equations of motion in flux form:

$$\begin{aligned}\frac{\partial u}{\partial t} + \frac{1}{a\cos\varphi}\frac{\partial}{\partial\lambda}(uu) + \frac{1}{a\cos\varphi}\frac{\partial}{\partial\varphi}(vu\cos\varphi) + \frac{\partial}{\partial p}(\omega u) \\ = \frac{uv\tan\varphi}{a} + fv - \frac{1}{a\cos\varphi}\frac{\partial\phi}{\partial\lambda} + g\frac{\partial F_u}{\partial p},\end{aligned} \tag{4}$$

$$\begin{aligned}\frac{\partial v}{\partial t} + \frac{1}{a\cos\varphi}\frac{\partial}{\partial\lambda}(uv) + \frac{1}{a\cos\varphi}\frac{\partial}{\partial\varphi}(vv\cos\varphi) + \frac{\partial}{\partial p}(\omega v) \\ = -\frac{u^2\tan\varphi}{a} - fu - \frac{1}{a}\frac{\partial\phi}{\partial\varphi} + g\frac{\partial F_v}{\partial p}.\end{aligned} \tag{5}$$

We also require mass continuity:

$$\frac{1}{a\cos\varphi}\frac{\partial u}{\partial\lambda} + \frac{1}{a\cos\varphi}\frac{\partial}{\partial\varphi}(v\cos\varphi) + \frac{\partial\omega}{\partial p} = 0. \tag{6}$$

Zonal averaging of (4)–(6) gives

$$\begin{aligned}&\frac{\partial \overline{u}^\lambda}{\partial t} + \frac{1}{a\cos\varphi}\frac{\partial}{\partial\varphi}\left\{\left(\overline{v}^\lambda\overline{u}^\lambda + \overline{v^*u^*}^\lambda\right)\cos\varphi\right\} + \frac{\partial}{\partial p}\left(\overline{\omega}^\lambda\overline{u}^\lambda + \overline{\omega^*u^*}^\lambda\right)\\ &= \left(\overline{v}^\lambda\overline{u}^\lambda + \overline{v^*u^*}^\lambda\right)\frac{\tan\varphi}{a} + f\overline{v}^\lambda + g\frac{\partial \overline{F}_u^\lambda}{\partial p},\end{aligned} \tag{7}$$

$$\begin{aligned}&\frac{\partial \overline{v}^\lambda}{\partial t} + \frac{1}{a\cos\varphi}\frac{\partial}{\partial\varphi}\left\{\left(\overline{v}^\lambda\overline{v}^\lambda + \overline{v^*v^*}^\lambda\right)\cos\varphi\right\} + \frac{\partial}{\partial p}\left(\overline{\omega}^\lambda\overline{v}^\lambda + \overline{\omega^*v^*}^\lambda\right)\\ &= \left(\overline{u}^\lambda\overline{u}^\lambda + \overline{u^*u^*}^\lambda\right)\frac{\tan\varphi}{a} - f\overline{u}^\lambda - \frac{1}{a}\frac{\partial\overline{\phi}^\lambda}{\partial\varphi} + g\frac{\partial \overline{F}_v^\lambda}{\partial p}\end{aligned} \tag{8}$$

and

$$\frac{1}{a\cos\varphi}\frac{\partial}{\partial\varphi}\left(\overline{v}^\lambda\cos\varphi\right) + \frac{\partial\overline{\omega}^\lambda}{\partial p} = 0. \tag{9}$$

We use the zonally averaged continuity equation, (9), to convert (7) and (8) to advective form:

$$\begin{aligned}&\frac{\partial \overline{u}^\lambda}{\partial t} + \frac{\overline{v}^\lambda}{a}\frac{\partial\overline{u}^\lambda}{\partial\varphi} + \overline{\omega}^\lambda\frac{\partial\overline{u}^\lambda}{\partial p} + \frac{1}{a\cos\varphi}\frac{\partial}{\partial\varphi}\left(\overline{v^*u^*}^\lambda\cos\varphi\right) + \frac{\partial}{\partial p}\overline{\omega^*u^*}^\lambda\\ &= \left(\overline{v}^\lambda\overline{u}^\lambda + \overline{v^*u^*}^\lambda\right)\frac{\tan\varphi}{a} + f\overline{v}^\lambda + g\frac{\partial \overline{F}_u^\lambda}{\partial p},\end{aligned} \tag{10}$$

$$\begin{aligned}&\frac{\partial \overline{v}^\lambda}{\partial t} + \frac{\overline{v}^\lambda}{a}\frac{\partial\overline{v}^\lambda}{\partial\varphi} + \overline{\omega}^\lambda\frac{\partial\overline{v}^\lambda}{\partial p} + \frac{1}{a\cos\varphi}\frac{\partial}{\partial\varphi}\left(\overline{v^*v^*}^\lambda\cos\varphi\right) + \frac{\partial}{\partial p}\overline{\omega^*v^*}^\lambda\\ &= -\left(\overline{u}^\lambda\overline{u}^\lambda + \overline{u^*u^*}^\lambda\right)\frac{\tan\varphi}{a} - f\overline{u}^\lambda - \frac{1}{a}\frac{\partial\overline{\phi}^\lambda}{\partial\varphi} + g\frac{\partial \overline{F}_v^\lambda}{\partial p}.\end{aligned} \tag{11}$$

Next, we multiply (10) by $[u]$ and (11) by $[v]$ and add the results to obtain

$$\begin{aligned}&\frac{\partial}{\partial t}KZ + \frac{\overline{v}^\lambda}{a}\frac{\partial}{\partial\varphi}KZ + \overline{\omega}^\lambda\frac{\partial}{\partial p}KZ\\ &+\frac{\overline{u}^\lambda}{a\cos\varphi}\frac{\partial}{\partial\varphi}\left(\overline{v^*u^*}^\lambda\cos\varphi\right) + \frac{\overline{v}^\lambda}{a\cos\varphi}\frac{\partial}{\partial\varphi}\left(\overline{v^*v^*}^\lambda\cos\varphi\right) + \overline{u}^\lambda\frac{\partial}{\partial\varphi}\overline{\omega^*u^*}^\lambda + \overline{v}^\lambda\frac{\partial}{\partial\varphi}\overline{\omega^*v^*}^\lambda\\ &= \overline{u}^\lambda\left(\cancel{\overline{v}^\lambda\overline{u}^\lambda} + \overline{v^*u^*}^\lambda\right)\frac{\tan\varphi}{a} - \overline{v}^\lambda\left(\cancel{\overline{u}^\lambda\overline{u}^\lambda} + \overline{u^*u^*}^\lambda\right)\frac{\tan\varphi}{a}\\ &-\frac{\overline{v}^\lambda}{a}\frac{\partial\overline{\phi}^\lambda}{\partial\varphi} + \overline{u}^\lambda g\frac{\partial \overline{F}_u^\lambda}{\partial p} + \overline{v}^\lambda g\frac{\partial \overline{F}_v^\lambda}{\partial p}.\end{aligned} \tag{12}$$

We use (9) to return to flux form and cancel the indicated terms of (12) to obtain

$$\begin{aligned}&\frac{\partial}{\partial t}KZ + \frac{1}{a\cos\varphi}\frac{\partial}{\partial\varphi}\left(\overline{v}^\lambda KZ\cos\varphi\right) + \frac{\partial}{\partial p}\left(\overline{\omega}^\lambda KZ\right)\\ &+\frac{\overline{u}^\lambda}{a\cos\varphi}\frac{\partial}{\partial\varphi}\left(\overline{v^*u^*}^\lambda\cos\varphi\right) + \frac{\overline{v}^\lambda}{a\cos\varphi}\frac{\partial}{\partial\varphi}\left(\overline{v^*v^*}^\lambda\cos\varphi\right) + \overline{u}^\lambda\frac{\partial}{\partial p}\overline{\omega^*u^*}^\lambda + \overline{v}^\lambda\frac{\partial}{\partial p}\overline{\omega^*v^*}^\lambda\\ &= \left(\overline{u}^\lambda\overline{v^*u^*}^\lambda - \overline{v}^\lambda\overline{u^*u^*}^\lambda\right)\frac{\tan\varphi}{a} - \frac{\overline{v}^\lambda}{a}\frac{\partial\overline{\phi}^\lambda}{\partial\varphi} + \overline{u}^\lambda g\frac{\partial \overline{F}_u^\lambda}{\partial p} + \overline{v}^\lambda g\frac{\partial \overline{F}_v^\lambda}{\partial p}.\end{aligned} \tag{13}$$

We can manipulate the terms on the second line of (13) as follows:

$$
\begin{aligned}
&\frac{\overline{u}^{\lambda}}{a\cos\varphi}\frac{\partial}{\partial\varphi}\left(\overline{v^{*}u^{*}}^{\lambda}\cos\varphi\right)+\frac{\overline{v}^{\lambda}}{a\cos\varphi}\frac{\partial}{\partial\varphi}\left(\overline{v^{*}v^{*}}^{\lambda}\cos\varphi\right)+\overline{u}^{\lambda}\frac{\partial}{\partial p}\overline{\omega^{*}u^{*}}^{\lambda}+\overline{v}^{\lambda}\frac{\partial}{\partial p}\overline{\omega^{*}v^{*}}^{\lambda}\\
&\quad=\frac{1}{a\cos\varphi}\frac{\partial}{\partial\varphi}\left[\left(\overline{u}^{\lambda}\overline{v^{*}u^{*}}^{\lambda}+\overline{v}^{\lambda}\overline{v^{*}v^{*}}^{\lambda}\right)\cos\varphi\right]-\frac{\overline{v^{*}u^{*}}^{\lambda}}{a}\frac{\partial\overline{u}^{\lambda}}{\partial\varphi}-\frac{\overline{v^{*}v^{*}}^{\lambda}}{a}\frac{\partial\overline{v}^{\lambda}}{\partial\varphi}\\
&\qquad+\frac{\partial}{\partial p}\left(\overline{u}^{\lambda}\overline{\omega^{*}u^{*}}^{\lambda}+\overline{v}^{\lambda}\overline{\omega^{*}v^{*}}^{\lambda}\right)-\overline{\omega^{*}u^{*}}^{\lambda}\frac{\partial\overline{u}^{\lambda}}{\partial p}-\overline{\omega^{*}v^{*}}^{\lambda}\frac{\partial\overline{v}^{\lambda}}{\partial p}.
\end{aligned}
\tag{14}
$$

Finally, we write

$$
\begin{aligned}
\frac{\overline{v}^{\lambda}}{a}\frac{\partial\overline{\phi}^{\lambda}}{\partial\varphi}&=\frac{\overline{v}^{\lambda}\cos\varphi}{a\cos\varphi}\frac{\partial\overline{\phi}^{\lambda}}{\partial\varphi}\\
&=\frac{1}{a\cos\varphi}\frac{\partial}{\partial\varphi}\left(\overline{v}^{\lambda}\overline{\phi}^{\lambda}\cos\varphi\right)-\frac{\overline{\phi}^{\lambda}}{a\cos\varphi}\frac{\partial}{\partial\varphi}\left(\overline{v}^{\lambda}\cos\varphi\right)\\
&=\frac{1}{a\cos\varphi}\frac{\partial}{\partial\varphi}\left(\overline{v}^{\lambda}\overline{\phi}^{\lambda}\cos\varphi\right)+\overline{\phi}^{\lambda}\frac{\partial\overline{\omega}^{\lambda}}{\partial p}\\
&=\frac{1}{a\cos\varphi}\frac{\partial}{\partial\varphi}\left(\overline{v}^{\lambda}\overline{\phi}^{\lambda}\cos\varphi\right)+\frac{\partial}{\partial p}\left(\overline{\omega}^{\lambda}\overline{\phi}^{\lambda}\right)-\overline{\omega}^{\lambda}\frac{\partial\overline{\phi}^{\lambda}}{\partial\varphi}\\
&=\frac{1}{a\cos\varphi}\frac{\partial}{\partial\varphi}\left(\overline{v}^{\lambda}\overline{\phi}^{\lambda}\cos\varphi\right)+\frac{\partial}{\partial p}\left(\overline{\omega}^{\lambda}\overline{\phi}^{\lambda}\right)+\overline{\omega}^{\lambda}\overline{\alpha}^{\lambda}.
\end{aligned}
\tag{15}
$$

Substituting back into (13) and combining terms, we find that

$$
\boxed{
\begin{aligned}
&\frac{\partial}{\partial t}KZ\\
&+\frac{1}{a\cos\varphi}\frac{\partial}{\partial\varphi}\left[\left(\overline{v}^{\lambda}KZ+\overline{u}^{\lambda}\overline{v^{*}u^{*}}^{\lambda}+\overline{v}^{\lambda}\overline{v^{*}v^{*}}^{\lambda}+\overline{v}^{\lambda}\overline{\phi}^{\lambda}\right)\cos\varphi\right]\\
&+\frac{\partial}{\partial p}\left(\overline{\omega}^{\lambda}KZ+\overline{u}^{\lambda}\overline{\omega^{*}u^{*}}^{\lambda}+\overline{v}^{\lambda}\overline{\omega^{*}v^{*}}^{\lambda}+\overline{\omega}^{\lambda}\overline{\phi}^{\lambda}\right)\\
&=\frac{\overline{v^{*}u^{*}}^{\lambda}}{a}\frac{\partial\overline{u}^{\lambda}}{\partial\varphi}+\frac{\overline{v^{*}v^{*}}^{\lambda}}{a}\frac{\partial\overline{v}^{\lambda}}{\partial\varphi}+\overline{\omega^{*}u^{*}}^{\lambda}\frac{\partial\overline{u}^{\lambda}}{\partial p}+\overline{\omega^{*}v^{*}}^{\lambda}\frac{\partial\overline{v}^{\lambda}}{\partial p}\\
&+\left(\overline{u}^{\lambda}\overline{v^{*}u^{*}}^{\lambda}-\overline{v}^{\lambda}\overline{u^{*}u^{*}}^{\lambda}\right)\frac{\tan\varphi}{a}\\
&-\overline{\omega}^{\lambda}\overline{\alpha}^{\lambda}\\
&+\overline{u}^{\lambda}g\frac{\partial\overline{F}_{u}^{\lambda}}{\partial p}+\overline{v}^{\lambda}\frac{\partial\overline{F}_{v}^{\lambda}}{\partial p}.
\end{aligned}
}
\tag{16}
$$

This is the zonal kinetic energy equation.

To derive the eddy kinetic energy equation, we return to the equations of motion. We use the continuity equation, (6), to convert (4)–(5) to advective form:

$$
\frac{\partial u}{\partial t}+\frac{u}{a\cos\varphi}\frac{\partial u}{\partial\lambda}+\frac{v}{a}\frac{\partial u}{\partial\varphi}+\omega\frac{\partial u}{\partial p}=\frac{uv\tan\varphi}{a}+fv-\frac{1}{a\cos\varphi}\frac{\partial\phi}{\partial\lambda}+g\frac{\partial F_{u}}{\partial p},\tag{17}
$$

$$
\frac{\partial v}{\partial t}+\frac{u}{a\cos\varphi}\frac{\partial v}{\partial\lambda}+\frac{v}{a}\frac{\partial v}{\partial\varphi}+\omega\frac{\partial v}{\partial p}=-\frac{u^{2}\tan\varphi}{a}-fu-\frac{1}{a}\frac{\partial\phi}{\partial\varphi}+g\frac{\partial F_{v}}{\partial p}.\tag{18}
$$

We subtract (10) and (11) from (17) and (18), respectively, to obtain

$$\begin{aligned}&\frac{\partial u^*}{\partial t}+\frac{u}{a\cos\varphi}\frac{\partial u^*}{\partial\lambda}+\left(\frac{v}{a}\frac{\partial u}{\partial\varphi}-\frac{\overline{v}^\lambda}{a}\frac{\partial\overline{u}^\lambda}{\partial\varphi}\right)+\left(\omega\frac{\partial u}{\partial p}-\overline{\omega}^\lambda\frac{\partial\overline{u}^\lambda}{\partial p}\right)\\&-\left[\frac{1}{a\cos\varphi}\frac{\partial}{\partial\varphi}\left(\overline{v^*u^*}^\lambda\cos\varphi\right)+\frac{\partial}{\partial p}\overline{\omega^*u^*}^\lambda\right]\\&=\frac{uv\tan\varphi}{a}-\frac{\left(\overline{v}^\lambda\overline{u}^\lambda+\overline{v^*u^*}^\lambda\right)\tan\varphi}{a}+fv^*-\frac{1}{a\cos\varphi}\frac{\partial\phi^*}{\partial\lambda}+g\frac{\partial F_u^*}{\partial p},\end{aligned}\tag{19}$$

$$\begin{aligned}&\frac{\partial v^*}{\partial t}+\frac{u}{a\cos\varphi}\frac{\partial v^*}{\partial\lambda}+\left(\frac{v}{a}\frac{\partial v}{\partial\varphi}-\frac{\overline{v}^\lambda}{a}\frac{\partial\overline{v}^\lambda}{\partial\varphi}\right)+\left(\omega\frac{\partial v}{\partial p}-\overline{\omega}^\lambda\frac{\partial\overline{v}^\lambda}{\partial p}\right)\\&-\left[\frac{1}{a\cos\varphi}\frac{\partial}{\partial\varphi}\left(\overline{v^*v^*}^\lambda\cos\varphi\right)+\frac{\partial}{\partial p}\overline{\omega^*v^*}^\lambda\right]\\&=-\frac{u^2\tan\varphi}{a}+\frac{\left(\overline{u}^\lambda\overline{u}^\lambda+\overline{u^*u^*}^\lambda\right)\tan\varphi}{a}-fu^*-\frac{1}{a}\frac{\partial\phi^*}{\partial\varphi}+g\frac{\partial F_v^*}{\partial p}.\end{aligned}\tag{20}$$

Expanding the nonlinear terms, we get

$$\begin{aligned}&\frac{\partial u^*}{\partial t}+\left(\frac{\overline{u}^\lambda+u^*}{a\cos\varphi}\right)\frac{\partial u^*}{\partial\lambda}+\frac{\left(\overline{v}^\lambda+v^*\right)}{a}\frac{\partial}{\partial\varphi}\left(\overline{u}^\lambda+u^*\right)+\left(\overline{\omega}^\lambda+\omega^*\right)\frac{\partial}{\partial p}\left(\overline{u}^\lambda+u^*\right)\\&-\left(\frac{\overline{v}^\lambda}{a}\frac{\partial\overline{u}^\lambda}{\partial\varphi}+\overline{\omega}^\lambda\frac{\partial\overline{u}^\lambda}{\partial p}\right)-\left[\frac{1}{a\cos\varphi}\frac{\partial}{\partial\varphi}\left(\overline{v^*u^*}^\lambda\cos\varphi\right)+\frac{\partial}{\partial p}\overline{\omega^*u^*}^\lambda\right]\\&=\left[\left(\overline{u}^\lambda+u^*\right)\left(\overline{v}^\lambda+v^*\right)-\left(\overline{v}^\lambda\overline{u}^\lambda+\overline{v^*u^*}^\lambda\right)\right]\frac{\tan\varphi}{a}+fv^*-\frac{1}{a\cos\varphi}\frac{\partial\phi^*}{\partial\lambda}+g\frac{\partial F_u^*}{\partial p},\end{aligned}\tag{21}$$

$$\begin{aligned}&\frac{\partial v^*}{\partial t}+\frac{\left(\overline{u}^\lambda+u^*\right)}{a\cos\varphi}\frac{\partial v^*}{\partial\lambda}+\frac{\left(\overline{v}^\lambda+v^*\right)}{a}\frac{\partial}{\partial\varphi}\left(\overline{v}^\lambda+v^*\right)+\left(\overline{\omega}^\lambda+\omega^*\right)\frac{\partial}{\partial p}\left(\overline{v}^\lambda+v^*\right)\\&-\left(\frac{\overline{v}^\lambda}{a}\frac{\partial\overline{v}^\lambda}{\partial\varphi}+\overline{\omega}^\lambda\frac{\partial\overline{v}^\lambda}{\partial p}\right)-\left[\frac{1}{a\cos\varphi}\frac{\partial}{\partial\varphi}\left(\overline{v^*v^*}^\lambda\cos\varphi\right)+\frac{\partial}{\partial p}\overline{\omega^*v^*}^\lambda\right]\\&=\left[-\left(\overline{u}^\lambda+u^*\right)^2+\left(\overline{u}^\lambda\overline{u}^\lambda+\overline{u^*u^*}^\lambda\right)\right]\frac{\tan\varphi}{a}-fu^*-\frac{1}{a}\frac{\partial\phi^*}{\partial\varphi}+g\frac{\partial F_v^*}{\partial p}.\end{aligned}\tag{22}$$

We rearrange and simplify:

$$\begin{aligned}&\left(\frac{\partial}{\partial t}+\frac{\overline{u}^\lambda}{a\cos\varphi}\frac{\partial}{\partial\lambda}+\frac{\overline{v}^\lambda}{a}\frac{\partial}{\partial\varphi}+\overline{\omega}^\lambda\frac{\partial}{\partial p}\right)u^*\\&+\left(\frac{u^*}{a\cos\varphi}\frac{\partial}{\partial\lambda}+\frac{v^*}{a}\frac{\partial}{\partial\varphi}+\omega^*\frac{\partial}{\partial p}\right)u^*+\frac{v^*}{a}\frac{\partial\overline{u}^\lambda}{\partial\varphi}+\omega^*\frac{\partial\overline{u}^\lambda}{\partial p}\\&-\left[\frac{1}{a\cos\varphi}\frac{\partial}{\partial\varphi}\left(\overline{v^*u^*}^\lambda\cos\varphi\right)+\frac{\partial}{\partial p}\overline{\omega^*u^*}^\lambda\right]\\&=\left[\left(\overline{u}^\lambda v^*+u^*\overline{v}^\lambda+v^*u^*\right)-\overline{v^*u^*}^\lambda\right]\frac{\tan\varphi}{a}+fv^*-\frac{1}{a\cos\varphi}\frac{\partial\phi^*}{\partial\lambda}+g\frac{\partial F_u^*}{\partial p},\end{aligned}\tag{23}$$

$$\begin{aligned}&\left(\frac{\partial}{\partial t}+\frac{\overline{u}^\lambda}{a\cos\varphi}\frac{\partial}{\partial\lambda}+\frac{\overline{v}^\lambda}{a}\frac{\partial}{\partial\varphi}+\overline{\omega}^\lambda\frac{\partial}{\partial p}\right)v^*\\&+\left(\frac{u^*}{a\cos\varphi}\frac{\partial}{\partial\lambda}+\frac{v^*}{a}\frac{\partial}{\partial\varphi}+\omega^*\frac{\partial}{\partial p}\right)v^*+\frac{v^*}{a}\frac{\partial\overline{v}^\lambda}{\partial\varphi}+\omega^*\frac{\partial\overline{v}^\lambda}{\partial p}\\&-\left[\frac{1}{a\cos\varphi}\frac{\partial}{\partial\varphi}\left(\overline{v^*v^*}^\lambda\cos\varphi\right)+\frac{\partial}{\partial p}\overline{\omega^*v^*}^\lambda\right]\\&=\left[-\left(2\overline{u}^\lambda u^*+u^*u^*\right)+\overline{u^*u^*}^\lambda\right]\frac{\tan\varphi}{a}-fu^*-\frac{1}{a}\frac{\partial\phi^*}{\partial\varphi}+g\frac{\partial F_v^*}{\partial p}.\end{aligned}\tag{24}$$

Now, we multiply (23) by u^* and (24) by v^*:

$$\begin{aligned}
&\left(\frac{\partial}{\partial t}+\frac{\overline{u}^{\lambda}}{a\cos\varphi}\frac{\partial}{\partial\lambda}+\frac{\overline{v}^{\lambda}}{a}\frac{\partial}{\partial\varphi}+\overline{\omega}^{\lambda}\frac{\partial}{\partial p}\right)\frac{u^{*2}}{2}\\
&+\left(\frac{u^*}{a\cos\varphi}\frac{\partial}{\partial\lambda}+\frac{v^*}{a}\frac{\partial}{\partial\varphi}+\omega^*\frac{\partial}{\partial p}\right)\frac{u^{*2}}{2}+\frac{u^*v^*}{a}\frac{\partial\overline{u}^{\lambda}}{\partial\varphi}+u^*\omega^*\frac{\partial\overline{u}^{\lambda}}{\partial p}\\
&-u^*\left[\frac{1}{a\cos\varphi}\frac{\partial}{\partial\varphi}\left(\overline{v^*u^*}^{\lambda}\cos\varphi\right)+\frac{\partial}{\partial p}\overline{\omega^*u^*}^{\lambda}\right]\\
&=u^*\left[\left(\overline{u}^{\lambda}v^*+u^*\overline{v}^{\lambda}+v^*u^*\right)-\overline{v^*u^*}^{\lambda}\right]\frac{\tan\varphi}{a}+fu^*v^*-\frac{u^*}{a\cos\varphi}\frac{\partial\phi^*}{\partial\lambda}+u^*g\frac{\partial F_u^*}{\partial p},
\end{aligned} \tag{25}$$

$$\begin{aligned}
&\left(\frac{\partial}{\partial t}+\frac{\overline{u}^{\lambda}}{a\cos\varphi}\frac{\partial}{\partial\lambda}+\frac{\overline{v}^{\lambda}}{a}\frac{\partial}{\partial\varphi}+\overline{\omega}^{\lambda}\frac{\partial}{\partial p}\right)\frac{v^{*2}}{2}\\
&+\left(\frac{u^*}{a\cos\varphi}\frac{\partial}{\partial\lambda}+\frac{v^*}{a}\frac{\partial}{\partial\varphi}+\omega^*\frac{\partial}{\partial p}\right)\frac{v^{*2}}{2}+\frac{v^*v^*}{a}\frac{\partial\overline{v}^{\lambda}}{\partial\varphi}+v^*\omega^*\frac{\partial\overline{v}^{\lambda}}{\partial p}\\
&-v^*\left[\frac{1}{a\cos\varphi}\frac{\partial}{\partial\varphi}\left(\overline{v^*v^*}^{\lambda}\cos\varphi\right)+\frac{\partial}{\partial p}\overline{\omega^*v^*}^{\lambda}\right]\\
&=v^*\left[-\left(2\overline{u}^{\lambda}u^*+u^*u^*\right)+\overline{u^*u^*}^{\lambda}\right]\frac{\tan\varphi}{a}-fv^*u^*-\frac{v^*}{a}\frac{\partial\phi^*}{\partial\varphi}+v^*g\frac{\partial F_v^*}{\partial p}.
\end{aligned} \tag{26}$$

Next, we use the zonal mean continuity equation, (4), to convert to flux form the terms of (25) and (26) that represent advection by the zonally averaged winds, and use the eddy continuity equation,

$$\frac{1}{a\cos\varphi}\frac{\partial u_*}{\partial\lambda}+\frac{1}{a\cos\varphi}\frac{\partial}{\partial\varphi}(v_*\cos\varphi)+\frac{\partial\omega_*}{\partial p}=0, \tag{27}$$

to similarly rewrite the terms that represent advection by the eddy winds. Taking the zonal means of the results, we get

$$\begin{aligned}
&\frac{1}{2}\frac{\partial}{\partial t}\overline{u^*u^*}^{\lambda}+\frac{1}{2}\frac{1}{a\cos\varphi}\frac{\partial}{\partial\varphi}\left[\left(\overline{v}^{\lambda}\,\overline{u^*u^*}^{\lambda}+\overline{v^*u^{*2}}^{\lambda}\right)\cos\varphi\right]+\frac{1}{2}\frac{\partial}{\partial p}\left(\overline{\omega}^{\lambda}\,\overline{u^*u^*}^{\lambda}+\overline{\omega^*u^{*2}}^{\lambda}\right)\\
&=-\frac{\overline{v^*u^*}^{\lambda}}{a}\frac{\partial\overline{u}^{\lambda}}{\partial\varphi}-\overline{\omega^*u^*}^{\lambda}\frac{\partial\overline{u}^{\lambda}}{\partial p}\\
&+\left(\overline{u}^{\lambda}\,\overline{v^*u^*}^{\lambda}+\overline{u^*u^*}^{\lambda}\,\overline{v}^{\lambda}+\overline{u^*v^*u^*}^{\lambda}\right)\frac{\tan\varphi}{a}+f\overline{v^*u^*}^{\lambda}-\overline{\frac{u^*}{a\cos\varphi}\frac{\partial\phi^*}{\partial\lambda}}^{\lambda}+\overline{u^*g\frac{\partial F_u^*}{\partial p}}^{\lambda},
\end{aligned} \tag{28}$$

$$\begin{aligned}
&\frac{1}{2}\frac{\partial\overline{v^*v^*}^{\lambda}}{\partial t}+\frac{1}{2}\frac{1}{a\cos\varphi}\frac{\partial}{\partial\varphi}\left[\left(\overline{v}^{\lambda}\,\overline{v^*v^*}^{\lambda}+\overline{v^*v^*v^*}^{\lambda}\right)\cos\varphi\right]+\frac{1}{2}\frac{\partial}{\partial p}\left(\overline{\omega}^{\lambda}\,\overline{v^*v^*}^{\lambda}+\overline{\omega^*v^{*2}}^{\lambda}\right)\\
&=-\frac{\overline{v^*v^*}^{\lambda}}{a}\frac{\partial\overline{v}^{\lambda}}{\partial\varphi}-\overline{\omega^*v^*}^{\lambda}\frac{\partial\overline{v}^{\lambda}}{\partial p}\\
&-\left(2\overline{u}^{\lambda}\,\overline{v^*u^*}^{\lambda}+\overline{u^*u^*v^*}^{\lambda}\right)\frac{\tan\varphi}{a}-f\overline{v^*u^*}^{\lambda}-\overline{\frac{v^*}{a}\frac{\partial\phi^*}{\partial\varphi}}^{\lambda}+\overline{gv^*\frac{\partial F_v^*}{\partial p}}^{\lambda}.
\end{aligned} \tag{29}$$

Now, we add (28) and (29) to obtain

$$\begin{aligned}
&\frac{\partial}{\partial t}KE+\frac{1}{a\cos\varphi}\frac{\partial}{\partial\varphi}\left\{\left[\overline{v}^{\lambda}KE+\frac{1}{2}\overline{v^*\left(u^{*2}+v^{*2}\right)}^{\lambda}\right]\cos\varphi\right\}+\frac{\partial}{\partial p}\left[\overline{\omega}^{\lambda}KE+\frac{1}{2}\overline{\omega^*\left(u^{*2}+v^{*2}\right)}^{\lambda}\right]\\
&=-\frac{\overline{v^*u^*}^{\lambda}}{a}\frac{\partial\overline{u}^{\lambda}}{\partial\varphi}-\frac{\overline{v^*v^*}^{\lambda}}{a}\frac{\partial\overline{v}^{\lambda}}{\partial\varphi}-\overline{\omega^*u^*}^{\lambda}\frac{\partial\overline{u}^{\lambda}}{\partial p}-\overline{\omega^*v^*}^{\lambda}\frac{\partial\overline{v}^{\lambda}}{\partial p}\\
&+\left(-\overline{u}^{\lambda}\,\overline{v^*u^*}^{\lambda}+\overline{u^*u^*}^{\lambda}\,\overline{v}^{\lambda}\right)\frac{\tan\varphi}{a}-\overline{\frac{u^*}{a\cos\varphi}\frac{\partial\phi^*}{\partial\lambda}}^{\lambda}-\overline{\frac{v^*}{a}\frac{\partial\phi^*}{\partial\varphi}}^{\lambda}+\overline{gu^*\frac{\partial F_u^*}{\partial p}}^{\lambda}+\overline{gv^*\frac{\partial F_v^*}{\partial p}}^{\lambda}.
\end{aligned} \tag{30}$$

Finally, by analogy with (15), we can rewrite the pressure-gradient terms of (30) as

$$\overline{\frac{u^*}{a\cos\varphi}\frac{\partial\phi^*}{\partial\lambda}}^{\lambda} + \overline{\frac{v^*}{a}\frac{\partial\phi^*}{\partial\varphi}}^{\lambda} = \frac{1}{a\cos\varphi}\frac{\partial}{\partial\varphi}\left(\overline{v^*\phi^*}^{\lambda}\cos\varphi\right) + \frac{\partial}{\partial p}\overline{\omega^*\phi^*}^{\lambda} + \overline{\omega^*\alpha^*}^{\lambda}. \tag{31}$$

Substitution of (31) into (30) gives

$$\boxed{\begin{aligned}
&\frac{\partial}{\partial t}KE \\
&+\frac{1}{a\cos\varphi}\frac{\partial}{\partial\varphi}\left[\left(\overline{v}^{\lambda}KE + \frac{1}{2}\overline{v^*\left(u^{*2}+v^{*2}\right)}^{\lambda} + \overline{v^*\phi^*}^{\lambda}\right)\cos\varphi\right] \\
&+\frac{\partial}{\partial p}\left(\overline{\omega}^{\lambda}KE + \frac{1}{2}\overline{\omega^*\left(u^{*2}+v^{*2}\right)}^{\lambda} + \overline{\omega^*\phi^*}^{\lambda}\right) \\
&= -\frac{\overline{v^*u^*}^{\lambda}}{a}\frac{\partial\overline{u}^{\lambda}}{\partial\varphi} - \frac{\overline{v^*v^*}^{\lambda}}{a}\frac{\partial\overline{v}^{\lambda}}{\partial\varphi} - \overline{\omega^*u^*}^{\lambda}\frac{\partial\overline{u}^{\lambda}}{\partial p} - \overline{\omega^*v^*}^{\lambda}\frac{\partial\overline{v}^{\lambda}}{\partial p} \\
&+\left(-\overline{u}^{\lambda}\overline{v^*u^*}^{\lambda} + \overline{u^*u^*}^{\lambda}\overline{v}^{\lambda}\right)\frac{\tan\varphi}{a} \\
&-\overline{\omega^*\alpha^*}^{\lambda} \\
&+\overline{gu^*\frac{\partial F_u^*}{\partial p}}^{\lambda} + \overline{gv^*\frac{\partial F_v^*}{\partial p}}^{\lambda}.
\end{aligned}} \tag{32}$$

All the transport terms, including the pressure-work terms, are combined on the second and third lines of (32). The terms on the fourth line represent gradient production, that is, the conversion between the kinetic energy of the mean flow and that of the eddies. This conversion increases the eddy kinetic energy when the eddy momentum flux is "downgradient," that is, when it is from higher mean momentum to lower mean momentum. The $[\omega^*\alpha^*]$ term represents conversion between eddy kinetic energy and eddy available potential energy.

The (possibly surprising) metric terms in the equations for KZ and KE arise because we defined "eddies" in terms of departures from the zonal mean, so that the latitude-longitude coordinate system is implicit in the very definition of KE. The metric terms cancel when we add the equations for KZ and KE to obtain the equation for the zonally averaged total kinetic energy, $[K]$:

$$\begin{aligned}
&\frac{\partial\overline{\mathrm{K}}^{\lambda}}{\partial t} \\
&+\frac{1}{a\cos\varphi}\frac{\partial}{\partial\varphi}\left[\left(\overline{v}^{\lambda}\overline{\mathrm{K}}^{\lambda} + \frac{1}{2}\overline{v^*\left(u^{*2}+v^{*2}\right)}^{\lambda} + \overline{u}^{\lambda}\overline{v^*u^*}^{\lambda} + \overline{v}^{\lambda}\overline{v^*v^*}^{\lambda} + \overline{v}^{\lambda}\overline{\phi}^{\lambda} + \overline{v^*\phi^*}^{\lambda}\right)\cos\varphi\right] \\
&+\frac{\partial}{\partial p}\left[\overline{\omega}^{\lambda}\overline{\mathrm{K}}^{\lambda} + \frac{1}{2}\overline{\omega^*\left(u^{*2}+v^{*2}\right)}^{\lambda} + \overline{u}^{\lambda}\overline{\omega^*u^*}^{\lambda} + \overline{v}^{\lambda}\overline{\omega^*v^*}^{\lambda} + \overline{\phi}^{\lambda}\overline{\omega}^{\lambda} + \overline{\omega^*\phi^*}^{\lambda}\right] \\
&= -\overline{\omega}^{\lambda}\overline{\alpha}^{\lambda} - \overline{\omega^*\alpha^*}^{\lambda} \\
&+\overline{gu^*\frac{\partial F_u^*}{\partial p}}^{\lambda} + \overline{gv^*\frac{\partial F_v^*}{\partial p}}^{\lambda} + \overline{u}^{\lambda}g\frac{\partial}{\partial p}\overline{F_u}^{\lambda} + \overline{v}^{\lambda}\frac{\partial}{\partial p}\overline{F_v}^{\lambda}.
\end{aligned} \tag{33}$$

APPENDIX G

Spherical Harmonics

The spherical surface harmonics are convenient functions for representing the distribution of geophysical quantities over the surface of the spherical Earth.

We look for solutions of Laplace's differential equation,

$$\nabla^2 S = 0, \tag{1}$$

in a three-dimensional space. The ∇^2 operator can be expanded in spherical coordinates as

$$\nabla^2 S = \frac{1}{r^2}\frac{\partial}{\partial r}\left(r^2 \frac{\partial S}{\partial r}\right) + \nabla_h^2 S = 0. \tag{2}$$

Here r is the distance from the origin, and $\nabla_h^2 S$ is the Laplacian on a two-dimensional surface of constant r, that is, on a spherical surface. We postpone showing the form of $\nabla_h^2 S$. Inspection of (2) suggests that S should be proportional to a power of r. We write

$$S = r^n Y_n. \tag{3}$$

Here we are assuming that the radial dependence of S is "separable," in the sense that the Y_n are independent of radius. They are called *spherical surface harmonics* of order n. The subscript n is attached to Y_n to remind us that it corresponds to a particular exponent in the radial dependence of S.

For S to remain finite as $r \rightarrow 0$, we need $n \geq 0$. Since $n = 0$ would mean that S was independent of radius, we conclude that n must be a *positive* integer. Using $(1/r^2)(\partial/\partial r)[r^2(\partial S/\partial r)] = [n(n+1)/r^2]S$, which follows immediately from (3), we can rewrite (2) as

$$\boxed{\nabla_h^2 Y_n + \frac{n(n+1)}{r^2} Y_n = 0}. \tag{4}$$

Before continuing with the separation of variables, it is important to point out that all the quantities appearing in (4) have meaning without the need for any particular coordinate system on the two-dimensional spherical surface. We are going to use longitude and latitude coordinates, but we do not need them to write (4).

At this point we make an analogy with trigonometric functions. Suppose that we have a "doubly periodic" function $W(x, y)$ defined on a plane, with the usual Cartesian coordinates x and y. As a particular example, we let

$$W(x,y) = A\sin(kx)\cos(ly), \tag{5}$$

where A is an arbitrary constant. In Cartesian coordinates the two-dimensional Laplacian of W is

$$\begin{aligned}\nabla_h^2 W &= \left(\frac{\partial^2}{\partial x^2} + \frac{\partial^2}{\partial y^2}\right)W \\ &= -(k^2 + l^2)W.\end{aligned} \tag{6}$$

Comparing (4) and (6), we see that they are closely analogous. In particular, $n(n+1)/r^2$ in (4) corresponds to (k^2+l^2) in (6). This shows $n(n+1)$ is proportional to a "total horizontal wavenumber" on the sphere.

We now write out the "horizontal Laplacian" as

$$\nabla_h^2 S = \frac{1}{r^2\cos\varphi}\frac{\partial}{\partial\varphi}\left(\cos\varphi\frac{\partial S}{\partial\varphi}\right) + \frac{1}{r^2\cos^2\varphi}\frac{\partial^2 S}{\partial\lambda^2}, \tag{7}$$

using the familiar spherical coordinate system in which r is the radial coordinate, λ is longitude, and φ is latitude. We can then rewrite (4) as

$$\frac{1}{\cos\varphi}\frac{\partial}{\partial\varphi}\left(\cos\varphi\frac{\partial Y_n}{\partial\varphi}\right) + \frac{1}{\cos^2\varphi}\frac{\partial^2 Y_n}{\partial\lambda^2} + n(n+1)Y_n = 0. \tag{8}$$

Factors of $1/r^2$ have canceled in (8), and as a result, r no longer appears. Nevertheless, its exponent, n, is still visible, like the smile of the Cheshire cat.

Next, we separate the longitude and latitude dependence in the $Y_n(\lambda,\varphi)$; that is,

$$Y_n(\varphi,\lambda) = \Phi(\varphi)\Lambda(\lambda), \tag{9}$$

where $\Phi(\varphi)$ and $\Lambda(\lambda)$ are to be determined. By substitution of (9) into (8), we find that

$$\frac{\cos^2\varphi}{\Phi}\left[\frac{1}{\cos\varphi}\frac{d}{d\varphi}\left(\cos\varphi\frac{d\Phi}{d\varphi}\right) + n(n+1)\Phi\right] = -\frac{1}{\Lambda}\frac{d^2\Lambda}{d\lambda^2}. \tag{10}$$

The left-hand side of (10) does not contain λ, and the right-hand side does not contain φ, so both sides must be a constant, c. Then, the longitudinal structure of the solution is governed by

$$\frac{d^2\Lambda}{d\lambda^2} + c\Lambda = 0. \tag{11}$$

It follows that $\Lambda(\lambda)$ must be a trigonometric function of longitude; that is,

$$\Lambda = A_S\exp(im\lambda),\text{ where } m = \sqrt{c}, \tag{12}$$

and A_S is an arbitrary complex constant. The cyclic condition $\Lambda(\lambda+2\pi) = \Lambda(\lambda)$ implies that $\sqrt{c}$ must be an integer, which we denote by m. We refer to m as the zonal wavenumber. Note that m is nondimensional and can be either positive or negative.

Notice that m has meaning only with respect to a particular spherical coordinate system. In this way m is *less fundamental* than n, which comes from the radial dependence of the three-dimensional function S, and has a meaning that is independent of any particular spherical coordinate system.

The equation for $\Phi(\varphi)$, which determines the meridional structure of the solution, is

$$\frac{1}{\cos\varphi}\frac{d}{d\varphi}\left(\cos\varphi\frac{d\Phi}{d\varphi}\right)+\left[n(n+1)-\frac{m^2}{\cos^2\varphi}\right]\Phi = 0. \tag{13}$$

Note that the *zonal* wavenumber, m, appears in this *meridional* structure equation, as does the radial exponent, n. The longitude and radius dependencies have disappeared, but the zonal wavenumber and the radial exponent are still visible.

For convenience, we define a new independent variable to measure latitude,

$$\mu \equiv \sin\varphi, \tag{14}$$

so that $d\mu \equiv \cos\varphi d\varphi$. We can then write (13) as

$$\frac{d}{d\mu}\left[(1-\mu^2)\frac{d\Phi}{d\mu}\right]+\left[n(n+1)-\frac{m^2}{1-\mu^2}\right]\Phi = 0. \tag{15}$$

Equation (15) is simpler than (13), in that (15) does not involve trigonometric functions of the independent variable. This added simplicity is the motivation for using (14). The solutions of (15) are called the *associated Legendre functions*, are denoted by $P_n^m(\mu)$, and are given by

$$\begin{aligned} P_n^m(\mu) = {} & \frac{(2n)!}{2^n n!(n-m)!}(1-\mu^2)^{\frac{m}{2}}\cdot\left[\mu^{n-m}-\frac{(n-m)(n-m-1)}{2(2n-1)}\mu^{n-m-2}\right. \\ & \left. +\frac{(n-m)(n-m-1)(n-m-2)(n-m-3)}{2\cdot 4(2n-1)(2n-3)}\mu^{n-m-4}-\ldots\right]. \end{aligned} \tag{16}$$

The subscript n and superscript m are just "markers" to remind us that n and m appear as parameters in (15), denoting the radial exponent and zonal wavenumber, respectively, of $S(r,\lambda,\varphi)$. The sum in (16) is continued as far as necessary to include all nonnegative powers of μ. The factor in brackets is therefore a polynomial of degree $n-m$, and so *we must require that*

$$n \geq m. \tag{17}$$

Substitution can be used to demonstrate that for $n \geq m$, the associated Legendre functions are indeed solutions of (15).

In view of the leading factor of $(1-\mu^2)^{m/2}$ in (16), the complete function $P_n^m(\mu)$ is a polynomial in μ for even values of m but not for odd values of m. The functions $P_n^m(\mu)$ are said to be of "order n" and "rank m." Figure G.1 gives some examples of associated Legendre functions, which you might wish to check for their consistency with (16). It can be shown that the associated Legendre functions are mutually *orthogonal*, i.e.,

$$\int_{-1}^{1} P_n^m(\mu)\cdot P_l^m(\mu)d\mu = 0, \text{ for } n \neq l, \text{ and } \int_{-1}^{1}[P_n^m(\mu)]^2 d\mu = \left(\frac{2}{2n+1}\right)\frac{(n+m)!}{(n-m)!}. \tag{18}$$

It follows that the functions

$$\sqrt{\left(\frac{2n+1}{2}\right)\frac{(n-m)!}{(n+m)!}}\,P_n^m(\mu),\ n=m,\ m+1,\ m+2,\ldots \tag{19}$$

are mutually *orthonormal* for $-1 \geq \mu \geq 1$.

Referring to (9), we see that a particular spherical surface harmonic can be written as

$$\boxed{Y_n^m(\mu,\lambda) = P_n^m(\mu)\exp(im\lambda)}. \tag{20}$$

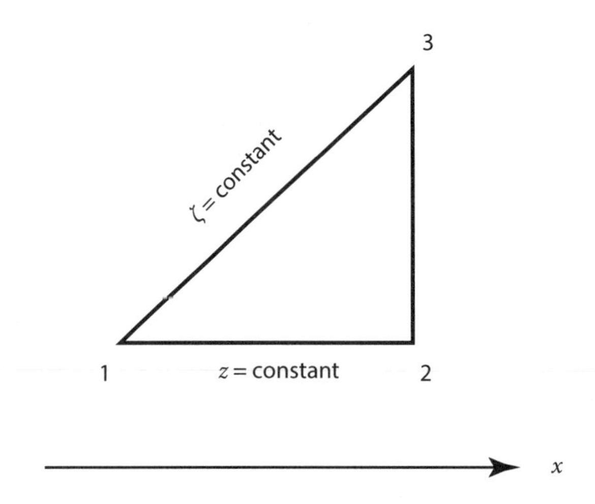

Figure G.1. Algebraic forms and plots of selected associated Legendre functions.

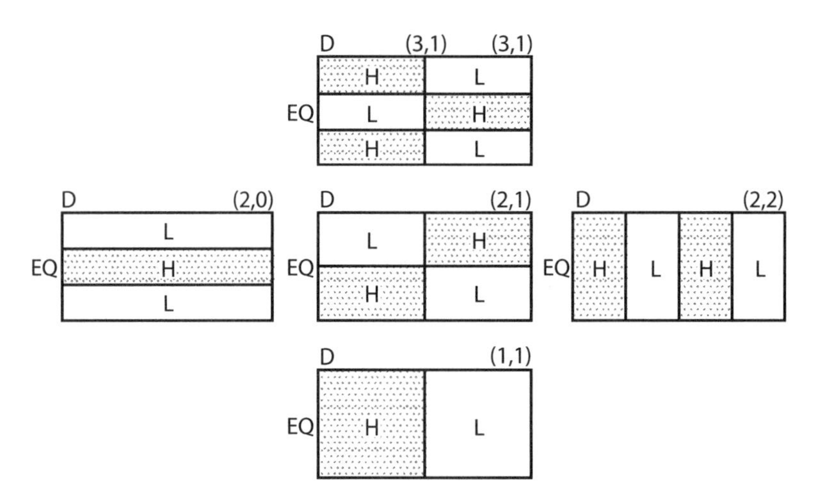

Figure G.2. Examples of low-resolution spherical harmonics, mapped onto the plane. The horizontal direction in each panel represents longitude, and the vertical direction represents latitude. The numbers in parentheses in each panel are the appropriate values of n and s, in that order. Recall that the number of nodes in the meridional direction is $n - s$. The shading in each panel represents the sign of the field (and all signs can be flipped arbitrarily). You may think of "white" as negative and "stippled" as positive. From Washington and Parkinson (1986).

It is the product of an associated Legendre function of μ with a trigonometric function of longitude. Note that the arbitrary constant has been set to unity.

Figure G.2 shows examples of spherical harmonics of low order, as mapped out onto the longitude-latitude plane. Figure G.3 gives similar diagrams for $n = 5$ and $m = 0, 1, 2, \ldots, 5$, plotted onto stretched spheres. Figure G.4 shows some low-order spherical harmonics mapped onto three-dimensional pseudospheres, in which the local radius of the surface of the pseudosphere is 1 plus a constant times the local value of the spherical harmonic.

By using the orthogonality condition (18) for the associated Legendre functions, and also the orthogonality properties of the trigonometric functions, we can show that

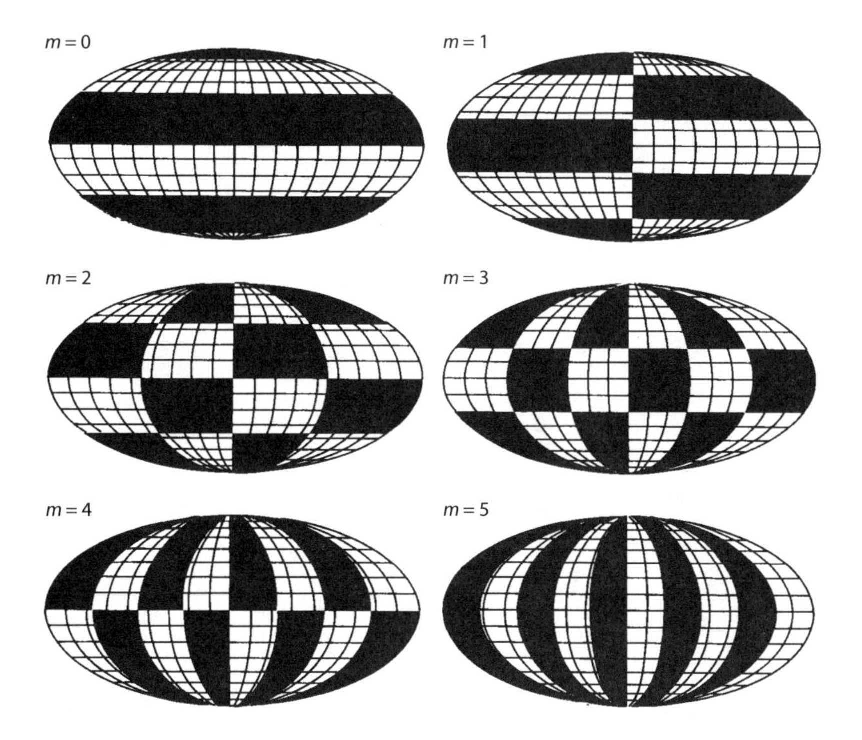

Figure G.3. Alternating patterns of positives and negatives for spherical harmonics with $n=5$ and $m=0,1,2,\ldots,5$. From Baer (1972). © American Meteorological Association. Used with permission.

$$\int_{-1}^{1}\int_{0}^{2\pi} P_n^m(\mu)\exp(im\lambda)P_l^m(\mu)\exp(im'\lambda)\,d\mu d\lambda = 0 \text{ unless } n=1 \text{ and } m=m'. \quad (21)$$

The mean value over the surface of a sphere of the *square* of a spherical surface harmonic is given by

$$\frac{1}{4\pi}\int_{-1}^{1}\int_{0}^{2\pi}[P_n^m(\mu)\exp(im\lambda)]^2\,d\mu d\lambda = \frac{1}{2(2n+1)}\frac{(n+m)!}{(n-m)!} \text{ for } s \neq 0. \quad (22)$$

For the special case $m = 0$, the corresponding value is $1/(2n+1)$.

For a given n, the mean values given by (22) vary greatly with m, which is inconvenient for the interpretation of data. For this reason, it is customary to use, instead of $P_n^m(\mu)$, the *seminormalized associated Legendre functions*, denoted by $\hat{P}_n^m(\mu)$. These functions are identical with $P_n^m(\mu)$ when $m = 0$. For $m > 0$, the seminormalized functions are defined by

$$\hat{P}_n^m(\mu) = \sqrt{2\frac{(n-m)!}{(n+m)!}}\cdot P_n^m(\mu). \quad (23)$$

The mean value over the sphere of the square of $\hat{P}_n^m(\mu)\exp(im\lambda)$ is then $(2n+1)^{-1}$, for any n and m.

The spherical harmonics can be shown to form a complete orthonormal basis and so can be used to represent an arbitrary function $F(\lambda,\varphi)$ of latitude and longitude:

$$F(\lambda,\varphi) = \sum_{m=-\infty}^{\infty}\sum_{n=|m|}^{\infty} F_n^m Y_n^m(\lambda,\varphi). \quad (24)$$

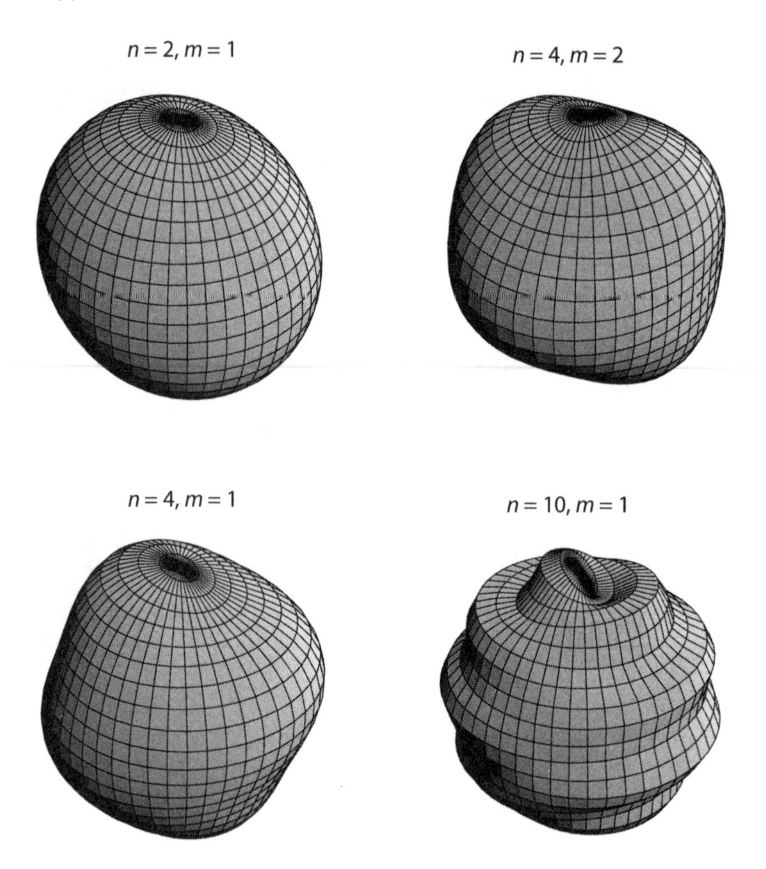

Figure G.4. Selected spherical harmonics mapped onto three-dimensional pseudospheres, in which the local radius of the surface of the pseudo-pseudosphere is 1 plus a constant times the local value of the spherical harmonic.

Here the T_n^m are the expansion coefficients. Note that the sum ranges over both positive and negative values and that the sum over n is taken so that $n-|m|\geq 0$.

The sums in (24) range over an infinity of terms, but in practice, of course, we have to truncate after a finite number of terms, so that (24) is replaced by

$$\overline{F} = \sum_{m=-M}^{M} \sum_{n=|m|}^{N(m)} F_n^m Y_n^m. \tag{25}$$

Here the overbar is a reminder that the sum is truncated. The sum over n ranges up to $N(m)$, which has to be specified somehow. The sum over m ranges from $-M$ to M. It can be shown that *this ensures that the final result is real*; this is an important result that you should prove for yourself.

The choice of $N(m)$ fixes what is called the *truncation procedure*. There are two commonly used truncation procedures. The first, called *rhomboidal*, takes

$$N(m) = M + |\, m \,|. \tag{26}$$

The second, called "*triangular*, takes

$$N(m) = M. \tag{27}$$

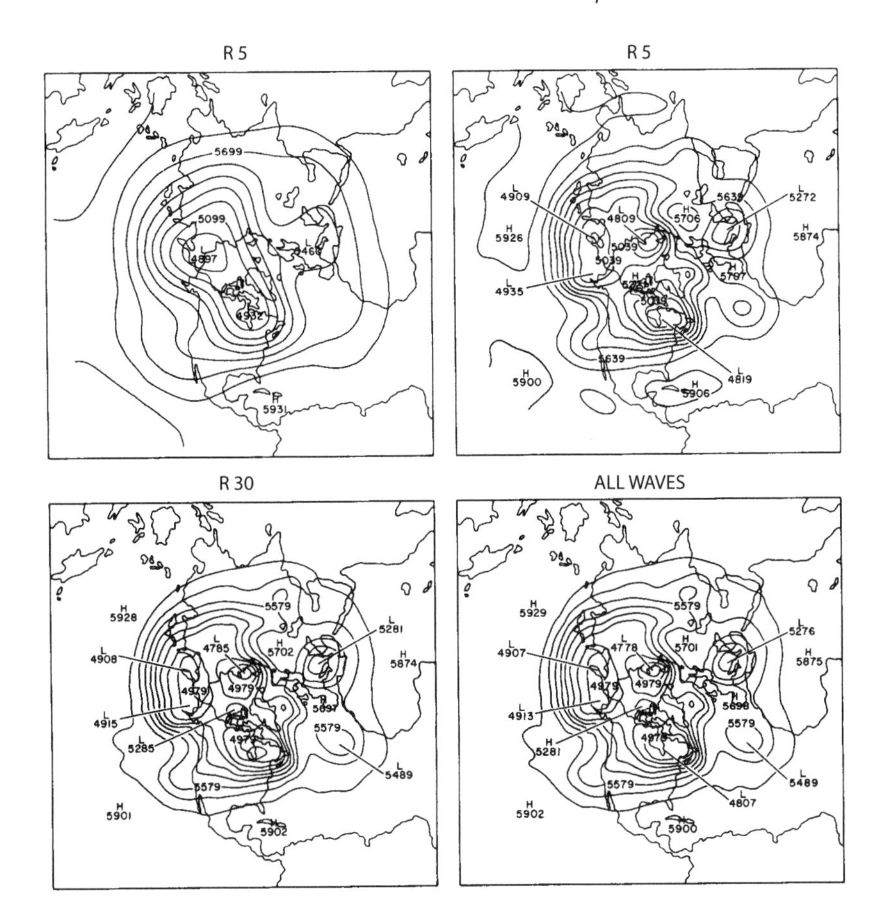

Figure G.5. Demonstration of the effects of various horizontal truncations of 500 m patterns of geopotential height (m) data provided originally on a 2.5° longitude-latitude grid. From Washington and Parkinson (1986).

Triangular truncation has the following beautiful property. To perform a spherical harmonic transform, it is necessary to adopt a spherical coordinate system (λ,φ). There are of course infinitely many such systems. There is no reason in principle that the coordinates have to be chosen in the conventional way, so that the poles of the coordinate system coincide with the Earth's poles of rotation. The choice of a particular spherical coordinate system is therefore somewhat arbitrary.

Suppose that we choose two different spherical coordinate systems (tilted with respect to one another in an arbitrary way), perform a triangularly truncated expansion in both, and then plot the results. It can be shown that the two maps will be identical. This means that the arbitrary orientations of the spherical coordinate systems used had no effect whatsoever on the results obtained. The coordinate system used "disappears" at the end. Triangular truncation is very widely used today, in part because of this nice property.

Figure G.5 shows an example based on 500 mb height data, provided originally on a 2.5° longitude-latitude grid. The figure shows how the data look when represented by just a few spherical harmonics (top left), a few more (top right), a moderate number (bottom left), and at full 2.5° resolution. The smoothing effect of severe truncation is clearly visible.

APPENDIX H

Hermite Polynomials

The Hermite polynomials are defined by

$$H_n(y) \equiv (-1)^n e^{y^2} \frac{d^n}{dy^n}(e^{-y^2}), \text{ for } n \geq 0. \tag{1}$$

The algebraic forms and plots of the first six Hermite polynomials are given in figure H.1. Note that the even-numbered Hermite polynomials are even functions, and the odd-numbered Hermite polynomials are odd functions.

The Hermite polynomials are orthogonal with respect to e^{-y^2}:

$$\int_{-\infty}^{\infty} H_n(y) H_m(y) e^{-y^2} dy = 0 \text{ for } n \neq m, \tag{2}$$

$$\int_{-\infty}^{\infty} H_n(y) H_m(y) e^{-y^2} dy = 2^n n! \sqrt{\pi} \text{ for } n = m. \tag{3}$$

Two useful recursion relations are

$$\frac{d}{dy} H_n(y) = 2n H_{n-1}(y) \text{ for } n \geq 1, \tag{4}$$

and

$$H_{n+1}(y) = 2y H_n(y) - 2n H_{n-1}(y). \tag{5}$$

We let

$$\psi_n(y) \equiv \frac{e^{-y^2} H_n(y)}{\sqrt{2^n n! \pi^{1/2}}}. \tag{6}$$

The $\psi_n(y)$ satisfy

$$\left\{\frac{d^2}{dy^2} + \left[(2n+1) - y^2\right]\right\} \psi_n = 0, \tag{7}$$

as can be verified by substitution. Note that for $n \geq 0$, the expression $2n+1$, which appears in (7), generates all positive odd integers.

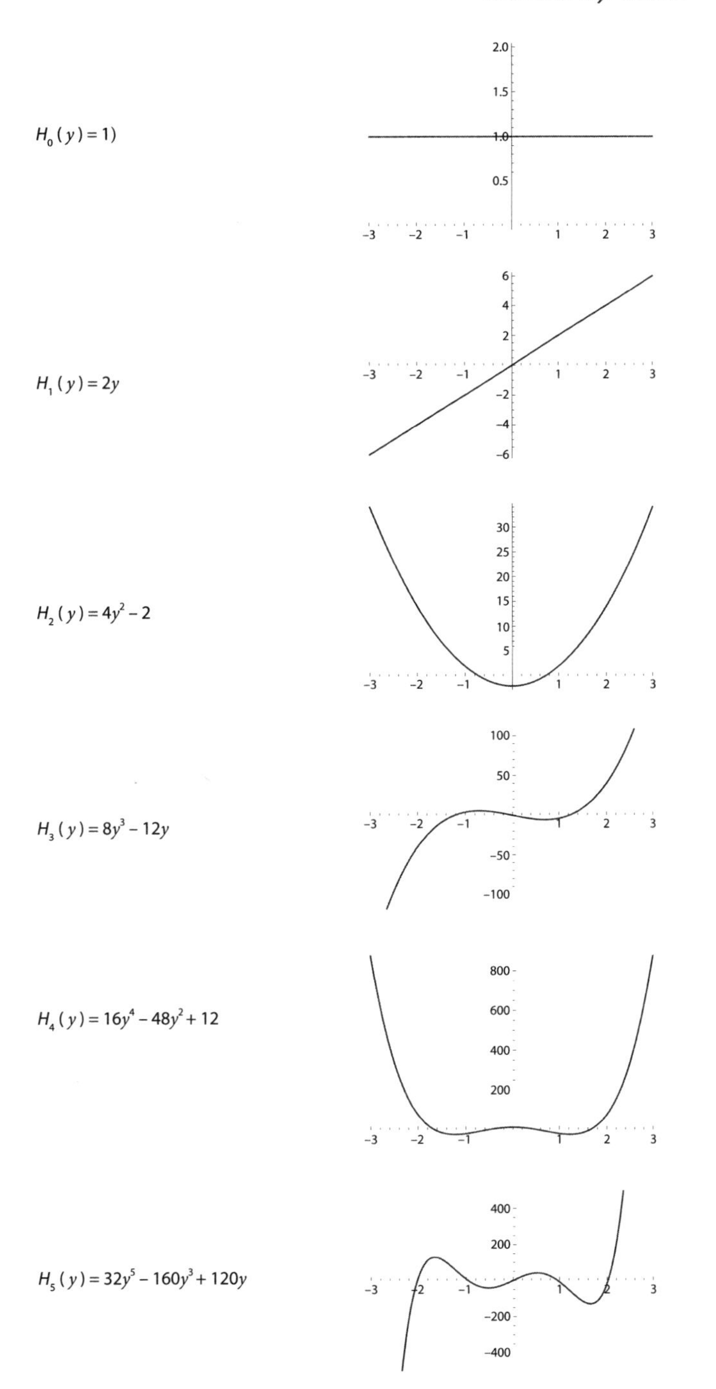

Figure H.1. The algebraic forms and plotted shapes of the first six Hermite polynomials.

Bibliography

Adler, R. F., G. J. Huffman, A. Chang, R. Ferraro, P. Xie, J. Janowiak, B. Rudolf, U. Schneider, S. Curtis, D. Bolvin, A. Gruber, J. Susskind, and P. Arkin. 2003. The Version-2 Global Precipitation Climatology Project (GPCP) Monthly Precipitation Analysis (1979–Present). *J. Hydrometeor.*, **4**, 1147–1167.

Albrecht, B. A., C. S. Bretherton, D. Johnson, W. H. Schubert, and A. S. Frisch. 1995. The Atlantic Stratocumulus Transition Experiment—ASTEX. *Bull. Amer. Meteor. Soc.*, **76**, 889–904.

Alexander, M. A., and C. Deser. 1995. A mechanism for the recurrence of wintertime midlatitude SST anomalies. *J. Phys. Oceanogr.*, **25**, 122–137.

Ambaum, M. 1997. Isentropic formation of the tropopause. *J. Atmos. Sci.*, **54**, 555–568.

Andersen, J. A., 2012. Investigations of the convectively coupled equatorial waves and the Madden-Julian oscillation. Ph.D. thesis, Harvard University.

Andersen, J. A., and Z. Kuang, 2012. Moist static energy budget of MJO-like disturbances in the atmosphere of a zonally symmetric aquaplanet. *J. Climate*, **25**, 2782-2804. doi:http://dx.doi.org/10.1175/JCLI-D-11-00168.1.

Anderson, D.L.T., and J. P. McCreary. 1985. Slowly propagating disturbances in a coupled ocean-atmosphere model. *J. Atmos. Sci.*, **42**, 615–629.

Anderson, J. L. 1995. A simulation of atmospheric blocking with a forced barotropic model. *J. Atmos. Sci.*, **52**, 2593–2608.

Anderson, J. R., and D. E. Stevens. 1987. The presence of linear wavelike modes in a zonally symmetric model of the tropical atmosphere. *J. Atmos. Sci.*, **44**, 2115–2127.

Andrews, D. G. 1983. Finite-amplitude Eliassen-Palm theorem in isentropic coordinates. *J. Atmos. Sci.*, **40**, 1877–1883.

Andrews, D. G., J. R. Holton, and C. B. Leovy. 1987. *Middle Atmosphere Dynamics*. International Geophysics Series, vol. 40. New York: Academic Press, 489 pp.

Añel, J. A., L. Gimeno, L. de la Torre, and R. Nieto. 2006. Changes in tropopause height for the Eurasian region from CARDS radiosonde data. *Naturwissenschaften*, **93**, 603–609. doi:10.1007/s00114-006-0147-5.

Annamalai, H., J. Hafner, K. P. Sooraj, and P. Pillai. 2013. Global warming shifts the monsoon circulation, drying South Asia. *J. Climate*, **26**, 2701–2718. doi:http://dx.doi.org/10.1175/JCLI-D-12-00208.1.

Arakawa, A. 1969. Parameterization of cumulus convection. *Proceedings of the WMO/IUGG Symposium on Numerical Weather Prediction*, Tokyo, November 26– December 4, 1968, Japan Meteor. Agency, **4** (8), 1–6.

Arakawa, A. 2004. The cumulus parameterization problem: Past, present, and future. *J. Climate*, **17**, 2493–2525.

Arakawa, A., and J.-M. Chen. 1987. Closure assumptions in the cumulus parameterization problem. *WMO/IUGG Symposium on Short- and Medium-Range Numerical Weather Prediction*, Tokyo, 107–130.

Arakawa, A., and M.-D. Cheng. 1993. The Arakawa-Schubert cumulus parameterization. In *The Representation of Cumulus Convection in Numerical Models*, edited by K. Emanuel and D. Raymond. *Meteor. Monogr.*, **24** (46), 1–246.

Arakawa, A., and C. Konor. 2009. Unification of the anelastic and quasi-hydrostatic systems of equations. *Mon. Wea. Rev.*, **137**, 710–726.

Arakawa, A., and W. H. Schubert. 1974. Interaction of a cumulus cloud ensemble with the large-scale environment, Part I. *J. Atmos. Sci.*, **31**, 674–701.

Arakawa, A., and K.-M. Xu. 1990. The macroscopic behavior of simulated cumulus convection and semiprognostic tests of the Arakawa-Schubert cumulus parameterization. *Proceedings of the Indo-U. S. Seminar on Parameterization of Sub-grid Scale Processes in Dynamical Models of Medium-Range Prediction and Global Climate*. Pune, India, August 6–10, pp. 3–18.

Archer, C. L., and K. Caldeira. 2008. Historical trends in the jet streams. *Geophys. Res. Lett.*, **35**, L08803. doi:10.1029/2008GL033614.

Arfken, G., H. Weber, and F. Harris. 2012. *Mathematical Methods for Physicists*, 7th ed. Waltham, MA: Academic Press, 1220 pp.

Aris, R. 1962. *Vectors, Tensors, and the Basic Equations of Fluid Mechanics*. New York: Dover, 286 pp.

Arnold, N. P., Z. Kuang, and E. Tziperman, 2013. Enhanced MJO-like variability at high SST. *J. Climate*, **26**, 988–1001. doi:http://dx.doi.org/10.1175/JCLI-D-12-00272.1.

Arnold, N., M. Branson, M. A. Burt, D. S. Abbot, Z. Kuang, D. A. Randall, and E. Tziperman. 2014. Significant consequences of explicit representation of atmospheric convection at high CO_2 concentration. *Proc. Nat. Acad. Sci. USA*, **111**, 10943–10948.

Arpé, K., C. Brankovic, E. Oriol, and P. Seth. 1986. Variability in time and space of energetics from a long series of atmospheric data produced by ECMWF. *Beitr. Phys. Atmos.*, **59**, 321–355.

Austin, J., and T. J. Reichler. 2008. Long-term evolution of the cold point tropical tropopause: Simulation results and attribution analysis. *J. Geophys. Res.*, **113**, D00B10. doi:10.1029/2007JD009768.

Bacon, S. 1997. Circulation and fluxes in the North Atlantic between Greenland and Ireland. *J. Phys. Oceanogr.*, **27**, 1420–1435.

Baer, F. 1972. An alternate scale representation of atmospheric energy spectra. *J. Atmos. Sci.*, **29**, 649–664.

———. 1974. Hemispheric spectral statistics of APE. *J. Atmos. Sci.*, **31**, 932–941.

———. 1981. Three-dimensional scaling and structure of atmospheric energetics. *J. Atmos. Sci.*, **38**, 53–68.

Baldwin, M. P., and T. J. Dunkerton. 2001. Stratospheric harbingers of anomalous weather regimes. *Science*, **294**, 581–584.

Baldwin, M. P., L. J. Gray, T. J. Dunkerton, K. Hamilton, P. H. Haynes, W. J. Randel, J. R. Holton, M. J. Alexander, I. Hirota, T. Horinouchi, D.B.A. Jones, J. S. Kinnersley, C. Marquardt, K. Sato, and M. Takahashi. 2001. The quasi-biennial oscillation. *Rev. Geophys.*, **39**, 179–229.

Ball, F. K. 1960. Control of inversion height by surface heating. *Quart. J. Roy. Meteor. Soc.*, **86**, 483–494.

Barenblatt, G. I. 2003. *Scaling*. Cambridge: Cambridge University Press, 171 pp.

Barkstrom, B. R. 1984. The Earth Radiation Budget Experiment (ERBE). *Bull. Amer. Meteor. Soc.*, **65**, 1170–1185

Barkstrom, B., E. F. Harrison, G. L. Smith, R. N. Green, J. Kibler, R. Cess, and the ERBE Science Team. 1989. Earth Radiation Budget Experiment (ERBE) archival and April 1985 results. *Bull. Amer. Meteor. Soc.*, **70**, 1254–1262.

Barnett, J. J., 1974. Mean meridional temperature behavior of the stratosphere from November 1970 to November 1971 derived from measurements by the Selective Chopper Radiometer on Nimbus IV. *Quart. J. Roy. Meteor. Soc.*, **100**, 505–530.

Barnett, T. P., 1991. On ENSO physics. *J. Climate*, **4**, 487–515.

Barnett, T. P., L. Bengtsson, K. Arpe, M. Flügel, N. Graham, M . Latif, J. Ritchie, E. Roeckner, U. Schlese, U. Schulzweida, and M. Tyree. 1994. Forecasting global ENSO-related climate anomalies. *Tellus*, **46A**, 367–380.

Barnett, T. P., L. Dümenil, U. Schlese, E. Roeckner, and M. Latif. 1989. The effect of Eurasian snow cover on regional and global climate variations. *J. Atmos. Sci.*, **46**, 661–685.

Barnett, T. P., M. Latif, N. Graham, N., and M. Flugel. 1995. On the frequency-wavenumber structure of the tropical ocean/atmosphere system. *Tellus*, **47A**, 998–1012.

Barriopedro, E., and N. Calvo. 2014. On the relationship between ENSO, stratospheric sudden warmings, and blocking. *J. Climate*, **27**, 4704–4720. doi:http://dx.doi.org/10.1175/JCLI-D-13-00770.1.

Barrow, J. D. 2002. *The Constants of Nature*. New York: Random House, 352 pp.

Barton, N. P., and A. W. Ellis. 2009. Variability in wintertime position and strength of the North Pacific jet stream as represented by re-analysis data. *Int. J. Climatol.*, **29**, 851–862.

Basdevant, C., B. Legras, R. Sadourny, and Me Béland. 1981. A study of barotropic model flows: Intermittency, waves, and predictability. *J. Atmos. Sci.*, **38**, 2305–2326.

Battisti, D. S. 1988. The dynamics and thermodynamics of a warming event in a coupled tropical atmosphere/ocean model. *J. Atmos. Sci.*, **45**, 2889–2919.

Battisti, D. S., and A. C. Hirst. 1989. Interannual variability in the tropical atmosphere/ocean system: Influence of the basic state, ocean geometry and nonlinearity. *J. Atmos. Sci.*, **46**, 1687–1712.

Battisti, D. S., and D. D. Ovens. 1995. The dependence of the low-level equatorial jet on Hadley and Walker circulations. *J. Atmos. Sci.*, **52**, 3911–3931.

Baytgin, K., and G. Laughlin. 2008. On the dynamical stability of the Solar System. *Astrophys. J.*, **683**, 1207–1216.

Bengtsson, L., M. Kanamitsu, P. Kallberg, and S. Uppala. 1982. FGGE 4-dimensional data assimilation at ECMWF. *Bull. Amer. Meteor. Soc.*, **63**, 29–43.

Bennett, A. F. 1978. Poleward heat fluxes in Southern Hemisphere Oceans. *J. Phys. Oceanogr.*, **8**, 785–798.

Benzi, R., S. Patarnello, and P. Santangelo. 1988. Self-similar coherent structures in two-dimensional decaying turbulence. *J. Phys. A: Math Gen.*, **21**, 1221–1237.

Berckmans, J., and T. Woollings, M.-E. Demory, P.-L. Vidale, and M. Roberts. 2013. Atmospheric blocking in a high resolution climate model: Influences of mean state, orography and eddy forcing. *Atmos. Sci. Lett.*, **14**, 34–40.

Berggren, R., B. Bolin, and C. G. Rossby. 1949. An aerological study of zonal motion, its perturbations and breakdown. *Tellus*, **1**, 13–47.

Betts, A. K. 1973. Nonprecipitating cumulus convection and its parameterization. *Quart. J. Roy. Meteor. Soc.*, **99**,178–196.

———. 1982. Saturation point analysis of moist convective overturning. *J. Atmos. Sci.*, **39**, 1484–1505.

———. 1985. Mixing line analysis of clouds and cloudy boundary layers. *J. Atmos. Sci.*, **42**, 2751–2763.

Betts, A. K., and W. Ridgway. 1989. Climatic equilibrium of the atmospheric convective boundary layer over a tropical ocean. *J. Atmos. Sci.*, **46**, 2621–2641.

Beyer, W. H. 1984. 27th ed. Boca Raton, FL: CRC Press, 301–305.

Beyers, N. J., B. T. Miers, and R. J. Reed. 1966. Diurnal tidal motions near the stratopause during 48 hours at White Sands Missile Range. *J. Atmos. Sci.*, **23**, 325–333.

Birner, T. 2010. Recent widening of the tropical belt from global tropopause statistics: Sensitivities. *J. Geophys. Res.*, **115**, D23109. doi:10.1029/2010JD014664.

Bister, M., and K. A. Emanuel. 1998. Dissipative heating and hurricane intensity. *Meteorol. Atmos. Phys.*, **65**, 233–240.

Bjerknes, J. 1938. Saturated-adiabatic ascent of air through dry-adiabatically descending environment. *Quart. J. Roy. Meteor. Soc.*, **64**, 325–330.

———. 1948. Practical application of H. Jeffrey's theory of the general circulation. *Programme et Résumé des Mémoires*. Réunion d'Oslo, Association de Météorolgie, Union de Géodesie et Géophysique International, 13–14.

———. 1966. A possible response of the atmospheric Hadley circulation to equatorial anomalies of ocean temperature. *Tellus*, **18**, 820–829.

———. 1969. Atmospheric teleconnections from the equatorial Pacific. *Mon. Wea. Rev.*, **97**, 163–172.

Blackmon, M. L., 1976. A climatological spectral study of the 500 mb geopotential height of the Northern Hemisphere. *J. Atmos. Sci.*, **33**, 1607–1623.

Blackmon, M. L., Y.-H. Lee, and J. M. Wallace. 1984a. Horizontal structure of 500 mb height fluctuations with long, intermediate and short time scales as deduced from lag-correlation statistics. *J. Atmos. Sci.*, **41**, 961–979.

Blackmon, M. L., Y.-H. Lee, J. M. Wallace and H.-H. Hsu, 1984b. Time variations of 500 mb height fluctuations with long, intermediate, and short time scales as deduced from lag-correlation statistics. *J. Atmos. Sci.*, **41**, 981–991.

Blackmon, M. L., and G. H. White. 1982. Zonal wavenumber characteristics of Northern Hemisphere transient eddies. *J. Atmos. Sci.*, **39**, 1985–1998.

Blade, I., and D. L. Hartmann. 1993. Tropical intraseasonal oscillations in a simple nonlinear model. *J. Atmos. Sci.*, **50**, 2922–2939.

Blazejewski, H., D. L. Cadet, and O. Marsal. 1986. Low-frequency sea surface temperature and wind variations over the Indian and Pacific Oceans. *J. Geophys. Res.*, **91**, 5129–5132.

Boer, G. J., and T. G. Shepherd. 1983. Large-scale two-dimensional turbulence in the atmosphere. *J. Atmos. Sci.*, **40**, 164–184.

Bolster, D., R. E. Hershberger, and R. J. Donnelly. 2011. Dynamic similarity, the dimensionless science. *Physics Today*, **64**, 42–47.

Bony, S., and K. A. Emanuel, 2005. On the role of moist processes in tropical intraseasonal variability: Cloud-radiation and moisture-convection feedbacks. *J. Atmos. Sci.*, **62**, 2770–2789. doi: http://dx.doi.org/10.1175/JAS3506.1.

Bony, S., R. Colman, V. M. Kattsov, R. P. Allan, C. S. Bretherton, J.-L. Dufresne, A. Hall, S. Hallegatte, M. M. Holland, W. Ingram, D. A. Randall, B. J. Soden, G. Tselioudis, and M. J. Webb, 2006. How well do we understand and evaluate climate change feedback processes? *J. Climate*, **19**, 3445–3482.

Booker, J. R., and F. P. Bretherton, 1967. The critical layer for internal gravity waves in a shear flow. *J. Fluid Mech.*, **27**, 513–539.

Bowman, K. P., and P. J. Cohen, 1997. Interhemispheric exchange by seasonal modulation of the Hadley circulation. *J. Atmos. Sci.*, **54**, 2045–2059.

Bowman, K. P., and G. D. Carrie, 2002. The mean-meridional transport circulation of the troposphere in an idealized GCM. *J. Atmos. Sci.*, **59**, 1502–1514.

Branstator, G. 1992. The maintenance of low-frequency atmospheric anomalies. *J. Atmos. Sci.*, **49**, 1924–1946.

———. 1995. Organization of storm track anomalies by recurring low-frequency circulation anomalies. *J. Atmos. Sci.*, **52**, 207–226.

Bretherton, C. S., and C. Schär. 1993. Flux of potential vorticity substance: A simple derivation and uniqueness property. *J. Atmos. Sci.*, **50**, 1834–1836.

Bretherton, F. P., 1969. Momentum transport by gravity waves. *Quart. J. Roy. Meteor. Soc.*, **95**, 213–243.

Broecker, W. S.1992. The great ocean conveyor. In *Global Warming: Physics and Facts*, edited by B. G. Levi, D. Hafemeister, and R. Scribner. New York: American Physical Society, 129–161.

Broecker, W. S., D. M. Peteet, and D. Rind. 1985. Does the ocean-atmosphere system have more than one stable mode of operation? *Nature*, **315**, 21–26.

Brown, R. G., and C. Zhang. 1997. Variability of midtropospheric moisture and its effect on cloud-top height distribution during TOGA COARE. *J. Atmos. Sci.*, **54**, 2760–2774.

Bryan, F. O. 1997. The axial angular momentum balance of a global ocean general circulation model. *Dyn. Atmos. Oceans*, **25**, 191–216.

Bryan, K., 1969. A numerical method for the study of the circulation of the world ocean. *J. Comput. Phys.*, **4**, 347–376.

Bryan, K., and J. L. Sarmiento. 1985. Modeling ocean circulation. *Adv. Geophys.*, **28A**, 433–459.

Bryden, H. L. 1993. Ocean heat transport across 24° N latitude. In *Interactions between Global Climate Subsystems: The Legacy of Hann*, edited by G. A. McBean and M. Hantel. *Geophys. Monogr.*, no. 75, 65–75.

Bukowinksi, M.S.T. 1999. Taking the core temperature. *Nature*, **401**, 432–433.

Bunker, A. F. 1971. Energy transfer and tropical cell structure over the central Pacific. *J. Atmos. Sci.*, **28**, 1101–1116.

Burrows, W. R. 1976. A diagnostic study of atmospheric spectral kinetic energetics. *J. Atmos. Sci.*, **33**, 2308–2321.

Businger, S., and J. A. Businger. 2001. Viscous dissipation of turbulence kinetic energy in storms. *J. Atmos. Sci.*, **58**, 3793–3796.

Butchart, N. 2014. The Brewer-Dobson circulation. *Rev. Geophys.*, **52**. doi:10.1002/2013RG000448.

Caballero, R., and M. Huber. 2010. Spontaneous transition to superrotation in warm climates simulated by CAM3. *Geophys. Res. Lett.*, **37**, L11701.

Cadet, D. L. 1986. Fluctuations of precipitable water over the Indian Ocean during the 1979 summer monsoon. *Tellus*, **38A**, 170–177.

Cadet, D. L., and S. Greco. 1987a. Water vapor transport over the Indian Ocean during the 1979 summer monsoon. Part I: Water vapor fluxes. *Mon .Wea. Rev.*, **115**, 653–663.

———. 1987b. Water vapor transport over the Indian Ocean during the 1979 summer monsoon. Part II: Water vapor budgets. *Mon. Wea. Rev.*, **115**, 2358–2366.

Cai, M., and M. Mak. 1990. Symbolic relation between planetary and synoptic-scale waves. *J. Atmos. Sci.*, **47**, 2953–2968.

Cane, M. A., and S. E. Zebiak. 1985. A theory for El Niño and the Southern Oscillation. *Science*, **228**, 1085–1087.

Clement, A., R. Seager, M. A. Cane, and S. E. Zebiak. 1996. An ocean thermostat. *J. Climate*, **9**, 2190–2196.

Cariolle, D., M. Amodei, M. Deque, J.-F. Mahfouf, P. Simon, and H. Teyssedre. 1993. A quasi-biennial oscillation signal in general circulation model simulations. *Science*, **261**, 1313–1316.

Carissimo, B. C., A. H. Oort, and T. H. Vonder Haar. 1985. Estimating the meridional energy transports in the atmosphere and ocean. *J. Phys. Oceanogr.*, **15**, 82–91.

Carnevale, G. F., J. C. McWilliams, Y. Pomeau, J. B. Weiss, and W. R. Young. 1991. Evolution of vortex statistics in two-dimensional turbulence. *Phys. Rev. Lett.*, **66**, 2735–2737.

Chandrasekhar, V. 1961. *Hydrodynamic and Hydromagnetic Stability*. Oxford: Clarendon Press, 652 pp.

Chang, C. P. 1977. Viscous internal gravity waves and low-frequency oscillations in the tropics. *J. Atmos. Sci.*, **34**, 901–910.

Chang, C. P., and H. Lim. 1988. Kelvin wave–CISK: A possible mechanism for the 30–50 day oscillations. *J. Atmos. Sci.*, **45**, 1709–1720.

Chang, E. K. M., Y. Guo, and X. Xia. 2012. CMIP5 multimodel ensemble projection of storm track change under global warming, *J. Geophys. Res.*, **117**, D23118. doi:10.1029/2012JD018578.

Chang, E. K.M., S. Lee, and K. L. Swanson. 2002. Storm track dynamics. *J. Climate*, **15**, 2163–2183.

Chang, C.-P., Y. Ding, N.-C. Lau, R. H. Johnson, Bin Wang, and T. Yasunari, eds. 2011. *The Global Monsoon System: Research and Forecast*. Hackensack, NJ: World Scientific, 608 pp.

Chang, P., L. Ji, and H. Li. 1997. A decadal climate variation in the tropical Atlantic Ocean from thermodynamic air-sea interactions. *Nature*, **385**, 516–518.

Chao, W. C., 1987. On the Origin of the Tropical Intraseasonal Oscillation. *J. Atmos. Sci.*, **44**, 1940–1949. doi:http://dx.doi.org/10.1175/1520-0469(1987)044<1940:OTOOTT>2.0.CO;2.

Charlton, A. J., and L. M. Polvani. 2007. A new look at stratospheric sudden warmings. Part I: Climatology and modeling benchmarks. *J. Climate*, **20**, 449–469. doi:http://dx.doi.org/10.1175/JCLI3996.1.

Charlton, A. J., L. M. Polvani, J. Perlwitz, F. Sassi, E. Manzini, K. Shibata, S. Pawson, J. E. Nielsen, and D. Rind. 2007. A new look at stratospheric sudden warmings. Pt. 2: Evaluation of numerical model simulations. *J. Climate*, **20**, 470–488. doi:http://dx.doi.org/10.1175/JCLI3994.1.

Charney, J. G. 1948. On the scale of atmospheric motions. *Geophys. Publ. Oslo*, **17**, 1–17.

———. 1963. A note on large-scale motions in the tropics. *J. Atmos. Sci.*, **20**, 607–609.

———. 1971. Geostrophic turbulence. *J. Atmos. Sci.*, **28**, 1087–1095.

———. 1973. Planetary fluid dynamics. In *Dynamic Meteorology*, edited by. P. Morel. Boston: D. Reidel, 97–352

Charney, J. G., and J. G. DeVore. 1979. Multiple flow equilibria in the atmosphere and blocking. *J. Atmos. Sci.*, **36**, 1205–1216

Charney, J. G., and P. G. Drazin. 1961. Propagation of planetary-scale disturbances from the lower into the upper atmosphere. *J. Geophys. Res.*, **66**, 83–109.

Charney, J. G., and A. Eliassen. 1949. A numerical method for predicting the perturbations of the middle latitude westerlies. *Tellus*, **1**, 38–54.

Charney, J. G., R. Fleagle, V. Lally, H. Riehl, and D. Wark.1966. The feasibility of a global observation and analysis experiment. *Bull. Amer. Meteor. Soc.*, **47**, 200–220.

Charney, J. G., and J. Shukla. 1981. Predictability of monsoons. In *Monsoon Dynamics*, edited by J. Lighthill and R. P. Pearce. Cambridge: Cambridge University Press, 99–109.

Charney, J. G., and M. E. Stern. 1962. On the stability of internal baroclinic jets in a rotating atmosphere. *J. Atmos. Sci.*, **19**, 159–172.

Charney, J. G., and D. M. Straus. 1980. Form-drag instability, multiple equilibria and propagating planetary waves in baroclinic, orographically forced planetary wave systems. *J. Atmos. Sci.*, **37**, 1157–1176.

Chen, T.-C., and A. Wiin-Nielsen. 1978. Nonlinear cascades of atmospheric energy and enstrophy in a two-dimensional spectral index. *Tellus*, **30**, 313–322.

Cheng, M.-D., and A. Arakawa. 1990. Inclusion of convective downdrafts in the Arakawa-Schubert cumulus parameterization. Tech. Rep., Dept. of Atmospheric Sciences, UCLA, 69 pp.

Cheng, M.-D., and A. Arakawa. 1997. Inclusion of rainwater budget and convective downdrafts in the Arakawa-Schubert cumulus parameterization. *J. Atmos. Sci.*, **54**, 1359–1378.

Chervin, R. M., and L. M. Druyan. 1984. Influence of ocean surface temperature gradient and continentality on the Walker circulation. Part I: Prescribed tropical changes. *Mon. Wea. Rev.*, **112**, 1510–1523.

Chikira, M. 2014. Eastward propagating intraseasonal oscillation represented by Chikira-Sugiyama cumulus parameterization. Part II: Understanding moisture variation under weak temperature gradient balance. *J. Atmos. Sci.*, **71**, 615–639.

Colucci, S. J. 1985. Explosive cyclogenesis and large-scale circulation changes: Implications for atmospheric blocking. *J. Atmos. Sci.*, **42**, 2701–2717.

———. 1987. Comparative diagnosis of blocking versus nonblocking planetary-scale circulation changes during synoptic-scale cyclogenesis. *J. Atmos. Sci.*, **44**, 124–139.

———. 2001. Planetary-scale preconditioning for the onset of blocking. *J. Atmos. Sci.*, **58**, 933–942.

Colucci, S. J., and T. L. Alberta. 1996. Planetary-scale climatology of explosive cyclogenesis and blocking. *Mon. Wea. Rev.*, **124**, 2509–2520.

Compo, G. P., G. N. Kiladis, and P. J. Webster. 1999. The horizontal and vertical structure of east Asian winter monsoon pressure surges. *Quart. J. Roy. Meteor. Soc.*, **125**, 29–54.

Cornejo-Garrido, A. G., and P. H. Stone. 1977. On the heat balance of the Walker circulation. *J. Atmos. Sci.*, **34**, 1155–1162.

Cornes, R. C., and P. D. Jones. 2011. An examination of storm activity in the northeast Atlantic region over the 1851–2003 period using the EMULATE gridded MSLP data series. *J. Geophys. Res. Atmos.*, **116**, D16110.

Courant, R., and D. Hilbert. 1989. *Methods of Mathematical Physics*, vol.1. New York: Wiley Interscience, 560 pp.

Craig, R. A. 1965. *The Upper Atmosphere: Meteorology and Physics.* New York: Academic Press, 509.

Cripe, D. G., and D. A. Randall. 2001. Joint variations of temperature and water vapor over the midlatitude continents. *Geophys. Res. Lett.*, **28**, 2613–2626.

Croci-Maspoli, M., and H. C. Davies. 2009. Key dynamical features of the 2005/06 European winter. *Mon. Wea. Rev.*, **137**, 664–678.

Crowley, T. J., and G. R. North. 1991. *Paleoclimatology.* New York: Oxford University Press, 339 pp.

Cushman-Roisin, B. 1982. Motion of a free particle on a beta-plane. *Geophys. Astrophys. Fluid Dyn.*, **22**, 85–102.

Cushman-Roisin, B., and J.-M. Beckers. 2011. *Introduction to Geophysical Fluid Dynamics: Physical and Numerical Aspects.* New York: Academic Press, 828 pp.

Danabasoglu, G., J. C. McWilliams, and P. R. Gent. 1994. The role of mesoscale tracer transports in the global ocean circulation. *Science*, **264**, 1123–1126.

Darnell, W. L., W. F. Staylor, S. K. Gupta, and F. M. Denn. 1988. Estimation of surface insolation using sun-synchronous satellite data. *J. Climate*, **1**, 820–835.

Darnell, W. L., W. F. Staylor, S. K. Gupta, N. A. Ritchey, and A. C. Wilber. 1992. Seasonal variation of surface radiation budget derived from International Satellite Cloud Climatology Project C1 data. *J. Geophys. Res.*, **97**, 15741–15760.

Defant, A.1921. Die Zirkulation der Atmosphäre in den gemässigten Breiten der Erde: Grundzüge einer Theorie der Klimaschwankungen. *Geograf. Ann.*, **3**, 209–266.

Del Genio, A. D., and M. S. Yao. 1993. Efficient cumulus parameterization for long-term climate studies: The GISS scheme. In *The Representation of Cumulus Convection in Numerical Models. Meteor. Monogr.* no. 46, 181–184.

Del Genio, A. D., M.-S. Yao, and J. Jonas. 2007. Will moist convection be stronger in a warmer climate? *Geophys. Res. Lett.*, **34**, L16703. doi:10.1029/2007GL030525.

Del Genio, A. D., M.-S. Yao, W. Kovari, and K. K.-W. Lo. 1996. A prognostic cloud water parameterization for global climate models. *J. Climate*, **9**, 270–304.

Deland, R. 1965. Some observations of the behavior of spherical harmonic waves. *Mon. Wea. Rev.*, **93**, 307–312.

Deser, C. 1993. Diagnosis of the surface momentum balance over the tropical Pacific Ocean. *J. Climate*, **6**, 64–74.

Deser, C., M. A. Alexander, and M. S. Timlin. 1996. Upper-ocean thermal variations in the North Pacific during 1970–1991. *J. Climate*, **9**, 1840–1855.

Deser, C., J. J. Bates, and S. Wahl. 1993. The influence of sea surface temperature gradients on stratiform cloudiness along the equatorial front in the Pacific Ocean. *J. Climate*, **6**, 1172–1180.

Deser, C, and M. L. Blackmon. 1993. Surface climate variations over the North Atlantic Ocean during winter: 1900–1989. *J. Climate*, **6**, 1743–1753.

———. 1995. On the relationship between tropical and North Pacific sea surface temperature variations. *J. Climate*, **8**, 1677–1680.

Deser, C, and M. S. Timlin. 1997. Atmosphere-ocean interaction on weekly timescales in the North Atlantic and Pacific. *J. Climate*, **10**, 393–408.

Deser, C., and J. M. Wallace. 1987. El Niño events and their relation to the Southern Oscillation: 1925–1986. *J. Geophys. Res.*, **92**, 14189–14196.

———. 1990. Large-scale atmospheric circulation features of warm and cold episodes in the tropical Pacific. *J. Climate*, **3**, 1254–1281.

DeMott, C. A. and S. R. Rutledge. 1998. The vertical structure of TOGA COARE convection. Part II: Modulating influences and implications for diabatic heating. *J. Atmos. Sci.*, **55**, 2748–2762.

Dickey, J. O., S. L. Marcus, J. A. Steppe, and R. Hide. 1992. The Earth's angular momentum budget on subseasonal time scales. *Science*, **255**, 321–324.

Dickinson, R. E. 1968. Planetary Rossby waves propagating vertically through weak westerly wind wave guides. *J. Atmos. Sci.*, **25**, 984–1002.

———. 1969. Theory of planetary wave-zonal flow interaction. *J. Atmos. Sci.*, **26**, 73–81.

Dijkstra, H. A., and D. J. Neelin. 1995. Ocean-atmosphere interaction and the tropical climatology. Part II: Why the Pacific cold tongue is in the east. *J. Climate*, **8**, 1343–1359.

Dima. I. M., and J. M. Wallace. 2003. On the Seasonality of the Hadley cell. *J. Atmos. Sci.*, **60**, 1522–1526.

Ding, P., and D. A. Randall. 1998. A cumulus parameterization with multiple cloud base levels. *J. Geophys. Res.*, **103**, 11341–11354.

Dole, R. M. 1986. The life cycles of persistent anomalies and blocking over the North Pacific. *Adv. Geophys.*, **29**, 31–69.

Dole, R. M., and M. D. Gordon. 1983. Persistent anomalies of the extratropical Northern Hemisphere wintertime circulation: Geographical distribution and regional persistence characteristics. *Mon. Wea. Rev.*, **111**, 1567–1586.

Dopplick, T. G. 1971a. Global radiative heating of the Earth's atmosphere. Planetary Circulation Project, Dept. of Meteorology, MIT, Report no. 24, 128 pp.

———. 1971b. The energetics of the lower stratosphere including radiative effects. *Quart. J. Roy. Meteor. Soc.*, **97**, 209–237.

Dong, D., R. S. Gross, and J. O. Dickey. 1996. Seasonal variations of the Earth's gravitational field: An analysis of atmospheric pressure, ocean tidal, and surface water excitation. *Geophys. Res. Lett.*, **23**, 725–728.

Douglass, A. R., P. A. Newman, and S. Solomon. 2014. The Antarctic ozone hole: An update. *Physics Today*, **67**, 42–48.

Dove, H. W. 1837. *Meteorologische Untersuchunger*. Berlin: Sandersche Buchhandlung, 344 pp.

Drazin, P. G. 1992. *Nonlinear Systems*. Cambridge: Cambridge University Press, 317 pp.

Durack, P. J., and S. E. Wijffels. 2010. Fifty-year trends in global ocean salinities and their relationship to broad-scale warming. *J. Climate*, **23**, 4342–4362.

Durack, P. J., S. E. Wijffels, and R. J. Matear. 2012. Ocean salinities reveal strong global water cycle intensification during 1950 to 2000. *Science*, **336**, 455–458.

Dutton, J. A. 1976. *The Ceaseless Wind*. New York: McGraw-Hill, 579 pp.

Eastman, R., and S. G. Warren. 2012. A 39-yr survey of cloud changes from land stations worldwide 1971–2009: Long-term trends, relation to aerosols, and expansion of the tropical belt. *J. Climate*, **26**, 1286–1303.

ECMWF (European Centre for Medium Range Weather Forecasts) .1997. ERA Description. *ECMWF Re-analysis Project Report Series*,1, 4.

Edmon, H. J., B. J. Hoskins, and M. E. McIntyre. 1980. Eliassen-Palm cross sections for the troposphere. *J. Atmos. Sci.*, **37**, 2600–2616. (See also Corrigendum, *J. Atmos. Sci.*, **38**, 1115.)

Edouard, S., R. Vautard, and G. Brunet, 1997. On the maintenance of potential vorticity in isentropic coordinates. *Quart. J. Roy. Meteor. Soc.*, **123**, 2069–2094.

Egger, J., and K.-P. Hoinka. 2014. Wave forcing of zonal mean angular momentum in various coordinate systems. *J. Atmos. Sci.*, **71**, 22212229. doi:http://dx.doi.org/10.1175/JAS-D-13-0111.1.

Egger, J., W. Metz, and G. Muller. 1986. Forcing of planetary-scale blocking anticyclones by synoptic-scale eddies. *Adv. Geophys.*, **29**, 183–198.

Ekman, V. W., 1905. On the influence of the earth's rotation on ocean currents. *Ark. Mat. Astron. Fys.*, 11: 52. Eliassen, A., and E. Kleinschmidt Jr. 1957. Dynamic meteorology. In *Handbuch der Physik*, Band 48: *Geophysik II*, 1–154, Berlin: Springer Verlag.

Eliassen, E., and B. Machenhauer. 1965. A study of the fluctuations of the atmospheric planetary flow patterns represented by spherical harmonics. *Tellus*, **17**, 220–238.

Eliassen, A., and E. Palm. 1961. On the transfer of energy in stationary mountain waves. *Geofys. Publ. Oslo*, **22**, 1–23.

Eliassen, A., and E. Raustein. 1968. A numerical integration experiment with a model atmosphere based on isentropic coordinates. *Meteor. Ann.*, **5**, 45–63.

———. 1970. A numerical integration experiment with a six-level atmospheric model with isentropic information surface. *Meteor. Ann.*, **5**, 429–449.

Emanuel, K. A. 1979. Inertial instability and mesoscale convective systems. Part I: Linear theory of inertial instability. *J. Atmos. Sci.*, **36**, 2425–2449.

———. 1982. Inertial instability and mesoscale convective systems. Part II: Symmetric CISK in a baroclinic flow. *J. Atmos. Sci.*, **39**, 1080–1097.

———. 1983a. The Lagrangian parcel dynamics of moist symmetric instability. *J. Atmos. Sci.*, **40**, 2368–2376.

———. 1983b. On assessing local conditional symmetric instability from atmospheric soundings. *Mon. Wea. Rev.*, **111**, 2016–2033.

———. 1987a. An air-sea interaction model of intraseasonal oscillations in the tropics. *J. Atmos. Sci.*, **44**, 2334–2340.

———. 1987b. The dependence of hurricane intensity on climate. *Nature*, **326**, 483–485.

———. 1991. A scheme for representing cumulus convection in large-scale models. *J. Atmos. Sci.*, **48**, 2313–2335.

———. 1994. *Atmospheric Convection*. New York: Oxford University Press, 580 pp.

———. 2005. Increasing destructiveness of tropical cyclones over the past 30 years. *Nature*, **436**, 686–688.

———. 2007. Environmental factors affecting tropical cyclone power dissipation. *J. Climate*, **20**, 5497–5509.

Emanuel, K. A., J. D. Neelin, and C. S. Bretherton. 1994. On large-scale circulations in convecting atmospheres. *Quart. J. Roy. Meteor. Soc.*, **120**, 1111–1143.

Emanuel, K., and D. Raymond, eds. 1993. The representation of cumulus convection in numerical models. *Meteor. Monogr.*, **24** (46).

Emanuel, K., R. Sundararajan, and J. Williams. 2008. Hurricanes and global warming: Results from downscaling IPCC AR4 simulations. *Bull. Amer. Meteor. Soc.*, **89**, 347–367.

England, M. H., S. McGregor, P. Spence, G. A. Meehl, A. Timmermann, W. Cai, A. S. Gupta, M. J. McPhaden, A. Purich, and A. Santoso. 2014. Recent intensification of wind-driven circulation in the Pacific and the ongoing warming hiatus. *Nature Climate Change*, **4**, 222–2227. doi:10.1038/nclimate2106.

Esbensen, S. K., and Y. Kushnir. 1981. The heat budget of the global ocean: An atlas based on estimates from surface marine observations. *Climatic Research Institute Report* no. 29, Oregon State University, Corvallis, OR.

Esbensen, S. K., and M. J. McPhaden. 1996. Enhancement of tropical ocean evaporation and sensible heat flux by mesoscale system. *J. Climate*, **9**, 2307–2325.

Fedorov, A. V., P. S. Dekens, M. McCarthy, A. C. Ravelo, P. B. deMenocal, M. Barreiro, R. C. Pacanowski, and S. G. Philander. 2006. The Pliocene paradox (mechanisms for a permanent El Niño). *Science*, **312**, 1485–1489. doi:10.1126/science.1122666.

Fels, S. B. 1985. Radiative-dynamical interactions in the middle atmosphere. *Adv. Geophys.*, **28A**, 277–300.

Feynman, R. P., R. B. Leighton, and M. Sands. 1963. *The Feynman Lectures on Physics*. Vol. 1: *Mainly Mechanics, Radiation, and Heat*. Reading, MA: Addison-Wesley.

Firestone, J. K., and B. A. Albrecht. 1986. The structure of the atmospheric boundary layer in the central equatorial Pacific during January and February of FGGE. *Mon. Wea. Rev.*, **114**, 2219–2231.

Fjortoft, R. 1953. On the changes in the spectral distribution of kinetic energy for a two-dimensional non-divergent flow. *Tellus*, **5**, 225–230.

Flasar, F. M., and 39 others. 2004. An intense stratospheric jet on Jupiter. *Nature*, **427**, 132–135.

Flatau, M., P. J. Flatau, P. Phoebus, and P. P. Niiler. 1997. The feedback between equatorial convection and local radiative and evaporative processes: The implications for intraseasonal oscillations. *J. Atmos. Sci.*, **54**, 2373–2386.

Flierl, G. 1978. Models of vertical structure and the calibration of two-layer models. *Dyn. Atmos. Oceans*, **2**, 341–382.

Fouchet, T., S. Guerlet, D. F. Strobel, A. A. Simon-Miller, B. Bézard, and F. Flasar. 2008. An equatorial oscillation in Saturn's middle atmosphere. *Nature*, **453**, 200–202. doi:10.1038/nature06912.

Fovell, R., D. Durran, and J. R. Holton. 1992. Numerical simulations of convectively generated stratospheric gravity waves. *J. Atmos. Sci.*, **49**, 1427–1442.

Fowler, L. D., and D. A. Randall. 1994. A global radiative-convective feedback. *Geophys. Res. Lett.*, **21**, 2035–2038.

———. 1996. Liquid and ice cloud microphysics in the CSU general circulation model. Part 3: Sensitivity tests. *J. Climate*, **9**, 561–586.

Fowler, L. D., D. A. Randall, and S. A. Rutledge. 1996. Liquid and ice cloud microphysics in the CSU general circulation model. Part 1: Model description and results of a baseline simulation. *J. Climate*, **9**, 489–529.

Friedson, A. J. 1999. New observations and modeling of a QBO-like oscillation in Jupiter's stratosphere. *Icarus*, **137**, 331–339.

Frierson, D. M. W., J. Lu, and G. Chen. 2007. Width of the Hadley cell in simple and comprehensive general circulation models. *Geophys. Res. Lett.*, **34**, L18804. doi:10.1029/2007GL031115.

Fu, C., and J. O. Fletcher. 1985. The relationship between Tibet-tropical ocean thermal contrast and interannual variability of Indian monsoon rainfall. *J. Climate Appl. Meteor.*, **24**, 841–847.

Fu, Q., C. M. Johanson, S. G. Warren, and D. J. Seidel. 2004. Contribution of stratospheric cooling to satellite-inferred tropospheric temperature trends. *Nature*, **429**, 55–58.

Fu, Q., and P. Lin. 2011. Poleward shift of subtropical jets inferred from satellite-observed lower stratospheric temperatures. *J. Climate*, **24**, 5597–5603.

Fu, R., A. D. Del Genio, W. B. Rossow, and W. T. Liu. 1992. Cirrus-cloud thermostat for tropical sea surface temperatures tested using satellite data. *Nature*, **358**, 394–397.

Fuchs, Ž., and D. J. Raymond, 2007: A simple, vertically resolved model of tropical disturbances with a humidity closure. *Tellus*, **59**A, 344–354.

Fulton, S. R., and W. H. Schubert. 1985. Vertical normal mode transforms: Theory and application. *Mon. Wea. Rev.*, **113**, 647–658.

Fultz, D. 1949. A preliminary report on experiments with thermally produced lateral mixing in a rotation hemispherical shell of liquid. *J. Atmos. Sci.*, **6**,17–33.

———. 1951. Non-dimensional equations and modeling criteria for the atmosphere. *J. Atmos. Sci.*, **8**, 262–267.

———. 1952. On the possibility of experimental models of the polar-front wave. *J. Atmos. Sci.*, **9**, 379–384.

———. 1959. A note on overstability and the elastoid-inertia oscillations of Kelvin, Solberg, and Bjerknes. *J. Atmos. Sci.*, **16**, 199–208.

———. 1991. Quantitative nondimensional properties of the gradient wind. *J. Atmos. Sci.*, **48**, 869–875.

Fultz, D., and Frenzen, P. 1955. A note on certain interesting ageostrophic motions in a rotating hemispherical shell. *J. Atmos. Sci.*, **12**, 332–338.

Fultz, D., and Murty, T. S. 1968. Effects of the radial law of depth on the instability of inertia oscillations in rotating fluids. *J. Atmos. Sci.*, **25**, 779–788.

Ganachaud, A., and C. Wunsch. 2000. Improved estimates of global ocean circulation, heat transport and mixing from hydrological data. *Nature*, **408**, 453–457.

Garcia, R. R., and M. L. Salby. 1987. Transient response to localized episodic heating in the tropics. Part II: Far-field behavior. *J. Atmos. Sci.*, **44**, 499–530.

Gates, W. Lawrence. 1992. AMIP: The Atmospheric Model Intercomparison Project. *Bull. Amer. Meteor. Soc.*, **73**, 1962–1970. doi:http://dx.doi.org/10.1175/1520-0477(1992)073<1962:ATAMIP>2.0.CO;2.

Geisler, J. E. 1981. A linear model of the Walker circulation. *J. Atmos. Sci.*, **38**, 1390–1400.

Geisler, J. E., and E. B. Kraus. 1969. The well-mixed Ekman boundary layer. Supplement, *Deep Sea Res.*, **16**, 73–84.

Gent, P. R., and J. C. McWilliams. 1990. Isopycnal mixing in ocean circulation models. *J. Phys. Ocean.*, **20**, 150–155.

———. 1996. Eliassen-Palm fluxes and the momentum equation in non-eddy-resolving ocean circulation models. *J. Phys. Oceanogr.*, **26**, 2539–2546.

Gettelman, A., T. Birner, V. Eyring, H. Akiyoshi, S. Bekki, C. Brühl, M. Dameris, D. E. Kinnison, F. Lefevre, F. Lott, E. Mancini, G. Pitari, D. A. Plummer, E. Rozanov, K. Shibata, A. Stenke, H. Struthers, and W. Tian. 2009. The tropical tropopause layer 1960–2100. *Atmos. Chem. Phys.*, **9**, 1621–1637. doi:10.5194/acp-9-1621-2009.

Gill, A. E. 1980. Some simple solutions for heat-induced tropical circulation. *Quart. J. Roy. Meteor. Soc.*, **106**, 447–462.

———. 1982a. *Atmosphere-Ocean Dynamics*. New York: Academic Press, 662 pp.

———. 1982b. Studies of moisture effects in simple atmospheric models: The stable case. *Geophys. Astrophys. Fluid Dyn.*, **19**, 119–152.

Gleckler, P. J., D. A. Randall, G. Boer, R. Colman, M. Dix, V. Galin, M. Helfand, J. Kiehl, A. Kitoh, W. Lau, X.-Z. Liang, V. Lykossov, B. McAvaney, K. Miyakoda, S. Planton, and W. Stern. 1995. Interpretation of ocean energy transports implied by atmospheric general circulation models. *Geophys. Res. Lett.*, **22**, 791–794.

Gleick, J. 1987. Chaos: *Making a New Science.* New York: Viking Press, 352 pp.

Goody, R. M., and Y. L. Yung. 1989. *Atmospheric Radiation Theoretical Basis*, 2nd ed. New York: Oxford University Press, 519 pp.

Grabowski, W., and M. W. Moncrieff, 2004. Moisture-convection feedback in the tropics. *Quart. J. Roy. Meteor. Soc.*, **130**, 3081–3104.

Graham, N. E., and W. B. White. 1990. The role of the western boundary in the ENSO cycle: Experiments with coupled models. *J. Phys. Oceanogr.*, **20**, 1935–1948.

Grassl, H. 1990. The climate at maximum entropy production by meridional and atmospheric heat fluxes. *Quart. J. Roy. Meteor. Soc.*, **107**, 153–166.

Gray, W. M. 1979. Hurricanes: Their formation, structure and likely role in the tropical circulation. In *Meteorology over the Tropical Oceans*, edited by D. B. Shaw. Bracknell, Berkshire: Royal Meteorology Society, 156–218.

Green, J.S.A. 1970. Transfer properties of the large-scale eddies and the general circulation of the atmosphere. *Quart. J. Roy. Meteor. Soc.*, **96**, 157–185.

———. 1977. The weather during July 1976: Some dynamical considerations of the drought. *Weather*, **32**, 120–126.

Gregory, D., and P. R. Rowntree. 1990. A mass flux convection scheme with representation of cloud ensemble characteristics and stability-dependent closure. *Mon. Wea. Rev.*, **118**, 1483–1506.

Grell, G. A., Y.-H. Kuo, and R. J. Pasch. 1991. Semiprognostic tests of cumulus parameterization schemes in the middle latitudes. *Mon. Wea. Rev.*, **119**, 5–31.

Grose, W. L., and B. J. Hoskins. 1979. On the influence of orography on large-scale atmospheric flow. *J. Atmos. Sci.*, **36**, 223–234.

Grotjahn , R. 1993. *Global Atmospheric Circulations.* New York: Oxford University Press, 430 pp.

Gu, D., and S. G. H. Philander. 1997. Interdecadal climate fluctuations that depend on exchanges between the tropics and extratropics. *Science*, **275**, 805–807.

Gupta, S. K. 1989. A parameterization of longwave surface radiation from sun-synchronous satellite data. *J. Climate*, **2**,305–320.

Gupta, S. K., W. L. Darnell, and A. C. Wilber. 1992. A parameterization for longwave surface radiation from satellite data: Recent improvements. *J. Appl. Meteor.*, **31**, 1361–1367.

Gupta, S. K., W. F. Staylor, W. L. Darnell, A. C. Wilber, and N. A. Ritchey. 1993. Seasonal variation of surface and atmospheric cloud radiative forcing over the globe derived from satellite data. *J. Geophys. Res.*, **98**, 20,761–20,778.

Hack, J. J. 1994. Parameterization of moist convection in the National Center for Atmospheric Research Community Climate Model (CCM2). *J. Geophys. Res.*, **99**, 5551–5568.

Hack, J. J., W. H. Schubert, and P. L. Silva Dias. 1984. A spectral cumulus parameterization for use in numerical models of the tropical atmosphere. *Mon. Wea. Rev.*, **112**, 704–716.

Hack, J. J., W. H. Schubert, D. E. Stevens, and H.-C. Kuo. 1989. Response of the Hadley circulation to convective forcing in the ITCZ. *J. Atmos. Sci.*, **46**, 2957–2913.

Haertel, P. T., and R. H. Johnson. 1998. Two-day disturbances in the equatorial western Pacific. *Quart. J. Roy. Meteor. Soc.*, **124**, 615–636.

Han, Y.-J., and S.-W. Lee. 1983. An analysis of monthly mean wind stress over the global ocean. *Mon. Wea. Rev.*, **111**, 1554–1566.

Hansen, A. R., and T.-C. Chen. 1982. A spectral energetics analysis of atmospheric blocking, *Mon. Wea. Rev.*, **110**, 1146–1165.

Haraguchi, P. Y. 1968. Inversions over the tropical eastern Pacific ocean. *Mon. Wea. Rev.*, **96**, 177–185.

Harman, P. M. 1998. *The Natural Philosophy of James Clerk Maxwell*. Cambridge: Cambridge University Press, 2323 pp.

Harshvardhan, D., A. Randall, T. G. Corsetti, and D. A. Dazlich. 1989. Earth radiation budget and cloudiness simulations with a general circulation model. *J. Atmos. Sci.*, **46**, 1922–1942.

Hart, J. E. 1979. Barotropic quasi-geostrophic flow over anisotropic mountains. *J. Atmos. Sci.*, **36**, 1736–1746.

Hartmann, D. L. 1994. *Global Physical Climatology*. New York: Academic Press, 411 pp.

Hartmann, D. L., and J. R. Gross. 1988. Seasonal variability of the 40–50 day oscillation in wind and rainfall in the tropics. *J. Atmos. Sci.*, **45**, 2680–2702.

Hartmann, D. L., and F. Lo. 1998. Wave-driven zonal flow vacillation in the Southern Hemisphere. *J. Atmos. Sci.*, **55**, 1303–1315.

Hartmann, D. L., and M. L. Michelsen. 1989. Intraseasonal periodicities in Indian rainfall. *J. Atmos. Sci.*, **46**, 2838–2862.

———. 1993. Large-scale effects on the regulation of tropical sea surface temperature. *J. Climate*, **6**, 2049–2062.

Hartmann, D. L., M. E. Ockert-Bell, and M. L. Michelson. 1992. The effect of cloud type on the Earth's energy balance: Global analysis. *J. Climate*, **5**, 1281–1304.

Hastenrath, S. 1971. On meridional circulation and heat budget of the troposphere over the equatorial central Pacific. *Tellus*, **23**, 60–73.

———. 1998. Contribution to the circulation climatology of the eastern equatorial Pacific: Lower atmospheric jets. *J. Geophys. Res.*, **D16**, 19433–19451.

Haurwitz, B., 1937. The oscillations of the atmosphere. *Gerlund's Beitr. zur Geophys.*, **51**, 195–233.

———. 1956. The geographical distribution of the solar semidiurnal pressure oscillation. *Meteor. Papers*, New York Univ. College of Engineering, **2** (5).

Haurwitz, B., and S. Chapman. 1967. Lunar air tide. *Nature*, **213**, 9–13.

Hay, G. E. 1953. *Vector and Tensor Analysis*. New York: Dover, 193 pp.

Hayashi, Y. 1970. A theory of large-scale equatorial waves generated by condensation heat and accelerating the zonal wind. *J. Meteor. Soc. Japan*, **48**, 140–160.

Hayashi, Y., and S. Miyahara. 1987. A three-dimensional linear response model of the tropical intraseasonal oscillation. *J. Met. Soc. Japan*, **65**, 843–852.

Hayashi, Y.-Y., and A. Sumi.1986. 30–40-day oscillations simulated in an "aqua planet" model. *J. Meteor. Soc. Japan*, **64**, 451–467.

Haynes, P. H., and M. E. McIntyre. 1987. On the evolution of potential vorticity in the presence of diabatic heating and frictional or other forces. *J. Atmos. Sci.*, **44**, 828–841.

Haynes, P. H., M. E. McIntyre, T. G. Shepherd, C. J. Marks, and K. P. Shine. 1991. On the "downward control" of extratropical diabatic circulations by eddy-induced mean zonal forces. *J. Atmos. Sci.*, **48**, 651–678. doi:http://dx.doi.org/10.1175/1520-0469(1991)048<0651:OTCOED>2.0.CO;2.

Heckley, W. A. 1985. Systematic errors of the ECMWF operational forecasting model in tropical regions. *Quart. J. Roy. Meteor. Soc.*, **111**, 709–738.

Held, I. 1999. Equatorial superrotation in Earth-like atmospheric models. Bernhard Haurwitz Memorial Lecture, AMS.

———. 2000. *The General Circulation of the Atmosphere*. 2000 Program in Geophysical Fluid Dynamics, Woods Hole Oceanographic Institute, Woods Hole, MA. (Available at http://gfd.whoi.edu/proceedings/2000/PDFvol2000.html).

Held, I., and M. Suarez. 1978. A two-level primitive equation atmospheric model designed for climatic sensitivity experiments. *J. Atmos. Sci.*, **35**, 206–229.

Held, I. M. 1983. Stationary and quasi-stationary eddies in the extratropical troposphere: Theory. In *Large-Scale Dynamical Processes in the Atmosphere*, edited by B. J. Hoskins and R. P. Pearce. New York: Academic Press, 397 pp.

Held, I. M., and B. J. Hoskins. 1985. Large-scale eddies and the general circulation of the atmosphere. *Adv. Geophys.*, **28A**, 3–31.

Held, I. M., and A. Y. Hou. 1980. Nonlinear axially symmetric circulations in a nearly inviscid atmosphere. *J. Atmos. Sci.*, **37**, 515–533.

Held, I. M., and B. J. Soden. 2000. Water vapor feedback and global warming. *Ann. Rev. Energy Environ.*, **25**, 441–475.

———. 2006. Robust responses of the hydrological cycle to global warming. *J. Climate*, **19**, 5686–5699. doi:http://dx.doi.org/10.1175/JCLI3990.1.

Held, I. M., M. Ting, and H. Wang. 2002. Northern winter stationary waves: Theory and modeling. *J. Climate*, **15**, 2125–2144

Hendon, H. H., and J. Glick. 1997. Intraseasonal air-sea interaction in the tropical Indian and Pacific Oceans. *J. Climate*, **10**, 647–661.

Hendon, H. H., and B. Liebmann. 1990. The intraseasonal (30–50 day) oscillation of the Australian summer monsoon. J. *Atmos. Sci.*, **47**, 2909–2923.

Hendon, H. H., and M. L. Salby. 1994. The life cycle of the Madden-Julian oscillation. *J. Atmos. Sci.*, **51**, 2225–2237.

Hess, S. L., R. M. Henry, C. B. Leovy, J. A. Ryan, and J. E. Tillman. 1977. Meteorological results from the surface of Mars: Viking 1 and 2. *J. Geophys. Res.*, **82**, 4559–4574.

Hicks, B. B. 1978. Some limitations of dimensional analysis and power laws. *Bound. Layer Meteor.*, **14**, 567–569.

Hide, R. 1969. Dynamics of the atmospheres of the major planets with an appendix on the viscous boundary layer at the rigid boundary surface of an electrically conducting rotating fluid in the presence of a magnetic field. *J. Atmos. Sci.*, **26**, 841–853.

Hide, R, and J. O. Dickey. 1991. Earth's variable rotation. *Science*, **253**, 629–637.

Highwood, E. J., and B. J. Hoskins. 1998. The tropical tropopause. *Quart. J. Roy. Meteor. Soc.*, **124**, 1579.

Highwood, E. J., B. J. Hoskins, and P. Berrisford. 2000. Properties of the Arctic tropopause. *Quart. J. Roy. Meteor. Soc.*, **126**, 1515.

Hirst, A. C. 1986. Unstable and damped equatorial modes in simple coupled ocean-atmosphere models. *J. Atmos. Sci.*, **43**, 606–630.

Holloway, C. E., and J. D. Neelin. 2009. Moisture vertical structure, column water vapor, and tropical deep convection. *J. Atmos. Sci.*, **66**, 1665–1683. doi:http://dx.doi.org/10.1175/2008JAS2806.1.

Holloway, C. E., S. J. Woolnough, and G.M.S. Lister. 2013. The effects of explicit versus parameterized convection on the MJO in a large-domain high-resolution tropical case study. Part I: Characterization of large-scale organization and propagation. *J. Atmos. Sci.*, **70**, 1342–1369. doi:http://dx.doi.org/10.1175/JAS-D-12-0227.1.

Holton, J. R. 1972. Waves in the equatorial stratosphere generated by tropical heat sources. *J. Atmos. Sci.*, **29**, 368–375.

———. 1992. *An Introduction to Dynamic Meteorology*, 3rd ed. International Geophysics Series, vol. 48. San Diego: Academic Press, 511 pp.

———. 1975. The dynamic meteorology of the stratosphere and mesosphere. *Meteor. Monogr.*, **15** (37), 1–218.

Holton, J. R. 1980. The dynamics of sudden stratospheric warmings. *Ann Rev. Earth Planet. Sci.*, **8**, 169–190.

———. 2004. *An Introduction to Dynamic Meteorology*, 4th ed. New York: Elsevier Academic Press, 535 pp.

Holton, J. R., and R. S. Lindzen. 1972. An updated theory for the quasi-biennial cycle of the tropical stratosphere. *J. Atmos. Sci.*, **29**, 1076–1080.

Holton, J. R., and T. Matsuno. 1984. *Dynamics of the Middle Atmosphere*. Advances in Earth and Planetary Sciences. Tokyo: Terra Scientific, 543 pp.

Hoskins, B. J. 1991. Towards a PV-theta view of the general circulation. *Tellus*, **43**, 27–35.

———. 1996. On the existence and strength of the summer subtropical anticyclones. *Bull. Amer. Meteor. Soc.*, **77**, 1287–1292.

———. 2003. Climate change at cruising altitude? *Science*, **301**, 469–470.

———. 2012. The potential for skill across the range of the seamless weather-climate prediction problem: A stimulus for our science. *Quart. J. Roy. Meteor. Soc.*, **139**, 573–584.

Hoskins, B. J., I. N. James, and G. H. White. 1983. The shape, propagation and mean-flow interaction of large-scale weather systems. *J. Atmos. Sci.*, **40**, 1595–1612.

Hoskins, B. J., and D. J. Karoly. 1981. The steady linear response of a spherical atmosphere to thermal and orographic forcing. *J. Atmos. Sci.*, **38**, 1179–1196.

Hoskins, B. J., M. E. McIntyre, and A. W. Robertson. 1985. On the use and significance of isentropic potential vorticity maps. *Quart. J. Roy. Meteor. Soc.*, **111**, 877–946.

Hoskins, B. J., and R. P. Pearce, eds. 1983. *Large-Scale Dynamical Processes in the Atmosphere*. New York: Academic Press, 397 pp.

Hou, A. Y., R. K. Kakar, S. Neeck, A. A. Azarbarzin, C. D. Kummerow, M. Kojima, R Oki, K. Nakamura, and T. Iguchi. 2014. The global precipitation measurement mission. Bull *Amer. Meteor. Soc.*, **95**, 701–722. doi:http://dx.doi.org/10.1175/BAMS-D-13-00164.1.

Hough, S. S. 1898. On the application of harmonic analysis to the dynamical theory of the tides. Part II: On the general integration of Laplace's dynamical equations. *Phil. Trans. Roy. Soc. A*, **191**, 139–185.

Hsu, Y.-J., and A. Arakawa. 1990. Numerical modeling of the atmosphere with an isentropic vertical coordinate. *Mon. Wea. Rev.*, **118**, 1933–1959.

Hsu, H.-H., and B. J. Hoskins. 1989. Tidal fluctuations as seen in ECMWF data. *Quart. J. Roy. Meteor. Soc.*, **115**, 247–264.

Hsu, H.-H., B. J. Hoskins, and F.-F. Jin. 1990. The 1985/86 intraseasonal oscillation and the role of the extratropics. *J. Atmos. Sci.*, **47**, 823–839.

Hsu, P.-C., and T. Li. 2012. Is "rich-get-richer" valid for Indian Ocean and Atlantic ITCZ? *Geophys. Res. Lett.*, **39**, L13705. doi:10.1029/2012GL052399.

Hsu, P.-C., T. Li, H. Murakami, and A. Kitoh. 2013. Future change of the global monsoon revealed from 19 CMIP5 models. *J. Geophys. Res. Atmos.*, **118**, 1247–1260. doi:10.1002/jgrd.50145.

Hu, Q., and D. A. Randall. 1994. Low-frequency oscillations in radiative-convective systems. *J. Atmos. Sci.*, **51**, 1089–1099.

Hu, Y., and Q. Fu. 2007. Observed poleward expansion of the Hadley circulation since 1979. *Atmos. Chem. Phys.*, **7**, 5229–5236.

Hu, Y., D. Yang, and J. Yang. 2008. Blocking systems over an aqua planet. *Geophys. Res. Lett.*, **35**, L19818.

Huang, H.-P., P. D. Sardeshmuk, and K. M. Weickmann. 1999. The balance of global angular momentum in a long-term atmospheric dataset. *J. Geophys. Res.*, **104**, 2031–2040.

Huang, H.-P., K. Weickmann, and C. Hsu. 2001. Trend in atmospheric angular momentum in a transient climate change simulation with greenhouse gas and aerosol forcing. *J. Climate*, **14**, 1525–1534.

Hung, M.-P., J.-L. Lin, W. Wang, Daehyun Kim, T. Shinoda, and S. J. Weaver. 2013. MJO and convectively coupled equatorial waves simulated by CMIP5 climate models. *J. Climate*, **26**, 6185–6214. doi:http://dx.doi.org/10.1175/JCLI-D-12-00541.1.

Huntley, H. E. 1967. *Dimensional Analysis*. New York: Dover, 158 pp.

Hurrell, J. W. 1995. Decadal trends in the North Atlantic Oscillation regional temperatures and precipitation. *Science*, **269**, 676–679.

Hurrell, J. W., and H. Van Loon. 1997. Decadal variations in climate associated with the North Atlantic Oscillation. *Climatic Change*, **36**, 301–326.

Huang, H.-P., and W. A. Robinson. 1998. Two-dimensional turbulence and persistent zonal jets in a global barotropic model. *J. Atmos. Sci.*, **55**, 611–632.

Huffman, G. J., R. F. Adler, P. Arkin, A. Chang, R. Ferraro, A. Gruber, J. Janowiak, A. McNab, B. Rudolf, and U. Schneider. 1997. The Global Precipitation Climatology Project (GPCP) combined precipitation dataset. *Bull. Amer. Meteor. Soc.*, **78**, 5–20.

Illari, L., 1982. Diagnostic study of the potential vorticity in a warm blocking anticyclone. *J. Atmos. Sci.*, **41**, 3518–3526.

Imbrie, J., and K. P. Imbrie. 1979. *Ice Ages: Solving the Mystery*. Cambridge, MA: Harvard University Press, 224 pp.

Ingersoll, A. P. 1990. Atmospheric dynamics of the outer planets. *Science*, **248**, 308–315.

Inness, P. M., J. M. Slingo, S. J. Woolnough, and R. B. Neale. 2001. Organization of tropical convection in a GCM with varying vertical resolution: Implications for the simulation of the Madden-Julian oscillation. *Climate Dynamics*, **17**, 777–793.

IPCC (Intergovernmental Panel on Climate Change). 2013. *Climate Change 2013: The Physical Science Basis. Contribution of Working Group I to the Fifth Assessment Report of the Intergovernmental Panel on Climate Change*, edited by T. F. Stocker, D. Qin, G.-K. Plattner, M. Tignor, S. K. Allen, J. Boschung, A. Nauels, Y. Xia, V. Bex, and P. M. Midgley. Cambridge: Cambridge University Press, 1535 pp.

Jacchia, L. G., and Z. Kopal. 1951. Atmospheric oscillations and the temperature of the upper atmosphere. *J. Meteorol.*, **9**, 13–23.

Jackson, D. L., and G. L. Stephens. 1995. A study of SSM/I derived precipitable water over the global oceans. *J. Climate*, **8**, 2025–2038.

James, I. N. 1994. *Introduction to Circulating Atmospheres*. Cambridge: Cambridge University Press, 422 pp.

Jarraud, M., and A. J. Simmons. 1983. The spectral technique. *Seminar on Numerical Methods for Weather Prediction*. European Centre for Medium Range Weather Prediction, Reading, England, 1–59.

Jeevanjee, N. 2011. An Introduction to Tensors and Group Theory for Physicists. New York: Birkhäuser, 258 pp.

Jeffreys, H. 1926. On the dynamics of geostrophic winds. *Quart. J. Roy. Met. Soc.*, **52**, 85–104

Jin, F.-F., J. D. Neelin, and M. Ghil. 1994. El Niño on the devil's staircase: Annual subharmonic steps to chaos. *Science*, **264**, 70–72.

Jin, F.-F., L.-L. Pan, and M. Watanabe. 2006a. Dynamics of synoptic eddy and low-frequency flow interaction. Part I: A linear closure. *J. Atmos. Sci.*, **63**, 1677–1694.

———. 2006b. Dynamics of synoptic eddy and low-frequency flow interaction. Part II: A theory for low-frequency modes. *J. Atmos. Sci.*, **63**, 1695–1708.

Jin, S. G., J. U. Park, J. H. Cho, and P. H. Park. 2007. Seasonal variability of GPS-derived zenith tropospheric delay (1994–2006) and climate implications. *J. Geophys. Res. Atmos.*, **112**, D09110.

Johnson, R. H. 1976. The role of convective-scale precipitation downdrafts in cumulus and synoptic-scale interaction. *J. Atmos. Sci.*, **33**, 1890–1910.

———. 2011. Diurnal cycle of monsoon convection. Chap. 15 of Chang et al. (2011).

Johnson, R. H., and R. A. Houze Jr., 1987. Precipitating cloud systems of the Asian monsoon. In Monsoon Meteorology, edited by C.-P. Chang and T. N. Krishnamurti. New York: Oxford University Press, 298–353.

Johnson, R. H., T. M. Rickenbach, S. A. Rutledge, P. E. Ciesielski, and W. H. Schubert. 1999. Trimodal characteristics of tropical convection. *J. Climate*, **12**, 2397–2418.

Jones, T. R., and D. A. Randall. 2011. Quantifying the limits of convective parameterizations. *J. Geophys. Res.*, **116**, D08210. doi:10.1029/2010JD014913.

Julian, P. R., and R. M. Chervin. 1978. Study of the Southern Oscillation and Walker circulation phenomenon. *Mon. Wea. Rev.*, **106**, 1433–1451.

Kållberg, P., P. Berrisford, B. Hoskins, A. Simmons, S. Uppala, S. Lamy-Thépaut, and R. Hine. 2005. ERA-40 Atlas. ERA-40 Project Report Series, no. 19, European Centre for Medium Range Weather Forecasts.

Kang, S. M., C. Deser, and L. M. Polvani. 2013. Uncertainty in climate change projections of the Hadley circulation: The role of internal variability. *J. Climate*, **26**, 7541–7554, doi:10.1175/JCLI-D-12-00788.1.

Kao, C.-Y. J., and Y. Ogura. 1987. Response of cumulus clouds to large-scale forcing using the Arakawa-Schubert cumulus parameterization. J. *Atmos. Sci.*, **44**, 2437–2458.

Karl, T. R., S. J. Hassol, C. D. Miller, and W. L. Murray, eds. 2006. *Temperature Trends in the Lower Atmosphere: Steps for Understanding and Reconciling Differences*. U.S. Climate Change Science Program and the Subcommittee on Global Change Research. Washington, DC: UNT Digital Library. http://digital.library.unt.edu/ark:/67531/metadc12017/.

Kasahara, A., 1974. Various vertical coordinate systems used for numerical weather prediction. *Mon. Wea. Rev.*, **102**, 509–522.

Kato, S. 1966. Diurnal atmospheric oscillation: 1. Eigenvalues and Hough functions. *J. Geophys. Res.*, 71, 3201–3209.

Kelly, M. A. 1998. A simple model of ocean-atmosphere interactions in the tropical climate system. Ph.D. thesis, Colorado State University.

Kelly, M. A., and D. A. Randall. 2001. A two-box model of a zonal atmospheric circulation in the tropics. *J. Climate*, **14**, 3944–3964.

Kelly, M. A., D. A. Randall, and G. L. Stephens. 1999. A simple radiative-convective model with a hydrologic cycle and interactive clouds. *Quart. J. Roy. Meteor. Soc.*, **125**, 837–869.

Kemball-Cook, S. R. and B. C. Weare. 2001. The onset of convection in the Madden-Julian oscillation. *J. Climate*, **5**, 780–793.

Khairoutdinov, M., and K. Emanuel. 2013. Rotating radiative-convective equilibrium simulated by a cloud-resolving model. *J. Adv. Model. Earth Syst.*, **5**, 816–825. doi:10.1002/2013MS000253.

Kiehl, J. T. 1994. On the observed near cancellation between longwave and shortwave cloud forcing in tropical regions. *J. Climate*, **7**, 559–565.

Kiehl, J. T., and K. E. Trenbert. 1997. Earth's annual global mean energy budget. *Bull. Amer. Meteor. Soc.*, **78**, 197–208.

Kiladis, G. N., M. C. Wheeler, P. T. Haertel, K. H. Straub, and P. E. Roundy. 2009. Convectively coupled equatorial waves. *Rev. Geophys.*, **47**, RG2003.

Killworth, P. D., 1983. Deep convection in the world oceans. *Rev. Geophys. Space Phys.*, **21**, 1–26.

Kim, D., K. Sperber, W. Stern, D. Waliser, I.-S. Kang, E. Maloney, W. Wang, K. Weickmann, J. Benedict, M. Khairoutdinov, M.-I. Lee, R. Neale, M. Suarez, K. Thayer-Calder, and G. Zhang. 2009. Application of MJO simulation diagnostics to climate models. *J. Climate*, **22**, 6413–6436.

Kim, D., P. Xavier, E. Maloney, M. Wheeler, D. Waliser, K. Sperber, H. Hendon, C. Zhang, R. Neale, Y.-T. Hwang, and H. Liu. 2014. Process-oriented MJO simulation diagnostic:

Moisture sensitivity of simulated convection. *J. Climate*, **27**, 5379–5395. doi:http://dx.doi.org/10.1175/JCLI-D-13-00497.1.

Kitoh, A. 2004. Effects of mountain uplift on East Asian summer climate investigated by a coupled atmosphere-ocean GCM. *J. Climate*, **17**, 783–802.

Kitoh, A., H. Endo, K. Krishna Kumar, I.F.A. Cavalcanti, P. Goswami, and T. Zhou. 2013. Monsoons in a changing world regional perspective in a global context. *J. Geophys. Res. Atmos.*, **118**, 3053–3065. doi:10.1002/jgrd.50258.

Klein, S. A., and D. L. Hartmann. 1993. The seasonal cycle of low stratiform clouds. *J. Climate*, **6**, 1587–1606.

Klemp, J. B., and D. K. Lilly. 1978. Numerical simulation of hydrostatic mountain waves. *J. Atmos. Sci.*, **35**, 78–107.

Knutson, T. R., and K. N. Weickmann. 1987. 30–60 day atmospheric oscillations: Composite life cycles of convection and circulation anomalies. *Mon. Wea. Rev.*, **115**, 1407–1436.

Koh, T. Y., and R. A. Plumb. 2004. Isentropic zonal average formalism and the near-surface circulation. *Quart. J. Roy. Meteor. Soc.*, **130**, 1631–1654.

Kolmogorov, A. N. 1941. The local structure of turbulence in incompressible viscous fluid for very large Reynolds numbers. *Dok. Akad. Mauk SSSR*, **30**, 301–305.

Konor, C. S., and A. Arakawa. 1997. Design of an atmospheric model based on a generalized vertical coordinate. *Mon. Wea. Rev.*, **125**, 1649–1673.

Konrad, C. E., II, and S. J. Colucci. 1988. Synoptic climatology of 500 mb circulation changes during explosive cyclogenesis. *Mon. Wea. Rev.*, **116**, 1431–1443.

Kraus, E. B., and L. D. Leslie. 1982. The interactive evolution of the oceanic and atmospheric boundary layers in the source regions of the trades. *J. Atmos. Sci.*, **39**, 2760–2772.

Kraus, E. B., and J. S. Turner. 1967. A one-dimensional model of the seasonal thermocline. II: The general theory and its consequences. *Tellus*, **19**, 98–105.

Krishnamurti, T. N., H. S. Bedi, and M Subramaniam. 1989. The summer monsoon of 1987. *J. Climate*, **2**, 321–340.

———. 1990. The summer monsoon of 1988. *Meteorol. Atmos. Phys.*, **42**, 19–37.

Krishnamurti, T. N., and D. Subrahmanyam. 1982. The 30–50 day mode at 850 mb during MONEX. *J. Atmos. Sci.*, **39**, 2088–2095

Krueger, A. J., and R. A. Minzner. 1976. A midlatitude ozone model for the 1976 U.S. standard atmosphere. *J. Geophys. Res.*, **81**, 4477–4481. doi:10.1029/JC081i024p04477.

Krueger, S. K. 1988. Numerical simulation of tropical cumulus clouds and their interaction with the subcloud layer. *J. Atmos. Sci.*, **45**, 2221–2250.

Kuang, Z. 2008. A moisture-stratiform instability for convectively coupled waves. *J. Atmos. Sci.*, **65**, 834–854.

Kunz, A., P. Konopka, R. Müller, and L. L. Pan. 2011. Dynamical tropopause based on isentropic potential vorticity gradients. *J. Geophys. Res.*, **116**, D01110. doi:10.1029/2010JD014343.

Kuo, H. L., 1965. On formation and intensification of tropical cyclones throughout latent heat release by cumulus convection. *J. Atmos. Sci.*, **22**, 40–63.

l'Heureux, M. L., S. Lee, and B. Lyon. 2013. Recent multidecadal strengthening of the Walker circulation across the tropical Pacific. *Nature Climate Change*, **3**, 571–576. doi:10.1038/nclimate1840.

Landu, K., and E. D. Maloney. 2011. Understanding intraseasonal variability in an aquaplanet GCM. *J. Meteor. Soc. Japan*, **89**, 195–210. doi:10.2151/jmsj.2011-302.

Laplace, P. S. 1832. *Mécanique Céleste*.Translated by N. Bowditch. Boston, 4 vols.; pt. 1, bk.4, sec. 3, p. 543.

Larson, K., D. L. Hartmann, and S. A. Klein. 1999. Climate sensitivity in a two-box model of the tropics. *J. Climate*, **12**, 2359–2374.

Laskar, J., 1994. Large-scale chaos in the Solar System. *Astron. Astrophys.*, **287**, L9–L12.

Laskar, J., and M. Gastineau. 2009. Existence of collisional trajectories of Mercury, Mars, and Venus with the Earth. *Nature*, **459**, 817–819.

Latif, M., and T. P. Barnett, 1994. Causes of decadal climate variability over the North Pacific and North America. *Science*, **266**, 634–637.

———. 1996. Decadal climate variability over the North Pacific and North America: Dynamics and predictability. *J. Climate*, **9**, 2407–2423.

Latif, M., R. Kleeman, R., and C. Eckert. 1997. Greenhouse warming, decadal variability, or El Niño? An attempt to understand the anomalous 1990s. *J. Climate*, **10**, 2221–2239.

Lau, K. M., and P. H. Chan. 1985. Aspects of the 40–50 day oscillation during northern winter as inferred from outgoing long wave radiation. *Mon. Wea. Rev.*, **113**, 1889–1909.

Lau, K.-M., and H. Lim, 1982. Thermally driven motions in an equatorial b-plane: Hadley and Walker circulations during the winter monsoon. *Mon. Wea. Rev.*, **110**, 336–353.

Lau, K. M., and L. Peng, 1987. Origin of the low-frequency (intraseasonal) oscillations in the tropical atmosphere. Part I: Basic theory. *J. Atmos. Sci.*, **44**, 950–972.

Lau, K.-M., and C.-H. Sui. 1997. Mechanisms of short-term sea surface temperature regulation: Observations during TOGA COARE. *J. Climate*, **10**, 465–472.

Lau, K.-M., C.-H. Sui, M.-D. Chou, and W.-K. Tao. 1994. An inquiry into the cirrus cloud thermostat effect for tropical sea surface temperatures. *Geophys. Res. Lett.*, **21**, 1157–1160.

Lau, K.-M., H.-T. Wu, and Bony, S. 1997. The role of large-scale circulation in the relationship between tropical convection and sea surface temperature. *J. Climate*, **10**, 381–392.

Lau, N-C. 1985. Modeling the seasonal dependence of the atmospheric response to observed El Niños in 1962–76. *Mon. Wea. Rev.*, **113**, 1970–1996.

———. 1997. Interactions between global SST anomalies and the midlatitude atmospheric circulation. *Bull. Amer. Meteor. Soc.*, **78**, 21–33.

Lau, N.-C., and M. J. Nath. 1991. Variability of the baroclinic and barotropic transient eddy forcing associated with monthly changes in the midlatitude storm tracks. *J. Atmos. Sci.*, **48**, 2589–2613.

Lavin, A., H. L. Bryden, and G. Parrilla. 1998. Meridional transport and heat flux variations in the subtropical North Atlantic. *Global Atmos. Ocean Syst.*, **6**, 269–293.

Legates, D. R. and C. J. Willmott. 1990. Mean seasonal and spatial variability in gauge-corrected global precipitation. *Int. J. Climatol.*, **10**, 111–127.

Leith, C. E. 1968. Diffusion approximation for two-dimensional turbulence. *Phys. Fluids*, **11**, 671–673.

Leovy, C. B., A. J. Friedson, and A. J. Orton. 1991. The quasi-quadrennial oscillation of Jupiter's equatorial stratosphere. *Nature*, **354**, 380–382.

Lesieur, M. 1995. *Turbulence in Fluids*. Dordrecht: Kluwer, 515 pp.

Levitus, S., J. I. Antonov, T. P. Boyer, O. K. Baranova, H. E. Garcia, R. A. Locarnini, A. V. Mishonov, J. R. Reagan, D. Seidov, E. S. Yarosh, and M. M. Zweng. 2012. World ocean heat content and thermosteric sea level change (0–2000 m), 1955–2010. *Geophys. Res. Lett.*, **39**, L10603, doi:10.1029/ 2012GL051106.

Levitus, S., J. I. Antonov, T. P. Boyer, R. A. Locarnini, H. E. Garcia, and A. V. Mishonov. 2009. Global ocean heat content 1955–2008 in light of recently revealed instrumentation problems. *Geophys. Res. Lett.*, **36**, L07608.

Levitus, S., J. Antonov, T. P. Boyer, and C. Stephens. 2000. Warming of the world ocean, *Science*, **287**, 2225-2229.

Lewis, J. S., and R. G. Prinn. 1984. *Planets and Their Atmospheres*. New York: Academic Press, 470 pp.

Li, T., and S. G. H. Philander. 1996. On the annual cycle of the eastern equatorial Pacific. *J. Climate*, **9**, 2986–2998.

———. 1997. On the seasonal cycle of the equatorial Atlantic Ocean. *J. Climate*, **10**, 813–817.

Lilly, D. K. 1968. Models of cloud-topped mixed layers under a strong inversion. *Quart. J. Roy. Meteor. Soc.*, **94**, 292–309.

———. 1972. Numerical simulation studies of two-dimensional turbulence. *Geophys. Fluid Mech.*, **3**, 289–319; **4**, 1–28.

———. 1983. Stratified turbulence and the mesoscale variability of the atmosphere. J. *Atmos. Sci.*, **40**, 749–761.

———. 1998. Stratified turbulence in the atmospheric mesoscales. *Theoret. Comput. Fluid Dyn.*, **11**, 139–153.

Lilly, D. K., and B. F. Jewett. 1990. Momentum and kinetic energy budgets of simulated supercell thunderstorms. *J. Atmos. Sci.*, **47**, 707–726.

Lim, H., and C. P. Chang. 1983. Dynamics of teleconnections and Walker circulations forced by equatorial heating. *J. Atmos. Sci.*, **40**, 1897–1915.

Limpasuvan, V., and D. L. Hartmann. 2000. Wave-maintained annular modes of climate Variability. *J. Climate*, **13**, 4414–4429. doi:http://dx.doi.org/10.1175/1520-0442(2000)013<4414:WMAMOC>2.0.CO;2.

Limpasuvan, V., D.W.J. Thompson, and D. L. Hartmann. 2004. The life cycle of the Northern Hemisphere sudden stratospheric warmings. *J. Climate*, **17**, 2584–2596.

Lin, C., and A. Arakawa. 1997. The macroscopic entrainment processes of simulated cumulus ensemble. Part I: Entrainment sources. *J. Atmos. Sci.*, **54**, 1027–1043.

Lin, X. and R. H. Johnson. 1996. Kinematic and thermodynamic characteristics of the flow over the western Pacific warm pool during TOGA COARE. *J. Atmos. Sci.*, **53**, 695–715.

Lindzen, R. S. 1966. On the theory of the diurnal tide. Mon. Wea. Rev., **94**, 295–301.

———. 1967. Planetary waves on beta planes. *Mon. Wea. Rev.*, **95**, 441–451.

Lindzen, R. S., and J. R. Holton. 1968. A theory of the quasi-biennial oscillation, *J. Atmos. Sci.*, **25**, 1095–1107.

———. 1974. Wave-CISK in the tropics. J. Atmos. Sci., **31**, 156–179.

———. 1990. Dynamics in Atmospheric Physics. Cambridge: Cambridge University Press, 310 pp.

Lindzen, R. S., and A. Y. Hou. 1988. Hadley circulations for zonally averaged heating centered off the equator. *J. Atmos. Sci.*, **45**, 2416–2427.

Lindzen, R. S., and S. Nigam.1987. On the role of sea surface temperature gradients in forcing low-level winds and convergence in the tropics. *J. Atmos. Sci.*, **44**, 2418–2436.

Liu, J., and T. Schneider. 2010. Mechanisms of jet formation on the giant planets. *J. Atmos. Sci.*, **67**, 3652–3672.

Liu, W. T. 1986. Statistical relation between monthly mean precipitable water and surface-level humidity over global oceans. *Mon. Wea. Rev.*, **114**, 1591–1602.

———. 2002. Progress in scatterometer application. *J. Oceanogr.*, **58**, 121–136.

Liu, Z. 1997. Oceanic regulation of the atmospheric Walker circulation. *Bull. Amer. Meteor. Soc.*, **78**, 407–412.

Liu, Z., and B. Huang. 1997. A coupled theory of tropical climatology: Warm pool, cold tongue, and Walker circulation. *J. Climate*, **10**, 1662–1679, 1997.

Loeb, N. G., J. M. Lyman, G. C. Johnson, R. P. Allan, D. R. Doelling, T. Wong, B. J. Soden, and G. L. Stephens. 2012. Observed changes in top-of-the-atmosphere radiation and upper-ocean heating consistent within uncertainty. *Nat. Geosci.*, **5**, 110–113. doi:10.1038/NGEO1375.

Longuet-Higgins, M. S. 1968. The eigenfunctions of Laplace's tidal equations over a sphere. *Phil. Trans. Roy. Soc. A*, **262**, 511–607.

Lord, S. J. 1982. Interaction of a cumulus cloud ensemble with the large-scale environment, Part III: Semi-prognostic test of the Arakawa-Schubert cumulus parameterization. *J. Atmos. Sci.*, **39**, 88–103.

Lord, S. J., and A. Arakawa, 1980. Interaction of a cumulus cloud ensemble with the large-scale environment, Part II. *J. Atmos. Sci.*, **37**, 2677–2692.

Lord, S. J., W. C. Chao, and A. Arakawa. 1982. Interaction of a cumulus cloud ensemble with the large-scale environment. Part IV: The discrete model. *J. Atmos. Sci.*, **39**, 104–113.

Lorenz, E. N. 1951. Seasonal and irregular variations of the Northern Hemisphere sea-level pressure profile. *J. Meteor.*, **8**, 52–29.

———. 1955. Available potential energy and the maintenance of the general circulation. *Tellus*, **7**, 157–167.

———. 1960a. Energy and numerical weather prediction. *Tellus*, **12**, 364–373.

———. 1960b. Generation of available potential energy and the intensity of the general circulation. In *Dynamics of Climate*, edited by R. L. Pfeffer. Oxford: Pergamon, 86–92.

———. 1963. Deterministic nonperiodic flow. *J. Atmos. Sci.*, **20**, 130–141.

———. 1967. *The Nature and Theory of the General Circulation of the Atmosphere*. Geneva: World Meteorological Organization, no. 218, TP115, 161 pp.

———. 1969a. Three approaches to atmospheric predictability. *Bull. Amer. Meteor. Soc.*, **50**, 345–349.

———. 1969b. The predictability of a flow which possesses many scales of motion. *Tellus*, **21**, 289–307.

———. 1976. Nondeterministic theories of climatic change. *Quat. Res.*, **6**, 495–506.

———. 1978. Available energy and the maintenance of a moist circulation. *Tellus*, **30**, 15–31.

———. 1979. Numerical evaluation of moist available energy. *Tellus*, **31**, 230–235.

———. 1982. Atmospheric predictability experiments with a large numerical model. *Tellus*, **34**, 505–513.

———. 1983. A history of prevailing ideas about the general circulation of the atmosphere. *Bull Amer. Meteor. Soc.*, **64**, 730–755.

———. 1984. Irregularity: A fundamental property of the atmosphere. *Tellus*, **36A**, 98–110.

———. 1990. Can chaos and intransitivity lead to interannual variability? *Tellus*, **42A**, 378–389.

———. 1993. *The Essence of Chaos*. Seattle: University of Washington Press, 227 pp.

———. 2001. Driven to extremes. *New Scientist*, **172**, 38–42.

Lorenz, D. J., and E. T. DeWeaver. 2007. Tropopause height and zonal wind response to global warming in the IPCC scenario integrations. *J. Geophys. Res.*, **112**: D10119.

Lorenz, D. J., and D. L. Hartmann. 2003. Eddy-zonal flow feedback in the Northern Hemisphere winter. *J. Climate*, **16**, 1212–1227.

Lorenz, R. D., J. I. Lunine, P. G. Withers, and C. P. McKay. 2001. Titan, Mars and Earth: Entropy production by latitudinal heat transport. *Geophys. Res. Lett.*, **28**, 415–418.

Lu J., C. Deser, and T. Reichler. 2009. Cause of the widening of the tropical belt since 1958. *Geophys. Res. Lett.*, **36**: L03803.

Lu, J., G. A. Vecchi, and T. Reichler. 2007. Expansion of the Hadley cell under global warming. *Geophys. Res. Lett.*, **34**, L06805, doi:10.1029/ 2006GL028443.

Luo, D., and Z. Chen. 2006. The role of land-sea topography in blocking formation in a block-eddy interaction model. *J. Atmos. Sci.*, **63**, 3056–3065. doi:http://dx.doi.org/10.1175 /JAS3774.1.

Ma, C.-C., C. R. Mechoso, A. Arakawa, and J. D. Farrara. 1994. Sensitivity of a coupled ocean-atmosphere model to physical parameterizations. *J. Climate*, **7**, 11883–1896.

Ma, C.-C., C. R. Mechoso, A. W. Robertson, A. Arakawa, 1996. Peruvian stratus clouds and the tropical Pacific circulation: A coupled ocean-atmosphere GCM study. *J. Climate*, **9**, 1635–1645.

MacDonald, A. M., and C. Wunsch, 1996. An estimate of global ocean circulation and heat fluxes. *Nature*, **382**, 436–439.

Madden, R., and P. R. Julian, 1971. Detection of a 40–50 day oscillation in the zonal wind in the tropical Pacific. *J. Atmos. Sci.*, **28**, 1109–1123.

———. 1972a. Description of global scale circulation cells in the tropics with a 40–50 day period. *J. Atmos. Sci.*, **29**, 1109–1123.

———. 1972b. Further evidence of global-scale 5-day pressure waves. *J. Atmos. Sci.*, **29**, 1464–1469.

———. 1994. Observations of the 40–50-day tropical oscillation: A review. *Mon. Wea. Rev.*, **122**, 814–837.

Malguzzi, P., 1993. An analytical study on the feedback between large- and small-scale eddies. *J. Atmos. Sci.*, **50**, 1429–1436.

Maloney, E. E., and D. L. Hartmann. 1998. Frictional moisture convergence in a composite life cycle of the Madden-Julian oscillation. *J. Climate*, **11**, 2387–2403.

Maloney, E. D. 2009. The moist static energy budget of a composite tropical intraseasonal oscillation in a climate model. *J. Climate*, **22,** 711–729.

Maloney, E. D., A. H. Sobel, and W. M. Hannah. 2010. Intraseasonal variability in an aquaplanet general circulation model. *J. Adv. Model. Earth Syst.*, **2**, 5. doi:10.3894/JAMES.2010.2.5.

Manabe, S., and K. Bryan. 1969. Climate calculation with a combined ocean-atmosphere model. *J. Atmos. Sci.*, **26**, 786–789.

Manabe, S., and F. Möller. 1961. On the radiative equilibrium and heat balance of the atmosphere. *Mon. Wea. Rev.*, **89**, 503–532.

Manabe, S., J. Smagorinsky and R. F. Strickler. 1965. Simulated climatology of a general circulation model with a hydrologic cycle. *Mon. Wea. Rev.*, **93**, 769–797.

Manabe, S., and R. J. Stouffer. 1988. Two stable equilibria of a coupled ocean-atmosphere model. *J. Climate*, **1**, 841–866.

Manabe, S., and R. F. Strickler, 1964. Thermal equilibrium of the atmosphere with a convective adjustment. *J. Atmos. Sci.*, **21**, 361–385.

Manabe, S., and T. Terpstra. 1974. The effects of mountains on the general circulation of the atmosphere as identified by numerical experiments. *J. Atmos. Sci.*, **31**, 3–42.

Manabe, S., and R. T. Wetherald. 1967. Thermal equilibrium of the atmosphere with a given distribution of relative humidity. *J. Atmos. Sci.*, **24**, 241–259.

Mapes, Brian E. 2000. Convective inhibition, subgrid-scale triggering energy, and stratiform instability in a toy tropical wave model. *J. Atmos. Sci.*, **57**, 1515–1535. doi:http://dx.doi.org/10.1175/1520-0469(2000)057<1515:CISSTE>2.0.CO;2.

Mapes, B. E., and J. T. Bacmeister. 2012. Diagnosing tropical biases and the MJO using patterns in MERRA's analysis tendencies. *J. Climate*, **25**, 6202–6214.

Margules, M. 1893. *Luftbewegungen in einer Rotierended Spharoidschale* (II. Teil). Sitzungsber. *Kais. Akad. Wiss. Wien, Math.-Nat. Cl.* **102**, Abt. IIA, 11–56. Air motion in a rotating spherical shell. Translated by B. Haurwitz. NCAR Tech. Note NCAR. TN-156+STR.

Marquardt, C. 1998. Die tropische QBO und dynamische Prozesse in der Stratosphäre. Ph.D. thesis, Met. Abh. FU-Berlin, Serie A, Band 9/Heft 4, Verlag Dietrich Reimer Berlin, 260 S.

Martius, O., L. M. Polvani, and H. C. Davies. 2009. Blocking precursors to stratospheric sudden warming events. *Geophys. Res. Lett.*, **36**, L14806.

Maruyama, T., and M. Yanai. 1967. Evidence of large-scale wave disturbances in the equatorial lower stratosphere. *J. Meteor. Soc. Japan*, **45**, 196–199.

Masuda, K. 1988. Meridional heat transport by the atmosphere and the ocean: Analysis of FGGE data. *Tellus*, **40A**, 285–302.

Matsuno, T. 1966. Quasi-geostrophic motions in the equatorial area. *J. Meteor. Soc. Japan*, **44**, 25–43.

Matsuno, T. 1970. Vertical propagation of stationary planetary waves in the winter Northern Hemisphere. *J. Atmos. Sci.*, **27**, 871–883.

Matsuno, T. 1971. A dynamical model of the stratospheric sudden warming. *J. Atmos. Sci.*, **28**, 1479–1494.

Matsuno, T., and K. Nakamura, 1979: The Eulerian- and Lagrangian-mean meridional circulations in the stratosphere at the time of a sudden warming. *J. Atmos. Sci.*, **36**, 640–654.

McCreary, J. P. 1981. A linear stratified ocean model of the equatorial undercurrent. *Phil. Trans. Roy. Soc. A*, **298**, 603–635.

McFarlane, N. A. 1987. Effect of orographically excited gravity wave drag on the general circulation of the lower stratosphere and troposphere. *J. Atmos. Sci.*, **44**, 1775–1800.

McWilliams, J. C. 1980. An application of equivalent modons to atmospheric blocking. *Dyn. Atmos. Oceans*, **5**, 43–66.

———. 1984. The emergence of isolated coherent vortices in turbulent flow. *J. Fluid Mech.*, **146**, 21–43.

McWilliams, J. C., G. R. Flierl, V. D. Larichev, and G. M. Reznik. 1981. Numerical studies of barotropic modons. *Dyn. Atmos. Oceans*, **5**, 219–238.

McWilliams, J. C., J. B. Weiss, and I. Yavneh. 1994. Anisotropy and coherent vortex structures in planetary turbulence. *Science*, **264**, 410–413.

Mechoso, C. R., A. W. Robertson, N. Barth, M. K. Davey, P. Delecluse, P. R. Gent, S. Ineson, S. B. Kirtman, M. Latif, H. Le Treut, T. Nagal, J. D. Neelin, S.G.H. Philander, J. Polcher, P. S. Schopf, T. Stockdale, M. J. Suarez, L. Terray, O. Thual, and J. J. Tribbia. 1995. The seasonal cycle over the tropical Pacific in coupled ocean-atmosphere general circulation models. *Mon. Wea. Rev.*, **123**, 2825–2838.

Meehl, G. A., G. N. Kiladis, K. M. Weickmann, M. Wheeler, D. S. Gutzler, and G. P. Compo. 1996. Modulation of equatorial subseasonal convective episodes by tropical-extratropical interaction in the Indian and Pacific Ocean regions. *J. Geophys. Res.*, **101**, 15033–15049.

Meehl, Gerald A., Aixue Hu, Julie M. Arblaster, John Fasullo, Kevin E. Trenberth. 2013. Externally forced and internally generated decadal climate variability associated with the interdecadal Pacific oscillation. *J. Climate*, **26**, 7298–7310. doi:http://dx.doi.org/10.1175/JCLI-D-12-00548.1.

Merilees, P. E., and T. Warn, 1972. The resolution implications of geostrophic turbulence. *J. Atmos. Sci.*, **29**, 990–991.

———. 1975. On energy and enstrophy exchanges in two-dimensional non-divergent flow. *J. Fluid Dyn.*, **69**, 625–630.

Miller, R. L., 1997. Tropical thermostats and low cloud cover. *J. Climate*, **10**, 409–440.

Miller, R. L., and X. Jiang. 1996. Surface energy fluxes and coupled variability in the tropics of a coupled general circulation model. *J. Climate*, **9**, 1599–1620.

Mitchell, H. L., and J. Derome. 1983. Blocking-like solutions of the potential vorticity equation: Their stability at equilibrium and growth at resonance. *J. Atmos. Sci.*, **40**, 2522–2536.

Mo, K., J. O. Dickey, and S. L. Marcus. 1997. Interannual fluctuations in atmospheric angular momentum simulated by the National Centers for Environmental Prediction medium range forecast model. *J. Geophys. Res.*, **102**, 6703–6713.

Mooley, D. A., and J. Shukla. 1987. Variability and forecasting of the summer monsoon rainfall over India. In *Monsoon Meteorology*, edited by C.-P. Chang and T. N. Krishnamurti. New York: Oxford University Press, 26–59.

Moorthi, S., and M. J. Suarez. 1992. Relaxed Arakawa-Schubert: A parameterization of moist convection for general circulation models. *Mon. Wea. Rev.*, **120**, 978–76.

Morel, P., ed. 1973. *Dynamic Meteorology*. Boston: D. Reidel, 622 pp.

Moura, A. D., and J. Shukla. 1981. On the dynamics of droughts in northeast Brazil: Observations, theory, and numerical experiments with a general circulation model. *J. Atmos. Sci.*, **38**, 2653–2675.

Mullen, S. L. 1987. Transient eddy forcing of blocking flows. *J. Atmos. Sci.*, **44**, 3–22.

Murakami, T. 1987a. Effects of the Tibetan Plateau. In *Monsoon Meteorology*, edited by C. P. Chang and T. N. Krishnamurti. New York: Oxford University, 235–270.

———. 1987b. Intraseasonal atmospheric teleconnection patterns during the Northern Hemisphere summer. *Mon. Wea. Rev.*, **115**, 2133–2154.

Murakami, T., L. X. Chen, A. Xie, and M. L. Shrestha. 1986. Eastward propagation of 30–60 day perturbations as revealed from outgoing longwave radiation data. *J. Atmos. Sci.*, **43**, 961–971.

Nakajima, K., and T. Matsuno. 1988. Numerical experiments concerning the origin of cloud clusters in the tropical atmosphere. *J. Meteor. Soc. Japan*, **66**, 309–329.

Nakamura, H. 1994. Rotational evolution of potential vorticity associated with a strong blocking flow configuration over Europe. *Geophys. Res. Lett.*, **21**, 2003–2006.

Nakamura, H., M. Nakamura, and J. L. Anderson. 1997. The role of high- and low-frequency dynamics in blocking formation. *Mon. Weather Rev.*, **125**, 2074–2093.

Nakazawa, T. 1986. Intraseasonal variations in OLR in the tropics during the FGGE year. *J. Meteor. Soc. Japan*, **64**, 17–34.

———. 1988. Tropical super clusters within intraseasonal variations over the western Pacific. *J. Meteor. Soc. Japan*, **66**, 823–839.

Nastrom, G. D., and K. S. Gage. 1985. A climatology of aircraft wavenumber spectra observed by commercial aircraft. *J. Atmos. Sci.*, **42**, 950–960.

Naujokat, B. 1986. An update of the observed quasi-biennial oscillation of the stratospheric winds over the tropics. *J. Atmos. Sci.*, **43**, 1873–1877.

Neelin, J. D., Battisti, D. S., Hirst, A. G., Jin, F.-F., Wakata, Y., Yamagata, T., Zebiak, S. E. 1998. ENSO theory, *J. Geophys. Res.*, **103**, 14261–14,290.

Neelin, J. D., and H. A. Dijkstra. 1995. Ocean-atmosphere interaction and the tropical climatology. Part I: The dangers of flux correction. *J. Climate*, **8**, 1325–1342.

Neelin, J. D., and I. M. Held. 1987. Modeling tropical convergence based on the moist static energy budget. *Mon. Wea. Rev.*, **115**, 3–12.

Neelin, J. D., I. M. Held, and K. H. Cook. 1987. Evaporation-wind feedback and low-frequency variability in the tropical atmosphere. *J. Atmos. Sci.*, **44**, 2241–2248.

Neelin, J. D., and F.-F. Jin. 1993. Modes of interannual tropical ocean-atmosphere interaction—a unified view. II: Analytical results in the weak coupling limit. *J. Atmos. Sci.*, **50**, 3504–3533.

Neelin, J. D., F.-F. Jin, and M. Latif. 1994. Dynamics of coupled ocean-atmosphere models: The tropical problem. *Ann. Rev. Fluid Mech.*, **26**, 617–659.

Neelin, J. D., M. Latif, M.A.F. Allaart, M. A. Cane, U. Cubasch, W. L. Gates, P. R. Gent, M. Ghil, C. Gordon, N. C. Lau, C. R. Mechoso, G. A. Meehl, J. M. Oberhuber, S.G.H. Philander, P. S. Schopf, K. R. Sperber, A. Sterl, T. Tokioka, J. Tribbia, and S. E. Zebiak. 1992. Tropical air-sea interaction in general circulation models. *Clim. Dyn.*, **7**, 73–104.

Neelin, J. D., and J.-Y. Yu. 1994. Modes of tropical variability under convective adjustment and the Madden-Julian oscillation. Part I: Analytical theory. *J. Atmos. Sci.*, **51**, 1876–1894.

Newell, R. E. 1963. Transfer through the tropopause and within the stratosphere. *Quart. J. Roy. Meteor. Soc.*, **89**, 167–204. doi:10.1002/qj.49708938002.

Newell, R. E., J. W. Kidson, D. G. Vincent and G. J. Boer. 1975. *The General Circulation of the Tropical Atmosphere*, vol. 2. Cambridge, MA: The MIT Press, 371 pp.

Newell, R. E., Y. Zhu, E. V. Browell, W. G. Read, J. W. Waters. 1996. Walker circulation and tropical upper tropospheric water vapor. *J. Geophys. Res.*, **101**, D1, 1961–1974.

Newman, P. A., and E. R. Nash. 2005. The unusual Southern Hemisphere stratosphere winter of 2002. *J. Atmos. Sci.*, **62**, 614–628. doi:http://dx.doi.org/10.1175/JAS-3323.1.

Newton, C. W. 1971. Mountain torques in the global angular momentum balance. *J. Atmos. Sci.*, **28**, 623–628.

———. 1972. Southern Hemisphere general circulation in relation to global energy and momentum balance requirements. *Meteor. Monogr.*, **35**, 215–246.

Nieto Ferreira, R. 1994. On the dynamics of the formation of multiple tropical disturbances. Atmospheric Science Paper No. 559, Dept. of Atmospheric Science, Colorado State University.

Niiler, P. P. 1975. Deepening of the wind-mixed layer. *J. Marine Res.*, **33**, 405–422.

Niiler, P. P., and E. B. Kraus. 1977. One-dimensional models of the upper ocean. In *Modelling and Prediction of the Upper Layers of the Ocean*, edited by E. B. Kraus. New York: Pergamon Press, 143–172.

Nitta, T. 1975. Observational determination of cloud mass flux distributions. *J. Atmos. Sci.*, **32**, 73–91.

Norris, J. R., and C. B. Leovy. 1994. Interannual variability in stratiform cloudiness and sea surface temperature. *J. Climate*, **7**, 1915–1925.

North, G. R., T. L. Bell, and R. F. Cahalan. 1982. Sampling errors in the estimation of empirical orthogonal function. *Mon. Wea. Rev.*, **110**, 669–706.

O'Gorman, P., and T. Schneider. 2008. The hydrological cycle over a wide range of climates simulated with an idealized GCM. *J. Atmos. Sci.*, **65**, 524–535.

O'Kane, T. J., J. S. Risbey, C. Franzke, I. Horenko, and D. P. Monselesan. 2013. Changes in the metastability of the midlatitude Southern Hemisphere circulation and the utility of nonstationary cluster analysis and split-flow blocking indices as diagnostic tools. *J. Atmos. Sci.*, **70**, 824–842. doi:http://dx.doi.org/10.1175/JAS-D-12-028.1.

Ohmura, H. and Ozuma, A. 1997. Thermodynamics of a global-mean state of the atmosphere: A state of maximum entropy increase. *J. Climate*, **10**, 441–445.

Oort, A. H. 1983. Global atmospheric circulation statistics, 1958–1973. *NOAA Prof. Paper* 14, 180 pp.

———. 1985. Balance conditions in the Earth's climate system. *Adv. in Geophys.*, **28A**, 75–98.

———. 1989. Angular momentum cycle in the atmosphere-ocean-solid earth system. *Bull. Amer. Meteor. Soc.*, **70**, 1231–1242.

Oort, A. H., and E. M. Rasmusson. 1971. Atmospheric circulation statistics. NOAA Prof. Paper, no. 5, U. S. Dept. of Commerce, Washington, DC, 323 pp.

Oort, A. H., and T. H. VonderHaar. 1976. On the observed annual cycle in the ocean-atmosphere heat balance over the Northern Hemisphere. *J. Phys. Oceanogr.*, **6**, 781–800.

Oort, A. H., and J. J. Yienger. 1996. Observed interannual variability in the Hadley circulation and its connection to ENSO. *J. Climate*, **9**, 2751–2767.

Ooyama, K. 1971. A theory on parameterization of cumulus convection. Special issue, *J. Meteor. Soc. Japan*, **49**, 744–756.

Orton, G. S., et al. 1991. Thermal maps of Jupiter: Spatial organization and time-dependence of stratospheric temperatures, 1980 to 1990. *Science*, **252**, 537–542.

Orton, G. S., A. J. Friedson, P. A. Yanamandra-Fisher, J. Caldwell, H. B. Hammel, K. H. Baines, J. T. Bergstrahl, T. Z. Martin, R. A. West, G. J. Veeder Jr., D. K. Lynch, R. Russell, M. E. Malcom, W. F. Golisch, D. M. Griep, C. D. Kaminski, A. T. Tokunaga, T. Herbst, and M. Shure. 1994. Spatial organization and time dependence of Jupiter's tropospheric temperatures, 1980–1993. *Science*, **265**, 625–631.

Paldor, N., and P. D. Killworth. 1988. Inertial trajectories on a rotating earth. *J. Atmos. Sci.*, **45**, 4013–4019.

Palmén, E., and C. W. Newton. 1969. *Atmospheric Circulation Systems*. New York: Academic Press, 603 pp.

Palmer, T. N. 1993. Extended range atmospheric prediction and the Lorenz model. *Bull. Amer. Meteor. Soc.*, **74**, 49–65.

———. 1999. A nonlinear dynamical perspective on climate prediction. *J. Climate*, **12**, 575–591.

Palmer, T. N., and D.L.T. Anderson. 1994. The prospects for seasonal forecasting: a review paper. *Quart. J. Roy. Meteor. Soc.*, **120**, 755–793.

Paltridge, G. W. 1975. Global dynamics and climate change: A system of minimum entropy exchange. *Quart. J. Roy. Meteor. Soc.*, **101**, 475–484.

Pan, D.-M., and D. A. Randall. 1998. A cumulus parameterization with a prognostic closure. *Quart. J. Roy. Meteor. Soc.*, **124**, 949–981.

Pan, H.-L., and W.-S. Wu. 1995. Implementing a mass flux convection parameterization package for the NMC medium-range forecast model. *NMC Office Note*, no. 409, 40 pp. (Available from the U. S. National Center for Environmental Prediction, 5200 Auth Road, Washington, DC 20233).

Pauluis, Olivier, Arnaud Czaja, Robert Korty. 2010. The global atmospheric circulation in moist isentropic coordinates. *J. Climate*, **23**, 3077–3093. doi:http://dx.doi.org/10.1175/2009JCLI2789.1.

Pauluis, O. M., and A. A. Mrowiec. 2013. Isentropic analysis of convective motions. *J. Atmos. Sci.*, **70**, 3673–3688. doi:http://dx.doi.org/10.1175/JAS-D-12-0205.1.

Peixóto, J. P. 1965. On the role of water vapor in the energetics of the general circulation of the atmosphere. *Portugalie Physica*, **4**, 135–170.

———. 1970. Water vapor balance of the atmosphere from five years of hemispheric data. *Nordic Hydrology*, **2**, 120–138.

Peixóto, J. P., and A. H. Oort. 1983. The atmospheric branch of the hydrological cycle and climate. In *Variations in the Global Water Budget*, edited by A. Street-Perrott et al. Boston: D. Reidel, 5–65.

———. 1992. *Physics of Climate*. New York: Springer-Verlag and American Institute of Physics, 520 pp.

Pennell, S. A., and K. L. Seitter. 1990. On inertial motion on a rotating sphere. *J. Atmos. Sci.*, **47**, 2032–2034.

Philander, S. G. 1990. *El Niño, La Niña, and the Southern Oscillation*. New York: Academic Press, 293 pp.

Philander, S.G.H., D. Gu, D. Halpern, G. Lambert, N.-C. Lau, T. Li, and R. Pacanowski. 1996. Why the ITCZ is mostly north of the equator. J. *Climate*, **9**, 2958–2972.

Philander, S.G.H., W. Hurlin, A. D. Siegal. 1987. A model of the seasonal cycle in the tropical Pacific ocean. *J. Phys. Oceanogr.*, **17**, 1986–2002.

Philander, S.G.H., R. C. Pacanowski, M.-C. Lau, and M. J. Nath. 1992. Simulation of ENSO with a global atmospheric GCM coupled to a high-resolution tropical Pacific Ocean GCM. *J. Climate*, **5**, 308–329.

Philander, S.G.H., T. Yamagata, and R. C. Pacanowski. 1984. Unstable air-sea interactions in the tropics. *J. Atmos. Sci.*, **41**, 604–613.

Phillips, N. A. 1966. The equations of motion for a shallow rotating atmosphere and the "traditional approximation." *J. Atmos. Sci.*, **23**, 626–628.

Pierrehumbert, R. T. 1995. Thermostats, radiator fins, and the local runaway greenhouse. *J. Atmos. Sci.*, **52**, 1784–1806.

Pierrehumbert, R. T., and P. Malguzzi. 1984. Forced coherent structures and local multiple equilibria in a barotropic atmosphere. *J. Atmos. Sci.*, **41**, 246–257.

Platzman, G. W. 1960. The spectral form of the vorticity equation. *J. Meteor.*, **17**, 635–644.

Plumb, R. A. 1984. The quasi-biennial oscillation. In *Dynamics of the Middle Atmosphere*, edited by J. R. Holton and T. Matsuno. Boston: D. Reidel, 217–251.

Plumb, R. A. and D. McEwan, 1978. The instability of a forced standing wave in a viscous stratified fluid: A laboratory analogue of the quasi-biennial oscillation. *J. Atmos. Sci.*, **35**, 1827–1839.

Poincaré, H. 1912. *Science et Méthode*. Paris: Flammarion. English translation: *Science and Method*. South Bend, IN: St. Augustine's Press, 288 pp.

Ponte, R. M., D. Stammer, and J. Marshall. 1998. Oceanic signals in observed motions of the Earth's pole of rotation. *Nature*, **391**, 476–479.

Previdi, M., and B. G. Liepert. 2007. Annular modes and Hadley cell expansion under global warming. *Geophys. Res. Lett.*, **34**, L22701, doi:10.1029/ 2007GL031243.

Pritchard, M. S., and Christopher S. Bretherton. 2014. Causal evidence that rotational moisture advection is critical to the superparameterized Madden–Julian Oscillation. *J. Atmos. Sci.*, **71**, 800–815. doi:http://dx.doi.org/10.1175/JAS-D-13-0119.1.

Provenzale, A., A. Babiano, A. Bracco, C. Pasquero, and J. B. Weiss. 2008. Coherent vortices and tracer transport. In *Transport and Mixing in Geophysical Flows*, vol. 744 in *Lecture Notes in Physics*, edited by Jeffrey B. Weiss and Antonello Provenzale. Berlin: Springer-Verlag.

Quiroz, R. S. 1986. The association of stratospheric warmings with tropospheric blocking. *J. Geophys. Res.*, **91**, 1723–1736.

Ramanathan, V., R. D. Cess, E. F. Harrison, P. Minnis, B. R. Barkstrom, E. Ahmad, and D. Hartmann. 1989. Cloud-radiative forcing and climate: Results from the Earth Radiation Budget Experiment. *Science*, **243**, 57–63.

Ramanathan, V., and J. A. Coakley, Jr. 1978. Climate modeling through radiative-convective models. *Rev. Geophys. Space Phys.*, **6**, 465–489.

Ramanathan, V., and W. Collins. 1991. Thermodynamic regulation of ocean warming by cirrus clouds deduced from observations of the 1987 El Niño. *Nature*, **351**, 27–32.

Randall, D. A. 1984. Buoyant production and consumption of turbulence kinetic energy in cloud-topped mixed layers. *J. Atmos. Sci.*, **41**, 402–413.

———. 2013. Beyond deadlock. *Geophys. Res. Lett.*, **40**, 1–7, doi:10.1002/2013GL057998.

Randall, D. A., J. A. Abeles, and T. G. Corsetti. 1985. Seasonal simulations of the planetary boundary layer and boundary-layer stratocumulus clouds with a general circulation model. J. Atmos. Sci., **42**, 641–676.

Randall, D. A., Curry, D. Battisti, G. Flato, R. Grumbine, S. Hakkinen, D. Martinson, R. Preller, J. Walsh, and J. Weatherly. 1998. Status of and outlook for large-scale modeling of atmosphere-ice-ocean interactions in the Arctic. *Bull. Amer. Meteor. Soc.*, **79**, 197–219.

Randall, D. A., P. Ding, and D.-M. Pan. 1997. The Arakawa-Schubert parameterization. In *The Physics and Parameterization of Moist Atmospheric Convection*, edited by R. K. Smith. Dordrecht: Kluwer Academic, 281–296.

Randall, D. A., Harshvardhan, and D. A. Dazlich. 1991. Diurnal variability of the hydrologic cycle in a general circulation model. *J. Atmos. Sci.*, **48**, 40–62.

Randall, D. A., Harshvardhan, D. A. Dazlich, and T. G. Corsetti, 1989. Interactions among radiation, convection, and large scale dynamics in a general circulation model. *J. Atmos. Sci.*, **46**, 1943–1970.

Randall, D. A., M. Khairoutdinov, A. Arakawa, and W. Grabowski. 2003. Breaking the cloud-parameterization deadlock. *Bull. Amer. Meteor. Soc.*, **84**, 1547–1564.

Randall, D. A., and D.-M. Pan. 1993. Implementation of the Arakawa-Schubert parameterization with a prognostic closure. In *The Representation of Cumulus Convection in Numerical Models*, edited by K. Emanuel and D. Raymond. *Meteor. Monogr.*, **24** (46), 1–246.

Randall, D. A., D.-M. Pan, and P. Ding. 1997. Quasi-equilibrium. In *The Physics and Parameterization of Moist Atmospheric Convection*, edited by R. K. Smith. Dordrecht: Kluwer Academic, 359–385.

Randall, D. A., and M. J. Suarez. 1984. On the dynamics of stratocumulus formation and dissipation. *J. Atmos. Sci.*, **41**, 3052–3057.

Randall, D. A., and J. Wang. 1992. The moist available energy of a conditionally unstable atmosphere. *J. Atmos. Sci.*, **49**, 240–255.

Randall, D. A., K.-M. Xu, R. J. C. Somerville, and S. Iacobellis. 1996. Single-column models and cloud ensemble models as links between observations and climate models. *J. Climate*, **9**, 1683–1697.

Randel, W. J., F. Wu, and D. J. Gaffen. 2000. Interannual variability of the tropical tropopause from radiosonde data and NCEP reanalyses. *J. Geophys. Res.*, **105**, 15509.

Rao, Y. P. 1976. Southwest monsoon. *Monograph 1/76*, India Meteorological Department, Pune, India.

Rasmusson, E. M. 1987. Tropical Pacific variations. *Nature*, **327**, 192.

Rasmusson, E. M., and T. H. Carpenter. 1983. The relationship between eastern equatorial Pacific sea surface temperatures and rainfall over India and Sri Lanka. *Mon. Wea. Rev.*, **111**, 517–528.

Rasmusson, E. M., and J. M. Hall. 1983. El Niño, the great equatorial Pacific Ocean warming event of 1982–1983. *Weatherwise*, **36**, 166–175.

Raval, A., and V. Ramanathan. 1989. Observational determination of the greenhouse effect. *Nature*, **342**, 758–761.

Raymond, D. J. 2000. The Hadley circulation as a radiative-convective instability. *J. Atmos. Sci.*, **57**, 1286–1297.

———. 2001. A new model of the Madden-Julian oscillation. *J. Atmos. Sci.*, **58**, 2807–2819.

Raymond, D. J., and A. M. Blyth. 1986. A stochastic mixing model for non-precipitating cumulus clouds. *J. Atmos. Sci.*, **43**, 2708–2718.

Reynolds, R. W., and T. M. Smith. 1994. Improved global sea surface temperature analyses using optimum interpolation. *J. Climate*, **7**, 929–948.

Reed, R. J. 1966. The present status of the 26-month oscillation. *Bull. Amer. Meteor. Soc.*, **46**, 374–387.

Reed, R. J., and M. J. Oard. 1969. A comparison of observed and theoretical diurnal tidal motions between 30 and 60 kilometers. *Mon. Wea. Rev.*, **97**, 456–459.

Ren, H.-L., F.-F. Jin, J.-S. Kug, J.-X. Zhao, and J. Park, 2009. A kinematic mechanism for positive feedback between synoptic eddies and NAO. *Geophys. Res. Lett.*, **36**, L11709, doi:10.1029/2009GL037294.

Rennó, N. O., K. A. Emanuel, and P. H. Stone. 1994. Radiative-convective model with an explicit hydrologic cycle, 1, Formulation and sensitivity to model parameters. *J. Geophys. Res.*, **99**, 14429–14442.

Rex, D. F. 1950a. Blocking action in the middle troposphere and its effect upon regional climate. Part I: An aerological study of blocking action. *Tellus*, **2**, 196–211.

———. 1950b. The effect of Atlantic blocking action upon European climate. *Tellus*, **3**, 199–212.

Rhines, P. 1975. Waves and turbulence on a b-plane. *J. Fluid Mech.*, **69**, 417–443.

Riehl, H., and J. S. Malkus. 1958. On the heat balance in the equatorial trough zone. *Geophysica*, **6**, 503–537.

Ringler, T. D., D. Jacobsen, M. Gunzburger, L. Ju, M. Duda, and W. Skamarock. 2011. Exploring a multiresolution modeling approach within the shallow-water equations. *Mon. Wea. Rev.*, **139**, 3348–3368. doi:10.1175/MWR-D-10-05049.1.

Robinson, T. D., and D. C. Catling. 2013. Common 0.1 bar tropopause in thick atmospheres set by pressure-dependent infrared transparency. *Nat. Geosci.*, **7**, 12–15. doi:10.1038/ngeo2020.

Roebber, P. J. 2009. Planetary waves, cyclogenesis, and the irregular breakdown of zonal motion over the North Atlantic. *Mon. Wea. Rev.*, **137**, 3907–3917. doi:http://dx.doi.org/10.1175/2009MWR3025.1.

Rogers, J. C., and Harry Van Loon. 1982. Spatial variability of sea level pressure and 500 m height anomalies over the Southern Hemisphere. *Mon. Wea. Rev.*, **110**, 1375–1392.

Romps, D. M. 2012. Weak pressure gradient approximation and its analytical solutions. *J. Atmos. Sci.*, **69**, 2835–2845.
Rosen, R. D., D. A. Salstein, T. M. Eubanks, J. O. Dickey, and J. A. Steppe. 1984. An El Niño signal in atmospheric angular momentum and Earth rotation. *Science*, **225**, 411–414.
Rosenlof, K. H. 1986. Walker circulation with observed zonal winds, a mean Hadley cell, and cumulus friction. *J. Atmos. Sci.*, **43**, 449–467.
Rossby, C. G. 1939. Relations between variations in the intensity of the zonal circulation and the displacements of the semi-permanent centers of action. *J. Mar. Res.*, **2**, 38–55.
———. 1941. The scientific basis of modern meteorology. In *Climate and Man*. Yearbook of Agriculture. Washington, DC: U.S. Government Printing Office, 599–655.
———. 1947. On the distribution of angular velocity in gaseous envelopes under the influence of large-scale horizontal mixing processes. *Bull. Amer. Meteor. Soc.*, **28**, 53–68.
Robinson, W. A. 1991. The dynamics of low-frequency variability in a simple model of the global atmosphere. *J. Atmos. Sci.*, **48**, 429–441.
Rueda, V.O.M. 1991. Tropical-extratropical atmospheric interactions. Ph.D. thesis, University of California, Los Angeles.
Rutledge, S. A., and R. A. Houze, Jr. 1987. A diagnostic modeling study of the trailing stratiform region of a midlatitude squall line. *J. Atmos. Sci.*, **44**, 2640–2656.
Sadourny, R., and C. Basdevant. 1985. Parameterization of subgrid scale barotropic and baroclinic eddies in quasi-geostrophic models: Anticipated potential vorticity method. *J. Atmos. Sci.*, **42**, 1353–1363.
Sakai, K., and W. R. Peltier. 1997. Dansgaard-Oeschger oscillations in a coupled atmosphere-ocean climate model. *J. Climate*, **10**, 949–970.
Salathé, E. P., Jr., and D. L. Hartmann. 1997. A trajectory analysis of tropical upper-tropospheric moisture and convection. *J. Climate*, **10**, 2533–2547.
Salby, M. L., and R. R. Garcia. 1987. Transient response to localized episodic heating in the tropics. Part I: Excitation and short-time near-field behavior. *J. Atm. Sci.*, **44**, 458–498.
Salby, M. L., R. R. Garcia, and H. Hendon. 1994. Planetary-scale circulations in the presence of climatological and wave-induced heating. *J. Atmos. Sci.*, **51**, 2344–2367.
Saltzman, B. 1970. Large-scale atmospheric energetics in the wavenumber domain. *Rev. Geophys. Space Phys.*, **8**, 289–302.
Santer, B. D., R. Sausen, T. M. L. Wigley, J. S. Boyle, K. AchutaRao, C. Doutriaux, J. E. Hansen, G. A. Meehl, E. Roeckner, R. Ruedy, G. Schmidt, and K. E. Taylor. 2003. Behavior of tropopause height and atmospheric temperature in models, reanalyses, and observations: Decadal changes. *J. Geophys. Res.*, **108**, 4002. doi:10.1029/2002JD002258.
Saravanan, R. 1990. Mechanisms of equatorial superrotation: Studies with two-level models. Ph.D. thesis, Princeton University.
Sasamori, T. 1982. Stability of the Walker circulation. *J. Atmos. Sci.*, **39**, 518–527.
Satoh, M., and Y.-Y. Hayashi. 1992. Simple cumulus models in one-dimensional radiative convective equilibrium problems. *J. Atmos. Sci.*, **49**, 1202–1220
Saunders, P. M., and B. A. King. 1995. Oceanic fluxes on the WOCE A11 section. *J. Phys. Oceanogr.*, 25, 1942–1958.
Savijärvi, H. I. 1988. Global energy and moisture budgets from rawinsonde data. *Mon. Wea. Rev.*, **116**, 417–430.
Sawyer, J. S. 1949. The significance of dynamic instability in atmospheric motions. *Quart. J. Roy. Meteor. Soc.*, **75**, 364–374.
———. 1965. The dynamical problems of the lower stratosphere. *Quart. J. Roy. Meteor. Soc.*, **91**, 407–416.

Scherhag, R. 1952. Die explosionsartigen Stratosphärenerwärmungen des Spätwinters 1951/52. *Ber. Deutsch. Wetterdienst* **38**, 51–63.

———. 1960. Stratospheric temperature changes and the associated changes in pressure distribution. *J. Meteor.*, **17**, 575–582.

Schey, H. M. 2004. *Div, Grad, Curl, and All That: An Informal Text on Vector Calculus*, 4th ed. New York: W. W. Norton, 163 pp.

Schilling, H.-D. 1982. A numerical investigation of the dynamics of blocking waves in a simple two-level model. *J. Atmos. Sci.*, **39**, 998–1017.

Schmidt, T., J. Wickert, G. Beyerle, and S. Heise. 2008. Global tropopause height trends estimated from GPS radio occultation data. *Geophys. Res. Lett.*, **35**, L11806, doi:10.1029/2008 GL034012.

Schneider, E. K. 1977. Axially symmetric steady-state models of the basic state for instability and climate studies. Part II: Nonlinear calculations. *J. Atmos. Sci.*, **34**, 280–296.

Schneider, E. K., and R. S. Lindzen. 1977. Axially symmetric steady-state models of the basic state for instability and climate studies. Part I: Linearized calculations. *J. Atmos. Sci.*, **34**, 263–279.

Schneider, T. 2006. The general circulation of the atmosphere. *Ann. Rev. Earth Planet. Sci.*, **34**, 655–688.

Schneider, T., and A. H. Sobel, eds. 2007. *The Global Circulation of the Atmosphere*. Princeton, NJ: Princeton University Press, 385 pp.

Schoeberl, M. R. 1978. Stratospheric warmings: Observations and theory. *Rev. Geophys.*, **16**, 5221–538.

Schopf, P. S., and M. J. Suarez. 1988. Vacillations in a coupled ocean-atmosphere model. *J. Atmos. Sci.*, **45**, 549–566.

Schubert, J. J., B. Stevens, and T. Crueger. 2013. The Madden-Julian oscillation as simulated by the MPI Earth System Model: Over the last and into the next millennium. *J. Adv. Model. Earth Syst.*, **5**, 71–84.

Schubert, W. H. 1976. Experiments with Lilly's cloud-topped mixed layer model. *J. Atmos. Sci.*, **33**, 436–446.

Schubert, W. H., Paul E. Ciesielski, C. Lu, and R. H. Johnson. 1995. Dynamical adjustment of the trade wind inversion layer. *J. Atmos. Sci.*, **52**, 2941–2952.

Schubert, W. H., and M. T. Masarik. 2006. Potential vorticity aspects of the MJO. *Dyn. Atmos. Oceans*, **42**, 127–151.

Schubert, W. H., J. S. Wakefield, E. J. Steiner, and S. K. Cox. 1979. Marine stratocumulus convection. Part I: Governing equations and horizontally homogeneous solutions. *J. Atmos. Sci.*, **36**, 1286–1307.

———. 1979. Marine stratocumulus convection. Part II: Horizontally inhomogeneous solutions. *J. Atmos. Sci.*, **36**, 1308–1324.

Schulman, L. L. 1973. On the summer hemisphere Hadley cell. *Quart. J. Roy. Meteor. Soc.*, **99**, 197–201.

Sclater, J. G., C. Jaupart, and D. Galson. 1980. The heat flow through oceanic and continental crust and the heat loss of the Earth. *Rev. Geophys. Space Phys.*, **18**, 269–311.

Screen, J. A., C. Deser, and I. Simmonds. 2012. Local and remote controls on observed Arctic warming. *Geophys. Res. Lett.*, **39**, L10709, doi:10.1029/2012GL051598.

Seager, R., and R. Murtugude. 1997. Ocean dynamics, thermocline adjustment and regulation of tropical SST. *J. Climate*, **10**, 521–534.

Seidel, D. J., Q. Fu, W. J. Randel, and T. J. Reichler. 2008. Widening of the tropical belt in a changing climate. *Nat. Geosci.*, **1**, 21–24. doi:10.1038/ngeo.2007.38.

Seidel, D. J., and W. J. Randel. 2006. Variability and trends in the global tropopause estimated from radiosonde data. *J. Geophys. Res.*, **111**, D21101. doi:10.1029/2006JD007363.

———. 2007. Recent widening of the tropical belt: Evidence from tropopause observations. *J. Geophys. Res.*, **112**, D20113. doi:10.1029/2007JD008861.

Seidel, D. J., R. J. Ross, J. K. Angell, and G. C. Reid. 2001. Climatological characteristics of the tropical tropopause as revealed by radiosondes. J. Geophys. Res. **106**, 7857. doi:10.1029/2000JD900837.

Seitter, K. L., and H.-L. Kuo. 1983. The dynamical structure of squall-line type thunderstorms. *J. Atmos. Sci.*, **40**, 2831–2854. doi:

Sellers, P. J., R. E. Dickinson, D. A. Randall, A. K. Betts, F. G. Hall, J. A. Berry, C. J. Collatz, A. S. Denning, H. A. Mooney, C. A. Nobre, and N. Sato. 1997. Modeling the exchanges of energy, water, and carbon between the continents and the atmosphere. *Science*, **275**, 502–509.

Sherwood, S. C. 1996. Maintenance of the free-tropospheric tropical water vapor distribution. Part I: Clear regime budget. J. Climate, **9**, 2903–2918.

———. 1999: Convective precursors and predictability in the tropical western Pacific. *Mon. Wea. Rev.*, **127**, 2977–2991.

Shinoda, T., H. H. Hendon, and J. Glick. 1998. Intraseasonal variability of surface fluxes and sea surface temperature in the tropical western Pacific and Indian Oceans. *J. Climate*, **11**, 1685–1702.

Showman, A., and L. Polvani. 2010. The Matsuno-Gill model and equatorial superrotation. *Geophys. Res. Lett.*, **37**, L18811.

Shukla, J. 1981. Dynamical predictability of monthly means. *J. Atmos. Sci.*, **38**, 2547–2572.

———. 1985. Predictability. *Adv. Geophys.*, **28B**, 87–122.

Shukla, J., and D. A. Paolino. 1983. The Southern Oscillation and long-range forecasting of the summer monsoon rainfall over India. *Mon. Wea. Rev.*, **111**, 1830–1837.

Shutts, G. J. 1983. The propagation of eddies in diffluent jet-streams: Eddy vorticity forcing of "blocking" flow fields. *Quart. J. Roy. Meteor. Soc.*, **109**, 737–761.

———. 1986. A case study of eddy forcing during an Atlantic blocking episode. *Adv. Geophys.*, **29**, 135–162.

Sikka, D. R., and S. Gadgil. 1980. On the maximum cloud zone and the ITCZ over Indian longitudes during the southwest monsoon. *Mon. Wea. Rev.*, **108**, 1840–1853.

Simmons, A., M. Hortal, G. Kelly, A. McNally, A. Untch, and S. Uppala. 2005. ECMWF analyses and forecasts of stratospheric winter polar vortex breakup: September 2002 in the Southern Hemisphere and related events. *J. Atmos. Sci.*, **62**, 668–689. doi:http://dx.doi.org/10.1175/JAS-3322.1.

Simmons, A. J., and B. J. Hoskins. 1978. The life cycles of some nonlinear baroclinic waves. *J. Atmos. Sci.*, **35**, 414–431.

Sjoberg, J. P., T. Birner. 2012. Transient tropospheric forcing of sudden stratospheric warmings. *J. Atmos. Sci.*, **69**, 3420–3432.

Slingo, J. M., K. R. Sperber, J. S. Boyle, J.-P. Ceron, M. Dix, B. Dugas, W. Ebisuzaki, J. Fyfe, D. Gregory, J.-F. Gueremy, J. Hack, A. Harzallah, P. Inness, A. Kitoh, W. K.-M. Lau, B. McAvaney, R. Madden, A. Matthews, T. N. Palmer, C.-K. Park, D. A. Randall, and N. Renno. 1996. Intraseasonal oscillations in 15 atmospheric general circulation models: Results from an AMIP diagnostic subproject. *Climate Dyn.*, **12**, 325–357.

Smith, R. K., ed. 1998. *The Physics and Parameterization of Moist Atmospheric Convection*. Dordrecht: Kluwer Academic.

Sobel, A. H., J. Nilsson, and L. M. Polvani. 2001. The weak temperature gradient approximation and balanced tropical moisture waves. *J. Atmos. Sci.*, **58**, 3650–3665.

Sobel, A., and E. Maloney, 2012: An idealized semi-empirical framework for modeling the Madden–Julian oscillation. *J. Atmos. Sci.*, **69**, 1691–1705.

Sobel, A., and E. Maloney, 2013: Moisture modes and the eastward propagation of the MJO. *J. Atmos. Sci.*, **70**, 187-192. doi: http://dx.doi.org/10.1175/JAS-D-12-0189.1.

Soden, B. J., and I. M. Held. 2006. An assessment of climate feedbacks in coupled ocean-atmosphere models. *J. Climate*, **19**, 3354–3360. doi:http://dx.doi.org/10.1175/JCLI3799.1.

Solberg, P. H. 1936. Le mouvement d'inertie de l'atmosphere stable et son role dans la theorie des cyclones. In *Proces Verbaux de l'Association de Météorologie*, International Union of Geodesy and Geophysics, 6th General Assembly, Edinburgh, **2**, 66–82.

Spence, T. W., and D. Fultz. 1977. Experiments on wave-transition spectra and vacillation in an open rotating cylinder. *J. Atmos. Sci.*, **34**,1261–1285.

Speranza, A. 1986. Deterministic and statistic properties of Northern Hemisphere, middle latitude circulation: Minimal theoretical models. *Adv. in Geophys.*, **29**, 199–225

Sohn, B.-J. 1994. Temperature-moisture biases in ECMWF analyses based on clear sky longwave simulations constrained by SSMI and MSU measurements and comparisons to ERBE estimates. *J. Climate*, **7**, 1707–1718.

Son, S.W., L. M. Polvani, E. W. Waugh, T. Birner, H. Akiyoshi, R. R. Garcia, A. Gettelman, D. A. Plummer, and E. Rozanov. 2009. The impact of stratospheric ozone recovery on tropopause height trends. *J. Climate*, **22**, 429–445. doi:http://dx.doi.org/10.1175/2008JCLI2215.1.

Soong, S.-T., and W.-K. Tao. 1980. Response of deep tropical cumulus clouds to mesoscale processes. *J. Atmos. Sci.*, **37**, 2016–2034.

Stacey, F. D., and P. M. Davis. 2008. *Physics of the Earth*. Cambridge: Cambridge University Press, 532 pp.

Stan, C., and D. A. Randall. 2007. Potential vorticity as a meridional coordinate. *J. Atmos. Sci.*, **64**, 621–633.

Starr, V. P. 1948. An essay on the general circulation of the Earth's atmosphere. *J. Meteor.*, **5**, 39–43.

Stephens, G. L., and D. O'Brien, 1993. Entropy and climate, I: ERBE observations of the entropy production of the earth. *Quart. J. Roy. Meteor. Soc.*, **119**, 1212–152.

Stern, W., and K. Miyakoda. 1995. Feasibility of seasonal forecasts inferred from multiple GCM simulations. *J. Climate*, **8**, 1071–1085.

Stockdale, T. N., D.L.T. Anderson, J.O.S. Alves, and M. A. Balmaseda. 1998. Global seasonal rainfall forecasts using a coupled ocean-atmosphere model. *Nature*, **392**, 370–373.

Stommel, H. 1961. Thermohaline convection with two stable regimes of flow. *Tellus*, **13**, 224–230.

Stone, P. H. 1972. A simplified radiative-dynamical model for the static stability of rotating atmospheres. *J. Atmos. Sci.*, **29**, 405–418.

———. 1973. The effects of large-scale eddies on climatic change. *J. Atmos. Sci.*, **30**, 521–529.

———. 1978. Constraints on dynamical transports of energy on a spherical planet. *Dyn. Atmos. Oceans*, **2**, 123–139.

Stone, P. H., and R. M. Chervin. 1984. Influence of ocean surface temperature gradient and continentality on the Walker circulation. Pt. 2: Prescribed global changes. *Mon. Wea. Rev.*, **112**, 1524–1534.

Strong, C., and R. E. Davis. 2007. Winter jet stream trends over the Northern Hemisphere. *Quart. J. Roy. Meteor. Soc.*, **133**, 2109–2115.

Stroeve, J. C., T. Markus, L. Voisvert, J. Miller, and A. Barrett. 2014. Changes in Arctic melt season and implications for sea ice loss. *Geophys. Res. Lett.*, **41**, 1216–1225. doi:10.1002/2013 GL058951.

Strutt, J. W. ("Lord Rayleigh"). 1916. On the dynamics of revolving fluids. *Proc. Roy. Soc.*, **A93**, 447–453.

Stull, R. B. 1988. *An Introduction to Boundary Layer Meteorology*. Dordrecht: Kluwer Academic, 666 pp.

Suarez, M., A. Arakawa, and D. A. Randall. 1983. Parameterization of the planetary boundary layer in the UCLA general circulation model: Formulation and results. *Mon. Wea. Rev.*, **111**, 2224–2243.

Suarez, M., and D. Duffy. 1992. Terrestrial superrotation: a bifurcation of the general circulation. *J. Atmos. Sci.*, **49**, 1541–1554.

Suarez, M. J., and P. S. Schopf. 1988. A delayed action oscillator for ENSO. *J. Atmos. Sci.*, **45**, 3283–3287.

Sugiyama, M. 2009. The moisture mode in the quasi-equilibrium tropical circulation model. Part I: Analysis based on the weak temperature gradient approximation. *J. Atmos. Sci.*, **66**, 1507–1523, doi:10.1175/2008JAS2690.1.

Sui, C.-H., and K.-M. Lau. 1992. Multiscale phenomena in the tropical atmosphere over the western Pacific. *Mon. Wea. Rev.*, **120**, 407–430.

Sun, D. Z. 1997. El Nino: A coupled response to radiative heating? *Geophys. Res. Lett.*, **24**, 2031–2034.

Sun, D.-Z., and R. S. Lindzen. 1994. A PV view of the zonal mean distribution of temperature and wind in the extratropical troposphere. *J. Atmos. Sci.*, **51**, 757–772.

Sun, D.-Z., and Z. Liu. 1996. Dynamic ocean-atmosphere coupling: A thermostat for the tropics. *Science*, **272**, 1148–1150.

Swanson, K. 2001. Blocking as a local instability to zonally varying flows. *Quart. J. Roy. Meteor. Soc.*, **127**, 1–15.

Takahashi, M. 1996. Simulation of the stratospheric quasi-biennial oscillation using a general circulation model. *Geophys. Res. Lett.*, **23**, 661–664.

Takahashi, M., and M. Shiobara. 1995. A note on a QBO-like oscillation in a 1/5 sector three-dimensional model derived from a GCM. *J. Meteor. Soc. Japan*, **73**, 131–137.

Taylor, E. S. 1974. *Dimensional Analysis for Engineers*. Oxford: Clarendon Press, 162 pp.

Taylor, G. I. 1950a. The formation of a blast wave by a very intense explosion. I: Theoretical discussion. *Proc. Roy. Soc. A*, **201**, 159–174.

———. 1950b. The formation of a blast wave by a very intense explosion. II: The atomic explosion of 1945. *Proc. Roy. Soc. A*, **201**, 175–186.

Thayer-Calder, K., and D. A. Randall, 2009: The Role of Convective Moistening in the Formation and Progression of the MJO. *J. Climate*, **66**, 3297–3312.

Thomas, R. A., and P. J. Webster. 1997. The role of inertial instability in determining the location and strength of near-equatorial convection. *Quart. J. Roy. Meteor. Soc.*, **123**, 1445–1482.

Thompson, D.W.J., M. P. Baldwin, and S. Solomon. 2005. Stratosphere-troposphere coupling in the Southern Hemisphere. *J. Atmos. Sci.*, **62**, 708–715. doi:http://dx.doi.org/10.1175/JAS-3321.1.

Thompson, D.W.J., and E. A. Barnes. 2014. Periodic variability in the large-scale southern hemisphere atmospheric circulation. *Science*, **343**, 641–645.

Thompson, D.W.J., D. J. Seidel, W. J. Randel, C.-Z. Zou, A.H. Butler, C. Mears, A. Osso, C. Long, R. Lin. 2012. The mystery of recent stratospheric temperature trends. *Nature*, **491**, 692–697. doi:10.1038/nature11579.

Thompson, D.W.J., and J. M. Wallace. 1998. The Arctic Oscillation signature in the wintertime geopotential height and temperature fields. *Geophys. Res. Lett.*, **25**, 1297–1300.

———. 2000. Annular modes in the extratropical circulation. Part I: Month-to-month variability. *J. Climate*, **13**, 1000–1016. doi:http://dx.doi.org/10.1175/1520-0442(2000)013<1000:AMITEC>2.0.CO;2.

Thompson, D.W.J., J. M. Wallace, and G. C. Hegerl. 2000. Annular modes in the extratropical circulation. Part II: Trends. *J. Climate*, **13**, 1018–1036. doi:http://dx.doi.org/10.1175/1520-0442(2000)013<1018:AMITEC>2.0.CO;2.

Thompson, D.W.J., and J. D. Woodworth. 2014. Barotropic and baroclinic annular variability in the Southern Hemisphere. *J. Atmos. Sci.*, **71**, 1480–1493.

Thual, O., and J. C. McWilliams. 1992. The catastrophe structure of thermohaline convection in a two-dimensional fluid model and a comparison with low-order box models. *Geophys. Astrophys. Fluid Dyn.*, **64**, 67–95.

Tiedtke, M. 1989. A comprehensive mass flux scheme for cumulus parameterization in large-scale models. *Mon. Wea. Rev.*, **117**, 1779–1800.

———. 1993. Representation of clouds in large-scale models. *Mon. Wea. Rev.*, **121**, 3040–3061.

Tokinaga, H., S.-P. Xie, C. Deser, Y. Kosaka, and Y. M. Okumura. 2012. Slowdown of the Walker circulation driven by tropical Indo-Pacific warming. *Nature*, **491**, 439–443, doi:10.1038/nature11576.

Townsend, R. D., and D. R. Johnson. 1985. A diagnostic study of the isentropic zonally averaged mass circulation during the first GARP global experiment. J. Atmos. Sci., **42**, 1565–1579.

Trenberth, Kevin E. 1990. Recent observed interdecadal climate changes in the Northern Hemisphere. *Bull. Amer. Meteor. Soc.*, **71**, 988–993.

Trenberth, K. E., and J. M. Caron. 2001. Estimates of meridional atmosphere and ocean heat transports. *J. Climate*, **14**, 3433–3443.

Trenberth, K. E., J. M. Caron, and D. P. Stepaniak. 2001. The atmospheric energy budget and implications for surface fluxes and ocean heat transports. *Clim. Dyn.*, **17**, 259–276.

Trenberth, K. E., J. R. Christy, and J. G. Olson. 1987. Global atmospheric mass, surface pressure, and water vapor variations. *J. Geophys. Res.*, **92**, 14815–14826.

Trenberth, K. E., J. T. Fasullo, and M. A. Balmaseda. 2014. Earth's energy imbalance. *J. Climate*, **27**, 3129–3144.

Trenberth, Kevin E., John T. Fasullo, Jeffrey Kiehl. 2009. Earth's global energy budget. *Bull. Amer. Meteor. Soc.*, **90**, 311–323. doi:http://dx.doi.org/10.1175/2008BAMS2634.1.

Trenberth, K. E., J. Fasullo, and L. Smith. 2005. Trends and variability in column-integrated water vapor. *Clim. Dyn.*, **24**, 741–758.

Trenberth, K. E., and C. J. Guillemot. 1995. Evaluation of the global atmospheric moisture budget as seen from analyses. *J. Climate*, **8**, 2255–2272.

Trenberth, K. E., and J. G. Olson. 1988. An evaluation and intercomparison of global analyses from the National Meteorological Center and the European Centre for Medium Range Weather Forecasts. *Bull. Amer. Meteor. Soc.*, **69**, 1047–1057.

Trenberth, K. E., and A. Solomon. 1994. The global heat balance: Heat transports in the atmosphere and ocean. *Clim. Dyn.*, **10**, 107–134.

Trenberth, K. E., D. P. Stepaniak, and J. M. Caron. 2000. The global monsoon as seen through the divergent atmospheric circulation. *J. Climate*, **13**, 3969, 399.

———. 2002. Accuracy of atmospheric energy budgets. *J. Climate*, **23**, 3343–3360.

Troup, A. J. 1965. The “Southern Oscillation. *Quart. J. Roy. Meteor. Soc.*, **91**, 490–506.

Tselioudis, G., W. B. Rossow, and D. Rind. 1992. Global patterns of cloud optical thickness variation with temperature. *J. Climate*, **5**, 1484–1495.

Tung, K. K. 1986. Nongeostrophic theory of zonally averaged circulation. Part I: Formulation. *J. Atmos. Sci.*, **43**, 2600–2618.

Tung, K.-K., and A. J. Rosenthal. 1985. The nonexistence of multiple equilibria in the atmosphere: Theoretical and observational considerations. *J. Atmos. Sci.*, **42**, 2804–2819.

Tziperman, E., and B. Farrell. 2009. Pliocene equatorial temperature: Lessons from atmospheric superrotation. *Paleoceanography*, **24**, PA1101.

Uppala, S. M., et al. 2005. The ERA-40 re-analysis. *Quart. J. Roy. Meteor. Soc.*, **131**, 2961–3012. doi:10.1256/qj.04.176.

Valdes, P. J., and B. J. Hoskins. 1989. Linear stationary wave simulations of the time-mean climatological flow. *J. Atmos. Sci.*, **46**, 2509–2527.

Vallis, G. K. 2006. *Atmospheric and Oceanic Fluid Dynamics*. Cambridge: Cambridge University Press, 745 pp.

Vautard, R., and B. Legras. 1988. On the source of midlatitude low-frequency variability. Part II: Nonlinear equilibration of weather regimes. *J. Atmos. Sci.*, **45**, 2845–2867.

Vecchi, G. A., and B. J. Soden. 2007. Global warming and the weakening of the tropical circulation. *J. Climate*, **20**, 4316–4340. doi: http://dx.doi.org/10.1175/JCLI4258.1.

Veronis, G. 1969. On theoretical models of the thermohaline circulation. *Deep-Sea Res.*, **16**, 301–323.

Vinnikov, K. Y., A. Robock, and A. Basist. 2002. Diurnal and seasonal cycles of trends of surface air temperature. *J. Geophys. Res.*, **107**, 4641. doi:10.1029/2001JD002007.

Vonder Haar, T. H., and A. H. Oort. 1973. A new estimate of annual poleward energy transport by Northern Hemisphere oceans. *J. Phys. Oceanogr.*, **2**, 169–172.

von Storch, H., and F. Zwiers. 1998. *Statistical Analysis in Climate Research*. Cambridge: Cambridge University Press, 528 pp.

Vose, R. S., D. R. Easterling, and B. Gleason. 2005. Maximum and minimum temperature trends for the globe: An update through 2004. *Geophys. Res. Lett.*, **32**, L23822.

Wahr, J. M., and A. H. Oort. 1984. Friction-and mountain-torque estimates from global atmospheric data. *J. Atmos. Sci.*, **41**, 190–204.

Waliser, D. E., and N. E. Graham. 1993. Convective cloud systems and warm-pool sea surface temperatures: Coupled interactions and self-regulation. *J. Geophys. Res.*, **98**, 12881–12893.

Waliser, D. E, K. M. Lau, and J. H. Kim. 1999. The influence of coupled sea surface temperatures on the Madden-Julian oscillation: A model perturbation experiment. *J. Atmos. Sci.*, **56**, 333–358.

Waliser, D. E., and R.C.J. Somerville. 1994. Preferred latitudes of the intertropical convergence zone. *J. Atmos. Sci.*, **51**, 1619–1639.

Walker, G. T., and E. W. Bliss. 1932. World weather V. *Mem. Roy. Meteor. Soc.*, **4**, 53–84.

Wallace, J. M. 1971. Spectral studies of tropospheric wave disturbances in the tropical western Pacific. *Rev. Geophys.*, **9**, 557–612.

———. 1983. The climatological mean stationary waves: Observational evidence. In *Large-Scale Dynamical Processes in the Atmosphere*, edited by B. J. Hoskins and R. P. Pearce. New York: Academic Press, 397 pp.

———. 1992. Effect of deep convection on the regulation of tropical sea surface temperature. *Nature*, **357**, 230–231.

Wallace, J. M., and D. S. Gutzler. 1981. Teleconnections in geopotential height field during the Northern Hemisphere winter. *Mon. Wea. Rev.*, **109**, 784–812.

Wallace, J. M., and F. R. Hartranft. 1969. Diurnal wind variations, surface to 30 kilometers. *Mon. Wea. Rev.*, **97**, 446–455.

Wallace, J. M., and V. E. Kousky. 1968. Observational evidence of Kelvin waves in the tropical stratosphere. *J. Atmos. Sci.*, **25**, 900–907.

Wallace, J. M., T. P. Mitchell, and C. Deser. 1989. The influence of sea-surface temperature on surface wind in the eastern equatorial Pacific: Seasonal and interannual variability. *J. Climate*, **2**, 1492–1499.

Wang, B. 1988. Dynamics of tropical low-frequency waves: An analysis of the moist Kelvin wave. *J. Atmos. Sci.*, **45**, 2051–2064.

Wang, B., and J. Chen. 1989. On the zonal-scale selection and vertical structure of equatorial intraseasonal waves. *Quart. J. Roy. Meteor. Soc.*, **115**, 1301–1323.

Wang, B., and X. Xie.1998. Coupled modes of the warm pool climate system. Part I: The role of air-sea interaction in maintaining Madden-Julian oscillation. *J. Climate*, **11**, 2116–2135.

Wang, J., and D. A. Randall. 1994. The moist available energy of a conditionally unstable atmosphere. II: Further analysis of the GATE data. *J. Atmos. Sci.*, **51**, 703–710.

Wang, J. S., D. J. Seidel, and M. Free. 2012. How well do we know recent climate trends at the tropical tropopause? *J. Geophys. Res. Atmos.*, **117**, D09118.

Warren, B. A. 1983. Why is no deep water formed in the North Pacific? *J. Marine Res.*, **41**, 327–347.

Washington, W. M., and G. A. Meehl. 1989. Climate sensitivity due to increased CO_2: Experiments with a coupled atmosphere and ocean general circulation model. *Clim. Dyn.*, **4**, 1–38.

Washington, W. M., and C. L. Parkinson. 1986. *An Introduction to Three-Dimensional Climate Modeling*. Mill Valley, NY: University Science Books, 422 pp.

Weaver, A. J., and E. S. Sarachik. 1991. The role of mixed boundary conditions in numerical models of the ocean's climate. *J. Phys. Oceanogr.*, **21**, 1470–1493.

Webster, P. J. 1972. Response of the tropical atmosphere to steady local forcing. *Mon. Wea. Rev.*, **100**, 518–541.

———. 1981. Monsoons. *Scientific American*, **245** (2), 108–118.

———. 1983. Mechanisms of monsoon low-frequency variability: Surface hydrological effects. *J. Atmos. Sci.*, **40**, 2110–2124.

———. 1987. The variable and interactive monsoon. In *Monsoons*, edited by J. S. Fein and P. L. Stephens. New York: Wiley, 269–330.

———. 1994. The role of hydrological processes in ocean-atmosphere interactions. *Rev. Geophys.*, **32**, 427–476.

Webster, P. J., V. O. Magana, T. N. Palmer, J. Shukla, R. A. Tomas, M. Yanai, and T. Yasunari. 1998. Monsoons: Processes, predictability, and the prospects for prediction. *J. Geophys. Res.*, **103**, 14,451–14,510.

Webster, P. J., and S. Yang. 1992. Monsoon and ENSO: Selectively interactive systems. *Quart. J. Roy. Meteor. Soc.*, **118**, 877–926.

Weickmann, K. M., and S.J.S. Khalsa. 1990. The shift of convection from the Indian Ocean to the western Pacific Ocean during a 30–60 day oscillation. *Mon. Wea. Rev.*, **118**, 964–978.

Weller, R. A., and S. P. Anderson. 1996. Surface meteorology and air-sea fluxes in the western equatorial Pacific warm pool during the TOGA coupled ocean-atmosphere response experiment. *J. Climate*, **9**, 1959–1990.

Wentz, F. J., L. Ricciardulli, K. Hilburn, and C. Mears. 2007. How much more rain will global warming bring? *Science*, **317**, 233–235.

Wheeler, M., and G. N. Kiladis. 1999. Convectively coupled equatorial waves: Analysis of clouds and temperature in the wavenumber-frequency domain. *J. Atmos. Sci.*, **56**, 374–399.

White, A. A., and R. A. Bromley. 1995. Dynamically consistent, quasi-hydrostatic equations for global models with a complete representation of the Coriolis force. *Quart. J. Roy. Meteor. Soc.*, **121**, 399–418.

Whitehead, J. A. 1995. Thermohaline ocean processes and models. *Ann. Rev. Fluid Mech.*, **27**, 89–113.

Whitlock, C. H., T. P. Charlock, W. F. Staylor, R. T. Pinker, I. Laszlo, R. C. DiPasquale, and N. A. Ritchey. 1993. WCRP surface radiation budget shortwave data product description: Version 1. 1. *NASA Technical Memorandum* 107747.

Wielicki, B. A., B. R. Barkstrom, B. A. Baum, T. P. Charlock, R. N. Green, D. P. Kratz, R. B. Lee, III, P. Minnis, G. L. Smith, T. Won, D. F. Young, R. D. Cess, J. A. Coakley Jr., D. A. H. Crommelynck, L. Donner, R. Kandel, M. D. King, A. J. Miller, V. Ramanathan, D. A. Randall, L. L.

Stowe, and R. M. Welch. 1998. Clouds and the Earth's Radiant Energy System (CERES): Algorithm overview. *IEEE Trans. Geosci. Remote Sens.* **36**, 1127–1141.

Wielicki, B. A., B. R. Barkstrom, E. F. Harrison, R. B. Lee III, G. L. Smith, and J. E. Cooper. 1996. Clouds and the Earth's Radiant Energy System (CERES): An Earth observing system experiment. *Bull. Amer. Meteor. Soc.*, **77**, 853–868.

Wiin-Nielsen, A. 1967. On the annual variation and spectral distribution of atmospheric energy. *Tellus*, **19**, 540–559.

———. 1972. A study of power laws in the atmospheric kinetic energy spectrum using spherical harmonic functions. *Meteor. Ann.*, **6**, 107–124.

Wiin-Nielsen, A., J. A. Brown, and M. Drake. 1963: On atmospheric energy conversions between the zonal flow and the eddies. *Tellus*, **15**, 261–279.

Wilczek, F. 2005: On absolute units. I: Choices. *Physics Today*, **58** (October).

———. 2006a: On absolute units. II: Challenges and responses. *Physics Today*, **59** (January).

———. 2006b: On absolute units. III: Absolutely not? *Physics Today*, **5** (May).

Abramowitz, M., and I. A. Segun, 1970: *Handbook of Mathematical Functions.* New York: Dover, 1046 pp.

Williams, G. P. 1988. The dynamical range of global circulations—I. *Clim. Dyn.*, **2**, 205–260.

Willoughby, H. E. 1998. Tropical cyclone eye thermodynamics. *Mon. Wea. Rev.*, **126**, 3053–3067.

Wong, T., G. L. Stephens, and P. W. Stackhouse Jr., and F.P. J. Valero. 1993. The radiative budgets of a tropical mesoscale convective system during the EMEX-STEP-AMEX experiment. 1. Observations. *J. Geophys. Res.*, **98**, 8683–8693.

Woollings, T., A. Charlton-Perez, S. Ineson, A. G. Marshall, and G. Masato. 2010. Associations between stratospheric variability and tropospheric blocking. J. Geophys. Res. Atmos., **115**, D06108.

Woollings, T., B. Hoskins, M. Blackburn, and P. Berrisford. 2008. A new Rossby wave-breaking interpretation of the North Atlantic Oscillation. *J. Atmos. Sci.*, **65**, 609–626.

Woolnough, S., J. Slingo, and B. Hoskins. 2000. The relationship between convection and surface fluxes on intraseasonal timescales. *J. Climate*, **13**, 2086–2104.

Wrede, R. C. 1972. *Introduction to Vector and Tensor Analysis.* New York: Dover, 418 pp.

Wyant, P. H., A. Mongroo, and S. Hammed. 1988. Determination of the heat-transport coefficient in energy-balance climate models by extremization of entropy production. *J. Atmos. Sci.*, **45**, 189–193.

Xie, P., and P. A. Arkin. 1996. Analyses of global monthly precipitation using gauge observations, satellite estimates, and numerical model predictions. *J. Climate*, **9**, 840–858.

Xie, S.-P., and S.G.H. Philander. 1994. A coupled ocean-atmosphere model of relevance to the ITCZ in the eastern Pacific. *Tellus*, **46A**, 340–350.

Xu, K.-M., and A. Arakawa. 1992. Semiprognostic tests of the Arakawa-Schubert cumulus parameterization using simulated data. *J. Atmos. Sci.*, **49**, 2421–2436.

Xu, K.-M., and K. A. Emanuel. 1989. Is the tropical atmosphere conditionally unstable? *Mon. Wea. Rev.*, **117**, 1471–1479.

Xu, K.-M., and D. A. Randall. 1996. Explicit simulation of cumulus ensembles with the GATE Phase III data: Comparison with observations. *J. Atmos. Sci.*, **53**, 3710–3736.

Yamagata, T., and Y. Hayashi. 1984. A simple diagnostic model for the 30–50 day oscillation in the tropics. *J. Met. Soc. Japan*, **62**, 709–717.

Yamagata, T., and Y. Masumoto. 1989. A simple ocean-atmosphere coupled model for the origin of warm El Niño Southern Oscillation event, *Philos. Trans. R. Soc. A*, **329**, 225–236.

Yanai, M., and C. Li. 1993. Mechanism of heating and the boundary layer over the Tibetan Plateau. *Mon. Wea. Rev.*, **122**, 305–323.

———. 1994. Interannual variability of the Asian summer monsoon and its relationship with ENSO, Eurasian snow cover and heating. In *Proceedings of the International Conference on Monsoon Variability and Prediction*, International Centre for Theoretical Physics, Trieste, Italy, May 9–13.

Yanai, M., C. Li, and Z. Song. 1992. Seasonal heating of the Tibetan Plateau and its effects on the Asian summer monsoon. *J. Meteor. Soc. Japan*, **70**, 319–351.

Yanai, M., and T. Maruyama. 1966. Stratospheric wave disturbance propagating over the equatorial Pacific. *J. Meteorol. Soc. Japan*, **44**, 227–243.

Yang, J., and J. D. Neelin. 1993. Sea-ice interaction with the thermohaline circulation. *Geophys. Res. Lett.*, **20**, 217–220.

Yasunari, T. 1979. Cloudiness fluctuations associated with the Northern Hemisphere summer monsoon. J. Met. Soc. Japan, **57**, 227–242.

Yeh, T.-C., and Y.-X. Gao. 1979. *The Meteorology of the Qinghai-Xizang (Tibet) Plateau*. Beijing: Science Press, 278 pp.

Yu, J.-Y., and J. D. Neelin. 1994. Modes of tropical variability under convective adjustment and the Madden-Julian oscillation. Part II: Numerical results. *J. Atmos. Sci.*, **51**, 1895–1914.

Zebiak, S. E., and M. A. Cane. 1987. A model of El Niño Southern Oscillation. *Mon. Wea. Rev.*, 115, 2262–2278.

Zent, A. P. 1996. The evolution of the Martian climate. *American Scientist*, **84**, 442–451.

Zhang, G. J., and N. A. McFarlane. 1995. Sensitivity of climate simulations to the parameterization of cumulus convection in the Canadian Climate Centre general circulation model. *Atmos.-Ocean*, **33**, 407–446.

Zhang, G. J., and M. J. McPhaden. 1995. The relationship between sea surface temperature and latent heat flux in the equatorial Pacific. *J. Climate*, **8**, 589–605.

Zhang, Y., J. M. Wallace, and D. S. Battisti. 1997. ENSO-like interdecadal variability: 1900–1993. *J. Climate*, **10**, 1004–1020.

Zhang, Z., W. Wang, and B. Qiu. 2014. Oceanic mass transport by mesoscale eddies. *Science*, **345**, 322–324. doi:10.1126/science.1252418.

Zhou, T., L. Zhang, and H. Li. 2008. Changes in global land monsoon area and total rainfall accumulation over the last half century. *Geophys. Res. Lett.*, **35**, L16707. doi:10.1029/2008 GL034881.

Index

Page numbers in italics refer to figures and tables